Theories of Spectral Line Shape

Theories of Spectral Line Shape

R. G. BREENE, JR.

A WILEY-INTERSCIENCE PUBLICATION

JOHN WILEY & SONS
New York • Chichester • Brisbane • Toronto

Library of Congress Cataloging in Publication Data:

Breene, Robert G

Theories of spectral line shape.

"A Wiley-Interscience publication."

Bibliography: p.

Includes index.

1. Spectral line broadening. I. Title.

QC467.B69 535.8′4 80-20664

ISBN 0-471-08361-5

Printed in the United States of America

10 9 8 7 6 5 4 3 2 1

To The General's Lady
In Poignant Memory of a More Gracious Day

Baccalaureus

A rogue, perhaps! What teacher will declare
The truth to us, exactly fair and square?
Each knows the way to lessen or exceed it....

Faust II, Act II, Scene I

Preface

During the last twenty to thirty years three general techniques for the theoretical description of spectral line broadening have risen to positions of dominance, these theory formulations appealing principally to the resolvant operator, to the operations of Liouville space, and to those descriptions involving Greens functions and Feynman diagrams. It is my purpose to describe these general formulations and their application to certain important broadening problems.

We begin with a study of that spectral line broadening that arises from the interaction of the radiator with its own electromagnetic field and that leads to the natural line shape. Although diagrammatic and Liouvillian treatments have been applied to this problem, it is perhaps the least dependent on them. After laying the groundwork requisite to perusal of the three techniques, I discuss each of them in turn and in detail. With this discussion as our preparation, I devote the last three chapters of the book to the principal specific problems which have their origin either in the motion of the radiator relative to the observer or in the presence of other particles in the neighborhood of the radiator.

Considerable advances in Doppler-pressure syntheses have taken place since Godfrey mounted the first attack on the problem in 1901. To a discussion of these advances I devote Chapter 7, in which I discuss specific applications of the Liouvillian and diagrammatic formulations.

Certain vagaries were first encountered in the resonance broadened spectral line around 1960. The problem has garnered attention from all three basic techniques in the interim, and I devote Chapter 8 to the results of this attention.

The spectral line satellites which are the consequence of certain radiator-broadener combinations can constitute a rather complex and quite intriguing phenomenon. In Chapter 9 we discuss the kaleidoscope of theories these satellites have evoked.

A number of individuals have furnished material for which the author assuredly wishes to express his deep appreciation. Here too I would like to thank Professor Abraham Ben-Reuven, Dr. Rosemary B. Higgins, and Dr. David W. Ross for their kindness in answering several, generally rather stupid, questions.

I cannot close without expressing my deep appreciation to the directors and staffs of the Physics Library of the University of Texas at Austin and the Library of Trinity University for the invaluable benefits which I have derived from the use of their libraries.

R. G. BREENE, JR.

San Antonio, Texas
August 1980

Contents

Theories of Spectral Line Shape

1

Introduction

Should we allow ourselves to lapse into the jingoism of this anything but scholarly age, we would probably call the period 1889–1958 the Era of Ideas, the period 1958–1978 the Era of Techniques. Although such high flown pretentiousness may be as gall and wormwood to the sensitive reader, the notion of ideas does indeed prove rather descriptive of the period that was ushered in by Lord Rayleigh's thoughts on the motional broadening of spectral lines and concluded by Henry Margenau's dissection of the various factors entering Stark broadening studies. With the Baranger initiation of the second temporal period, the ideas were all expounded on; actually, any number of these ideas had been floating about for considerably longer than many authors appear to realize. However, in an ever burgeoning flood of sometimes rather intricate expositions, the techniques began to pour forth, and we have found ourselves inundated by an ever increasing tide of density matrices, Liouvillians, diagrams, cumulants—in a word, techniques. Our purpose in this book is to investigate these techniques. First, however, it behooves us to review some ideas from the book to which this book is a sequel.

We begin with a radiating atom. Now we may view this atom as consisting of a photoelectron harmonically oscillating against a more or less stationary core and emitting or absorbing radiation as a consequence of the accelerations which accompany such motions. On the other hand, we might simply consider the atom as some sort of structure for which the laws (rules?) of the quantum theory predict (at least) two levels, transitions between which lead to emission or absorption of radiation. (The reader is of course aware that the latter model corresponds to the world picture which has proven to be by far the most productive. The question as to whether such a world picture is unique constitutes the subject matter of another of the author's books (a book in which the publishing milieu evinces an impressive lack of interest and with which the reader will not be burdened again). Using either model, our first impression is that a spectral "line" would be observed as—in an intensity-frequency plot—a widthless spike at the characteristic frequency. Why this is never the case is the subject of line broadening theory.

One obviously never deals with a single radiator, but occasionally it is helpful for us to behave as if a single radiator were feasible. Therefore, let us remove all other atoms, either of the same or different species, from the neighborhood of our radiator to infinity. Let us further suppose the translational temperature of our radiator to be 0 °K, the radiator at rest. Under these conditions the spectral line will consist, not of a widthless spike, but of an apparently symmetrical distribution of intensities within a band about an intensity maximum. Such is the natural line shape. Since we did not discuss this subject in the earlier book, we shall devote the second chapter of this one to it.

Now let us raise the translational temperature of the radiator to some finite value. The consequence is a spectral line broadening effect called "Doppler broadening," first explained to us by Lord Rayleigh (259). If the radiator is moving toward us at the time of emission, the emitted radiation will suffer an augmentation of its frequency corresponding to the increase in the number of waves which reach us per unit time. If the radiator is moving away, there will be a corresponding diminution in frequency and so on for an assemblage of radiators carrying out random motions governed by a Boltzmann distribution of velocities. The result will be a distribution of intensities in the spectral line which is exponentially dependent on the square of the frequency separation from line center. Now let us bring our neighboring atoms in from infinity and ignore the radiator's thermal motion.

As to the broadening effects that we may expect, from the beginning in 1895 the resulting phenomenon has tended to divide into two limiting cases: what we have chosen to call "interruption broadening" and what Margenau originally dubbed "statistical broadening." (That interruption broadening is often called "impact broadening" is worthy of note.) We first consider interruption broadening, the development of which by Michelson (200), Lorentz (186), Lenz (178), Weisskopf (279, 280), and Lindholm (181, 182) we have considered *in extenso*.

Let us suppose that our classical oscillator photoelectron is behaving as classical oscillators will and oscillating. It is thus emitting electromagnetic radiation that corresponds to the characteristic frequency of its oscillation. Were it not for the damping effect of the necessity for it to radiate away energy, it would of course emit a monochromatic wave of infinite length. We suppose that it would be able to do so were is not for perturbing collisions that it experiences with its atomic neighbors. These collisions occur very rapidly, so rapidly in fact that the only effect on the radiator is an instantaneous change in the phase of the emitted radiation. For distant collisions these changes in phase are slight, and an observer would be essentially unaware of their existence in the monochromatic wave that he was observing. However, the close collisions would induce large phase changes, so large indeed that the resulting before/after incoherency would make it appear as if the wave train were suddenly lopped off at the point and time of collision. Between two such close collisions then, it would seem that a wave train of finite length corresponding to the intercollision time is emitted. We observe, however, wave trains of infinite length. The manner in which a wave train of finite length may be constructed from an infinite number of wave trains of infinite length and differing frequencies is detailed for us by the Fourier transform. A particular frequency will thus correspond to a particular Fourier amplitude. Thus the intensity at that frequency in the spectral line will be proportional to the square of the appropriate Fourier amplitude.

The theory in this elementary form yields a symmetrical, unshifted spectral line having an inverse square dependence on the frequency separation from line center. The definition of a cutoff phase shift must perforce be arbitrary if sufficient to yield fore and aft incoherency. Whatever it is, however, we find that phase shifts of this order and greater are the important contributors to the linewidth. That the neglected phase shifts of lesser magnitude are largely responsible for the line shift is qualitatively reasonable. Our wave train of finite length we suppose established by the large phase shifts, and, as observers, we would be looking at the line center frequency in looking at this train. Now, however, the small phase shifts enter the picture and somewhat distort the frequency of the finite length train. They nudge the frequency from that of the unperturbed spectral line center to the maximum of the shifted line. At this point our rapid-collision interruption theory has provided a broadened and shifted spectral line which has often been found useful for situations where the time of collision is small compared to the time between collisions. (We may, for example, define a "collision sphere" having a radius of the order of the range of the broadening force. Then the time of collision can be defined as the time of perturber residence within the collision sphere, the time between collisions, the time between sphere penetrations.) This time restriction leads us to a line portion restriction that is frequently placed on the interruption theory.

What we shall consistently refer to as the optical collision diameter is the distance of closest approach of the radiator-perturber collision pair. The sort of cerebration involved in interruption theory ruminations does not tell us the following, but we shall accept it for the present. In a very qualitative way the collisions of the sort leading to intensities in the wing of the line have smaller optical collision diameters the farther we progress into that wing. This means that, the farther out in the wings we are, the longer the perturber will remain in the interaction sphere. At some point then we may no longer say that the collision times are small compared to the intercollision times. Therefore, the interruption theory cannot be expected to hold "far" out in the line wing.

How far is far? How small is small? How much greater than is "much greater than?" Obviously the questions cannot be answered with any precision except in reference to an experiment which may not exist. Although we shall not attempt to emphasize their occurrence, there are a number of articles wherein "much greater than" really amounts to "equal to." It does seem proper, however, that such restrictions are often rendered almost meaningless in application.

The interruption theory which we have so far discussed is "classical" if we consider a wave train classical and a light particle quantal. Otherwise it need not be so considered. However, we can arrive at a Fourier transform another way. Suppose we consider the electronic structure of our radiator and the translational structure of radiator plus perturbers and describe the whole thing quantally, that is, by a wave function. Then a contribution to the intensity corresponding to a particular frequency will be the product of the matrix elements of, say, the dipole moment at the beginning and that at the end of some temporal period, preferably the period of collision. The total intensity will be a sum over all such possible contributions multiplied by a delta function to assure that we deal only with the desired frequency. When we go from a sum over all such contributions to an integral over them, and when we utilize the integral definition of the Dirac delta function, we find that we have arrived at precisely the Fourier transform of these moment matrix elements, which is what Anderson (55) did, appealing to the time development operator (TDO) and the S matrix. (Foley (122) should be parenthetically inserted here.)

Now we suppose that the collisions are slow, the time of perturber residence in the collision sphere sizeable compared to the time between sphere penetrations. It is easier here to think in terms of energy levels rather than oscillating photoelectrons, but one can utilize the latter. With all due deference to Holtsmark, who did indeed develop a probability for a perturber distribution in considering Stark and related broadening, the statistical theory is the brainchild of Margenau (188, 191). Margenau said, The probability for the emission of a particular frequency—and hence the intensity corresponding to that frequency—is proportional to the probability of a perturber distribution such that the resultant level distortion will yield that frequency. A chain of reasoning, through which we have already proceeded, tells us that such a theory is a close collision, long time collision, line wing theory. Having thus very briefly remarked on what are generally considered to be the two limiting cases in line broadening theory, we mention that their synthesis has been attempted in various ways. Perhaps the most often encountered method amounts to a folding of the two limiting cases, the smearing of an interruption distribution over a statistical one. (We do not necessarily refer here to those later theories which reduce in the appropriate limits to the proper theory.)

We have made an incidental comment on the equivalence of van der Waals and foreign gas broadening and on the existence of resonance or self broadening. There is really not a great deal more to be said about foreign gas broadening in general—except of course that the potentials yielding it are not simply asymptotic van der Waals potentials. However, there is something to be said about resonance broadening, the broadening phenomenon wherein the broadener is of the same species as the radiator. Suppose the radiator is in an excited state, the broadeners, all of the same species, in their ground states, and suppose further that this excited state is linked to the ground state by a large oscillator strength. Then prior to emission the quantum of excitation may be exchanged by the radiator with one of the perturbers, by this perturber with another, and so on. All of which leads, if you wish to consider the matter this way, to a shortened lifetime for the excited state and hence, through the Heisenberg uncertainty principle, to a broadened upper state. (As the reader is probably aware, the Fourier transform is a statement of the Heisenberg uncertainty principle and of course relates Heisenberg-conjugate variables as $f(t)=\int \exp[iEt]f(E)\,dE$. Of course the Dirac delta function does the same thing.) This in turn leads to a spectral line broader than that induced by an equal density of foreign gas. This was classically recognized by Holtsmark (145) in 1925. Intriguingly enough, various authors were still rediscovering the effect until relatively recently.

We have considered foreign-gas and resonance broadening. Another broadening phenomenon which has been the subject of a great deal of discussion is that of Stark broadening, which is often, as in a plasma, induced by the rapid motions of the free electrons and the more sedate motions of the positive ions. Although Holtsmark's original treatment (144) was not restricted to what we now know as Stark broadening, he was the first to consider the positive ions as a broadening source. He computed the probability, at the emitter, for a given electric field strength with a corresponding spectral line split into its Stark components. Averaging over the various possible distributions led to a broadened spectral line. This was of course a statistical theory, but it was not the statistical theory of Margenau, the level-distortion theory. Most authors seem to behave as if it were. This is important only insofar as one considers accuracy of quotation important. In any event, we would suppose that the rapid motion of the electrons might be treated by a theory of the interruption type, and indeed this has proved to be the case to a rather good approximation. In order to determine the broadening effect of a plasma of electrons and positive ions, interruption theories have been folded into statistical distributions. With considerable success. (We are not going to consider Stark broadening in this work. Since the appearance of our last book on line shape, Griem (18, 19) has published two books whose coverage of the subject is quite comprehensive. The reader will find that, by and large, the principles are all here in any case.)

We earlier included the Jablonski theory in a chapter entitled "Statistical Broadening." We shall defend this categorization after we recall the outlines of the Jablonski theory, which will play a role of some consequence in our satellite studies. In a word, this theory establishes and utilizes the analogy between the broadened electronic spectral line of an atom and the electronic-vibrational band system of a diatomic molecule. (Originally Jablonski postulated a polyatomic molecule, but, since no one has overmuch concerned themselves with more than a diatomic, there is no particular reason for us to maintain the fiction.) This of course amounts to considering the electronic-vibrational potentials which the radiator-broadener pair may combine to produce. Assuming the applicability of the Born-Oppenheimer approximation

—to the effect that the total wave function is a product of the electronic and vibrational wave functions—the square of the vibrational matrix element, averaged over the initial energy and momentum of the pair, will tell us the intensity corresponding to a particular electronic-vibrational transition energy, that is, to a particular frequency in the spectral line. This vibrational matrix element has generally been approached through the Wentzel-Kramers-Brillouin (WKB) wave function. Classifying this theory as a statistical one may be justified with the following argument.

The statistical theory is usually viewed as relating to close collisions. Constructing a quasimolecule of the type crucial to the Jablonski hypothesis would seem a somewhat bizarre procedure for distant collisions during which the collision partners rather well maintain their atomic character as opposed to taking on the molecular character of even a dissociated diatomic. To this argument we appeal in making the admittedly tenuous assignment of the Jablonski theory to the statistical category. In our remarks to this point we have not made much distinction as to whether our discussion is set against a classical or quantal background, but the point has occasionally arisen.

Whether the background is classical or quantal is of course of some concern to questions of adiabaticity and of considerable concern to the matter of the classical path. We may consider a collision to be adiabatic if the collision partners emerge from a collision in the states in which they entered into it. (Exchanges of translational energy are not generally considered examples of diabaticity.) A classic example (but not the only one) of diabaticity is provided by Spitzer's rotational adiabaticity (255), which we shall have occasion to encounter in what follows. Initial contact between emitter and perturber is established by their interaction field which generates a direction in space and hence renders the magnetic quantum number of the radiator meaningful. On contact establishment then we find the radiator in a state defined by a particular value of the magnetic quantum number. If the radiator swings round with perturber passage so that it remains in the state specified by that value, the collision is adiabatic. If the radiator does not swing with perturber passage, the magnetic quantum number will change during that passage, and the collision is diabatic. Obviously the optical collision diameter is going to determine whether such swinging, such adiabaticity, is possible.

If the translational motion of the particles undergoing a broadening collision may be treated classically, the so-called classical path approximation is applicable. This approximation may be more or less broken down into two subapproximations, one relating to a definition of the separation of the two interacting particles, the other to whether or not a region of space is classically accessible to, say, the perturber. If we consider the wave packet associated with the perturber, we may develop a criterion for the classical path validity. If, during the time of residence of the perturber within the collision sphere, that is, during the time of collision, the perturber packet diffusion is "much less than" the collision sphere radius, the classical path approximation may be taken as valid. In practice "much less than" has often been taken as "equal to." Two considerations obtrude: (1) the duration of the collision and (2) the rate of diffusion. Electrons, for example, generally spend a short time in the collision sphere, but, at the same time, these particles diffuse more rapidly than do the comparatively massive atoms.

A chapter of the book that we have been more or less reviewing was devoted to the broadening of polyatomic, as opposed to monatomic, spectral lines. This was really more devoted to those special forces which arise in interactions between such molecules because of the extra degrees of freedom which are theirs as a consequence of their possessing more than one nucleus. We hardly need tell the reader that the field of intermolecular forces is an immense one. We will tell him, however, that we have no intention of cluttering up our study of line shape theory with specific details of intermolecular forces, with one exception which we shall encounter in our chapter dealing with satellites.

In Chapter 2, and because we did not touch on the subject previously, we shall discuss the natural line shape, really a subject more or less unto itself. Let us preface our sketch of what is to be anticipated in the chapters which follow Chapter 2 with some very general remarks on what has been happening in line broadening in the last twenty years. In doing so we begin by writing down five sets of words (1) overlapping spectral lines, (2) resolvant operators, (3) density matrices, (4) Liouville spaces, and (5) Feynman diagrams.

Since Anderson's development of a more or less quantal theory for the broadening of an isolated spectral line, the TDO played an important role in furnishing the time dependence of the dipole moment operator. This put the Hamiltonian into an exponential with the frequency from the Fourier transform, which in turn meant that the temporal integration would yield a denominator containing the difference between Hamiltonian and frequency. This denominator became known as the "resolvant" and was obviously rather frequently encountered. We have said that Anderson carefully treated the isolated line.

A line that is not isolated must perforce be overlapped by other spectral lines associated with the same molecule. This subject was treated carefully by Baranger in 1958 (63) and also by Kolb and Griem (164). (That the Baranger work had to a certain extent been anticipated by Leslie (179) is quite properly pointed out by Roney (36). That Leslie has not received the recognition due him is one of those injustices about which we can do very little.) The subject of overlapping lines would be of no consequence were the resultant profile simply the sum of a set of isolated profiles. However, this is not the case, and we have a subject worth at least some of our attention. The same may be said of the density matrix, although we have encountered this entity previously. The density matrix generally enters treatments whose main theme is independent of them. For example, in the Roney treatment, which appeals to the reduced density matrix, the general thrust of the study is based on the Liouville operator. This brings us to the main line of advance of spectral line broadening theory during the last twenty years.

The intensity in a spectral line—and hence the shape—is proportional to the probability for the atomic transition which leads to the line. This probability in turn is proportional to the matrix element of the perturbing portion of the Hamiltonian which induces the transition. This perturbing portion is almost universally approximated by the dipole term in the multipole expansion for the radiator together, of course, with the vector potential for the radiation field. (This vector potential leads to Doppler effects, which we ignore for the moment.) At this point the main stream branches, the first offshoot yielding the Liouville space treatment, the second the Feynman diagram. In a word, the Liouville operator arises immediately from the expression of the dipole operator in a form of Heisenberg representation, the Feynman diagram from that of the current commutator in second quantized form which is equivalent to a function of one- and two-particle Greens functions.

The Liouville space formulation of spectral line broadening theory was initiated by Fano in 1963 (115) and advanced, through the efforts of a number of authors, to the well rounded formulation of Ben-Reuven (71). The Feynman diagram approach was utilized in natural line shape studies as early as 1951 (187). Its general application to the problem of our interest, however, was initiated by Ross in 1966 (236). At present it appears that the main line of immediate advance will be provided by the Liouville formulation, the diagrammatic formulation continuing to provide greater physical insight and understanding.

It would be rather difficult for us to overemphasize the importance of Greens functions, Liouville space, and the operators therein. These entities we discuss in a rather general—as opposed to broadening oriented—way in Chapter 3. Every line broadening treatment perforce begins with some expression for the intensity in the spectral line, be it a squared state growth coefficient (SGC) as in Chapter 2, the susceptibility or something similar. We also discuss these starting expressions in this chapter.

Chapter 4 is devoted, first, to the Baranger-Kolb-Griem treatment of the broadening of overlapping spectral lines, the treatment involving the resolvant operator to the extent that we are justified in using it as a description. In this chapter also we have our first direct, albeit preliminary, encounter with the density matrix. Here too we consider evaluations of the S matrix, which is very important to broadening theory, first by truncation, then in its entirety by an appeal to the classical path. The chapter concludes with a brief remark on the Boltzmann factor.

In Chapters 5 and 6 we study the Liouvillian and the Greens function formulation of line broadening, respectively. Since one encounters Greens functions in the product space of the Liouville operator, such an abbreviated description can obviously be misleading. But I hope it will serve for the sort of broad brush description at which we aim here.

Chapters 4 to 6 are chiefly concerned with broad general formulations, whether of resonance or foreign-gas broadening, high pressures or low, and so on. The last three chapters are devoted to more specialized treatments of specific phenomena. Chapter 7 is principally devoted to syntheses of Doppler and pressure broadening, although the subject matter does stray somewhat from a rigid restriction to this area. Chapter 8 is devoted to self or resonance broadening and included within it are studies rooted in the Liouvillian and Greens function theory of the earlier chapters as well as a certain amount of work which cannot really be so classified. The book concludes with Chapter 9, a rumination on the satellites which sometimes appear as secondary maxima in the wings of pressure-broadened spectral lines.

We do not include a chapter on molecular—polyatomic as opposed to monatomic—spectral line broadening. Important to such broadening is of course the theory of overlapping spectral lines and the unique phenomena which are encountered when we stray into that comparatively low frequency region, the microwave. In the main, the ideas important to molecular broadening are here, and the author felt that, as in the case of intermolecular forces, he would be exceeding the scope of what he might sensibly hope to encompass by a serious treatment of molecular broadening per se.

We have, we hope, covered the principal articles in the areas which we have considered. However, our bibliography makes no claim to completeness. Fuhr et al. (1) have published an extensive bibliography plus three supplements which certainly seem to include the principal references on spectral line broadening. A word now on units would probably be in order.

We shall make use of "natural units." These units may be referenced to the cgs system by taking the unit of length as one centimeter. The unit of time will be taken as the time required for light to travel one centimeter, the unit of mass that mass for which the Compton wavelength ($\lambda = h/mc$ cgs) is 2π. The conversion between the two systems is given in Table 1.1.

It is probably obvious that $\hbar = c = 1$.

The following are a few abbreviations which we will save a certain amount of space by utilizing: (1) effective interaction tetradic (EIT), (2) bath operator (BO), (3) state growth coefficient (SGC), (4) time development operator (TDO), (5) time ordering operator (TOO), (6) vector coupling coefficient (VCC), and (7) self energy part (SEP).

Table 1.1

	Natural units	CGS units
Length	1	1 cm
Time	1	$1/c$ sec
Mass	1	$\hbar/c$ g
Energy	1	$\hbar c$ erg
Charge	1	$\sqrt{\alpha hc}$
Mass of electron	2.59×10^{-10}	9.108×10^{-28}
Radius of first Bohr orbit	$1/m\alpha$	$\hbar^2/me^2$

2

Natural Line Shape

1 Introduction

We begin by supposing our radiator at rest relative to the observer, thus eliminating the Doppler broadening of the spectral line. Next, we isolate the radiator by removing its molecular neighbors to infinity. Now when our isolated, dormant molecule makes a radiant transition between two discrete energy levels, the resulting radiation, viewed as an energy versus intensity plot, will be an apparently symmetrical distribution in the neighborhood of the unperturbed level separation. From a qualitative point of view, the fact that an unbroadened, delta-function spectral line is not observed may be attributed either to radiation damping or to finite energy state width, the first a more or less classical, the second a quantum point of view.

Radiation damping results from the retarding self-force which the electromagnetic field, produced by the electron within the molecule, exerts on that same electron. The calculation of the amplitudes and hence the intensities of the emitted electric vector is straightforward [see § 4 of Heitler (21)] and leads to a symmetrical distribution about the position of the unperturbed spectral line. This is indeed a reasonable approximation. However, the width of the spectral line at half maximum intensity bears little relationship to the observed width. Regardless of emitter type, the width proves to be 0.188×10^{-12} cm. For example, Minkowski (203) and Weingeroff (278) have both found, for the NaD line, a width of 0.62×10^{8} $\sec^{-1}$, which of course corresponds, since $\delta\lambda=c\delta\nu/\nu^2$, to a width in wavelength units of 3.65×10^{-12}. (We shall consistently use λ for wavelength, ν for frequency, and c for the velocity of light.)

The Heisenberg uncertainty principle tells us that energy and time can at most be specified to within $\delta E\delta t>\hbar$, a relation which may be applied to a set of discrete energy levels. Let us suppose that we place our radiator in its ground state, isolated from any sort of perturbing fields. We may quite properly expect our system to remain in that state for an infinite period of time. Since the time is infinite, we may anticipate a zero uncertainty in the measurement of the state energy, not a sort of probability distribution of energies as we would otherwise expect. In other words, this ground state will have zero width. Such is not the case for an excited state, even a metastable one, for the excited state will have a finite lifetime. From the uncertainty relation it follows that the width of the state is inversely proportional to the state lifetime. A very useful approximation is provided by the assumption that the width of the spectral line resulting from an emitting transition to the ground state is equal to the width of the upper state, $\delta=\Delta E=\hbar(\Delta t)^{-1}$, where δ is the spectral line width at half intensity. Now the lifetime of the upper state is the inverse of the probability, w_{ug}, of a radiating transition between the upper and ground states. Therefore,

$$\delta=\hbar w_{ug} \tag{1.1}$$

This relation is of some consequence; indeed, the most esoteric of natural width theories have only been able to improve slightly on it (slightly from a percentage point of view but enormously from a phenomenological point of view). In the following we shall concern ourselves with this improvement.

The Weisskopf-Wigner (281, 282) approach to the natural line shape problem provides the very preponderant first approximation and will be considered from the Källén (27) viewpoint. A few preliminaries, however, are appropriate.

The radiant process is instigated by the interaction between the electromagnetic field, into which photons are being injected or from which they are being withdrawn, and the bound electrons responsible for emission or absorption. The classic treatment of this interaction and its effect on the spectral line is that of Weisskopf and Wigner, a treatment which is basically a first approximation to the two-state emitter problem and which we shall consider in Section 2. In Sections 3 and 4 the Heitler-Ma and the Bali-Higgins studies of the multistate emitter are discussed, followed by a study of Low's Feynman diagram approach (Section 5) and Mizushima et al.'s Liouville approach (Section 6). The chapter concludes with Arnous-Heitler's self energy transformation approach (Section 7) to the inclusion of the effects of higher-order corrections on the natural line shape.

We shall suppose the reader to be familiar with basic Hilbert space concepts [cf. Friedriches (15) and von Neumann (35)] and hence we shall devote no specific attention to these concepts. There will, however, be considerable occasion for us to concern oursleves with tensor products of Hilbert spaces, so that it behooves us to write down a few defining equations. These product spaces will be designated by $H_1\otimes\cdots\otimes H_n$, where the $H_1,\ldots, H_n$ are Hilbert spaces, and the inner product of two elements in the product space is given by

$$\langle f|g\rangle=\langle f^{(1)}\cdots f^{(n)}|g^{(1)}\cdots g^{(n)}\rangle=\prod_{k=1}^{n}\langle f^{(k)}|g^{(k)}\rangle_k \tag{1.2}$$

From Eq. (1.2) it is clear that a vector in the product space may be formed from the vectors in the individual spaces of which the product space is constituted as follows. Given the vectors $h^{(1)},\ldots, h^{(n)}$ in the Hilbert spaces

$H_1, \ldots, H_n$, then the vector in the product space $H_1 \otimes \cdots \otimes H_n$ that corresponds to these vectors is

$$h^{(1)} \cdots h^{(n)} = \left\{ \frac{1}{c} h^{(1)} x \cdots x(ch^{(k)}) x \cdots x h^{(n)} : c \in \mathbb{C}^1, |c| > 0, k = 1, \ldots, n \right\}$$

$\{ch: c \in \mathbb{C}^1\}$ reads the set of all ch such that c belongs to the set of complex numbers.

We shall require certain matrix elements and matrix elements of products in one or more product spaces. Equation (1.2) allows us to draw up the prescriptions for such products. The Liouville operator will be of interest to us so that it may be appropriate to use this particular operator as our example. One way in which we may obtain the operator is as follows.

The equation of motion for an operator A we will agree to be given by

$$\mathrm{i}\dot{A} = [H, A] \tag{1.3}$$

where H is the Hamiltonian, and $[H, A] = HA - AH$. It is often convenient to write this in terms of a Liouville operator, $\mathfrak{L}$, where

$$\mathrm{i}\dot{A} = \mathfrak{L} A \tag{1.4}$$

The matrix element of $\mathfrak{L}A$ must be the same as that of the commutator $[H, A]$ which fact affords us a method of specificizing $\mathfrak{L}$. In order to do so let us take the matrix element of the commutator thus:

$$\begin{aligned}
[H, A]_{im} &= \sum_j \langle i|H|j\rangle\langle j|A|m\rangle - \sum_j \langle i|A|j\rangle\langle j|H|m\rangle \\
&= \sum_{jk} \langle j|A|k\rangle\left[\langle i|H|j\rangle\delta_{km} - \delta_{ij}\langle k|H|m\rangle\right] \\
&= \sum_{jk} \left[\langle i|H|j\rangle\delta^*_{mk} - \delta_{ij}\langle m|H^*|k\rangle\right]\langle j|A|k\rangle
\end{aligned} \tag{1.5}$$

where $\langle k|H|m\rangle = \langle m|H^*|k\rangle$ and $\delta^*_{mk} = \delta_{km}$ since H and δ_{km} are Hermitian. Now let us look back at Eq. (1.2) in order to compare it to the first term, $\langle i|H|j\rangle\delta_{mk}$ in Eq. (1.5). Surely $\delta_{mk} = \langle m|I^*|k\rangle = \langle k|I|m\rangle$, where I is the identity operator. Therefore, if we had a product of two Hilbert spaces, H an operator in one and I^* an operator in the other, we would have an operator HI^* in the product space with the matrix element in that space given by $\langle i|H|j\rangle\delta^*_{mk}$. The second term within the square bracket of Eq. (1.5) would similarly arise from an operator IH^* in the product space. Therefore, if we choose a Liouville operator in the product space,

$$\mathfrak{L} = HI^* - IH^* \tag{1.6}$$

and define matrix multiplication of two operators, one (A) in the simple space (H_1) and one ($\mathfrak{L}$) in the product space ($H_1 \otimes H_2$), as

$$[\mathfrak{L}A]_{im} = \sum_{jk} \langle im|\mathfrak{L}|jk\rangle\langle j|A|k\rangle \tag{1.7a}$$

where

$$\langle im|\mathfrak{L}|jk\rangle = \langle i|H|j\rangle\delta^*_{mk} - \delta_{ij}\langle m|H^*|k\rangle \tag{1.7b}$$

we obtain the desired result.

2 Natural Line Shape to First Approximation

In speaking of the shift or broadening of a spectral line, we are, consciously or unconsciously, basing our discussion on a Dirac delta function sort of line which could never exist in the real world but which we suppose would exist in the absence of interactions (interactions between the reference frames of the observer and the emitter, between the emitter and its neighbors, between the emitting atom and its own electromagnetic field). As a consequence of the interaction of the electron with its own field what "would" have been its unperturbed level location is shifted. This is the Lamb shift. On the basis that it is absurd to treat this as a spectral line shift—it has been discussed *in extenso* in any event [cf. § 37 of Källén (27), S5-3 of Jauch and Rohrlich (26)] —we shall make no attempt to discuss this, restricting ourselves to the broadening, which it may be equally absurd to discuss as if the spectral line could be any other way.

The familiar TDO has the effect of inducing the requisite modifications of a wave function at some temporal boundary, t_0, so as to yield the proper wave function at a later time, t. If we take $\mathfrak{U}(t, t_0)$ as our TDO, we recall the defining equations,

$$|\Psi(t)\rangle = \mathfrak{U}(t,t_0)|\Psi(t_0)\rangle \Rightarrow \dot{\mathfrak{U}}(t,t_0) = -\mathrm{i}H_1\mathfrak{U}(t,t_0) \tag{2.1}$$

where H_1 is that portion of the Hamiltonian responsible for the changes occurring in the interval $[t_0, t]$. We shall suppose the reader familiar with the development leading to [cf. Källén (27), Kirzhnits (28), Schweber (40), etc.]:

$$\mathfrak{U}(t,t_0) = \mathsf{T}_{\leftarrow}\exp\left(-\mathrm{i}\int_{t_0}^{t} dt'\,H_1(t')\right) \tag{2.2}$$

where $\mathsf{T}_{\leftarrow}$ is the TOO whereby time is ordered from right to left. Time ordering from left to right would be indicated by $\mathsf{T}_{\rightarrow}$. Allowing t_0 to recede into the past, t into the future such that $\mathfrak{U}(t, t_0) \rightarrow S(\infty, -\infty)$ yields the S matrix. In general then we may say

$$|\Psi(t)\rangle = S(t, t_0)|\psi_j(t_0)\rangle = \sum \mathsf{c}_i(t)|\psi_i(t_0)\rangle \tag{2.3}$$

since the general principle of superposition may be expected to hold [cf. Schiff (39)]. We know that the probability, w_{ij}, for a transition from an initial state j to a final state i will be given by the square of the SGC, c_i:

$$w_{ij} = |\mathsf{c}_i|^2 = |\langle i|S|j\rangle|^2 \tag{2.4}$$

From Eq. (2.2) we see that the matrix elements of S will correspond to matrix elements of various ordered powers of H_1 so that this operator constitutes the next area of our concern. For natural line broadening this operator describes the particle-field interaction. Section III.13(12) of Heitler (21) provides an expression for this particular interaction Hamiltonian, although we specifically appeal to Eq. (17.22) of Källén in order to obtain:

$$H'(A,\psi) = -\tfrac{1}{2}ie\int d^3x A_i(x)[\Psi(x), \gamma_i\psi(x)] \tag{2.5a}$$

Here and hereafter the square brackets will indicate the commutator; the curly brackets the anticommutator. $A_i(x)$ is the second quantized electromagnetic field operator which is given by Eq. (11.20) of Källén as:

$$A_i(x) = \sum_k \sum_{\lambda=1}^{4} e_i^{(\lambda)} \frac{\left[e^{ikx}\mathsf{a}^{(\lambda)}(k) + e^{-ikx}\mathsf{a}^{*(\lambda)}(k)\right]}{\sqrt{2V\omega}} \tag{2.5b}$$

Here kx is the scalar product of two four-vectors, the four-momentum, $k(=\mathbf{k}, \omega)$, $\mathbf{k}$ the photon momentum, ω the photon frequency, and the position vector x in four-space. The index, λ, relates to polarization directions in four-space. $\mathsf{a}^{(\lambda)}(k)$ annihilates a photon having polarization λ and momentum k. $\mathsf{a}^{*(\lambda)}(k)$ is the analogous creation operator. From, say, § 15g of Schweber (40), we may obtain the sort of Furry particle field operators that we must utilize for ψ.

$$\psi(x) = \sum_{E>0} \{\mathsf{A}_n\psi_i^{(n)}(x)e^{-iE_nt} + B_n^*\psi_i^{(n)}(x)e^{iE_nt}\} \tag{2.5c}$$

where A_n is an electron annihilation operator, B_n^* a positron creation operator. For the cases of interest to us we may simply drop the summation and second term in Eq. (2.5c). The field operator conjugate to $\psi(x)$ is surely

$$\bar{\psi}(x) = \mathsf{A}_n^*\bar{\psi}_i^{(n)}(x)e^{iE_nt}$$

where A_n^* is an electron creation operator. $\psi_i^{(n)}(x)$ is, of course, the wave function for the bound electron. It is obvious that, in evaluating powers of Eq. (2.5a) by means of Eqs. (2.5b) and (2.5c), we shall be creating and annihilating both photons and electrons. We shall take b_n^* as creating a state configuration of photons and electrons, an operator that might create an electron at the same time it is annihilating a photon, create both photon and electron, or, in general, create and annihilate these two entities in whatever fashion is requisite to the creation of the state n. We take b_n as an analogous annihilation operator and return to Eq. (2.4). Since the probability given by this equation is frequency dependent, it is obvious that this probability will yield a spectral line shape.

It is customary to denote the constant term in Eq. (2.2) by $S^{(0)}$, the term containing a product of n of the H_1 as $S^{(n)}$. Let us now write the matrix element of $S^{(n)}$ in terms of the vacuum state function, Φ_0, that is, the second-quantized wave function corresponding to no photons or electrons in the field. In order to do this we recall that a state, n', having some specified number of photons and electrons in the field is expressible as:

$$|\Psi_{n'}\rangle = \mathsf{b}_{n'}^{*}|\Phi_0\rangle$$

Let us now suppose that our radiant transition is one from a state in which the atom is in a state n' and there are no photons in the field to a state in which the atom is in the state n and there is one photon in the field. The matrix element over the scattering matrix of nth order will then be:

$$\langle\Psi_n|S^{(n)}|\Psi_n\rangle = \langle\Phi_0|\mathsf{b}_n S^{(n)}\mathsf{b}_{n'}^{*}|\Phi_0\rangle \tag{2.6}$$

In Eq. (2.6) Φ_0 is the vacuum function, b_n annihilates an electron in a state n, $\mathsf{b}_{n'}^{*}$ creates an electron in a state n' and a photon in an appropriate state. In the zeroth approximation then ($S^{(0)}=1$) we have $\mathsf{b}_{n'}^{*}$ creating an electron in n' and a photon, then b_n attempting to eliminate an electron in state n, an electron which is not there. The result is, of course, zero. In the first approximation ($S^{(1)}=H_1$), $\mathsf{b}_{n'}^{*}$ creates as before, but now Ψ, ψ, and A furnish another A^{*}, A, and a. If $\mathsf{A}\equiv\mathsf{A}_{n'}$, the initially created electron is annihilated; if $\mathsf{A}^{*}\equiv\mathsf{A}_{n}^{*}$, an electron is created in state n that the operator in Eq. (2.5) may annihilate. The a of H_1 has, in the meantime, annihilated the photon originally created. The point of all this is that the matrix element of $S^{(1)}$ will not disappear for this emission event, at least not as a consequence of these operators. Also, as may analogously be seen, no $S^{(n)}$ will disappear for n odd. In like manner all $S^{(n)}$ disappear for n even. As we shall see, these even n relate to scattering processes where the initial and final states are the same, there being photons and electrons present in both states. They may also relate to the probability for the existence of a no-photon initial state after time t. Let us return to Eq. (2.1) where, of course, $\mathfrak{U}$ may correspond to S.

In this equation $|\Psi(t)\rangle$ is the system state function (vector) at some time t, Ψ_n by our own assumption being the "initial state" of the system. At what time should we take this "initial state" to have existed? In the S-matrix treatment the initial time is usually taken to be $-\infty$, but such an assumption will find the radiator as yet unexcited and hence not in a particularly appropriate condition. So we take this initial time as $t=0$ and thus begin following the treatment of Weisskopf and Wigner (281, 282) as adapted to the TDO formalism by Källén (see § 28). This all important boundary condition is expressed as

$$\mathfrak{U}(0)=1. \tag{2.7}$$

Let us now use Eqs. (2.5) in Eq. (2.4) in order to write down the matrix element

$$\langle\Psi_n|H_1|\Psi_{n'}\rangle = -\frac{1}{2}\mathrm{i}e\sum_{\lambda i}\int d^3x\left[\psi_n(x)\gamma_i e_i^{(\lambda)}\psi_{n'}(x)\right]e^{-\mathrm{i}\mathbf{k}\cdot\mathbf{x}}e^{\mathrm{i}(E_n+\omega-E_{n'})t}$$
$$\cdot\frac{\langle\Phi_0|b_n b_n^{*} b_{n'} b_{n'}^{*}|\Phi_0\rangle}{\sqrt{2V\omega}}$$

The $\bar{\psi}_n$ and $\psi_{n'}$ are row and column matrices, respectively [see Bethe and Salpeter (7)], originating in the relativistic Dirac equation for the bound electron. These wave functions are supposed to describe the radiating "atom," but, practically speaking, they do so only if the atom possesses but a single electron or if a photoelectron and core are sufficiently well defined so that this is effectively the case. The γ_i is a 4×4 matrix.

The matrix element on the right is, of course, unity. We use Källén's shorthand

$$\langle \Psi_n | H_1 | \Psi_{n'} \rangle = C_k \exp(i\Delta\omega t) \tag{2.8a}$$

$$\Delta\omega = E_n + \omega - E_{n'} \tag{2.8b}$$

$$C_k = -\tfrac{1}{2} ie \sum_{\lambda i} \int \left[\psi_n(x) \gamma_i e_i^{(\lambda)} \psi_{n'}(x) \right] e^{-i\mathbf{k}\cdot\mathbf{x}} d^3x / \sqrt{2V\omega} \tag{2.8c}$$

From Eq. (2.4) we know that, if $\mathsf{a}(t)$ is the SGC for the upper or initial state, $\mathsf{b}_k(t)$ that for the lower or final state with a photon of momentum $\mathbf{k}$ in the field,

$$\mathsf{a}(t) = \langle \Psi_{n'} | \mathcal{U}(t) | \Psi_{n'} \rangle \tag{2.8d}$$

$$\mathsf{b}(t) = \langle \Psi_n | \mathcal{U}(t) | \Psi_{n'} \rangle \tag{2.8e}$$

The elementary time-dependent perturbation theory [cf. § 29 of Schiff (39)] tells us that

$$i\dot{\mathsf{a}}(t) = \sum_k c_k^* \exp(-i\Delta\omega t) \mathsf{b}_k(t) \tag{2.9a}$$

$$i\dot{\mathsf{b}}_k(t) = c_k \exp(i\Delta\omega t) \mathsf{a}(t) \tag{2.9b}$$

Now if $\mathcal{U}(0) = 1$ [Eq. (2.7)], Eqs. (2.8d) and (2.8e) tell us that

$$\mathsf{a}(0) = 1, \qquad \mathsf{b}_k(0) = 0 \tag{2.10}$$

Equation (2.9b) yields an expression for b_k which we substitute into Eq. (2.9a) in order to obtain

$$\mathsf{b}_k(t) = -ic_k \int_0^t dt' \, e^{i\Delta\omega t'} a(t') \Rightarrow$$

$$\dot{\mathsf{a}}(t) = -\sum_k |\mathsf{c}_k|^2 \int_0^t dt' \, e^{-i\Delta\omega(t-t')} \mathsf{a}(t') \tag{2.11a}$$

Equation (2.11a) may now be solved by Laplace transforming $\mathsf{a}(E)$ $(= \int dt \exp(-Et)\mathsf{a}(t))$, integrating and inverting the transform. The left side of Eq. (2.11a) may be integrated by parts as

$$\int_0^\infty e^{-Et} \dot{\mathsf{a}}(t) \, dt = e^{-Et} \mathsf{a}(t) |_0^\infty + E \int_0^\infty dt \, e^{-Et} \mathsf{a}(t) = E\mathsf{a}(E) - 1 \tag{2.11b}$$

since the integrand on the right is, save for $a(t)$, a delta function. Using a well-known property of definite double integrals we evaluate the right side of Eq. (2.11a) as

$$\int_0^\infty dt\, e^{-Et}\int_0^t dt'\, e^{-\mathrm{i}\Delta\omega(t-t')}a(t') = \int_0^\infty dt'\, \mathsf{a}(t')\int_{t'}^\infty dt\, e^{-(E+\mathrm{i}\Delta\omega)t+\mathrm{i}\Delta\omega t'}\frac{\mathsf{a}(E)}{(E+\mathrm{i}\Delta\omega)} \tag{2.11c}$$

From Eqs. (2.11) we then obtain

$$\mathsf{a}(E) = \left(E + \sum_k \frac{|c_k|^2}{(E+\mathrm{i}\Delta\omega)}\right)^{-1} \tag{2.12}$$

The familiar inverson formula for the Laplace transform is readily available [cf. § 29 of Abramowitz and Stegun (3)] and yields the SGC:

$$\mathsf{a}(t) = (2\pi\mathrm{i})^{-1}\int_{\epsilon-\mathrm{i}\infty}^{\epsilon+\mathrm{i}\infty} dE\, e^{Et}\Bigg/\left[E + \frac{\sum_k |c_k|^2}{(E+\mathrm{i}\Delta\omega)}\right] \tag{2.13}$$

and we are left with a rather involved problem in complex integration. The path is given by the limits lying parallel to and to the right of the imaginary axis in the E plane. Taking E as $\epsilon+\mathrm{i}z$, Eq. (2.13) may be written

$$\mathsf{a}(t) = (2\pi\mathrm{i})^{-1}\int_{-\infty}^{\infty} dz\, e^{\mathrm{i}zt}e^{\epsilon t}\Bigg/\left(z - \mathrm{i}\epsilon - \sum_k \frac{|c_k|^2}{(z+\Delta\omega-\mathrm{i}\epsilon)}\right)$$

$$= (2\pi\mathrm{i})^{-1}\int_{-\infty}^{\infty} dz\, e^{\mathrm{i}zt}\Bigg/\left(z - \mathcal{P}\sum \frac{|c_k|^2}{(z+\Delta\omega)} - \mathrm{i}\pi\sum_k |c_k|^2\delta(z-\Delta\omega)\right) \tag{2.14}$$

where we have let $\epsilon\to 0$ and used $\lim(a-\mathrm{i}\epsilon)^{-1} = \mathcal{P}a^{-1} + \mathrm{i}\pi\delta(a)$.

As Källén has remarked, for small t, transients would be present as a consequence of switching on the interaction at $t=0$. (We shall later encounter transients so-called in pressure broadening. These, however, are a physical phenomenon related to the time of collision and not an artificial consequence of this form of adiabatic switching.) These transients are hardly of interest since our observation of the spectral line can only be expected to take place at a considerably later time. Therefore, we are quite justified in simplifying matters by considering only times such that $tE_n \gg 1$ and $tE_{n'} \gg 1$. Remark the conditions for the integral of Eq. (2.14) over a semicircular path in the upper half-plane are such as to fulfill Jordan's lemma [cf. § 6.222 of Whittaker and Watson (47)] so that the integral over this path is zero. Therefore, the integral on the right of Eq. (2.14) is simply the residue at whatever poles are provided by the denominator. When we drop the z from the principal value and the delta function, there is a simple pole and a ready result:

$$\mathsf{a}(t)=(2\pi\mathsf{i})^{-1}\int_{-\infty}^{\infty} dz\, e^{\mathsf{i}zt}\Big/\left(z-\mathcal{P}\sum_k \frac{|c_k|^2}{\Delta\omega}-\mathsf{i}\pi\sum_k |c_k|^2\delta(\Delta\omega)\right)$$

$$=\exp\left[-\mathsf{i}\Delta t-\tfrac{1}{2}\delta t\right] \tag{2.15a}$$

$$\delta=2\pi\sum_k |c_k|^2\delta(\Delta\omega) \tag{2.15b}$$

$$\Delta=-\mathcal{P}\sum_k \frac{|c_k|^2}{\Delta\omega} \tag{2.15c}$$

The conditions for lemma fulfillment are derived from the fact that $|\oint \exp(\mathsf{i}zt)Q(z)\,dz|<\epsilon/t$ where $|Q(z)|<\epsilon$ for $|z|>r_0$, ϵ and r_0 being chosen.

Therefore, the probability that our atom is still, say, in its upper state at time, t, is

$$|\mathsf{a}(t)|^2=\exp(-\delta t) \tag{2.16}$$

the assumption originally made by Weisskopf and Wigner and the sort of decay one has come to expect. The state lifetime is δ^{-1}; δ is the transition probability per unit time. Now we are in a position to evaluate $\mathsf{b}_k(t)$ from Eq. (2.11a),

$$\mathsf{b}_k(t)=-c_k\frac{\left\{\exp\left[\left(\mathsf{i}\Delta\omega-\mathsf{i}\Delta-\tfrac{1}{2}\delta\right)t\right]-1\right\}}{\Delta\omega-\Delta E+\tfrac{1}{2}\mathsf{i}\delta}$$

The spectral line shape for large times is then

$$I(\omega)=|\mathsf{b}_k(t)|^2=\frac{|c_k|^2}{(\Delta\omega-\Delta)^2+(\delta/2)^2}\left[1+e^{-\delta t}f(t)\right]$$

$$=\frac{|c_k|^2}{(\Delta\omega-\Delta)^2+(\delta/2)^2} \tag{2.17}$$

for t large as assumed. In Eq. (2.17) δ is the width of the line, that is, the frequency width at half maximum intensity and Δ is the "shift" of the intensity maximum from a fictitious unperturbed location. Finally, $\Delta\omega$ is the frequency separation from the unshifted intensity maximum.

The result Eq. (2.17) is general for (1) large times and (2) a two-state atom. Now if we wish to use Eqs. (2.8) for a evaluation of the c_k, we are applying first-order S-matrix theory, the walking-stick diagram (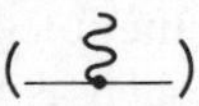) . This appears to agree quite well with experiment, although one could go to higher approximations. Of this more later, for now let us investigate the widths as a function of atomic number, an investigation largely concerned with hydrogenic atoms.

Sufficient complications will be found in the emitting transition employing the simplest wave function, this transition being from $2p: P_{3/2}$ or $P_{1/2}$ to the $1s: S_{1/2}$. We shall consider only the spin-up transition. The wave functions for the upper and lower states are given by Eqs. (14.38) through (14.42) of Bethe and Salpeter (7). As an example, we write down the upper state functions:

$$G_{2p}(\mathfrak{r}_3)=Z_1^{3/2}\sqrt{\frac{1+\epsilon_3}{2\Gamma(2\gamma_2+1)}}\; e^{-r/2}3\mathfrak{r}_3^{\gamma_2-1} \tag{2.18a}$$

$$F_{2p}(\mathfrak{r}_3)=-\sqrt{\frac{1-\epsilon_3}{(1+\epsilon_3)}}\; G_{2p}(\mathfrak{r}_3) \tag{2.18b}$$

$$\gamma_2=\sqrt{4-\alpha^2 Z^2}\,,\qquad N_3=2,\qquad \epsilon_3=\left[1+(\alpha Z/\gamma_2)^2\right]^{-1/2},\qquad \mathfrak{r}_3=Z_1 r \tag{2.18c}$$

$$\bar{\psi}_{2p}=\sqrt{3/8\pi}\,\big(G_{2p}(\mathfrak{r}_3)\sin\vartheta e^{i\varphi},\quad 0,\quad iF_{2p}(\mathfrak{r}_3)\sin\vartheta\cos\vartheta e^{i\varphi},$$
$$-iF_{2p}(\mathfrak{r}_3)\sin^2\vartheta e^{2i\varphi}\big) \tag{2.18d}$$

We are very nearly in a position to evaluate Eq. (2.8c). However, base vectors, $e_i^{(\lambda)}$, are required. The base vectors as given by Källén [Eqs. (5.30)] are quite convenient for evaluating certain commutation relations, but the $e_i^{(1)}$ and $e_i^{(2)}$, $i=1,2$ are somewhat complex. We therefore develop a different set which we display as Table 2.1.

Table 2.1

i \ λ	1	2	3	4	
1	(k_1+ik_2)	0	0	0	$\frac{1}{\omega}$
2	0	ω	0	0	
3	k_3	0	(k_1+ik_2)	0	
4	0	0	0	ω	

That this set of basis vectors yields the commutation relations desired is immediately demonstrable.

We consider the exponential factor in Eq. (2.8a). The wavelength of the radiation is inversely proportional to k, the region of integration, and hence to the maximum value of x of the order of the dimensions of the emitting atom. Therefore, in the visible region the exponent $\mathbf{k}\cdot\mathbf{x}$ will be of the order of 10^{-4}, and the exponential may surely be taken as unity. We shall so take it, but with the understanding that, if we move to sufficiently high frequencies (x ray and gamma) our result will be too large as a consequence. In order to write down the matrix element we recall that $|\mathbf{k}|=\omega$ so that

$$\frac{\mathbf{k}_3}{\omega}=\cos\vartheta,\qquad \frac{\mathbf{k}_1+\mathrm{i}\mathbf{k}_2}{\omega}=\sin\vartheta e^{i\varphi} \tag{2.19}$$

We now apply Eqs. (2.18), an analogous expression for the S function, Eq. (2.19), and Table 2.1 to the evaluation of

$$\int d^3x\Big(\bar{\psi}_{1s}(x)\gamma_i\sum_\lambda e_i^{(\lambda)}\psi_{2p}(x)\Big)=\frac{1}{4\pi}\sqrt{\tfrac{3}{2}}\int d^3x$$

$$\times\left[(G_{1s},0,-\mathrm{i}\cos\vartheta G_{1s},-\mathrm{i}\sin\vartheta e^{i\varphi}G_{1s})\sum_\lambda e_1^{(\lambda)}\begin{pmatrix}G_{2p}\sin^2\vartheta e^{-\mathrm{i}2\varphi}\\ G_{2p}\sin\vartheta\cos\vartheta e^{-\mathrm{i}\varphi}\\ 0\\ G_{2p}\sin\vartheta e^{-\mathrm{i}\varphi}\end{pmatrix}+\cdots\right]$$

$$=\frac{1}{4\pi}\sqrt{\tfrac{3}{2}}\int d^3x\,G_{1s}[k_1+\mathrm{i}k_2-k_3]G_{2p}\sin^2\vartheta e^{-\mathrm{i}2\varphi}/\omega+\cdots$$

$$=\frac{\sqrt{6}}{5}\int G_{1s}G_{2p}r^2\,dr \tag{2.20}$$

Using Eqs. (2.18a) and a corresponding equation for G_{1s} we may evaluate the integral in Eq. (2.20) as

$$\int G_{1s}G_{2p}r^2\,dr=2^{\gamma_1-1/2}\mathrm{Z}^{\gamma_1+\gamma_2+1}$$

$$\times\left\{\frac{\left[1+\sqrt{1-\alpha^2\mathrm{Z}^2}\,\right]\left[1+\sqrt{1-\alpha^2\mathrm{Z}^2/4}\,\right]}{\Gamma(2\gamma_1+1)\Gamma(2\gamma_2+1)}\right\}^{1/2}\int r^{\gamma_1+\gamma_2}e^{-3\mathrm{Z}_1 r/2}\,dr \tag{2.21}$$

For the Schrödinger hydrogen case $\gamma_1=\mathrm{Z}=1$, $\gamma_2=2$, and Eq. (2.21) reduces to $6^{-1/2}\int r^3e^{-3r/2}\,dr$ as it should. Carrying out the integration in Eq. (2.21) yields

$$\int G_{1s}G_{2p}r^2\,dr=\frac{2^{2\gamma_1+\gamma_2+1/2}}{3^{\gamma_1+\gamma_2+1}}\Gamma(\gamma_1+\gamma_2+1)$$

$$\times\left\{\frac{\left[1+\sqrt{1-\alpha^2\mathrm{Z}^2}\,\right]\left[1+\sqrt{1-\alpha^2\mathrm{Z}^2/4}\,\right]}{\Gamma(2\gamma_1+1)\Gamma(2\gamma_2+1)}\right\}^{1/2} \tag{2.22}$$

We remark that, for the nonrelativistic case, this integral has the value $16\sqrt{6}/81$. Thus since the spectral line width is directly proportional to the square of the integral, we may readily calculate the variation of line width with atomic number. We do so for $\mathrm{Z}=1, 10, 50, 100$. The results are displayed as Table 2.2.

Table 2.2

Z	I_r	I_{nr}	δ_r	δ_{nr}
1	0.483	0.483	1.000	1.000
10	0.483	0.483	1.000	1.000
50	0.435	0.483	0.814	1.000
100	0.299	0.483	0.384	1.000

In Table 2.2 the I_r column lists the relativistic integral, the I_{nr} the nonrelativistic. As we shall see, we cannot list the line widths as a function of Z since they depend on the spectral line frequency. However, the widths will be proportional to the squares of the integrals, and we thus list the normalized squares of the integrals as δ_r (relativistic) and δ_{nr} (nonrelativistic). We may certainly remark that the relativistic effects on the widths have become pronounced for Z of the order of 100.

If we call the value of the integral Eq. (2.22) A, then, from Eqs. (2.15b) and (2.8c), we may find the linewidth as

$$\delta = \frac{2\pi e^2 A^2}{2V}\,\frac{2}{3}\sum_k \frac{\delta(\Delta\omega)}{\omega} = \frac{e^2A^2}{3(2\pi)^2}\int \frac{\delta(\Delta\omega)}{\omega}\,d^3\mathsf{k}$$

where we have multiplied Eq. (2.8c) by two to account for both spin directions. Using Eq. (2.8b) and the fact that $\omega = |\mathbf{k}|$, we find

$$\delta = \frac{e^2A^2}{3(2\pi)^2}\int \frac{\delta[\omega-(E_{n'}-E_n)]}{\omega}\,\omega^2\,d\omega\,d\Omega = \frac{e^2A^2(E_{n'}-E_n)}{3\pi} \tag{2.23a}$$

For the nonrelativistic $\mathsf{Z}=1$ case this becomes

$$\delta = 6.55\times10^{-7}(E_{n'}-E_n) \tag{2.23b}$$

The constant is dimensionless, although computed in natural units, so that δ will be in whatever units are chosen for E_n. Suppose we consider the NaD line, $\nu = 5.1\times10^{14}\,\text{sec}^{-1}$. Then $\delta = 9.2\times10^{8}$, hardly in good agreement with the observed 0.62×10^{8}, but the D line does arise from a $3p\to3s$ transition which we have approximated by a $2p\to1s$. The s orbit will certainly be a penetrating one for which the electron will hardly see a Z of one. The latter consideration is demonstrably more serious than the fact that we are using the wrong orbitals in computing the matrix element. Let us take a $3p$ orbital with $\mathsf{Z}=1$ and a $3s$ orbital of arbitrary Z from, say, Bethe and Salpeter [see § 3β but remark the error in Eq. (3.17)]. A Z value of about nine (for Na, $\mathsf{Z}=11$) yields the correct result. This is certainly substantial penetration, and at least some of it could and should be eliminated by increasing Z in the $3p$ orbital. Be this as it may, the broadening theory and level of approximation fare rather well.

3 The Multistate Emitter. Heitler Approach

The results which we have obtained involve two obvious approximations: (1) we have supposed the emitting atom to have only two energy states and (2) in computing the SGC we have evaluated the TDO only through first approximation. Apparently then, improvements in the theory could be anticipated if the atomic state structure chosen more realistically reflected that of the radiator and if the higher-order corrections to the TDO were considered. As we shall see, (1) must be modified for close lying states, but (2) is an excellent approximation.

A level lying close to the upper level of a transition pair can noticeably distort the natural shape of the spectral line, but fairly distant levels which are not coupled to the final state by large transition probabilities are essentially unimportant, as are the higher-order corrections to the TDO. That the original Weisskopf-Wigner result is improved almost not at all by these corrections was demonstrated some years ago by Arnous and Heitler (58), and we shall consider that demonstration later. We now consider what we may call the Heitler approach to the three-level atom which was developed by Morozov and Shorygin (MS) (209).

We may write the state function for our system, atom plus radiation field, and the equation for the SGC's as usual:

$$|\Psi(t)\rangle = \sum_i \mathsf{a}_i(t)|\psi_i\rangle \qquad \dot{\mathsf{a}}_j = \sum_i \mathsf{a}_i(t)H_{ji}(t) \tag{3.1}$$

where now the index i runs over a number of atomic states. We solve Eq. (3.1) iteratively as we may for the TDO, beginning with $\mathsf{a}_j^0 = 1$ and obtaining

$$\mathsf{a}_j^{(1)} = \sum_i \int H_{ji}(t)\,dt$$

for the first approximation,

$$\mathsf{a}_j^{(2)} = \sum_{ik} \int\int H_{jk}(t')H_{ki}(t)\,dt\,dt' = \sum_i \int\int [H(t')H(t)]_{ji}\,dt\,dt'$$

and so on. The final result is

$$\mathsf{a}_j = \mathsf{T}_{\leftarrow}\exp\left(\sum_i \int H_{ij}(t)\,dt\right) \tag{3.2}$$

Equation (3.2) differs from the two-state result only in the summation within the exponential. Therefore, we see that, where the line breadth was equal to the transition probability between the two states in the two-state case, it is now, for emission, equal to the sum of the various transition probabilities to the various lower lying states. (This is intuitively obvious from our elementary uncertaintly principle discussion of Section 1.) Practically speaking, one can often obtain a reasonable idea as to the importance of multistate considerations. In the case of the NaD line, for example [cf. § 28 of Allen (5)] the

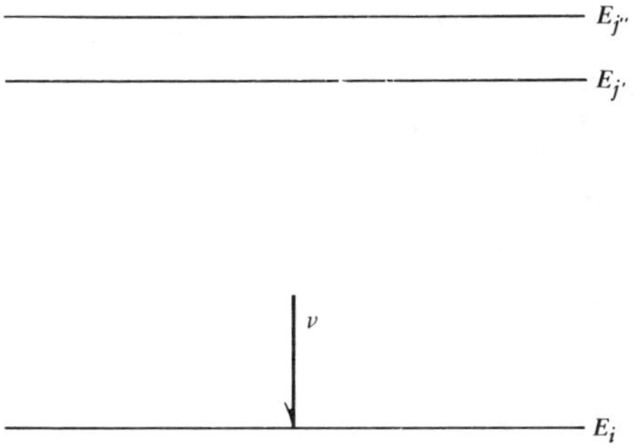

Figure 3.1 After Morozov and Shorygin (209).

oscillator strength for the multiplet ($3p^2P \to 2s^2S$) is large (1.93); there are no other emitting transitions, and the two-state approximation should be excellent. For the He $1s2p^1P$ level such should not be the case, however, for, from this level, allowed transitions may proceed to the $1s2s^1S$ ($gf=0.375$) or to the $1s^2\,{}^1S$ (0.35).

MS have demonstrated the asymmetrizing effect on the spectral line shape of a level lying close to the upper state of an emitting transition and interacting with that state. These authors begin with Eq. (2.15b) for the line width, and we reproduce their energy level scheme as Fig. 3.1. The atom is originally in a state j'' with no photons in the field, ultimately in a state i with a photon of frequency ν in the field. From Eq. (3.1) we may write

$$\mathrm{i}\dot{a}_{i\nu} = a_{j'} H_{ij'} e^{\mathrm{i}(\omega_\nu - \omega_{j'i})t} + a_{j''} H_{ij''} e^{\mathrm{i}(\omega_\nu - \omega_{j''i})t} \tag{3.3a}$$

$$\mathrm{i}\dot{a}_{j'} = \sum_\nu a_{i\nu} H_{j'i} e^{\mathrm{i}(\omega_{j'i} - \omega_\nu)t}, \qquad \mathrm{i}a_{j'} = \sum_\nu a_{i\nu} H_{j''i} e^{\mathrm{i}(\omega_{j''i} - \omega_\nu)t} \tag{3.3b}$$

wherein the H operator is in the interaction representation and gives rise to the exponentials. Although the authors do not so state, they rather obviously have supposed $H_{j'j''}=0$ in order to obtain Eq. (3.3b). For, say, electric dipole transitions this would surely be the case. Their method of solution for the desired $a_{i\nu}(\infty)$ has been given by Heitler (see § 16).

Our boundary conditions are again $a_{j''}(+0)=1$, $a_i(0)=a_{j'}(0)=0$. Although the a's are not physically meaningful for $t<0$, it is convenient to extend formally their definition to this region. We do this by letting $a_j(t)=a_i(t)=0$ for $t<0$. Now $a_{j''}(+0)=1$ so that there will be a singularity at $t=0$. This means that the time derivative, zero for -0 and zero for $+0$, will be infinite there, so that $a_{j''}$ will be a Dirac delta function in the neighborhood of $t=0$. (By the time $+0$ what we mean is the limiting value of whatever the function of time may be as time approaches zero from positive values.) Equation (3.3a) may be written for $-\infty < t < \infty$:

$$\mathrm{i}\dot{a}_{i\nu} = \sum_j a_j H_{ij} e^{\mathrm{i}(\omega_\nu - \omega_{ji})t} + \mathrm{i}\delta_{ij}\delta(t) \tag{3.4}$$

In solving Eq. (3.4) it is convenient to take the Fourier transform of a_k, thus

$$\mathsf{a}_k(t) = -(2\pi \mathrm{i})^{-1}\int_{-\infty}^{\infty} dE\, G_{kj''}(E)e^{-\mathrm{i}(E_k - E)t} \tag{3.5a}$$

where $G_{kj''}$ is to be determined. Also

$$\mathrm{i}\delta(t) = -(2\pi \mathrm{i})^{-1}\int_{-\infty}^{\infty} dE\, e^{\mathrm{i}(E_{j''} - E)t} \tag{3.5b}$$

We substitute Eqs. (3.5) into Eq. (3.4) and obtain

$$(E - E_i)G_{ij''}(E) = \sum_j H_{ij}G_{jj''}(E) + \delta_{kj''} \tag{3.6}$$

Dividing through by $E - E_k$ is of course complicated by the fact that $E - E_k$ may be zero. For $xf(x) = g(x)$, $g(0)$ nonsingular, $f(x) = g(x)[1/x + \alpha\delta(x)]$, α arbitrary, since $x\delta(x) = 0$. We then choose $\alpha = -\mathrm{i}\pi$, which will satisfy the boundary conditions, and divide through by $(E - E_k)$ in Eq. (3.6) in order to obtain a result from which we separate $G_{j'j''}$ by introducing $U_{ij''}$ as follows:

$$G_{ij''}(E) = U_{ij''}(E)G_{j''j''}(E)\zeta(E - E_i),\quad U_{j''j''} \equiv 0 \tag{3.7a}$$

$$\zeta(E - E_n) = \mathcal{P}\left(\frac{1}{(E - E_n)}\right) - \mathrm{i}\pi\delta(E - E_n) \tag{3.7b}$$

Remarking that $(E - E_i)\zeta(E - E_i) = 1$ and substituting Eqs. (3.7) into Eq. (3.6) yields:

$$U_{ij''}(E)G_{j''j''}(E) = H_{ij''}G_{j'j''}(E) + H_{ij'}G_{j'j''}(E)$$

$$= H_{ij''}G_{j''j''}(E) + H_{ij'}U_{j'j''}(E)G_{j''j''}(E)\zeta(E - E_{j'})$$

$$U_{ij''}(E) = H_{ij''} + H_{ij'}U_{j''j''}(E)\zeta(E - E_{j'}) \tag{3.8a}$$

$$U_{j'j''}(E) = \sum_{\nu} H_{j'i_\nu}\zeta(E - E_i)U_{i_\nu j''}(E) \tag{3.8b}$$

From this point, $i_\nu \equiv i$.

Next we may obtain, from Eq. (3.6), the following equation:

$$(E - E_{j''})G_{j''j''}(E) = 1 + G_{j''j''}(E)\left\{H_{j''j''} + \sum_k H_{j''k}\zeta(E - E_k)U_{kj''}(E)\right\}$$

of solution

$$G_{j''j''} = \left[E - E_{j''} + \tfrac{1}{2}\mathrm{i}\Gamma(E)\right]^{-1} \tag{3.9a}$$

where

$$\tfrac{1}{2}\Gamma(E) = \mathrm{i}H_{j''j''} + \mathrm{i}\sum_k H_{j''k}\zeta(E - E_k)U_{kj''}(E) \tag{3.9b}$$

Now Eqs. (3.9a) and (3.7a) may be substituted into Eq. (3.5a) with the result

$$\mathsf{a}_k(t) = -(2\pi \mathrm{i})^{-1} \int_{-\infty}^{\infty} dE\, U_{kj''}(E) \zeta(E-E_k) \frac{e^{\mathrm{i}(E_k - E)t}}{\left[E - E_{j''} + \frac{1}{2}\mathrm{i}\Gamma(E)\right]} \tag{3.10a}$$

$$\mathsf{a}_{j''}(t) = -(2\pi \mathrm{i})^{-1} \int_{-\infty}^{\infty} dE \frac{e^{\mathrm{i}(E_{j''} - E)t}}{\left[E - E_{j''} + \frac{1}{2}\mathrm{i}\Gamma(E)\right]} \tag{3.10b}$$

When $t=0$, the exponential in Eq. (3.10a) becomes unity so that, for $k \neq j$,

$$\mathsf{a}_k(0) = \int_{-\infty}^{\infty} dE\, U_{kj''}(E) G_{j''j''}(E) \zeta(E-E_k) = 0 \tag{3.11}$$

according to our boundary conditions. This means that the sum of all residues in the upper half of the complex plane (ζ regular there) is zero. Equations (3.11) and (3.10a) may be combined as

$$\begin{aligned}\mathsf{a}_k(t) &= -(2\pi \mathrm{i})^{-1} \int_{-\infty}^{\infty} dE\, U_{kj''}(E) \zeta(E-E_k) \frac{\left[e^{\mathrm{i}(E_k - E)t} - 1\right]}{\left[E - E_0 + \frac{1}{2}\mathrm{i}\Gamma(E)\right]} \\ &= -(2\pi \mathrm{i})^{-1} \int_{-\infty}^{\infty} dE\, U_{kj''}(E) \frac{\left[e^{\mathrm{i}(E_k - E)t} - 1\right]}{\left[E - E_0 + \frac{1}{2}\mathrm{i}\Gamma(E)\right]\left[E - E_k\right]}\end{aligned} \tag{3.12}$$

since the bracketed factor vanishes for $E = E_k$ and eliminates the singularity with which $\zeta(E - E_k)$ is concerned. Having eliminated ζ, we shall now reintroduce it, the net effect of this apparently circuitous procedure being the elimination of the exponential in Eq. (3.10a). Let us recall (§ 8 of Heitler):

$$\zeta(x) = -\mathrm{i} \int_0^{\infty} e^{\mathrm{i}xt}\, dt = \lim_{t \to \infty} \frac{\left[1 - e^{-\mathrm{i}xt}\right]}{x} = \mathcal{P}\left(\frac{1}{x}\right) - \mathrm{i}\pi\delta(x) \tag{3.13}$$

Therefore, when we take t very large, Eq. (3.12) becomes

$$\mathsf{a}_k(\infty) = -(2\pi \mathrm{i})^{-1} \int_{-\infty}^{\infty} dE\, U_{kj''}(E) \frac{\zeta(E_k - E)}{\left[E - E_{j'} + \frac{1}{2}\mathrm{i}\Gamma(E)\right]} \tag{3.14}$$

When we compare Eq. (3.14) with Eq. (3.11), we find that the former differs from the latter by the replacement of $\zeta(E_k - E)$ by $\zeta(E - E_k)$. This of course is the crucial step in the derivation. For (1) $\zeta(x) + \zeta(-x) = -2\pi\mathrm{i}\delta(x)$, (2) zero in the form of Eq. (3.11) may be added to Eq. (3.44) with no effect, and (3) the integration has now become trivial.

$$\begin{aligned}\mathsf{a}_k(\infty) &= \int_{-\infty}^{\infty} dE\, U_{kj''}(E) \frac{\delta(E - E_k)}{\left[E - E_{j''} + \frac{1}{2}\mathrm{i}\Gamma(E)\right]} \\ &= \frac{U_{kj''}(E_k)}{\left[E_k - E_{j''} + \frac{1}{2}\mathrm{i}\Gamma(E_k)\right]}\end{aligned} \tag{3.15}$$

The probability that our atom will eventually be found in the lower state with a photon of frequency ν in the field is thus

$$I(\nu)=|\mathrm{a}_\nu(\infty)|^2=\frac{|U_{ij''}(E_i)|^2}{(E_i-E_{j''}-\Delta)^2+(\frac{1}{2}\delta)^2} \tag{3.16a}$$

$$\delta=\Re[\Gamma(E_i)], \qquad \Delta=\Im[\Gamma(E_i)] \tag{3.16b}$$

Equation (3.16a) is the Heitler-Ma solution with which MS began. From Eqs. (3.8) we may obtain expressions for the $U_{ij''}$ and $U_{ij'}$ in terms of $H_{ij''}$ and $H_{ij'}$. Since the only real point here is the asymmetry introduced into the spectral line by the neighboring level, j', and since the numerator of Eq. (3.16a) is responsible for asymmetry, our principal concern is with $U_{ij''}$. This may be obtained from Eqs. (3.8) as

$$U_{ij''}=H_{ij''}-\frac{\frac{1}{2}\mathrm{i}\gamma H_{ij'}\zeta(E-E_{j'})}{\left[1+\frac{1}{2}\mathrm{i}\gamma_j\zeta(E-E_{j'})\right]}$$

$$=H_{ij''}-\frac{\frac{1}{2}\mathrm{i}\gamma H_{ij'}}{\left[E-E_{j'}+\frac{1}{2}\mathrm{i}\gamma_j\right]} \tag{3.17a}$$

where

$$\gamma_j=2\mathrm{i}\sum_\nu |H_{ij'}|^2\zeta(E-E_{j''}) \tag{3.17b}$$

$$\gamma=2\mathrm{i}\sum_\nu H_{j'i}H_{ij''}\zeta(E-E_{ij}) \tag{3.17c}$$

It is obvious that the $E-E_{j'}$ factor in the denominator of Eq. (3.17a) will lead to asymmetries in the spectral line shape Eq. (3.16a). We see that, if the j' level is eliminated, $H_{ij'}=0$, and the asymmetry disappears.

4 The Multistate Emitter. Transition Operator Approach

The MS result of Section 3 has been obtained by Bali and Higgins (BH) (61) using a considerably different approach. In their approach, we first encounter the density matrix. We may recall that the density matrix was first applied to spectral line broadening over thirty years ago by Karplus and Schwinger [159 and see § 2.13 of Breene (9)]. Lenitzky (174–176) used this type of treatment in order to deal with the natural linewidth to be anticipated in the radiation from a two-level atom. Mollow and Miller (207) likewise used the density matrix—in their case, the reduced density operator which we shall encounter in Section 36—in order to study the two-level atom as a driven system that is damped by its coupling to a "bath" of radiation field oscillators, their treatment being restricted to second order in the perturbation. At about the same time Lehmberg (173) treated the multilevel atom using the density matrix together with a so-called transition operator, his treatment not being restricted to any particular order of perturbation. Finally, BH used this transition operator together with certain of the techniques introduced by Lehmberg in order to study the problem considered by MS and arrive at the same result. This we now consider.

In so doing we follow the nomenclature of their predecessors by terming the radiator a "system," the field oscillators a "reservoir." We construct the Hamiltonian for our system-reservoir ensemble in a second-quantized form which simply means that the state vectors are number vectors, the functionals associated with the axes being integral parts of the operators. The system portion of our Hamiltonian is

$$H_s=\sum_\alpha \epsilon_\alpha \mathsf{A}_\alpha^+\mathsf{A}_\alpha \Rightarrow H_s|i\rangle=\epsilon_i\mathsf{A}_i^+\mathsf{A}_i|i\rangle=\epsilon_i|i\rangle \tag{4.1a}$$

where A_α^+ creates an atom in a state α, A_α annihilates one.

In like manner the reservoir will have the Hamiltonian

$$H_R=\sum_\nu \omega_\nu \mathsf{a}_\nu^+\mathsf{a}_\nu \Rightarrow H_R|n\rangle=\omega_n\mathsf{a}_n^+\mathsf{a}_n|n\rangle=\omega_n|n\rangle \tag{4.1b}$$

where the a_ν^+ creates a photon of frequency ν, a_ν annihilates one.

The nonrelativistic particle-field interaction is of course

$$H_{SR}=\frac{\mathbf{p}\cdot\mathbf{A}}{\mathrm{m}}=\sum_\nu e\mathbf{p}\cdot\mathbf{e}\left[e^{i\mathbf{k}\cdot\mathbf{x}}\mathsf{a}_\nu+e^{-i\mathbf{k}\cdot\mathbf{x}}\mathsf{a}_\nu^+\right]/\mathrm{m}\sqrt{2\omega V} \tag{4.2}$$

wherein e is the direction of polarization.

Using the arguments following Eq. (2.18) we drop the exponential factors. In order to attain the desired form for H_{SR} we first recall that, for $j=j',j''$, p may induce transitions from i to j or vice versa. If the transition is $i\rightarrow j$, absorbing, a_ν will extract the photon from the field while $\mathsf{A}_j^+\mathsf{A}_i$ changes the atomic state, the term being $\mathsf{A}_j^+\mathsf{A}_i\mathsf{a}_\nu$. For the $j\rightarrow i$ transition the corresponding term will be $\mathsf{a}_\nu^+\mathsf{A}_i^+\mathsf{A}_j$. Thus

$$\begin{aligned}H_{SR}&=\sum_{\nu j} e\mathbf{e}\cdot\langle i|\mathbf{p}|j\rangle\left[\mathsf{A}_j^+\mathsf{A}_i\mathsf{a}_\nu+\mathsf{a}_\nu^+\mathsf{A}_i^+\mathsf{A}_j\right]/\mathrm{m}\sqrt{2\omega V}\\&=-\sum_{\nu j} g_{\nu j}\left[\mathsf{A}_j^+\mathsf{A}_i\mathsf{a}_\nu+\mathsf{a}_\nu^+\mathsf{A}_i^+\mathsf{A}_j\right]\end{aligned} \tag{4.1c}$$

Now we agree that $\langle i|\mathbf{p}|j\rangle/\mathrm{m}=\langle i|\dot{\mathbf{x}}|j\rangle=-\omega_\nu\langle i|\mathbf{x}|j\rangle$ so that

$$g_{\nu j}=-\mathbf{e}\cdot\langle i|e\mathbf{x}|j\rangle\sqrt{\omega/2V} \tag{4.1d}$$

Obviously

$$H=H_S+H_R+H_{SR}. \tag{4.1e}$$

We now introduce the transition operator, P_{ij}, which induces a transition from j to i by annihilating state j and creating state i:

$$\mathsf{P}_{ij}=|i\rangle\langle j| \tag{4.2a}$$

Our Hamiltonian may now be written as

$$H=\sum_{\alpha=i,j}\epsilon_\alpha \mathsf{P}_{\alpha\alpha}+\sum_q \omega_q \mathsf{a}_q^+\mathsf{a}_q-\sum_{qj} g_{jq}(\mathsf{P}_{ji}\mathsf{a}_q+\mathsf{a}_q^+\mathsf{P}_{ij}) \tag{4.2b}$$

We are now in a position to write out the equations of motion, $\dot{\mathsf{a}}_n=\mathrm{i}[H,\mathsf{a}_n]$ and $\dot{\mathsf{P}}_{ij}=\mathrm{i}[H,\mathsf{P}_{ij}]$ using the familiar relations:

$$[\mathsf{a}_i,\mathsf{a}_j^+]=\{\mathsf{A}_i,\mathsf{A}_j^+\}=\delta_{ij};\ [\mathsf{a}_j,\mathsf{A}_j^+]=[\mathsf{a}_i,\mathsf{a}_j]=[\mathsf{A}_i,\mathsf{A}_j]=\cdots=0 \tag{4.3}$$

We remark that

$$\mathsf{a}_\nu^+\mathsf{a}_\nu\mathsf{a}_n-\mathsf{a}_n\mathsf{a}_\nu^+\mathsf{a}_\nu=\begin{cases}0 & \text{for } n\neq\nu\\ -\mathsf{a}_\nu & \text{for } n=\nu\end{cases}$$

The equations of motion are

$$\dot{\mathsf{a}}_q=-\mathrm{i}\omega_q\mathsf{a}_q+\mathrm{i}\sum_j g_{jq}\mathsf{P}_{ij} \tag{4.4a}$$

$$\dot{\mathsf{P}}_{ij''}=-\mathrm{i}\omega_{j''}\mathsf{P}_{ij''}-\mathrm{i}P_{j'j''}\sum_q g_{j'q}\mathsf{a}_q-\mathrm{i}(\mathsf{P}_{j''j''}-\mathsf{P}_{ii})\sum_q g_{j''q}\mathsf{a}_q \tag{4.4b}$$

As an example of the manipulation involved we work out the first term in Eq. (4.4b).

$$[H_s,\mathsf{P}_{ij''}]=\left[\sum\epsilon_\alpha\mathsf{P}_{\alpha\alpha},\mathsf{P}_{ij''}\right]=\left[\sum\epsilon_\alpha\mathsf{P}_{\alpha\alpha}\mathsf{P}_{ij''}-\sum\epsilon_\alpha\mathsf{P}_{ij''}\mathsf{P}_{\alpha\alpha}\right]$$

From Eq. (4.2a) we see that $\mathsf{P}_{\alpha\alpha}\mathsf{P}_{ij''}=0$ unless $\alpha=i$, $\mathsf{P}_{ij''}\mathsf{P}_{\alpha\alpha}=0$ unless $j''=\alpha$. Therefore

$$\begin{aligned}[H_S,\mathsf{P}_{ij''}]&=\epsilon_i\mathsf{P}_{ii}\mathsf{P}_{ij''}-\epsilon_{j''}\mathsf{P}_{ij''}\mathsf{P}_{j''j''}=[\epsilon_i-\epsilon_{j''}]\mathsf{P}_{ij''}\\ &=\omega_{j''}\mathsf{P}_{ij''}\end{aligned}$$

The solution to Eq. (4.4a) may be written down as

$$\mathsf{a}_q(t)=\exp(-\mathrm{i}\omega_n t)\mathsf{a}_q(0)+\sum_j g_{jq}\int_0^t dt'\exp[-\mathrm{i}\omega_q(t-t')]\,\mathsf{P}_{ij}(t') \tag{4.5}$$

When Eq. (4.5) is substituted into Eq. (4.4b) we obtain terms of the form

$$\sum_q g_{j_1q}g_{j_2q}\int_0^t dt'\exp[-\mathrm{i}\omega_q(t-t')]\mathsf{P}_{ij_2}(t') \tag{4.6}$$

There will effectively be a continuous distribution of the ω_q and hence of the q so that we may make the following replacement:

$$\sum_q g_{j_1q}g_{j_2q}\to4\pi\int_0^\infty d\omega_q f(\omega_q)g_{j_1}(\omega_q)g_{j_2}(\omega_q) \tag{4.7}$$

where $f(\omega_q)\,d\omega_q\,d\Omega=4\pi\omega_q^2\,d\omega$ is the density function for the frequencies. Equation (4.6) becomes

$$4\pi\int_0^\infty d\omega_q f(\omega_q)g_{j_1}(\omega_q)g_{j_2}(\omega_q)\int_0^t dt'\exp[-\mathrm{i}\omega_q(t-t')]\mathsf{P}_{ij_2}(t') \tag{4.8}$$

We will agree that

$$\mathsf{P}_{ij}(t')=e^{\mathrm{i}Ht'}\mathsf{P}_{ij}e^{-\mathrm{i}Ht'}=e^{\mathrm{i}(H_0+H_{SR})t'}\mathsf{P}_{ij}e^{-\mathrm{i}(H_0+H_{SR})t'} \tag{4.9}$$

By definition $H_0 \gg H_{SR}$ so that, for small t', we may ignore the H_{SR} in the exponential. Under these circumstances also $\exp(\mathrm{i}H_0 t)|i\rangle = \exp(\mathrm{i}E_i t)|i\rangle$. From Eq. (4.8) obviously $t > t'$ so that

$$\begin{aligned}\mathsf{P}_{ij}(t') &= e^{\mathrm{i}E_i t' + \mathrm{i}H_{SR}t}\mathsf{P}_{ij}e^{-\mathrm{i}E_j t' - \mathrm{i}H_{SR}t}\\ &= e^{\mathrm{i}Ht}e^{-\mathrm{i}E_i(t-t')}\mathsf{P}_{ij}e^{\mathrm{i}E_j(t-t')}e^{-\mathrm{i}Ht}\\ &= \mathsf{P}_{ij}(t)e^{\mathrm{i}E_{ji}(t-t')} \end{aligned} \tag{4.10}$$

and Eq. (4.8) becomes

$$4\pi\mathsf{P}_{ij_2}\int_0^\infty d\omega_q f(\omega_q)g_{j_1}(\omega_q)g_{j_2}(\omega_q)\int_0^t dt' \exp\left[-\mathrm{i}(\omega_q - E_{j_2 i})(t-t')\right] \tag{4.11}$$

We are going to encounter terms involving $E_{j'i}$ and $E_{j''i}$. In the energy region between j'' and j'. $(\omega_q - E_{ji})$ will have opposite sign for j' and j'' leading to $a + \mathrm{i}\sin(\omega_q - E_{j'i})$ and $a - \mathrm{i}\sin(\omega_q - E_{j''i})$ which we shall suppose will eventually cancel assuming approximate equality of coupling coefficients. A similar argument for ω_q greater or lesser than both $E_{j'i}$ and $E_{j''i}$ leads to the replacement of the exponentials by the cosine.Thus Eq. (4.11) becomes

$$\begin{aligned} &4\pi\mathsf{P}_{ij_2}\int_0^\infty d\omega_q f(\omega_q)g_{j_1}(\omega_q)g_{j_2}(\omega_q)\sin(\omega_q - E_{j_2 i})t/\mathrm{i}(\omega_q - E_{j_2 i})\\ &\qquad \xrightarrow[t\to\infty]{} 4\pi^2 P_{ij_2}\int_0^\infty d\omega_q f(\omega_q)g_{j_1}(\omega_q)g_{j_2}(\omega_q)\delta(\omega_q - E_{j_2 i})\\ &\qquad = 4\pi^2\mathsf{P}_{ij_2}f(E_{j_2 i})g_{j_1}(E_{j_2 i})g_{j_2}(E_{j_2 i})\\ &\qquad = 4\pi^2\gamma_{j_1 j_2}\mathsf{P}_{ij_2}\omega_{j_2 i}^3 \end{aligned} \tag{4.12}$$

[We should emphasize that we are here anticipating the fact that we will eventually take the limit of infinite time. We will not actually take this limit until Eq. (4.22).]

We now substitute the equivalents of Eq. (4.12) into Eq. (4.5), the result into Eq. (4.4b). Keeping in mind that $\mathsf{P}_{ik}\mathsf{P}_{ij} = 0$ for $k = i, j', j''$, we find

$$\begin{aligned}\dot{\mathsf{P}}_{ij''} = &-(\mathrm{i}\omega_{j''} + \gamma_{j''}\omega_{j''}^3)\mathsf{P}_{ij''} - \gamma_{j'j''}\omega_{j'}^3\mathsf{P}_{ij'} - \Big(\mathrm{i}\mathsf{P}_{j''j''}\sum_q g_{j'q}\\ &+ \mathrm{i}(\mathsf{P}_{j'j''} - \mathsf{P}_{ii})\sum_q g_{j''q}\Big)e^{-\mathrm{i}\omega_3 t}\mathsf{a}_q(0)\end{aligned} \tag{4.13}$$

Next we operate with Eq. (4.13) on the photon vacuum, $|0\rangle$. Obviously $\mathsf{a}_q(0)|0\rangle = 0$ except for $\omega_q = 0$. For $\omega_q = 0$, $g_{j'q} = g_{j''q} = 0$, so that the term involving $\mathsf{a}_q(0)$ in Eq. (4.13) falls out leaving

$$\left\{\dot{\mathsf{P}}_{ij''} = -(\mathrm{i}\omega_{j''} + \gamma_{j''}\omega_{j''}^3)\mathsf{P}_{ij''} - \gamma_{j'j''}\omega_{j'}^3\mathsf{P}_{ij'}\right\}|0\rangle \tag{4.14a}$$

$$\left\{\dot{\mathsf{P}}_{ij'} = -\gamma_{j'j''}\omega_{j''}^3\mathsf{P}_{ij''} - (\mathrm{i}\omega_{j'} + \gamma_{j'}\omega_{j'}^3)\mathsf{P}_{ij'}\right\}|0\rangle \tag{4.14b}$$

where Eq. (4.14b) follows from Eq. (4.14a) by exchanging j' and j''.

Let us now make the substitution, $\mathsf{P}_1 = \mathsf{P}_{ij''} + \mathsf{P}_{ij'}$, $\mathsf{P}_2 = \mathsf{P}_{ij''} - \mathsf{P}_{ij'}$, a substitution obviously inferred by the form of Eq. (4.14). In the resulting equations we shall suppose that $\omega'-\omega''=\omega'^3-\omega''^3=0$, $\gamma_{j'j''}=\gamma_{j'}=\gamma_{j''}$. The solution is then immediate

$$\mathsf{P}_1=\mathsf{P}_1^0\exp\left\{-\tfrac{1}{2}\left[\mathrm{i}(\omega'+\omega'')+(\omega'^3+\omega''^3)\right]t\right\},\mathsf{P}_2=\mathsf{P}_2^0\exp\left[-\tfrac{1}{2}(\omega'+\omega'')t\right] \tag{4.15}$$

Now we will agree that $\mathsf{P}_1^0=c_1\mathsf{P}_{ij''}(0)+c_2\mathsf{P}_{ij'}(0)$, and so on, so that our solutions will have the form

$$\mathsf{P}_{ij''}=\left[ae^{At}+be^{Bt}\right]\mathsf{P}_{ij''}(0)+\left[ce^{At}+de^{Bt}\right]\mathsf{P}_{ij'}(0) \tag{4.16a}$$

$$\mathsf{P}_{ij'}=\left[a'e^{At}+b'e^{Bt}\right]\mathsf{P}_{ij''}(0)+\left[c'e^{At}+d'e^{Bt}\right]\mathsf{P}_{ij'}(0) \tag{4.16b}$$

although they will not, of course, contain the exponentials of Eq. (4.15) unless we make the approximations $\omega'=\omega''$, and so on, which we do not desire.

In order to obtain the exponent A as an example, we substitute Eqs. (4.16) into Eqs. (4.14). In so doing we make the following substitution:

$$\sigma=\mathrm{i}\omega+\gamma\omega^3, \qquad \Gamma'=\gamma_{j'j''}\omega''^3 \tag{4.17}$$

Equating coefficients of $\exp(At)\mathsf{P}_{ij''}(0)$ on both sides of the equations yields:

$$a'A=-(\sigma''a+\Gamma'a'), \qquad a'A=-(\Gamma''a+\sigma'a')$$

from which we find

$$A=-\tfrac{1}{2}[(\sigma'+\sigma'')\pm m], \qquad m=\sqrt{(\sigma'-\sigma'')^2+4\Gamma'\Gamma''} \tag{4.18}$$

A similar solution yields the same result for B. We therefore choose opposite signs of m for A and B.

We return to the boundary conditions in order to evaluate the coefficients in Eqs. (4.16). At $t=0$, $\mathsf{P}_{ij''}(t)=\mathsf{P}_{ij''}(0)$ so that $a+b=1$, $c+d=0$. Also $\mathsf{P}_{ij'}(t)=\mathsf{P}_{ij'}(0)$ so $a'+b'=0$. We may now use the equations arising from the coefficients of $\exp(At)\mathsf{P}_{ij''}(0)$ and $\exp(Bt)\mathsf{P}_{ij''}(0)$ together with these coefficient relations in order to obtain

$$a=\frac{\sigma'-\sigma''+m}{2m}, \qquad b=\frac{m-\sigma'+\sigma''}{2m}, \qquad b'=-a'=\frac{-\Gamma''}{m} \tag{4.19}$$

As always, we may express any state vector in terms of SGC's, $\mathsf{b}_q(t)$, thus

$$|\psi\rangle=\sum \mathsf{b}_q(t)|\{n_q\}\rangle \tag{4.20}$$

With a_q again referring to the annihilation operator which annihilates a photon in the qth oscillator, we see that

$$\mathsf{b}_q(t)=\langle\{n_q\}|\psi\rangle=\mathsf{a}_q(t)|\{n_q\}\rangle$$

Therefore, since $I(\omega_q)=|\mathsf{b}_q(\infty)|^2$, we obviously desire a_q which we may obtain from Eq. (4.5) with the appropriate P_{ij}. Since we are interested in that situation where the system is ultimately in its ground state, a photon in the field, the appropriate parts of Eqs. (4.16) are $\mathsf{P}_{ij''}(0)$ and $\mathsf{P}_{ij'}(0)$.

We therefore find

$$\mathsf{P}_{ij''}|j''0\rangle = \frac{1}{2m}\left[(\sigma'-\sigma''+m)e^{At}+(m-\sigma'+\sigma'')e^{Bt}\right]|i0\rangle \qquad (4.21a)$$

$$\mathsf{P}_{ij'}|j'0\rangle = \frac{\Gamma''}{m}\left[e^{At}-e^{Bt}\right]|i0\rangle \qquad (4.21b)$$

From Eqs. (4.5) and (4.21) we see that

$$\mathsf{b}_q=\mathsf{a}_q=g_q'\int_0^t dt'\exp(-\mathrm{i}\omega_q(t-t'))\Gamma''\left[e^{At}-e^{Bt}\right]/m$$
$$+g_q''\int_0^t dt'\exp\left[\mathrm{i}\omega(t-t')\right]\left[(\sigma'-\sigma''+m)e^{At}+(m-\sigma'+\sigma'')e^{Bt}\right]/2m$$

Since negative real exponents occur in all exponentials, the upper limits for $t\to\infty$ all zero out.

$$\begin{aligned}\mathsf{b}_q(\infty)&=(g_q'\Gamma''/m)\left[(A+\mathrm{i}\omega)^{-1}-(B+\mathrm{i}\omega)^{-1}\right]\\&\quad+(g_q''/2m)\left[(\sigma'-\sigma''+m)/(A+\mathrm{i}\omega)+(m-\sigma'+\sigma'')/(B+\mathrm{i}\omega)\right]\\&=(g_q'\Gamma''/m)\left[m/(A+\mathrm{i}\omega)(B+\mathrm{i}\omega)\right]\\&\quad+(g_q''/2m)\{m(\sigma'-\sigma'')+m\left[2\mathrm{i}\omega-(\sigma'+\sigma'')\right]\}/(A+\mathrm{i}\omega)(B+\mathrm{i}\omega)\end{aligned} \qquad (4.22)$$

Now let us first suppose there to be but a single upper level so that $\nu'-\nu''=0$. Then Eq. (4.22) collapses to the symmetrical form,

$$\mathsf{b}_q(\infty)=\frac{\frac{1}{2}g_q\mathrm{i}(2\omega-\omega'-\omega'')}{-\frac{1}{4}(2\omega-\omega'-\omega'')^2+\frac{1}{2}\gamma\mathrm{i}(2\omega-\omega'-\omega'')}$$

$$=\mathrm{i}g\left[-(\omega-\omega')+\gamma\mathrm{i}\right]^{-1}$$

which yields the interruption line shape. (The familiar linewidth is of course $\delta=2\gamma$.) With the two upper levels we obtain

$$\mathsf{b}_q(\infty)=\frac{\frac{1}{2}g_q\mathrm{i}2(\omega-\omega'')}{-(2\omega-\omega'-\omega'')^2/4+\frac{1}{2}\gamma\mathrm{i}(2\omega-\omega'-\omega'')}$$

$$\doteq\frac{g_q\mathrm{i}(\omega-\omega'')}{-\frac{1}{2}(\omega-\omega')(\omega-\omega'')+\frac{1}{2}\gamma\mathrm{i}(2\omega-\omega'-\omega'')} \qquad (4.23)$$

where we have supposed the levels close together so that only terms first order in $(\omega-\omega')$ and $(\omega-\omega'')$ have been retained in $(2\omega-\omega'-\omega'')^2$.

We therefore find that, as in the MS case, two upper levels yield the asymmetrical spectral line:

$$I(\omega)=|\mathsf{b}(\infty)|^2=\frac{4|g|^2(\omega-\omega'')^2}{(\omega-\omega')^2(\omega-\omega'')^2+\gamma^2(2\omega-\omega'-\omega'')^2} \qquad (4.24)$$

We may carry the BH result an obvious and faltering step further. Suppose that, instead of two upper levels, there are three of them. The sum in Eq. (4.4a) will now run over three terms, and so on. The solution will be of the form Eqs. (4.16) with three pairs of exponentials instead of two. The exponent, A, for example, will now be the solution of a cubic equation in the σ_i and Γ_i which can only be dealt with numerically. Other equations may be solved for the coefficients, and Eq. (4.21a) and the final result follow. A total of n levels leads to an nth order equation for A, the point at which a general solution breaks down, although one could, of course, continue the problem numerically. It appears doubtful that the result would differ substantially.

5 Higher-Order Corrections. The S-Matrix Approach

Let us consider a photon scattered by an atom. At time $t=-\infty$ we shall suppose the atom to be in its ground state, the photon in state k. The physical process may then be considered to be the absorption of the photon by the atom followed by the emission of the photon in state k' and the return of the atom to its ground state. The initial and final states of the atom will be the same, those of the photon k and k'. Let us consider Eq. (2.6), in particular, $\mathsf{b}_{n'}S^{(i)}\mathsf{b}_n^*$. In all cases, b_n^* creates an atom in its ground state; $\mathsf{b}_{n'}$ must ultimately destroy the same atom. Therefore, i may take on any even value as we may now demonstrate. Let $i=0$. No additional operators enter, and $k=k'$ since $\mathsf{b}_{n'}$ must destroy precisely the photon which b_n^* created. For $i=2$, six additional operators are introduced so that we have $\mathsf{b}_{n'}\mathsf{A}_{n'}^*\mathsf{a}_{k'}^*\mathsf{A}_i\mathsf{A}_i^*\mathsf{a}_k\mathsf{A}_n\mathsf{b}_n^*$, A the atomic operators and a the photon operators. Now a_k will destroy the k-state photon which b_n^* created, $\mathsf{a}_{k'}^*$ create the k' photon that $\mathsf{b}_{n'}$ will destroy. The operator A_n must destroy the ground state atom created by b_n^*, but A_i^* may create an atom in any state rendered accessible to it by H_1 and the photon momentum. Higher even, but only even $S^{(i)}$ are analogous. It can be seen that the scattered radiation will have a spectral distribution of intensities or a line shape, and this shape has been studied by Low (187). Before discussing the Low work, however, it behooves us to recall certain generalities relating to Feynman diagrams.

A great many authors have devoted themselves to the explanation of Feynman diagrams of whom we remark Abrikosov et al. (4), Jauch and Rohrlich (26), Källén (27), Kirzhnits (28), Mattuck (29), Schweber (40), and Ziman (52). We therefore need do little more than display the familiar, general rules which of course calls for the examination of a term in a matrix element of the S-matrix expansion, Eq. (2.6).

The first two terms (1 and $S^{(1)}$) are too simple to be instructive; the fourth ($S^{(3)}$) is needlessly complicated. Therefore, we consider $S^{(2)}$. We write down the matrix element of $S^{(2)}$ for a transition between an initial state having n'_a

electrons in state a' and n'_k photons in a state k' to a state having n_a electrons and n_k photons in a and k.

$$
\begin{aligned}
&\langle n'_a\{n'_k\}|S^{(2)}|n_a\{n_k\}\rangle \\
&\qquad = \tfrac{1}{2}e^2\int\int d^4x\,d^4x'\langle a'k'|\mathsf{T}\{\mathsf{N}[\bar{\psi}^{(0)}(x)\gamma_i\psi^{(0)}(x)] \\
&\qquad\quad \cdot \mathsf{A}_i^{(0)}(x)\mathsf{N}[\bar{\psi}^{(0)}(x')\gamma_i\psi^{(0)}(x')]\mathsf{A}_j^{(0)}(x')\}|ak\rangle \\
&\qquad = \tfrac{1}{2}e^2\int\int d^4x\,d^4x'\,\langle a'|\mathsf{T}\{\mathsf{N}[\bar{\psi}^{(0)}(x)\gamma_i\psi^{(0)}(x)] \\
&\qquad\quad \cdot \mathsf{N}[\bar{\psi}^{(0)}(x')\gamma_i\psi^{(0)}(x')]\}|a\rangle\langle k'|\mathsf{T}[\mathsf{A}_i^{(0)}(x)\mathsf{A}_i^{(0)}(x')]|k\rangle
\end{aligned}
\tag{5.1}
$$

where N indicates a normal product, that is, the product of operators arranged so that all annihilation operators are to the right of all creation operators. We have here also introduced the symbol $|\{n_k\}\rangle$ for the wave function or state vector of the electromagnetic field, the symbol referring to the set, $\{n_k\}$, of occupation numbers for the field oscillators. We shall suppose the reader to recall (cf. § 21 of Källén, § 82 of Abrikosov et al., Appendix F of Mattuck) the way in which a time-ordered normal product may be expanded and write out a few terms of Eq. (5.1) as:

$$
\begin{aligned}
&\tfrac{1}{2}e^2\int\int d^4x\,d^4x'\{\langle a'|\mathsf{N}[\bar{\psi}^{(0)}(x)\gamma_i\psi^{(0)}(x)\bar{\psi}^{(0)}(x')\gamma_j\psi^{(0)}(x')]|a\rangle\langle k'|\mathsf{N}[\mathsf{A}_i^{(0)}(x) \\
&\qquad \cdot\mathsf{A}_j^{(0)}(x')]|k\rangle + \langle a'|\mathsf{N}\Big[\bar{\psi}^{(0)}(x)\gamma_i\overbrace{\psi^{(0)}(x)\bar{\psi}^{(0)}}(x')\gamma_j\psi^{(0)}(x')\Big]|a\rangle \\
&\cdot\langle k'|\mathsf{N}\Big[\overbrace{\mathsf{A}_i^{(0)}(x)\mathsf{A}_j^{(0)}}(x')\Big]|k\rangle + \langle a'|\mathsf{N}\Big[\overbrace{\bar{\psi}^{(0)}(x)\gamma_i\overbrace{\psi^{(0)}(x)\bar{\psi}^{(0)}}(x')\gamma_j\psi^{(0)}}(x')\Big]|a\rangle \\
&\qquad\qquad \cdot\langle k'|\mathsf{N}\Big[\overbrace{\mathsf{A}_i^{(0)}(x)\mathsf{A}_j^{(0)}}(x')\Big]|k\rangle + \cdots\}
\end{aligned}
$$

$$
\begin{aligned}
&= \tfrac{1}{2}e^2\int\int d^4x\,d^4x'\{\langle a'|\mathsf{N}[\cdots]|a\rangle\langle k'|\mathsf{N}[\mathsf{A}_i^{(0)}\mathsf{A}_j^{(0)}]|k\rangle \\
&\quad + \langle a'|\mathsf{N}[\bar{\psi}^{(0)}(x)\gamma_i\langle 0|\psi^{(0)}(x)\bar{\psi}^{(0)}(x')|0\rangle\gamma_j\psi^{(0)}(x')]|a\rangle\langle 0|\mathsf{A}_i^{(0)}\mathsf{A}_j^{(0)}|0\rangle \\
&\quad + \langle 0|\bar{\psi}^{(0)}(x)\psi^{(0)}(x')|0\rangle\langle 0|\psi^{(0)}(x)\bar{\psi}^{(0)}(x')|0\rangle\gamma_i\gamma_j\langle 0|\mathsf{A}_i^{(0)}(x)\mathsf{A}_j^{(0)}(x')|0\rangle + \cdots\}
\end{aligned}
$$

$$
\begin{aligned}
&= \tfrac{1}{2}e^2\int\int d^4x\,d^4x'\{\langle a'|\mathsf{N}[\cdots]|a\rangle\langle k'|\mathsf{N}[\mathsf{A}_i^{(0)}\mathsf{A}_j^{(0)}]|k\rangle \\
&\quad + \langle a'|\bar{\psi}^{(0)}(x)|0\rangle\gamma_i\mathbb{S}_F(x-x')\gamma_i\langle 0|\psi^{(0)}(x')|a\rangle\mathfrak{D}_F(x-x') \\
&\quad + \mathbb{S}_F(x-x')\mathbb{S}_F(x'-x)\gamma_i\gamma_i\mathfrak{D}_F(x'-x) + \cdots\}
\end{aligned}
\tag{5.2}
$$

In Eq. (5.2) we have indicated the contraction of a pair of operators by $\overline{A\cdots B}$. The $\mathfrak{D}_F$ is one form of the photon Greens function defined, for example, by Eq. (7.29) of Källén. The particle analogy to $\mathfrak{D}$, $\mathfrak{S}$, the reader will find treated *in extenso* in § 15 of the same author. Remark that, for $a = a' =$ vacuum state, all the normal products disappear leaving only the fully contracted term.

We now draw the familiar Feynman diagrams using the Boson ladder method. There will be one point for each γ_i (corresponding to the order of the approximation), a straight line for each electron operator, $\psi^{(0)}$ or $\mathfrak{D}_F$, a wavy line for each photon operator, $\mathsf{A}^{(0)}$ or $\mathfrak{S}_F$. These lines will connect with one point for each free operator $\psi^{(0)}$ or $\mathsf{A}^{(0)}$, with two points for each contraction $\mathfrak{D}_F$ or $\mathfrak{S}_F$. There will be two electron and one photon lines which connect with each point. We draw the graphs which correspond to the terms in Eq. (5.2) as Fig. 5.1. There are obviously other graphs in second order (Fig. 5.2), and the construction of fourth (Fig. 5.3), sixth (Fig. 5.4), and so on, pose no particular problems. The method of drawing such diagrams is obvious from the diagrams. As is well known, one may begin by drawing such graphs and then reverse the procedure in order to obtain the integrals corresponding, the rules being inverse to those leading to the graphs.

A diagram such as Fig. 5.1*c*, having no external lines, is called an "unlinked" diagram. Now remark Figs. 5.3*g* and 5.4*e*. Remark them and recall the factorials which are to be anticipated in the expansion of Eq. (2.2).

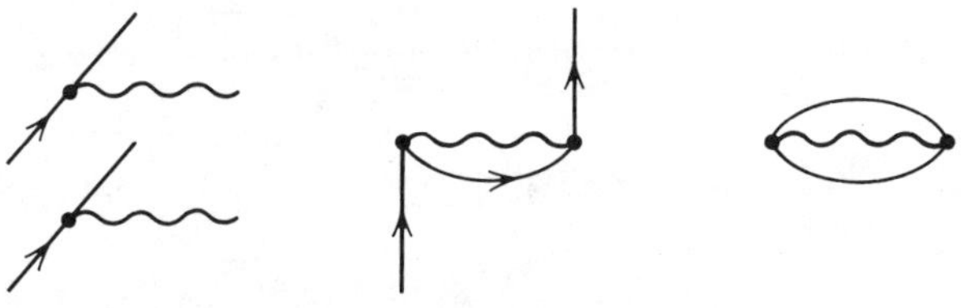

(*a*) (*b*) (*c*)

Figure 5.1

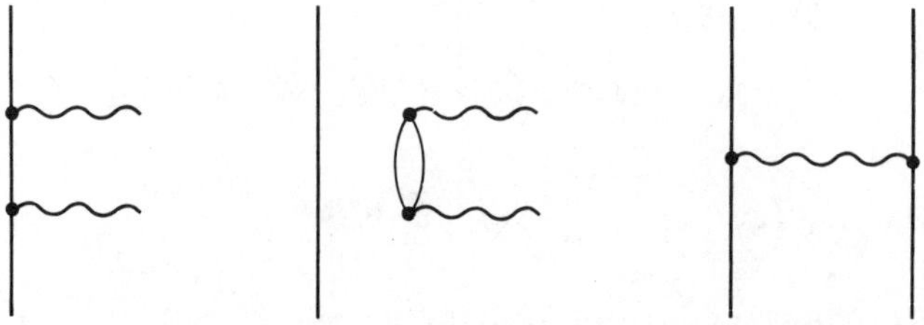

(*a*) (*b*) (*c*)

Figure 5.2

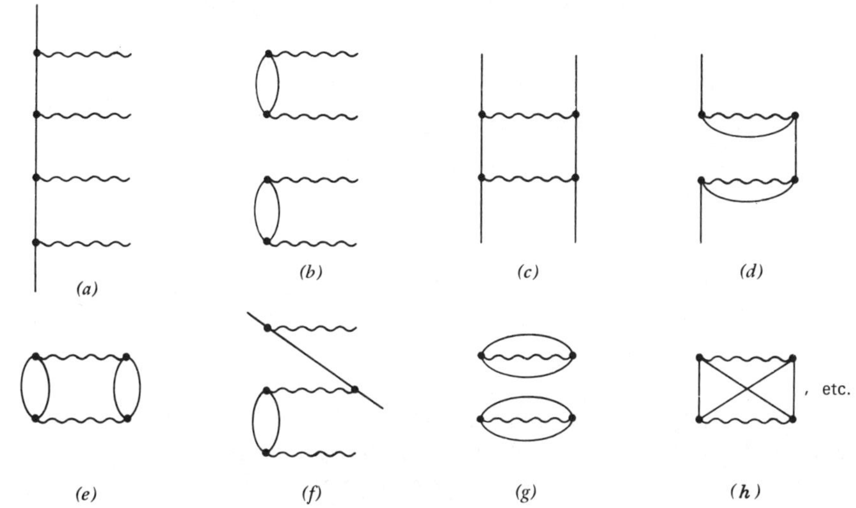

Figure 5.3

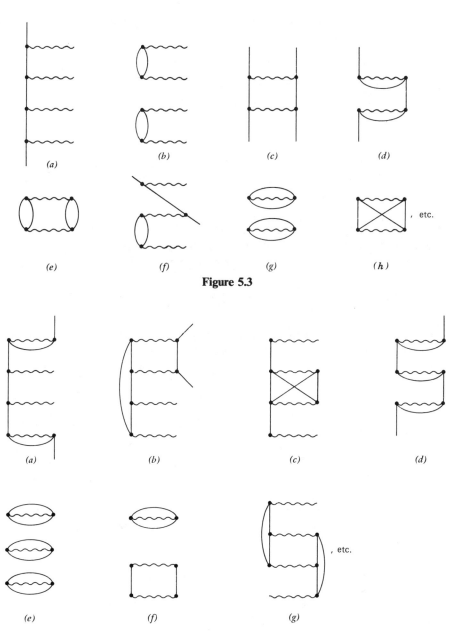

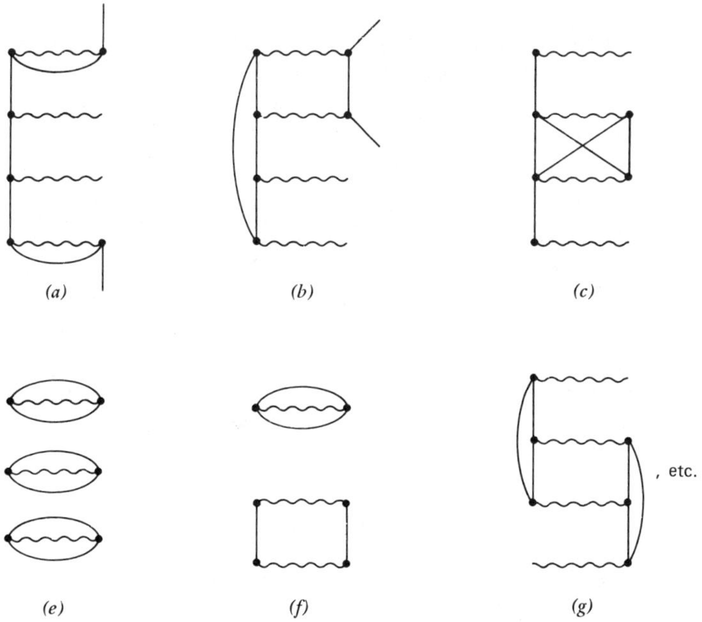

Figure 5.4

From such as these we will eventually obtain the following infinite series:

$$1 + \text{〰} + \frac{1}{2!}(\text{〰})^2 + \frac{1}{3!}(\text{〰})^3 + \cdots = \exp[\text{〰}]$$

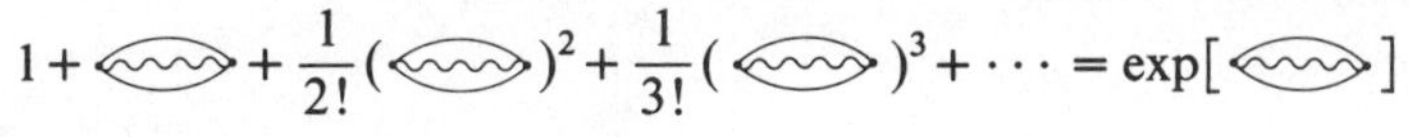

But diagrams like Fig. 5.3h will lead to

$$\exp\left[\boxtimes\right]$$

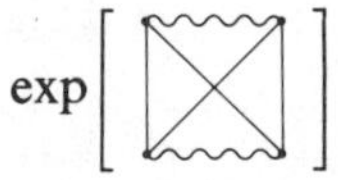

and diagrams like Fig. 5.4f will lead, when combined with the earlier series, to

$$\exp\left[\text{〰} + \boxtimes\right]$$

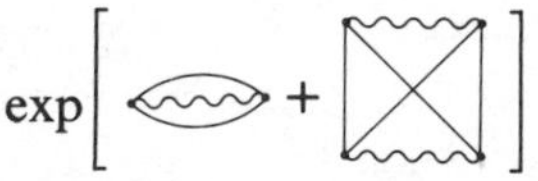

As by this time has become apparent, all unlinked diagrams sum to a single exponential which, being imaginary, has the form $\exp(i\eta)$. The following elementary graph collection leads to the desired conclusion:

$$\left[1 + \text{〰} + \boxtimes + \cdots + \frac{1}{2!}(\text{〰} + \boxtimes + \cdots)^2 + \cdots\right]$$

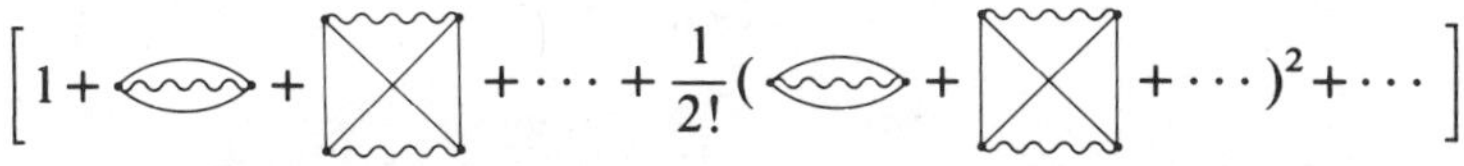

For we see that each term in the infinite series of linked diagrams is thus multiplied by the same phase factor, $\exp(i\delta)$, so that the phase factor itself and hence all the unlinked diagrams may be neglected. (Wave functions are always indeterminate to within a phase factor in any event.) Now we return to the scattering and line shape study of Low.

In doing so it may be of interest to point out that Fig. 5.2a corresponds to Heisenberg-Kramers scattering, Fig. 5.2b to Delbruck scattering. These figures appear as Fig. 1 in Low, who obviously did not use the Boson ladder method of drawing diagrams. Similarly, our Fig. 5.4b is his Fig. 2f, our Fig. 5.3f his Fig. 2b, and so on.

Low's treatment of emission line shape occupies him but briefly in the last section of his paper, the approach being analogous to the photon incident on an atom in its ground state. This time, however, the photon is replaced by an "effective thermal excitation function,"

$$V(x,t) = \int I(\omega) e^{-i\omega t} d\omega V(x) \tag{5.3}$$

which will excite the atom. We remark that $I(\omega)$ is the energy distribution in the incident "excitation function," the integral and exponential merely providing the Fourier transform. The practical result is that the factor $\exp(ikx)$ which the incident photon had introduced into the matrix element is now replaced by Eq. (5.3). The intensity distribution is the spectral line will of course be given by $|\langle k|S|0\rangle|^2$. We consider the $S^{(2)}$ term as the first of interest, our method not being precisely that of Low.

We write down Eq. (5.1), dropping the T and N designations—we shall be careful to see that $t > t'$—and introducing obvious matrix multiplication:

$$\langle 0k'|S^{(2)}|0k\rangle = \tfrac{1}{2}e^2 \sum_n \iiint d^4x\, d^4x'\, d\omega \langle 0k'|\bar{\psi}_0^{(0)}(x)\gamma_i \psi_n^{(0)}(x) V(x)|nk'\rangle$$

$$\cdot \langle nk'|\psi_n^{(0)}(x')\gamma_j \psi_0^{(0)}(x')\mathsf{A}_j^{(0)}(x')|0k\rangle I(\omega) e^{-i\omega t} \tag{5.4}$$

where we are designating the ground state by 0. The ψ are in the interaction representation, a fact which we now explicitly introduce. We know that the upper state will decay as $\exp(-\Gamma t)$. (We also know that it had best do so or we will obtain an embarrassingly divergent expression for the spectral line shape.) This factor we also introduce into Eq. (5.4). Remembering that the spatial integrations will practically lead to matrix elements over the wave functions, we may write Eq. (5.4) as

$$\langle 0k'|S^{(2)}|0k\rangle = \tfrac{1}{2}e^2 \sum_n \iiint dt\, dt'\, d\omega\, e^{i(E_b - E_n)t} e^{i(E_0 - E_n)t'} e^{ikt'} e^{-i\omega t}$$

$$\cdot e^{-\Gamma(t-t')} \langle 0|\gamma_i \mathsf{V}|n\rangle \langle n|\gamma_j|0\rangle \langle k'|\mathsf{A}_j^{(0)}(x')|k\rangle I(\omega) \tag{5.5}$$

We now change from t, t' to $t - t' = t_1$, $t = t_2$. The limits on $t_1, (0, \infty)$, assure us of time ordering, the limits on t_2 being $(-\infty, \infty)$. Therefore

$$\iiint d\omega\, dt\, dt' \exp[-i(E_0 - E_n - i\Gamma)(t - t')] \exp(ikt') \exp(-i\omega t) I(\omega)$$

$$= \iiint d\omega\, dt_1\, dt_2 \exp[-i(E_0 - E_n + k - i\Gamma)t_1] \exp[i(k - \omega)t_2] I(\omega)$$

$$= 2\pi \iint d\omega\, dt_1 \exp[-i(E_0 - E_n + k - i\Gamma)t_1] \delta(k - \omega)$$

$$= -2\pi i \int d\omega \frac{\delta(\mathsf{k} - \omega) I(\omega)}{E_0 - E_n + k - i\Gamma} = \frac{-2\pi i I(k)}{E_0 - E_n + k - i\Gamma} \tag{5.6}$$

For the two-state case the summation in Eq. (4.2) disappears, and that equation in conjunction with Eq. (4.4) yields the familiar form for the spectral line. The same result essentially eventuates for widely separated levels.

For two levels, $n = 1, 2$, lying closely enough together so that $E_1 \doteq E_2 = E$ and having transition probabilities approximately the same, $\Gamma_1 \doteq \Gamma_2 = \Gamma$, we find the line shape proportional to

$$\frac{(E_0 - E + \mathsf{k})^2 - \Gamma^2}{\left[(E_0 - E + \mathsf{k})^2 - \Gamma^2\right]^2 + 4\Gamma^2 (E_0 - E + \mathsf{k})^2} \tag{5.7}$$

In the line wing where $\Gamma \ll E_0 - E + \mathsf{k}$, we see that our shape is the familiar symmetric one while asymmetry sets in when Γ is of the order of separation from line center.

The higher orders could be treated in the same fashion although this has apparently not been done. $S^{(4)}$ would be approached with an equation which is the direct extrapolation of Eq. (5.2), this one having four matrix elements, an additional integration over frequency due to matrix multiplication, and an additional $\exp(-i\omega t)$. A complicated form of Eq. (5.5) eventuates. Now suppose that, from left to right, the times are designated t, t', t'', t'''. Then we satisfy the time-ordering requirement by adopting the new coordinates $t_4 = t$, $t_3 = t - t'$, $t_2 = t' - t''$, $t_1 = t'' - t'''$ with integration limits $-\infty < t_4 < \infty$, $0 < t_i < \infty$ for $i = 1, 2, 3$. We find that our integrand is of the form

$$\sum_{nmp} \exp[-\mathrm{i}(E_0 - E_p - k - \mathrm{i}\Gamma_p)t_1] \exp[-\mathrm{i}(E_0 - E_m - k - \mathrm{i}\Gamma_m)t_2]$$

$$\exp[-\mathrm{i}(E_0 - E_n - k - \mathrm{i}\Gamma_n)t_3] \exp[-\mathrm{i}(k - \omega)t_4] \quad (5.8)$$

In Eq. (5.6) the summation has arisen from the matrix multiplications. For the two-state case $m \equiv 0$, $n \equiv p$. This fourth-order correction to the line shape has the form

$$\left| \left[k(E_0 - E_n - \omega - \mathrm{i}\Gamma)^2 \right]^{-1} \right|^2 \quad (5.9)$$

6 The Graphed Liouville Approach to Natural Line Shape

Mizushima, Robert, and Galatry (MRG) (206) utilized certain, to us at least, important new tools in their intriguing treatment of natural line shape, although once again we shall see the Weisskopf-Wigner result given a rousing round of applause. First of all, working through the density matrix, these authors introduced the Liouville operator into the problem, it then being the operator in terms of which the propagator for the density operator was expanded. The Liouville operator, ushered into the line broadening arena by Fano (115), was to see considerable service in the line-broadening cause, as we shall see. Another widely used technique is that of the Feynman diagram, which MRG also use. We will recall that the unlinked graphs may be collected, all aleph-null (20) of them, into an exponential. This is what MRG were able to do, and, although they obviously could not consider all varieties of graph, those they did consider constitute contributions through aleph-null order when placed in the exponential. We now turn to the details of their treatment.

The intensity corresponding to a frequency ω_i in the spectral line will assuredly relate to the probability, $p(t)$, that a photon of frequency ω_i is present in the field. As we have seen, this probability is often expressed as the square of the SGC. MRG, however, express it as the difference of density operators.

The equation of motion for the density matrix, $\Upsilon(t)$, [Eq. (2.128) of Breene (9)] may be recalled as

$$\mathrm{i}\dot{\Upsilon} = [H, \Upsilon] \quad (6.1)$$

Now suppose we construct the operator

$$\Upsilon = \sum_{mn} |m\rangle \Upsilon_{mn} \langle n| \tag{6.2}$$

where Υ_{mn} is an element of the density matrix. If we multiply through on the left by $\langle m|$ and on the right by $|n\rangle$ we find

$$\langle m|\Upsilon|n\rangle = \Upsilon_{mn}$$

Therefore, the operator which we have constructed as Eq. (6.2) must be that associated with the density matrix, that is, it must be the density operator. When we turn from our appeals of the previous sections to this appeal, we are shifting our viewpoint from that of particle mechanics to that of statistical mechanics. Generally speaking, when we make such a shift it is because we are interested, not in a single system, but in a collection or ensemble of systems. Our mental vision is to be focused on phase points, ergodic hypotheses, and so on. This amounts, in a manner of speaking, to a focus on the smoothed behavior of a single system.

Since it obeys an equation of motion, the density matrix must be a function of time which introduces the idea of boundary conditions. MRG quite sensibly used the boundary conditions of their predecessors, that is to say, at time $t=0$ they supposed the atom to be in one of its unperturbed states, the radiation oscillators to be distributed in some fashion over their unperturbed states. Such a boundary condition corresponds to adiabatic switching or whatever other label one may choose and simply amounts to saying that there is no interaction at time zero. However, from the density matrix viewpoint, it means that there is no probability for any mixing, so that the density matrix is diagonal.

$$\langle i|\Upsilon(0)|j\rangle = \delta_{ij}\langle i|\Upsilon(0)|i\rangle \tag{6.3}$$

Since $\Upsilon_{ij}(0)=0$, we see that there is zero probability for a time zero transition from i to j, a simple repetition of our earlier boundary condition on the SGC's. Suppose now that we desire the intensity in the spectral line which results from the transition from state a in the two-state case. This will be equal to the probability for the change in the number of systems, $|i\rangle = |a\{n_i\}\rangle$, in the initial state. This probability is

$$p(t) = \sum_{n_i=N} \langle\{n_i\}a|\Upsilon(0)-\Upsilon(t)|a\{n_i\}\rangle \tag{6.4}$$

We now suppose that both $\Upsilon(t)$ and $H(=H_0+H_1)$ are in the interaction representation. If we substitute $\Upsilon(t)$ and H into Eq. (6.1), terms involving H_0 drop out with the result:

$$i\dot{\Upsilon} = [H_1', \Upsilon'] \tag{6.5}$$

It is important to emphasize that Υ' and H_1' are now in the interaction representation. The first crucial step in the MRG study consists in the introduction of the Liouville operator:

$$\mathfrak{L}_1 A = [H_1', A] \tag{6.6}$$

where $\mathfrak{L}_1$ is given by Eq. (1.6) with H replaced by H_1'.

From Eqs. (6.5) and (6.6) it will certainly be formally true that

$$\Upsilon'(t) = \Upsilon(0) - \mathrm{i}\int_0^t \mathfrak{L}_1(t')\Upsilon'(t')\,dt' \tag{6.7}$$

When Eq. (6.7) is substituted into Eq. (6.4), we find

$$p(t) = \sum_{n_i = N} \left\langle i \middle| \mathrm{i}\int_0^t \mathfrak{L}_1(t_1')\Upsilon'(t_1')\,dt_1' \middle| i \right\rangle \tag{6.8}$$

At which point MRG take another step crucial to their development.

By our heretofore standard assumptions H_1 came into being at $t=0$ so that $H_1 \neq 0$ for $t>0$. A new time, t_2, is now introduced such that $H_1 = 0$ for $t<t_2$, the time t_2 being such that $t_2<t_1$. Obviously then, $H_1 \neq 0$ for $t_2<t<t_1$; in fact, the authors require a radiative transition between $t=t_2$ and $t=t_2+dt_2$. This has the effect of moving the temporal origin to that

$$\Upsilon'(t_2) = \Upsilon'(0) = \Upsilon(0) \tag{6.9}$$

We now introduce the temporal propagator, $\mathsf{G}'(t_1, t_2)$, which we suppose capable of transforming the density matrix at t_2 into the density matrix at t_1, thus

$$\Upsilon'(t_1) = \mathsf{G}'(t_1, t_2)\Upsilon'(t_2) \tag{6.10}$$

If we substitute Eq. (6.10) into Eq. (6.5), where we now suppose the differentiation with respect to t_1, the fact that $\partial[\Upsilon(t_1)]/\partial t_1 = 0$ yields:

$$\mathrm{i}\frac{\partial}{\partial t}\mathsf{G}'(t_1, t_2) = \mathfrak{L}_1(t_1)\mathsf{G}'(t_1, t_2) \tag{6.11a}$$

$$\mathsf{G}'(t_1, t_2) \equiv 1, \qquad t_1 = t_2 \tag{6.11b}$$

The iterative solution to this equation is:

$$\mathsf{G}'(t_1, t_2) = 1 + \sum_{n=0}^{\infty} (-\mathrm{i})^{n+1} \int_{t_2}^{t_1} \cdots \int_{t_2}^{t^{(n)}} \mathfrak{L}_1(t') \cdots \mathfrak{L}_1(t^{(n+1)})\,dt' \cdots dt^{(n+1)} \tag{6.12}$$

where times increase from right to left. Now suppose the first radiative transition occurs between t_2 and $t_2 + dt_2$. This means that

$$\int_{t_2}^{t^{(n)}} \mathfrak{L}_1(t^{(n+1)})\,dt^{(n+1)} \to \mathfrak{L}_1(t_2)\,dt_2$$

so that

$$
\begin{aligned}
&\mathsf{G}'_{t_2}(t_1,t_2)\\
&=1+\left(1+\sum_{n=0}^{a}(-\mathrm{i})^{(n+1)}\int_{t_2}^{t_1}\cdots\int_{t_2}^{t^{(n)}}\mathfrak{L}_1(t')\cdots\mathfrak{L}_1(t^{(n+1)})\,dt'\cdots dt^{(n+1)}\right)\\
&\quad\cdot(-\mathrm{i})\mathfrak{L}_1(t_2)\,dt_2\\
&=1-\mathrm{i}\mathsf{G}'(t_1,t_2)\mathfrak{L}_1(t_2)\,dt_2
\end{aligned}
\tag{6.13}
$$

We integrate Eq. (6.13) over t_2 in order to obtain Eq. (6.10) which we substitute into Eq. (6.8) with the result:

$$
\begin{aligned}
p(t)=&\sum_{n_i=N}\mathrm{i}\int_0^t dt_1\int_0^{t_1}dt_2\langle i|\mathfrak{L}_1(t_1)\Upsilon(0)|i\rangle\\
&-\sum_{n_i=N}\mathrm{i}\int_0^t dt_1\int_0^{t_1}dt_2\langle i|\mathfrak{L}_1(t_1)\mathsf{G}'(t_1,t_2)\mathfrak{L}_1(t_2)\Upsilon(0)|i\rangle
\end{aligned}
\tag{6.14}
$$

Since $\langle i|\Upsilon(0)|j\rangle=\langle i|\Upsilon(0)|i\rangle\delta_{ij}$, the first term in Eq. (6.14) is

$$
\begin{aligned}
\langle i|\mathfrak{L}_1(t_1)\Upsilon(0)|i\rangle&=\sum_{jk}\langle ii|\mathfrak{L}_i|jk\rangle\langle j|\Upsilon_0|k\rangle\\
&=\sum_{j}\langle ii|\mathfrak{L}_1|jj\rangle\langle j|\Upsilon_0|j\rangle=0
\end{aligned}
$$

We consider the second integral where we let $\mathsf{G}'(t_1,t_2)\mathfrak{L}_1(t_2)\Upsilon(0)=\mathbf{X}$:

$$
\begin{aligned}
&\langle i|\mathfrak{L}_1\mathbf{X}|i\rangle=\mathrm{i}\sum_{jk}\langle ii|\mathfrak{L}_1|jk\rangle\langle j|\mathbf{X}|k\rangle\\
&=\mathrm{i}\sum_{jk}\big[\langle i|H|j\rangle\langle j|\mathbf{X}|k\rangle\delta_{ik}e^{\mathrm{i}(E_i-E_j)t_1}\\
&-\langle i|H^*|k\rangle\langle j|\mathbf{X}|k\rangle\delta_{ij}e^{\mathrm{i}(E_k-E_i)t_1}\big]\\
&=\mathrm{i}\sum_{j}\big[\langle i|H|j\rangle\langle j|\mathbf{X}|k\rangle e^{\mathrm{i}(E_1-E_j)t_1}\\
&-\langle i|H^*|j\rangle\langle i|\mathbf{X}|j\rangle e^{-\mathrm{i}(E_i-E_j)t_1}\big]\\
&=\mathrm{i}\sum_{j}\big[e^{\mathrm{i}(E_i-E_j)t_1}-e^{-\mathrm{i}(E_i-E_j)t_1}\big]\langle i|H|j\rangle\langle j|\mathbf{X}|i\rangle=2\Re\sum\langle i|H_1(t_1)|j\rangle\langle j|\mathbf{X}|i\rangle
\end{aligned}
$$

so that

$$
p(t)=2\Re\sum_{n_i=N}\int_0^t dt_1\int_0^{t_1}dt_2\langle i|H_1(t_1)|a\rangle\langle a|\mathsf{G}'(t_1,t_2)\mathfrak{L}_1(t_2)\Upsilon(0)|i\rangle
\tag{6.15}
$$

As a consequence of Eq. (1.5)

$$\langle a|\mathsf{G}'(t_1,t_2)\mathfrak{L}_1(t_2)\Upsilon(0)|i\rangle=\sum_{bj}\langle ai|\mathsf{G}'(t_1,t_2)|bj\rangle\langle b|\mathfrak{L}_1(t_2)\Upsilon(0)|j\rangle \tag{6.16}$$

Taking only the first term in Eq. (6.13) for $\mathsf{G}'_{t_2}(t_1,t_2)$ in Eq. (6.14), we find for the limit of large times

$$\lim_{t\to\infty}\mathfrak{R}\int_0^t dt_1\langle i|H_1(t_1)|a\rangle\int_0^{t_1}dt_2\langle a|\Upsilon(0)|a\rangle$$

$$=\langle a|\Upsilon(0)|a\rangle\langle i|H_1|a\rangle\lim_{t\to\infty}\mathfrak{R}\left(\int_0^t dt_1\exp[\mathrm{i}(E_i-E_a)t_1]\int_0^{t_1}dt_2\right)$$

$$=\langle a|\Upsilon(0)|a\rangle\langle i|H_1|a\rangle t\cdot\delta(\omega_{ia}),\ \omega_{ia}=E_i-E_a \tag{6.17}$$

which demonstrates the absence of line breadth with the absence of the propagator.

On the other hand, a Weisskopf-Wigner type results eventuates when the diagonal elements of the entire propagator are considered, the propagator furnishing the damping which, of course, is the essence of the result. That only diagonal elements need be considered comes about as follows. Take the term linear in the Liouville in Eq. (6.12) and keep in mind that $H_1(t)=\exp(\mathrm{i}H_0t)H_1\exp(-\mathrm{i}H_0t)$. Then the diagonal and off-diagonal elements of the Liouville operator are

$$\langle ab|\mathfrak{L}_1(t)|ab\rangle=\langle ab|H_1(t)I-IH_1^*(t)|ab\rangle=\langle a|H_1|a\rangle-\langle b|H_1|b\rangle \tag{6.18a}$$

$$\begin{aligned}\langle ab|\mathfrak{L}_1(t)|cd\rangle=&\langle a|H_1|c\rangle\delta_{bd}\exp[\mathrm{i}(E_a-E_c)t]\\&-\langle d|H_1|b\rangle\delta_{ac}\exp[\mathrm{i}(E_d-E_b)t]\end{aligned} \tag{6.18b}$$

We will eventually deal with large times, $t\to\infty$, albeit a different but related t, and the rapid oscillations of the imaginary exponentials of Eq. (6.18b) will tend to cancel the contribution of these off-diagonal elements to the time integral.

As we can see from Eq. (6.11a), we are dealing, in G and $\mathfrak{L}$, with the product space analogy of $\mathfrak{U}$ (the TDO) and H_1 in Hilbert space. Therefore, MRG were able to apply the S-matrix machinations of the simple space to their $H\otimes H$ with, however, some rather interesting innovations, the first of which is a redefinition of the vacuum with a very useful objective for so doing.

We recall that (cf. Appendix F of 29) matrix elements of S—or of a propagator—over the vacuum contain only completely contracted terms or diagrams. As we have seen, the complete perturbation series of such terms will be an exponential containing all single unlinked diagrams as indicated in the discussion of Section 5. Thus

$$\langle O|\mathsf{G}'(t_1,t_2)|0\rangle=\exp[\mathsf{G}'_0] \tag{6.19}$$

where MRG consider only diagrams through second order in the propagator expansion thus obtaining

exp 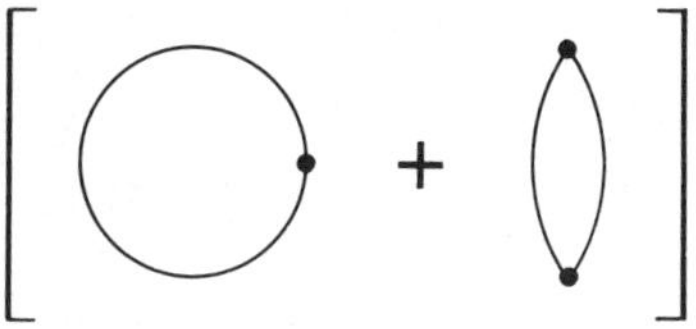

Although the Liouville has diagonal elements in the product space, it is important to remember that the perturbing Hamiltonian $(\mathbf{p}\cdot\mathbf{A})$ has only off-diagonal elements. Indeed, this accounts for the disappearance of the first-order term in the propagator expansion:

$$
\begin{aligned}
&= -\mathrm{i}\int_{t_2}^{t_1}\langle O|\mathfrak{L}_1(t')|O\rangle\, dt' \\
&= -\mathrm{i}\int_{t_2}^{t_1}dt'[\langle a|H_1|a\rangle - \langle b|H_1|b\rangle] = 0,
\end{aligned}
\tag{6.20}
$$

and the form of the second-order term:

$$
\begin{aligned}
&= -\sum_A \int_{t_2}^{t_1}dt'\int_{t_2}^{t'}dt''\langle O|\mathfrak{L}_1(t')|A\rangle\langle A|\mathfrak{L}_1(t'')|O\rangle \\
&= -\sum_{cd}\int_{t_2}^{t}dt'\int_{t_2}^{t'}dt''\left[e^{\mathrm{i}\omega_{ac}t'}H_{ac}\delta_{bd} - \delta_{ac}e^{\mathrm{i}\omega_{db}t'}H_{bd}^*\right] \\
&\quad\times\left[e^{-\mathrm{i}\omega_{ac}t''}H_{ca}\delta_{db} - \delta_{ca}e^{-\mathrm{i}\omega_{db}t''}H_{db}^*\right] \\
&= -\sum_c \int_{t_2}^{t_1}dt'\int_{t_2}^{t'}dt''\left[|\langle a|H_1|c\rangle|^2 e^{\mathrm{i}\omega_{ac}(t'-t'')}\right. \\
&\quad\left.+|\langle b|H_1|c\rangle|^2 e^{\mathrm{i}\omega_{bc}(t''-t')}\right]
\end{aligned}
\tag{6.21}
$$

where $\omega_{ij} = E_i - E_j$.

The prescription for diagram interpretation is admittedly somewhat different here, but the relation should be reasonably apparent. Integrating, say, the first factor in Eq. (6.21) over t'' yields

$$
\begin{aligned}
-\mathrm{i}\int_{t_2}^{t_1}dt'\frac{1-e^{\mathrm{i}\omega_{ac}(t'-t)}}{\omega_{ac}} &= \mathrm{i}\int_{t_2}^{t_1}dt'\zeta(\omega_{ac}) \\
&= (t_1 - t_2)\left[\mathrm{i}\mathcal{P}\left(\frac{1}{\omega_{ac}}\right) + \pi\delta(\omega_{ac})\right]
\end{aligned}
\tag{6.22a}
$$

where

$$
\zeta(\omega_{ac}) = \lim_{K\to\infty}\frac{(1-e^{\mathrm{i}\omega_{ac}K})}{\omega_{ac}} = \mathcal{P}\left(\frac{1}{\omega_{ac}}\right) - \mathrm{i}\pi\delta(\omega_{ac}) \tag{6.22b}
$$

when we suppose $t'-t_2$ becomes very large (cf. § 8 of Heitler) as the authors do. Obviously then

$$\langle ab|\mathsf{G}'(t_1,t_2)|ab\rangle=\exp[-\Gamma_{ab}(t_1-t_2)] \tag{6.23a}$$

where

$$\Gamma_a=\sum_c\left[\pi\delta(\omega_{ac})+i\mathcal{P}\left(\frac{1}{\omega_{ac}}\right)\right]|\langle a|H_1|c\rangle|^2=\gamma_a+i\gamma_a' \tag{6.23b}$$

$$\Gamma_{ab}=\Gamma_a+\Gamma_b=\gamma_{ab}+i\gamma_{ab}'=\gamma_a+\gamma_b+i(\gamma_a'-\gamma_b') \tag{6.23c}$$

Obtaining the line shape is now virtually immediate, involving familiar matrix elements of the propagator and the Liouville. Thus from Eq. (6.15),

$$p(t)=2\mathcal{R}\sum_{n_{i_a}=N}\int_0^t dt_1\int_0^{t_1}dt_2\{\langle i|H_1(t_1)|a\rangle\langle ia|\mathsf{G}'(t_1,t_2)|ia\rangle\langle ia|\mathfrak{L}|ii\rangle\Upsilon_i\}$$

$$=2\mathcal{R}\sum_{n_{i_a}=N}|\langle i|H_1|a\rangle|^2\int_0^t dt_1\int_0^{t_1}dt_2\exp\{-(i\omega_{ia}+\Gamma_{ia})(t_1-t_2)\}\Upsilon_i \tag{6.24}$$

where $\Upsilon_i=\langle i|\Upsilon|i\rangle$.

In Eq. (6.24) we have the probability after time t. We desire the probability per unit time, that is, $p(t)/t$. For t very large, only the term involving t will survive in $p(t)/t$ so that:

$$P=2\mathcal{R}\sum_{n_{i_a}=N}|\langle i|H_1|a\rangle|^2\Upsilon_i/(i\omega_{ia}+\Gamma_{ia})$$

$$=2\sum_{n_{i_a}=N}|\langle i|H_1|a\rangle|^2\Upsilon_i\frac{\gamma_{ia}}{(\omega_{ia}+\gamma_{ia}')^2+\gamma_{ia}^2} \tag{6.25}$$

Once again we have obtained the Weisskopf-Wigner result, and we might remark that once again the linewidth corresponds to the damping constant, that is, to the real part of Γ_{ia}.

We have been dealing with that portion of the interaction Hamiltonian which is linear in the vector potential, **A**. MRG also considered that portion of the interaction quadratic in **A**. Perhaps the basic difference between the two cases lies in the fact that, while Eq. (6.20) is zero in the linear case, it is nonzero in the quadratic case. This leads to a somewhat different displacement of the energy levels but has no effect on the linewidth.

7 Higher-Order Contributions. The Self Energy Transformation

We conclude our considerations of natural line shape with what was a most enlightening work, the study of the effects of higher-order perturbations by Arnous and Heitler (AH)(58). The success of the treatment was vouchsafed by the fact that almost no subsequent attention was paid to it. For AH demonstrated the venerable Weisskopf-Wigner first approximation to be such an excellent one that the higher orders and hence their careful treatment of them could be quite justifiably ignored.

We will recall that, in classical mechanics, Hamilton's canonical equations relating coordinates and conjugate momenta are invariant under a canonical transformation (sometimes called a contact transformation); in quantum mechanics the commutation relations among these quantities are invariant under a canonical (unitary) transformation. That we remark on this transformation is appropriate here since it plays an important role in the treatment of the effects of the higher orders of perturbation on the natural line shape as developed by Heitler and his associates, in particular by AH. We shall do more than remark shortly, but it is first necessary to recall a few elementary facts about self energy.

We recall (cf. § 14 of Heitler) the equations for the SGC's in first order for, say, the absorption or emission of a single photon

$$i\dot{\mathrm{b}}_n = H'_{nO}\mathrm{b}_O e^{\mathrm{i}(E_n - E_0)t}$$

$$i\dot{\mathrm{b}}_O = H'_{On}\mathrm{b}_n e^{\mathrm{i}(E_0 - E_n)t} \xrightarrow[\mathrm{b}_0(0)=1,\,\mathrm{b}_n(0)=0]{} \mathrm{b}_n(t) = H'_{nO}\frac{e^{\mathrm{i}(E_n - E_0)t} - 1}{E_O - E_n} \tag{7.1a}$$

H' being, as usual, the particle-field interaction. Now let us suppose there to be states n' within one photon of both O and n. Then

$$\mathrm{i}\dot{\mathrm{b}}_{n'} = H'_{n'O}\mathrm{b}_O e^{\mathrm{i}(E_{n'} - E_0)t}$$

$$\mathrm{i}\dot{\mathrm{b}}_n = \sum_{n'} H'_{nn'}\mathrm{b}_{n'} e^{\mathrm{i}(E_n - E_{n'})t} \xrightarrow[\mathrm{b}_0(0)=1,\,\mathrm{b}_n(0)=0]{} \mathrm{i}\mathrm{b}_n = K_{nO} e^{\mathrm{i}(E_n - E_0)t} \tag{7.1b}$$

where

$$K^{(2)}_{nO} = \sum_{n'} \frac{H'_{nn'}H'_{n'O}}{(E_O - E_{n'})} \tag{7.1c}$$

K_{nO} being dubbed a "compound matrix element" by Heitler, the n' being virtual states. When three instead of two photons are involved the doubly compound matrix element would be replaced by a triply compound one,

$$K^{(3)}_{n0} = \sum_{n'n''} \frac{H'_{nn'}H'_{n'n''}H''_{n''O}}{(E_O - E_{n'})(E_O - E_{n''})} \tag{7.1d}$$

and so on.

H' has no diagonal elements, but this is not to infer that it cannot contribute to the diagonal elements for the complete Hamiltonian in orders of perturbation theory higher than the first. One may use Eq. (7.1b) to arrive at the elementary second-order perturbation theory result

$$\Delta E_n = \sum_{n'} \frac{H'_{nn'}H'_{n'n}}{(E_n - E_{n'})} \tag{7.2}$$

Equation (7.2) tells us the change of energy of the electron in state n due to its interaction with radiation; this is the self energy of the electron. A rather common way of dealing with it is to introduce a term into the

Hamiltonian, say $\delta_m H'$, so that the diagonal elements of the electron Hamiltonian, now $H_0+\delta_m H'$, include the self energy. To (α, p) we add $\beta\delta\mu$ which, however, does not commute with (α, p) so that the off-diagonal elements of the operator with respect to the eigenvectors of H_0, exist.

AH begin with the wave equation,

$$\left(i\frac{\partial}{\partial t}-K_0-K\right)\psi(t)=0 \tag{7.3}$$

where K_0 and K are related to H_0 and H' through a canonical transformation which we now consider.

Our mass "renormalized," unperturbed Hamiltonian is now taken as $H_0\rightarrow H_0+H_s$ which means that, in order that $H(=H_0+H')$ remain the same, $H'\rightarrow H'-H_s$. The eigenvalues of H_0+H_s are still the undisplaced (by the self energy) levels, a situation which we account for by adding H_L to H_0+H_s. H_L has the characteristics: (1) it describes the level shifts and (2) it commutes with H_0+H_s. Since H_L commutes with H_0+H_s, the eigenfunctions of the latter are those of the former. Again we assure the invariance of H by adding $-H_L$ to H'. Finally, AH work in the Coulomb gauge, which means that the interaction now has components H_t for that with the transverse field, H_c for the Coulomb interaction. Thus

$$H_0=H_0+H_S+H_L, \qquad H'=H_t+H_c-H_s-H_L, \qquad H=H_0+H' \tag{7.4}$$

We take a state $|\Psi\rangle=|0\rangle$ to second order in ordinary perturbation theory:

$$|0\rangle=|\Psi_0^0\rangle+\mathcal{P}(E-H_0)^{-1}H'|\Psi_0^0\rangle \tag{7.5}$$

and write out the matrix element of H', $\langle\Psi_0^0|\mathrm{H}'|0\rangle$, to this order. First of all, we recall that H_t has no diagonal elements, so that $\langle\Psi_0^0|H_t|\Psi_0^0\rangle$ disappears. Therefore,

$$\langle\Psi_0^0|H'|0\rangle=\langle\Psi_0^0|H_c-H_s-H_L|\Psi_0^0\rangle+\langle\Psi_0^0|H_t\mathcal{P}(E_0-H_0)^{-1}H_t|\Psi_0^0\rangle \tag{7.6}$$

where we have supposed H_c, H_s, and H_L to contribute to the second matrix element only in orders higher than the second. Therefore,

$$H_{2L00}=(H_c-H_{2S})_{00}+\left(H_t\mathcal{P}(E_0-H_0)^{-1}H_t\right)_{00} \tag{7.7}$$

where the shorthand should be obvious, and

$$H'=H_t+H_c-H_s-H_L\equiv H_t+H_2+H_4+\cdots \tag{7.8}$$

$$H_2=H_c-H_{2S}-H_{2L}, \; H_4=-(H_{4S}+H_{4L}) \tag{7.9}$$

What AH call the "crucial problem of the theory," the elimination of the virtual states, may now be essayed. This procedure is begun by introducing a transformation which is intended to redefine the ground state in the absence of real photons, the vacuum state, and the state of a single free particle in such a way that the virtual states, which are sometimes referred to as "clothing" the particles, are included in the induced definition. We suppose that this may be accomplished as follows:

$$T^{-1}(H_0+H')T=K_0+K \tag{7.10a}$$

$$T=1+T_1+T_2+T_3+\cdots \tag{7.10b}$$

$$K=K_1+K_2+K_3+\cdots \tag{7.10c}$$

Eq. (7.10a) of course implies a transformation from the eigenvectors $|\Psi\rangle$ of (H_0+H') to the eigenvectors $|\Psi'\rangle=T^{-1}|\Psi\rangle$ of (K_0+K) so that

$$\mathrm{i}|\dot{\Psi}\rangle=\mathrm{i}(\dot{T}|\Psi'\rangle+T|\dot{\Psi}'\rangle)+(H_0+H')T|\Psi'\rangle.$$

Or

$$\mathrm{i}|\dot{\Psi}'\rangle=\left(T^{-1}H_0T-\mathrm{i}T^{-1}\dot{T}+T^{-1}H'T\right)|\Psi'\rangle=(K_0+K)|\Psi'\rangle \tag{7.11}$$

Since $TT^{-1}=1$, Eq. (7.10b) tells us that

$$T^{-1}=I-T_1+T_1^2-T_2+T_1T_2+T_2T_1-T_1^3-T_3+\cdots \tag{7.10d}$$

and we shall require that

$$K_0=H_0, \tag{7.10e}$$

Equations (7.11) and (7.10e) tell us that

$$TK=H'T-\mathrm{i}\dot{T} \tag{7.12}$$

Substituting from Eqs. (7.10b) and (7.10c) into Eq. (7.12) and equating terms of the same order (i.e., $H'T_3$ is of the same order as T_2K_2) we obtain:

$$T_1=H'-\mathrm{i}\dot{T}_1 \tag{7.13a}$$

$$T_2=H'T_1-T_1K_1-\mathrm{i}\dot{T}_2 \tag{7.13b}$$

$$T_3=H'T_2-T_1K_2-T_2K_1-\mathrm{i}\dot{T}_3 \tag{7.13c}$$

$$T_4=H'T_3-T_1K_3-T_2K_2-T_3K_1-\mathrm{i}\dot{T}_4 \tag{7.13d}$$

We have not really had to concern ourselves to this point with the elimination of the virtual states, but it is time that we do so. The customary method is the introduction of something called the "energy shell." Now our transformation to K is to be such that the total energy of the system is preserved, a point that soon introduces difficulties when we begin discussing levels of finite width, that is, excited levels. Now in theory this total energy conservation may be accomplished by requiring that K have elements only "on the energy shell," that is, within a "thickness," in energy space defined by requiring that $K_{nm}=0$ for $|E_n-E_m|>\epsilon$, ϵ being the energy shell thickness, then letting ϵ go to zero. (We see that this eliminates virtual states since there can be no terms of the form $K_{Om}K_{mn}$ [cf. Eq. (7.1c)] whereby such states are defined.) This is a very straightforward procedure for the ground state, which we know to have zero thickness or width, but leaves us rather in a quandary for the excited states which, as we know, tail off exponentially toward infinity insofar as their energy spread is concerned. What do we do?

First, we put off doing anything by claiming that the transformation carries within it the seeds of the explanation which we will examine later. When later arrives, as unfortunately it must, Heitler and his earlier associates referred back to the excitation phenomenon leading to the state and explained that the exciting radiation was a band of frequencies of twice the thickness of the energy shell. Such an assumption leaves us with a rather obvious dilemma, one of the horns relates to the restriction of the spectral line extension, the other to the elimination of the virtual states. However, in the paper that we are considering, a major part of this difficulty was eliminated by an ingenious development which eliminated the specific appearance of the transformation details through fourth order (an ample one) in the linewidth.

We shall use Heitler's subscripts (§ 15 of Heitler) *d* for on-shell elements, *nd* for off-shell elements, and *D* for diagonal with respect to all variables. The final requirement or boundary condition is on the element for *T*. Since $|\Psi'\rangle = T^{-1}|\Psi\rangle$, we agree that

$$|\Psi'_m\rangle = \sum_n T^{-1}_{mn}|\Psi_n\rangle$$

In this case elements of *T* on the energy shell are declared zero, only *D* diagonal elements of *T* being allowed· $T_d=0$; $T_D, T_{nd}\neq 0$.

Canonical transformations are à priori unitary which means that $T^{-1}=T^{+}$. From Eq. (7.10b) then $T^{+}=I^{+}+T_1^{+}+T_2^{+}+T_3^{+}+\cdots$, where I is of course the unit matrix, and $I^{+}=I$. By comparing the expression for T to Eq. (7.10d) we see that

$$T_1^{+}+T_1=0, \qquad T_2^{+}+T_2=T_1^{2}, \qquad T_3+T_3^{+}=T_1T_2+T_2T_1-T_1^{3} \quad (7.14)$$

Equations (7.10b), (7.10c), (7.9b), and (7.14) are substituted into Eq. (7.10a) and use is made of Eq. (7.10e). On equating terms of the same order on both sides, we obtain

$$K_1=H_t+[H_0,T_1] \quad (7.15a)$$

$$K_2=H_tT_1-T_1K_1+[H_0,T_2]+H_2 \quad (7.15b)$$

$$K_3=H_tT_2-T_1K_2-T_2K_1+H_2T_1+[H_0,T_3] \quad (7.15c)$$

$$K_4=H_tT_3-T_1K_3-T_2K_2-T_3K_1+[H_0,T_4]+H_2T_2+H_4 \quad (7.15d)$$

We see from Eq. (7.14) that $\langle a|T_1|b\rangle=-\langle a|T_1^{+}|b\rangle$ which is acceptable for $a\neq b$, but which must mean, for $a=b$ on the energy shell, $T_{1d}=0$. On the other hand, if we suppose T_{2d} not zero, then $T_{2d}^{+}=T_{2d}$, and, from Eq. (7.14),

$$T_{1d}=0, \qquad T_{2d}=\tfrac{1}{2}(T_1^2)_d \quad (7.16a)$$

We know $(H_t)_d=0$, and, by Eqs. (7.10a) and (7.10e), T_m and H_0 commute so that $[H_0,T_m]_d=0$. From Eqs. (7.15a) then,

$$K_{1d}=0 \quad (7.16b)$$

Again from Eq. (7.14) we find ($T_{3d}^{*}=T_{3d}$)

$$T_{3d}=\tfrac{1}{2}\left(T_1T_2+T_2T_1-T_1^3\right)_d \quad . \tag{7.16c}$$

With which we leave the virtual-state-eliminating transformation to its own devices for the moment. Using our transformed Hamiltonian operators, K_0+K, we may write a wave equation

$$\left(\mathrm{i}\frac{\partial}{\partial t}-K_0-K\right)|\Psi(t)\rangle=0 \tag{7.17a}$$

of solution

$$|\Psi(t)\rangle=R(t)|\Psi(t_0)\rangle \tag{7.17b}$$

$$R(t)=\begin{cases}\exp[-\mathrm{i}(t-t_0)(K_0+K)], & t>t_0\\ 0, & t<t_0\end{cases} \tag{7.17c}$$

where

$$\left(\mathrm{i}\frac{\partial}{\partial t}-K_0-K\right)R(t)=\mathrm{i}\delta(t-t_0) \tag{7.17d}$$

As is often the case in electrodynamics, the Dirac delta function or one of its relations—in this case, the zeta function of, Eq. (3.13)—is called on to play an important role. Let us rewrite Eq. (7.17c) slightly:

$$\begin{aligned}R(t)&=\tfrac{1}{2}\exp[-\mathrm{i}(t-t_0)(K_0+K)]+\tfrac{1}{2}\exp[-\mathrm{i}(t-t_0)(K_0+K)]\\ &=-\frac{\mathrm{i}}{2\pi}\int_{-\infty}^{\infty}dE\,e^{-\mathrm{i}(t-t_0)E}\mathrm{i}\pi\delta(E-K_0-K)\\ &\quad+\frac{\mathrm{i}}{2\pi}\lim_{R\to\infty}\int_{-R}^{R}dE\frac{e^{-\mathrm{i}(t-t_0)E}}{E-(K_0+K)}\end{aligned} \tag{7.18}$$

The quantity (K_0+K) is in the first quadrant of the complex plane so that $R(t)/2$, the latter term on the right of Eq. (7.18), is actually the principal value of $[E-(K_0+K)]^{-1}$. Therefore, by Eq. (3.13) we obtain

$$R(t)=\frac{\mathrm{i}}{2\pi}\int_{-\infty}^{\infty}dE\,e^{-\mathrm{i}(t-t_0)E}\zeta(E-K_0-K) \tag{7.19}$$

For $t<t_0$, $R(t)$ has the desired zero value, which discontinuously jumps to one for $t=t_0+0$ as it should.

The alternate definition for ζ provides a convenient expansion for that function:

$$\begin{aligned}\zeta(E-K_0-K)&=\lim_{\sigma\to0}\{[(E-K)+\mathrm{i}\sigma]-K\}^{-1}\\ &=\zeta+\zeta K\zeta+\zeta K\zeta K\zeta+\cdots, \qquad \zeta=\zeta(E-K)\end{aligned} \tag{7.20}$$

Such an expansion converges for $K^2<[(E-K_0)+\mathrm{i}\sigma]^2$.

There will now be matrix elements of the zeta function, and we designate the on-shell elements as

$$[\zeta(E-K_0-K)]_d=L(E)\equiv L_d \tag{7.21}$$

The ζ matrix may surely be written as the sum of a diagonal and an off-diagonal matrix. (By an "off-diagonal" matrix we mean a matrix having zeros on the principal diagonal.) Remarking that $\zeta(E-K_0)$ and $L(E)$ are diagonal, we see that, if $U^s(E)$ is off-diagonal, $L(E)U^s(E)N(E)$ will be off-diagonal. We may therefore write

$$\zeta(E-K_0-K)=L(E)+L(E)U^s(E)L(E) \tag{7.22a}$$

$$\zeta(E-K_0-K)=L(E)+\zeta(E-K_0)U(E)L(E) \tag{7.22b}$$

where

$$LU^s=\zeta(E-K_0)U, \qquad U^s\equiv U^s_{nd}, \qquad U\equiv U_{nd} \tag{7.22c}$$

Multiplying Eqs. (7.22b) and (7.22a) through on the left by $(E-K_0-K)$ and on the right by L^{-1} yields

$$L^{-1}=E-K_0-K+U-K\zeta U \tag{7.23a}$$

$$L^{-1}=E-K_0-K+(E-K_0-K)LU^s \tag{7.23b}$$

Let us call the operator which is the right side of Eq. (7.23a) A. The $(E-K_0)$ portion of A will be diagonal, the U portion off-diagonal, the $-(K+K\zeta U)$ portion diagonal and off diagonal. *In toto*, A will be $d+nd$. This means, of course, that A^{-1} must be $d+nd$. Remark however, that the *nd* portions of A^{-1} will only arise from U and from the *nd* parts of $K+K\zeta U$. Therefore, if we desire only the *d* parts of A^{-1}, we drop U and take only the *d* parts of $K+K\zeta U$. Multiplying Eq. (7.23a) through on the right by such an A^{-1} and on the left by L yields

$$L=\left[E-K_0+\tfrac{1}{2}\mathrm{i}\Gamma(E)\right]^{-1} \tag{7.24a}$$

$$\tfrac{1}{2}\mathrm{i}\Gamma(E)=(K+K\zeta U)_d \tag{7.24b}$$

In Eq. (7.24a) we have encountered something which should be somewhat familiar. For, as we shall see, $E-K_0$ is at least reminiscent of a frequency location in a spectral line, $\Gamma(E)$ of a damping factor in whose real part we may perhaps seek the line breadth. We substitute Eq. (7.24a) into Eqs. (7.23) in order to obtain

$$U=K+K\zeta U+\tfrac{1}{2}\mathrm{i}\Gamma=(K+K\zeta U)_{nd} \tag{7.25a}$$

$$U^s=K+\tfrac{1}{2}\mathrm{i}\Gamma+\left(K+\tfrac{1}{2}\mathrm{i}\Gamma\right)LU^s \tag{7.25b}$$

The next step is to obtain expansions for $\Gamma(E)$ and $U^s(E)$; the former expansion is obtained by substituting repeatedly for U from Eq. (7.25a) into $\Gamma(E)$ from Eq. (7.24b).

$$\begin{aligned}\tfrac{1}{2}\mathrm{i}\Gamma(E)&=\left[K+K\zeta(K+K\zeta U)_{nd}\right]_d=\cdots\\&=\left[K+K\zeta K_{nd}+K\zeta(K\zeta K_{nd})_{nd}+\cdots\right]_d\end{aligned} \tag{7.26}$$

In order to obtain the U^s expansion we begin with a rewritten Eq. (7.24a),

$$(E-K_0)L=1-\tfrac{1}{2}\mathrm{i}\Gamma L$$

which, when multiplied through on the right by $(1-\frac{1}{2}\mathrm{i}\Gamma L)^{-1}$, keeping in mind that $\zeta\cdot(E-K_0)=1$, yields

$$\zeta=L\left(1-\tfrac{1}{2}\mathrm{i}\Gamma L\right)^{-1}=L\left(1+\tfrac{1}{2}\mathrm{i}\Gamma L+\tfrac{1}{2}\mathrm{i}\Gamma L\tfrac{1}{2}\mathrm{i}\Gamma L+\cdots\right) \tag{7.27}$$

Equation (7.27) is substituted into Eq. (7.26), and Eq. (7.26) is substituted into the result so that we obtain

$$\begin{aligned}\tfrac{1}{2}\mathrm{i}\Gamma=\{&K+K_{nd}LK_{nd}+KL(K_{nd}LK_{nd})_{nd}+KL[K_{nd}L(K_{nd}LK_{nd})_{nd}]_{nd}\\&-KL(KLK_{nd})_dLK_{nd}+\cdots\}_d\end{aligned} \tag{7.28}$$

This puts us in a position to obtain the expansion for U^s in terms of K and L from Eq. (7.25b). The result is

$$\begin{aligned}U^s=\{&K+KLK_{nd}+[K_{nd}L(K_{nd}]_{nd}LK_{nd})_{nd}-\{K_{nd}L(K_{nd}\}_dLK_{nd})_d\\&+[\{K_{nd}L\{(K_{nd}\}_{nd}\,L[K_{nd})_{nd}]_{nd}LK_{nd}]_{nd}\}_{nd}\\&-\{K_{nd}L(K_{nd}\}_dLK_{nd})_dLK_{nd}-K_{nd}L(K_{nd}L[K_{nd})_dLK_{nd}]_d\\&-(K_{nd}LK_{nd}L[K_{nd})_dLK_{nd}]_d-[K_{nd}L(K_{nd}]_dLK_{nd}LK_{nd})_d\\&-[K_{nd}L(K_{nd}LK_{nd}]_dLK_{nd})_d+\cdots\}_{nd}\end{aligned} \tag{7.29}$$

In obtaining Eq. (7.29) AH made use of relations such as $(K_{nd}LK_d)_d=0$, justifiable as follows: Since L and K_d contain only diagonal elements, $K_{nd}LK_d$ contains only off-diagonal elements. Therefore, the diagonal elements of the product are zero. We may readily show that all but the first three terms and the fifth term of Eq. (7.29) vanish. Consider the fourth term.

$K_{nd}L$ is *nd*, say A_{nd}, so that we have $\{A_{nd}(K_{nd}\}_dA_{nd})_d$. There is simply no way in which these *nd* elements can furnish a *d* element, so that the result is zero. Eq. (7.29) exhibits "all possible resonances," as AH put it, through the operator whose resonances or poles are obvious from Eq. (7.24a).

The diagonal elements of the TDO are obtained from Eq. (7.19) and (7.22) as

$$R_d=\tfrac{1}{2}\mathrm{i}\pi^{-1}\int dE\,\exp[-\mathrm{i}(t-t_0)E]L(E) \tag{7.30}$$

The off-diagonal elements of the TDO in the interaction representation,

$$S(t)=e^{\mathrm{i}K_0t}R(t)e^{-\mathrm{i}K_0t},\qquad |\psi'(t)\rangle=e^{\mathrm{i}K_0t}|\psi(t)\rangle=S(t)|\psi'(t)\rangle \tag{7.31}$$

are obtained from the same equations as

$$S_{nd}(t)=\tfrac{1}{2}\mathrm{i}\pi^{-1}\int dE\,e^{-\mathrm{i}(E-K_0)t}\zeta ULe^{\mathrm{i}(E-K_0)t_0} \tag{7.32a}$$

$$=\tfrac{1}{2}\mathrm{i}\pi^{-1}\int dE\,e^{-\mathrm{i}(E-K_0)t}LU^sLe^{\mathrm{i}(E-K_0)t} \tag{7.32b}$$

Eq. (7.32a) corresponding to Eq. (7.22b), Eq. (7.32b) to Eq. (7.32a).

As usual, we shall be interested in the situation after infinite time, $t \to \infty$, and S_{nd} may be written down for this condition using the ζ relation,

$$\tfrac{1}{2}\mathrm{i}\pi^{-1}\int dx\,\zeta(x)e^{\mathrm{i}xt} = \begin{cases} 1 & \text{for } t>0 \\ 0 & \text{for } t<0 \end{cases} \tag{7.33}$$

If U has no singularity for $E=E_F\equiv K_{OF}$, F the final state, we obtain the following from Eq. (7.32a):

$$S_F A(\infty) = U_{FA}(E_A)L_{AA}(E_F)\exp[\mathrm{i}(E_F - E_A)t_0] \tag{7.34a}$$

It is probable that the atom will be in the ground state for $t\to\infty$, and we suppose this state to have zero width. If such is the case, the real part of Γ that corresponds to this level width will be zero. Under the assumption—which will be modified later—that the imaginary part of Γ is also zero, Eq. (7.24a) tells us that L is given by $\zeta(E-K_0)$. When this is substituted into Eq. (7.23b) we find

$$S_{FA}(\infty) = U_{FA}^{s}L_{AA}(E_F)\exp[\mathrm{i}(E_F - E_A)t_0] \tag{7.34b}$$

Now we are in a position to return to Eqs. (7.16) and those preceding and recall the objectives of the T transformation under discussion there. Our purpose in carrying out the transformation is to obtain a transformed ground state G, a vacuum V, and a single free-particle state which do not change with time under the influence of the interaction K. Our digression on S has put us in a position to state these requirements partially through the expression

$$S_{GG}(t) = \tfrac{1}{2}\mathrm{i}\pi^{-1}\int dE\,\exp[-\mathrm{i}(E-H_0)(t-t_0)]L_{GG}(E) = 1 \tag{7.35}$$

which we obtain from Eq. (7.30), the factor $\exp[iH_0(t-t_0)]$ entering via the interaction representation.

Now if Eq. (7.35) is to hold as we desire, Eq. (7.33) tells us that

$$L_{GG} = \zeta(E - H_{0GG}) \tag{7.36}$$

Eq. (7.36) in conjunction with Eq. (7.24a) tells us further that

$$\Gamma_{GG}(E) = 0 \tag{7.37}$$

Into the expansion Eq. (7.26) we substitute the expansion Eq. (7.10c) and keep only terms through second order with the result

$$-\tfrac{1}{2}\mathrm{i}\Gamma_2(E) = K_{2d} + (K_1\zeta K_1)_d \tag{7.38}$$

where K_{1d} has disappeared because of Eq. (7.16b). We write out the GG matrix element of Γ in detail:

$$\Gamma_{GG} = K_{2GG} + \sum_A K_{1GA}\zeta_{AA}K_{1AG} = 0 \tag{7.39}$$

since ζ is diagonal. Therefore, $K_{2GG}=0$. Since $K_{1AG}=0$ for every A, it must be true that $K_1|G\rangle = 0$. We may therefore determine $T_1|G\rangle$ directly from Eq. (7.15a), and this T_1 will induce the transformation results which we have been seeking.

At this point AH establish certain ground rules for their subsequent procedure which it behooves us to adopt also. $|G\rangle$, $|V\rangle$, and $|k\rangle$ represent the atomic ground state, the vacuum, and the photon in state k, respectively. In considering a photon incident on an atom in its ground state—which is going to be excited for future emission—we are dealing with $|G+k\rangle$. We suppose, however, that at some time, t_0, there is no interaction between G and k so that T_1 transforms $|G\rangle$, $|k\rangle$, and $|V\rangle$ independently. As a consequence, T_1 may be represented by $T_{1G}+T_{1V}+T_{1k}$ acting on these states. This brings up a rather vital definition.

AH call transitions taking place from $|G\rangle$ alone, $|k\rangle$ alone, or $|V\rangle$ alone virtual transitions, those involving both G and k real transitions. As examples of virtual transitions they display:

$$G+k\rightarrow G+k+k'+\text{pair},\qquad G+k\rightarrow G+\text{pair},\qquad G+k\rightarrow z+k+k'$$

while as a real transition they cite $G+k\rightarrow z$. In both the virtual and real cases z is an arbitrary atomic state other than the ground state. Evidently then we may decompose the interaction Hamiltonian as follows:

$$H|G+k\rangle=H_r|G+k\rangle+H_v|G+k\rangle \tag{7.40}$$

where the subscript r refers to real, the subscript v to virtual transitions. Now K_1 is required to induce real transitions,

$$K_1|G+k\rangle=H_r|G+k\rangle \tag{7.41a}$$

Therefore, Eqs. (7.40) and (7.15a) tell us that

$$[H_0,T_1]|G+k\rangle=(K_1-H_t)|G+k\rangle=-H_v|G+k\rangle \tag{7.41b}$$

We are now in a position to consider the spectral line shape, that is, we are in a position to study Eq. (7.24a),

$$L=\left[E-H_0+\tfrac{1}{2}i\Gamma(E)\right]^{-1} \tag{7.24a}$$

wherein $K_0=H_0$ by Eq. (7.10e).

Through Eqs. (7.34), the absolute square of L will of course tell us the spectral line contour, the real part of Γ yielding the linewidth and the imaginary part, the line shift. What we rather obviously must study then are the various orders in the expansion of Γ [Eq. (7.26)] as modified by the K expansion [Eq. (7.10c)]. The lowest order we have already written down as Eq. (7.38); this we consider first. In doing so we use the AH shorthand,

$$\zeta=\zeta(E-H_0),\qquad \mathcal{P}=\mathcal{P}(E-H_0)^{-1},\qquad \delta=\delta(E-H_0) \tag{7.42a}$$

$$\zeta_z=\zeta(E_z-H_0),\qquad \mathcal{P}_z=\mathcal{P}(E_z-H_0)^{-1},\qquad \delta_z=\delta(E_z-H_0) \tag{7.42b}$$

$$Q_m=[H_0,T_m]=[T_m,E-H_0],\qquad Q_{md}=0 \tag{7.42c}$$

That

$$\zeta=\zeta_0-(E-E_0)\zeta\zeta_0 \tag{7.43a}$$

we demonstrate using $x\zeta(x)=1$. Multiply Eq. (7.34a) through on the left by

$(E-H_0)$, the argument of ζ. Then the right side will equal one if the equation holds.

$$(E-H_0)\zeta(E_0-H_0)-(E-H_0)(E-E_0)\zeta(E-H_0)\zeta(E_0-H)=1 \quad (7.44)$$

We multiply through on the left by $(E-H_0)^{-1}=\zeta(E-H_0)$, on the right by $\zeta^{-1}(E_0-H_0)=(E_0-H_0)$ in order to obtain

$$1-(E-E_0)\zeta(E-H_0)=\zeta(E-H_0)(E_0-H_0)$$

Both sides of this equation are diagonal. Taking the OO element of both sides yields

$$1-(E-E_0)\zeta(E-E_0)=0$$

an equation which holds since $x\zeta(x)=1$. Eq. (7.43a) is thus valid. Since it is also true that $(E_O-H_O)|O\rangle=0$, $|O\rangle$ an arbitrary state, we may obtain certain other useful relations:

$$\langle O|Q_m\delta_0=\delta_0 Q_m|O\rangle=0 \quad (7.43b)$$

$$\langle O|Q_m\zeta_0=\langle O|T_m, \qquad \zeta_0 Q_m|O\rangle=-T_m|O\rangle \quad (7.43c)$$

We substitute Eq. (7.7) into Eq. (7.9) in order to obtain an expression for a diagonal element of H_2 which in turn we substitute into Eq. (7.15b) to obtain ($[H_0, T_m]_d=0$)

$$K_{2d}=(H_tT_1-T_1K_1-H_t\mathcal{P}_0H_t)_d=-(H_0\mathcal{P}_0Q_1+Q_1\mathcal{P}_0K_1+H_t\mathcal{P}_0H_t)_d \quad (7.45)$$

wherein we use the fact that H_tT_1 is replaceable by $H_0\mathcal{P}_0Q$ by Eqs. (7.43), for

$$(H_tT_1)_d=T_{1ab}=(\mathcal{P}_0)_{aa}Q_{ab}=(\zeta_0+\mathrm{i}\pi\delta_0)_{aa}Q_{ab}=-(T_m)_{ab}$$

$(H_0\mathcal{P}_0Q_1)_d$ will disappear since $H_0\mathcal{P}_0$ has only diagonal elements while Q_1 has no diagonal elements. By Eq. (7.15a) then,

$$K_{2d}=-(K_1\mathcal{P}_0K_1)_d \quad (7.46)$$

and Eq. (7.38) becomes

$$\tfrac{1}{2}\mathrm{i}\Gamma(E)=(K_1\zeta K_1-K_1\mathcal{P}_0K_1)_d \quad (7.47)$$

From the definition of the zeta function we find the real and imaginary parts to be

$$(2\pi)^{-1}\mathcal{R}[\Gamma_2(E)]=(K_1\delta K_1)_d \quad (7.48a)$$

$$\tfrac{1}{2}\mathcal{I}[\Gamma_2(E)]=(K_1\mathcal{P}K_1-K_1\mathcal{P}_0K_1)_d \quad (7.48b)$$

As was desired, the second-order level shift has been eliminated, for

$$\mathcal{I}[\Gamma_2(E_0)_{OO}]=0$$

At this point we have obtained Γ to second order, real, and imaginary parts as indicated. However, we could write Eq. (7.47) somewhat differently, replacing $\mathcal{P}_0$ by a sum of ζ_0 and δ_0 in order to obtain

$$\Gamma_2(E)=2\pi(H_t\delta_0H_t)+2\mathrm{i}(K_1\zeta K_1-K_1\zeta_0K_1)_d$$

with $(K_1\delta_0 K_1)=(H_t\delta_0 H_t)_d$ by Eq. (7.43b). Thus

$$\Gamma_2(E)=\gamma_2+(E-H_0)2\mathrm{i}\Delta_2(E) \tag{7.49a}$$

$$\gamma_2=2\pi(H_t\delta_0 H_t)_d \tag{7.49b}$$

$$\Delta_2(E)=-(K_1\zeta\zeta_0 K_1)_d \tag{7.49c}$$

by Eq. (7.43a).

In obtaining the form of Eq. (7.49a) we have split off or segregated what we may call the familiar second-order level width, γ_2, but keep in mind that there are still real parts lurking in Δ_2. Let us specify these as $\Gamma_4(E)$. Then Eq. (7.24a) may be written

$$L=\{[E-H_0][1-\Delta_2(E)]+\tfrac{1}{2}\mathrm{i}[\gamma_2+\Gamma_4(E)]\}^{-1}$$

Multiply the numerator and the denominator on the right by

$$(1-\Delta_2)^{-1}=1+\Delta_2+\Delta_2^2+\cdots$$

in order to obtain

$$L=\frac{1+\Delta_2(E)+\cdots}{E-H_0+\frac{1}{2}\mathrm{i}[\gamma_2+\bar{\Gamma}_4(E)]} \tag{7.50a}$$

where

$$\bar{\Gamma}_4=\Gamma_4(E)+\gamma_2\Delta_2(E) \tag{7.50b}$$

the imaginary series in the denominator having been cut off at fourth order. A precise analogy to Eq. (7.49a) is assumed for $\bar{\Gamma}_4$:

$$\bar{\Gamma}_4(E)_{OO}=\left(\bar{\Gamma}_4(E_0)+(E-H_0)2\mathrm{i}\Delta_4(E)\right)_{OO} \tag{7.51}$$

We insert this in the expression for L and obtain an analogy to Eq. (7.50a) having a sixth-order contribution to Γ from $\Delta_4(E)$ and a fourth-order contribution to the numerator, neither of which were considered by AH.

Perhaps this is an appropriate point for a remark on the object of all these exercises. γ_2 is the "classical"—as AH term it—Weisskopf-Wigner result for the linewidth or level width for a transition to the ground state. The question, how important are the contributions of the higher orders of perturbation, will surely be answered by determining $\bar{\Gamma}_4(E_0)$ in terms of γ_2. When AH made precisely this determination they found that γ_2 was some six orders of magnitude (137^3) greater than $\bar{\Gamma}_4(E_0)$. The general *modus operandi*, to the details of which we now turn our attention, is a by now familiar one.

We begin with Eq. (7.26) for Γ, substituting Eq. (7.10c) into it and, appropriately, keeping only fourth-order terms:

$$\begin{aligned}\tfrac{1}{2}\mathrm{i}\Gamma_4(E_0)=\{&K_4+K_1\zeta_0K_3+K_3\zeta_0K_1+K_2\zeta_0K_{2nd}+K_1\zeta_0K_2\zeta_0K_1\\&+K_1\zeta_0K_1\zeta_0K_{2nd}+K_2\zeta_0(K_1\zeta_0K_1)_{nd}+K_1\zeta_0K_1\zeta_0(K_1\zeta_0K_1)_{nd}\}_d\end{aligned} \tag{7.52}$$

It may be demonstrated that

$$\zeta_0 Q_m\zeta_0=\zeta_0 T_m-T_m\zeta_0, \qquad \zeta_0 F_d=F_d\zeta_0 \tag{7.53}$$

where F is an arbitrary but diagonal operator. It is now possible to use Eqs. (7.15), (7.43), and (7.53) in order to obtain

$$\tfrac{1}{2}\mathrm{i}\Gamma_4(E) = -\{(H_t\zeta_0 H_t + H_2)\zeta_0(H_t\zeta_0 H_t + H_2)_{nd} + H_t\zeta_0 H_2\zeta_0 H_t - \mathrm{i}\pi(H_t\zeta_0 H_t)(K_1\zeta_0^2 K_1 - H_t\zeta_0^2 H_t)_d + H_4\}_d \tag{7.54}$$

Note that T_2 and T_3 have dropped out, K_1 or T_1 remaining only in the third term.

As AH put it, "it may happen that" the central $H_t\zeta_0 H_t$ is diagonal in $H_t\zeta_0 H_t\zeta_0(H_t\zeta_0 H_t)_{nd}$, and so may H_2 in the second term. Whether it happens or not, these two cases may be written as

$$H_t\zeta_0 H_t\zeta_0(H_t\zeta_0 H_t)_{nd} + H_t\zeta_0 H_2\zeta_0 H_t = (H_t\zeta_0\{H_t)_{nd}\zeta_0 H_t + H_2\}_{nd}\zeta_0 H_t + (H_t\zeta_0\{H_t)_{nd}\zeta_0 H_t + H_2\}_d\zeta_0 H_t \tag{7.55}$$

Next, we may use Eq. (7.43a) in order to write the diagonal element,

$$(\zeta_0 H_t\zeta_0)_{zz} = \zeta_0(\zeta_z H_t)_{zz} - (H_t\zeta_0\zeta_z H_t)_{zz} \tag{7.56}$$

where we are going to suppose z the state for which $\{H_t\zeta_z H_t + H_2\}_d$ is the diagonal element. Remember that

$$H_t\zeta_z H_t + H_2 = H_t\frac{\mathcal{P}}{E_z - H_0}H_t - \mathrm{i}\pi\delta(E_z - H_0) + H_t\frac{\mathcal{P}}{E_0 - H_0}H_t = -\mathrm{i}\delta_z$$

since $E \equiv E_z$ for our diagonal assumption, and Eq. (7.55) is simplified accordingly. This element then becomes

$$-\mathrm{i}\pi H_t\zeta_0(H_t\delta_z H_t)_{zz}\zeta_0 H_t = -\tfrac{1}{2}\mathrm{i}(H_t\zeta_0\gamma_z\zeta_0 H_t)_{zz}, \qquad \gamma_z = \gamma_2(E_z)_{zz} \tag{7.57}$$

It may perhaps amuse the reader to collect the results of the last few equations and insert them into Eq. (7.54) in order to obtain

$$\tfrac{1}{2}\mathrm{i}\Gamma_4(E)_{OO} = -\{(H_t\zeta_0\{H_t)_{nd}\zeta_0 H_t + H_2\}_{nd}\zeta_0 H_t + H_2\zeta_0(H_t\zeta_0 H_t + H_2)_{nd} + H_t\zeta_0 H_t\zeta_0 H_{2nd} - \tfrac{1}{2}\mathrm{i}H_t\zeta_0\gamma_2\zeta_0 H_t - H_t(H_t\zeta_0\zeta_z H_t)_{zz}\zeta_0 H_t - \tfrac{1}{2}\mathrm{i}\gamma_2(K_1\zeta_0^2 K_1 - H_t\zeta_0^2 H_t) + H_4\}_{OO} \tag{7.58}$$

With which we have almost obtained the desired result, $\overline{\Gamma}_4$, if not quite. According to Eq. (7.50b), we must add to Eq. (7.58) the product of γ_2 as given by Eq. (7.49b) and Δ_2 as given by Eq. (7.49c) in order to obtain Γ_4.

$$\tfrac{1}{2}\mathrm{i}\overline{\Gamma}_4(E_0)_{OO} = -\{(H_t\zeta_0\{H_t)_{nd}\zeta_0 H_t + H_2\}_{nd}\zeta_0 H_t + H_2\zeta_0(H_t\zeta_0 H_t + H_2)_{nd} + H_t\zeta_0 H_t\zeta_0 H_{2nd} - \tfrac{1}{2}\mathrm{i}H_t\zeta_0\gamma_2\zeta_0 H_t - H_t(H_t\zeta_0\zeta_z H_t)_{zz}\zeta_0 H_t + \tfrac{1}{2}\mathrm{i}\gamma_2 H_t\zeta_0^2 H_t + H_4\}_{OO} \tag{7.59}$$

It is immensely important for us to remark the absence of any K or any T in Eq. (7.59). This means that the result is independent of the canonical transformation T. Therefore, to this order, our line breadth is not a function of the mode of excitation of the upper state in the radiating transition, a fact which considerably generalizes the result. On the other hand, since Δ_2 in Eq. (7.50a) is not independent of the transformation, the asymmetry which this introduces through the numerator will not be independent of the excitation mechanism.

AH now write

$$L = L(1+\Delta(E)) = \frac{1+\Delta_2(E)+\cdots}{(E-H_0)+i(\gamma/2)} \tag{7.60a}$$

where

$$\Delta(E) = \Delta_2(E)+\cdots, \qquad \gamma = \gamma_2 + \bar{\Gamma}_4(E_0)+\cdots \tag{7.60b}$$

We shall not study the numerator in detail but simply refer the reader to Section 4 of AH where these energy-shell-dependent terms are considered. Insofar as the actual calculation of $\mathfrak{R}[\bar{\Gamma}_4(E_0)]$ from Eq. (7.59) is concerned, it is, as we might suspect, lengthy and has been discussed by Arnous (57). It is approximately given by

$$\mathfrak{R}\left[\bar{\Gamma}_4(E_0)\right] \doteq \gamma_2 p^2/137\mathrm{m}^2 \doteq \gamma_2/137^3 \tag{7.61}$$

where p is the average momentum of the electron in the atom. AH point out that the accurate result may be as much as a factor of 20 larger than this. Under any conditions, the AH study clearly demonstrates that higher-order corrections to the natural line width are negligible.

Asymmetries in the natural line shape due to the higher order terms in the numerator can, however, become important in the microwave, as AH emphasize.

3

Greens Functions, Liouvillians, and Shape Bases

That the Greens function cannot be ignored in contemporary line broadening theory the reader is doubtless aware. In Section 8, we look at it briefly and develop its relationship to the Liouville operator. Then we briefly review certain familiar scattering results to include a brush with the optical theorem.

Without attempting the probably impossible assessment of this technique's value in contemporary broadening theory, one may certainly classify the manipulations of the Liouville space as immensely important. In preparation for later work we discuss various Liouville space operators and manipulations in Section 10. This we follow with a particular three-particle scattering treatment whose results we shall require.

We have begun our (natural) line shape treatments with an expression for the intensity which consisted chiefly of the squared SGC. There are other, equally important starting expressions, and these we discuss in Section 12.

8 The Greens Function and the Liouville Operator

Reichsfreiherr Gottfried Wilhelm Leibniz occupied himself with a bewildering variety of concerns ranging from the promulgation of the first—if we except Louis IX—French Egyptian expedition and the reconciliation of the Catholic and Protestant faiths to the creation of the monad. Unfortunately, a careful survey shows that the once popular monad is currently remembered only by an obscure witch doctor in Upper Volta, the French Egyptian expedition by no one at all. A few of us do recall the delightful—in spite of the rather dour Moritz Cantor—hubbub involving the Reichsfreiherr, Sir Isaac, and their devotees. All of us immediately bring to mind his invention of the calculus notation and the second differential, the latter so important to the spur which developed the calculus, the search for a method of determining maxima and minima. Insofar as this second differential is concerned, Christian Huygens' almost immediate declaration that it was meaningless can hardly be construed as a highlight of his otherwise illustrious career. However, our concern with the famous lens grinder has little to do with differentials although it has a great deal to do with the waves to which he devoted so much of his attention.

Let us suppose that a traveling wave is described by a wave function $\varphi(x,t)$ at a time t. We may, by Huygens' principle, consider each point of space, x, as the source of a spherical wave, this source having as its generator the original traveling wave. Therefore, at some later time, t', we would encounter a wave at point, x', whose amplitude, $\varphi(x',t)$, is a superposition of those of all the wavelets emitted throughout space at the time, t, the constant of proportionality being assumed to be $\mathrm{i}\mathsf{G}(x,x')$ $[x\equiv(\mathbf{x},t)]$. Huygens' principle then tells us:

$$\varphi(x')=\mathrm{i}\int d\mathbf{x}\,\mathsf{G}(x',x)\varphi(x), \qquad t'>t \tag{8.1}$$

$\mathsf{G}(x',x)$ is of course the Greens function. It is important to remark that, if we concern ourselves more directly with a particle rather than a wave, $\mathsf{G}(x',x)$ is the probability that, if we place a particle at space-time point x, we will find it at space-time x'. Let us determine the free-particle propagator.

In order to do so we shall (1) determine the relationship of the propagator to the Schrödinger equation and (2) Fourier transform the free-particle propagator.

When we insert $\epsilon(t'-t)\varphi(x')$, ϵ assuring that $t'>t$, in the Schrödinger equation, we obtain

$$\left[\mathrm{i}\frac{\partial}{\partial t}-H(x')\right]\epsilon(t'-t)\varphi(x')=\mathrm{i}\delta(t'-t)\varphi(x') \tag{8.2}$$

because (1) the bracketed expression operating on $\varphi(x')$ yields zero and (2) $[\cdots]\epsilon(t'-t)=\mathrm{i}\delta(t'-t)$ by analogy to Eq. (3.4). Therefore, operating on Eq. (8.1) yields

$$\left[\mathrm{i}\frac{\partial}{\partial t}-H(x')\right]\epsilon(t'-t)\varphi(x')=\mathrm{i}\delta(t'-t)\varphi(x')$$
$$=\mathrm{i}\int dx\left[\mathrm{i}\frac{\partial}{\partial t}-H(x')\right]\mathsf{G}(x',x)\varphi(x) \tag{8.3}$$

which tells us that

$$\left[\mathrm{i}\frac{\partial}{\partial t}-H(x')\right]\mathsf{G}(x',x)=\delta(x'-x) \tag{8.4}$$

where $\delta(x-x')$ is of course a four-dimensional delta function. That Eq. (8.4) is a solution of Eq. (8.3) may be verified by multiplying through on the right by $\varphi(x)$ and integrating over all x. This completes our determination of the relationship between the Greens function and the Schrödinger equation.

We consider the free-particle Greens function, that is, the Greens function for which $H(x')=\nabla'^2/2\mathrm{m}$, which we Fourier transform as follows:

$$\mathrm{G}^0(x',x)=\mathrm{G}^0(x'-x)=\int d\mathbf{k}\,d\omega\frac{\exp[\mathrm{i}\mathbf{p}\cdot(\mathbf{x}'-\mathbf{x})-\mathrm{i}\omega(t'-t)]\mathrm{G}^0(\mathbf{p},\omega)}{(2\pi)^4} \tag{8.5}$$

We operate on both sides of Eq. (8.5) with the bracketed expression on the left of Eq. (8.4).

$$\left(\mathrm{i}\frac{\partial}{\partial t'}+\frac{\nabla'^2}{2\mathrm{m}}\right)\mathrm{G}^0(x',x)=\int d\mathbf{k}\,d\omega\left(\omega-\frac{\mathrm{k}^2}{2\mathrm{m}}\right)\mathrm{G}^0(\mathbf{p},\omega)$$
$$\cdot\exp[-\mathrm{i}\omega(t'-t)+\mathrm{i}\mathbf{p}\cdot(\mathbf{x}'-\mathbf{x})]=\delta(x'-x)$$
$$\equiv\int d\mathbf{k}\,d\omega\frac{\exp[-\mathrm{i}\omega(t'-t)+\mathrm{i}\mathbf{p}\cdot(\mathbf{x}'-\mathbf{x})]}{(2\pi)^4} \tag{8.6}$$

Equation (8.6) tells us that

$$\mathrm{G}^0(\mathbf{p},\omega)=[\omega-p^2/2\mathrm{m}]^{-1} \tag{8.7a}$$

for $\omega\neq p^2/2\mathrm{m}$.

This result may also be obtained as follows. The free-particle wave function is of course $\exp[\mathrm{i}\mathbf{p}\cdot\mathbf{r}-\mathrm{i}\omega t]$. Therefore, for the free-particle case we may rewrite Eq. (8.4) as

$$[\omega-p^2/2\mathrm{m}]\mathrm{G}^0=1\Rightarrow\mathrm{G}^0=[\omega-p^2/2\mathrm{m}]^{-1}=[E-H_0]^{-1} \tag{8.7b}$$

as a consequence of multiplying through on the left by $[\omega-p^2/2\mathrm{m}]^{-1}$ or $[E-H_0]^{-1}$. In order to obtain the Greens function in projection operator form we may then begin by expressing it in the basis which solves the Schrödinger equation corresponding to Eq. (8.4), thus:

$$\mathrm{G}^0(\mathbf{r},\mathbf{r}';E)=\sum_{ij}g_{ij}\varphi_i(\mathbf{r})\varphi_j^*(\mathbf{r}') \tag{8.8}$$

where $[E_i-H_0]\psi_i=0$. When Eq. (8.8) is substituted into the equivalent of Eq. (8.4), we obtain

$$-\delta(\mathbf{r}-\mathbf{r}')=(H_0-E)\mathrm{G}^0=\sum_{ij}g_{ij}(E_i-E)\varphi_i(\mathbf{r})\varphi_j^*(\mathbf{r}')$$

Multiplying this equation through on the right by $\varphi_i(\mathbf{r}')$ and integrating over all space leads to the solution,

$$g_{ij}=\delta_{ij}/(E-E_i)$$

so that

$$\mathrm{G}^0(\mathbf{r},\mathbf{r}';t)=\sum_i\varphi_i(\mathbf{r})\varphi_j^*(\mathbf{r}')/(E-E_i)=\sum_i|m\rangle[E-E_i]^{-1}\langle m| \tag{8.9}$$

in the denominator of which an $\mathrm{i}\delta$ is, as usual, added or subtracted in order to eliminate the indeterminacy at infinity [cf. § 9.4 and 10.1 of Kirzhnits (28)].

It will be of interest to us to express certain wave function important to scattering problems in terms of Greens functions. In order to do this we shall begin by expressing the wave function for time, t, in terms of that for some later time, t', by means of the TDO $\exp[-\mathrm{i}H(t-t')]$: $\Psi_n(t)=\exp[-\mathrm{i}H(t-t')]\Phi_n(t')$. The transients which arise are eliminated by "feeding in the incident beam over a period of time," as Watson and Goldberger (16) put it, an infinite period of time in order that the uncertainty principle may allow a sharply defined energy. Then a state $\Psi_n^{(\epsilon)}(t)$,

$$\Psi_n^{(\epsilon)}(t)=\epsilon\int dt'\, e^{\epsilon t'}e^{-\mathrm{i}H(t-t')}\Phi_n(t')$$

is defined whose limit is the desired state. Thus

$$\Psi_n=\lim_{\epsilon\to0+}\Psi_n^{(\epsilon)}(0)=\lim_{\epsilon\to0+}(-\epsilon)\int_{-\infty}^{0}dt'\, e^{\epsilon t'}e^{\mathrm{i}(H-E_n)t'}\Phi_n$$

$$=\lim_{\epsilon\to0+}(-\mathrm{i}\epsilon)\left[H-E_n-\mathrm{i}\epsilon\right]^{-1}\Phi_n=\lim_{\epsilon\to0+}(-\mathrm{i}\epsilon)\mathrm{G}(E_n+\mathrm{i}\epsilon)\Phi_n \quad (8.10)$$

Let us remark that, for $\mathrm{G}^0=[H^0-E_n^0-\mathrm{i}]^{-1}$,

$$\lim_{\epsilon\to0+}(-\mathrm{i}\epsilon)\mathrm{G}(E_0^0+\mathrm{i}\epsilon)\Phi_n=\lim_{\epsilon\to0+}\frac{-\mathrm{i}\epsilon}{H^0-E_n^0-\mathrm{i}\epsilon}\Phi_n$$

$$=\left(1+\frac{1}{H^0-E_n^0-\mathrm{i}\epsilon}(E_n^0-H^0)\right)\Phi_n=\Phi_n \quad (8.11)$$

since $(E_n^0-H^0)\Phi_n=0$ by definition.

Let us use Eq. (8.7a) in order to carry out the integration of Eq. (8.5), adding the imaginary infinitesimal of the last paragraph to the denominator of the former which has here the effect of removing the ω pole below the path of integration along the real axis in the complex plane,

$$\mathrm{G}^0(x'-x)=(2\pi)^{-3}\int d\mathbf{p}\, e^{\mathrm{i}\mathbf{p}\cdot(\mathbf{x}'-\mathbf{x})}\int_{-\infty}^{\infty}d\omega\frac{e^{-\mathrm{i}\omega(t'-t)}}{2\pi\left[\omega-p^2/2\mathrm{m}+\mathrm{i}\epsilon\right]} \quad (8.12)$$

We replace ω by $\Omega=\omega-p^2/2\mathrm{m}$ so that the ω integral in Eq. (8.12) becomes

$$\lim_{\epsilon\to0}\exp\left(\frac{-\mathrm{i}p^2(t'-t)}{2\mathrm{m}}\right)\int_{-\infty}^{\infty}d\Omega\frac{e^{-\mathrm{i}\Omega(t'-t)}}{2\pi(\Omega+\mathrm{i}\epsilon)}\equiv-\mathrm{i}\exp\left(\frac{-\mathrm{i}p^2(t'-t)}{2\mathrm{m}}\right)\epsilon(t'-t) \quad (8.13)$$

Therefore, Eq. (8.12) becomes

$$\mathrm{G}^0(x'-x)=-\mathrm{i}\epsilon(t'-t)\int d\mathbf{p}\,\varphi_k(x')\varphi_k^*(x) \quad (8.14)$$

which is to be compared to the Dirac electron Greens function, S, of Chapter 2.

We shall suppose the Greens function to have the same form as the free-particle Greens function, the φ being appropriately modified. Should the state vectors be a discrete instead of a continuous set, a summation will replace the integration of Eq. (8.14) so that

$$\mathsf{G}(x', x) = -\mathrm{i}\sum_n \varphi_n(x')\varphi_n^*(x)\epsilon(t'-t) \tag{8.15}$$

The φ_n are wave functions, that is, simple or not so simple functions of the configuration coordinates. (They could be eigenvectors and, in a sense, are, but this is of no particular consequence at the moment.) It behooves us to rewrite the equation in operator form, which means that we shall proceed to a second-quantized expression for G. We write G as

$$\mathsf{G}(x'x) = -\mathrm{i}\langle\Psi_0|\mathsf{T}_{\leftarrow}[\psi(x')\psi^+(x)]|\Psi_0\rangle \tag{8.16a}$$

$$\psi(x) = \sum_i \mathsf{A}_i\varphi_i(x) \tag{8.16b}$$

$$\psi^+(x) = \sum_i \mathsf{A}_i^+\varphi_i^*(x) \tag{8.16c}$$

In Eqs. (8.16) the φ_i and φ_i^* are still the wave functions of Eq. (8.15) and previous. The A_i and A_i^+ annihilate and create, respectively, a particle in state i. Ψ_0 is the second quantized vacuum state, that is to say, the state corresponding to no particle present in the field. Thus,

$$\sum_i \mathsf{A}_i\varphi_i|\Psi_0\rangle = 0, \qquad \sum_i \mathsf{A}_i^+\varphi_i^*(x)|\Psi_0\rangle = \sum_i \varphi_i^*(x)|\Psi_{OO\cdots i_1\cdots}\rangle$$

where $\Psi_{OO\cdots i_1\cdots}$ is the function corresponding to no particles in any state save i wherein there is one particle. The reader will therefore see that Eq. (8.16a) reproduces Eq. (8.15). A shorthand form is generally preferred:

$$\mathsf{G}(x', x) = -\mathrm{i}\langle\psi(x')\psi^+(x)\rangle \tag{8.17}$$

The point of immediate importance here is that G is in operator form in the Schrödinger representation. It may be put in the Heisenberg representation as follows:

$$\psi(x) = e^{\mathrm{i}Ht}\psi(\mathbf{x})e^{-\mathrm{i}Ht} \tag{8.18}$$

For simplicity let us take $t' = t$, $t = 0$. Therefore, Eq. (8.17) becomes

$$\begin{aligned}\mathsf{G}_{ij}(x', x) &= -\mathrm{i}\langle\Psi_0|\mathsf{T}\left[e^{\mathrm{i}Ht}\psi_i(\mathbf{x}')e^{-\mathrm{i}Ht}\psi_j^+(\mathbf{x})\right]|\Psi_0\rangle \\ &= -\mathrm{i}\sum_{\alpha\beta\gamma}\langle\Psi_0|e^{\mathrm{i}Ht}|\Psi_\alpha\rangle\langle\Psi_\alpha|\psi_i(\mathbf{x}')|\Psi_\beta\rangle \\ &\quad\cdot\langle\Psi_\beta|e^{-\mathrm{i}Ht}|\Psi_\gamma\rangle\langle\Psi_\gamma|\psi_j^+(\mathbf{x})|\Psi_o\rangle \\ &= -\mathrm{i}\sum_{\alpha\beta\gamma} e^{\mathrm{i}[H_{0\alpha}\delta_{\beta\gamma}-\delta_{0\alpha}H^*_{\gamma\beta}]t}\langle\Psi_\alpha|\psi_i(\mathbf{x}')|\Psi_\beta\rangle\langle\Psi_\alpha|\psi_j^+(\mathbf{x})|\Psi_0\rangle\end{aligned} \tag{8.19}$$

Consider the last matrix element on the right of Eq. (8.19). $\psi_j^+(\mathbf{x})|\Psi_0\rangle = |\Psi_j\rangle$ so that this matrix element disappears unless $\alpha=0$. Similarly, $\beta=i$ and $\alpha=0$ or the next element to the left disappears. Using Eq. (1.7) then we see that

$$\mathrm{G}_{ij}(x',x) = -\mathrm{i}\sum_{\alpha\beta\gamma}\langle O\gamma|e^{i\mathfrak{L}t}|\alpha\beta\rangle\langle\alpha|\psi_i|\beta\rangle\langle\gamma|\psi_j^+|O\rangle$$

$$= -\mathrm{i}\langle Oj|e^{i\mathfrak{L}t}|Oi\rangle$$

or

$$\mathrm{G}(x',x) = -\mathrm{i}e^{\mathrm{i}\mathfrak{L}t} \tag{8.20}$$

If we take the Fourier transform, we see that

$$\mathrm{G}(\omega) = -\mathrm{i}\int_0^\infty e^{\mathrm{i}\omega t}e^{-\mathrm{i}\mathfrak{L}t}\,dt = -\mathrm{i}\lim_{\epsilon\to 0+}\left[\omega-\mathfrak{L}-\mathrm{i}\epsilon\right]^{-1} \tag{8.21}$$

which temporarily concludes the discussion of the Greens function and the Liouville operator.

9 Certain Well-Known Scattering Results

The asymptotic form of the wave function for a scattered particle can logically be taken to be superposition of an incident plane wave and a spherical scattered wave having the form

$$\psi(r) = e^{\mathrm{i}\mathbf{p}\cdot\mathbf{r}} + \frac{e^{\mathrm{i}pr}f(\vartheta)}{r} \tag{9.1}$$

where ϑ is the scattering angle measured with respect to the initial line of flight of the scattered particle. By the nature of the wave function, the probability that the particle will be scattered through an angle, ϑ, will be proportional to the absolute square of the scattering amplitude, $f(\vartheta)$. The scattering cross section or total probability that the particle is scattered at all will then be

$$\sigma = \int |f(\vartheta)|^2\,d\Omega = 2\pi\int |f(\vartheta)|^2\sin\vartheta\,d\vartheta \tag{9.2}$$

where Ω is the solid angle.

Next we will recall the partial wave expansion of Faxen and Holtsmark (116) which satisfies the Schrödinger equation for a spherically symmetric scattering potential, $\mathrm{V}(r)$,

$$f(\vartheta) = \sum_l (2l+1)\left[\exp(2\mathrm{i}\delta_l)-1\right]\frac{P_l(\cos\vartheta)}{2\mathrm{i}p} \tag{9.3}$$

where δ_l are the phase shift induced in the wave by the scattering center. If we substitute Eq. (9.3) into Eq. (9.2), we obtain

$$\sigma = 4\pi\sum_l (2l+1)\frac{\sin^2\delta_l}{p^2} \tag{9.4}$$

From Eq. (9.3) it is obvious that

$$I[f(0)] = \sum_l (21+1)\frac{\sin^2\delta_l}{2ip} \tag{9.5}$$

Combining Eqs. (9.4) and (9.5) leads to the optical theorem,

$$I[f(0)] = [p/4\pi]\sigma \tag{9.6}$$

which we shall obtain in a slightly different fashion. Before doing so, however, let us recall what we shall presume to be familiar facts about the scattering amplitude.

Given the wave equation for a spherically symmetric potential, $V(r)$,

$$\nabla^2\psi + [p^2 - U(r)]\psi = 0, \qquad U(r) = 2mV(r) \tag{9.7}$$

the solution [cf. Chapter 7 of Mott and Massey (32)] is

$$\psi = e^{i\mathbf{p}\cdot\mathbf{r}} - \frac{1}{4\pi}\int \frac{e^{ipR}}{R} U(r')\psi(\mathbf{r}')\,d\mathbf{r}', \qquad \mathbf{R} = \mathbf{r} - \mathbf{r}' \tag{9.8a}$$

of asymptotic form

$$\psi = e^{i\mathbf{p}\cdot\mathbf{r}} - \frac{e^{ipr}}{4\pi r}\int e^{-i\mathbf{p}\cdot\mathbf{r}'} U(r')\psi(\mathbf{r}')\,d\mathbf{r}' \tag{9.8b}$$

A comparison of Eqs. (8.1) and (8.8) tells us that the scattering amplitude is given by

$$f(\vartheta) = \frac{1}{4\pi}\int e^{-i\mathbf{p}\cdot\mathbf{r}'} U(r')\psi(\mathbf{r}')\,dr' = \frac{m\langle\psi_p^0|V(r)|\psi_{p'}\rangle}{2\pi} \tag{9.9}$$

where ψ_p^0 is the plane wave, $\psi_{p'}$ the exact solution. The first Born approximation for this amplitude is obtained by replacing $\psi_{p'}$ by ψ_p^0 in Eq. (9.9). With which we turn to an alternate derivation of the optical theorem which will be of practical value to us at a later time.

Perhaps it is fair to say that the following relation between the S and T matrices is well known [cf. § 4.14 of Ziman (52)]:

$$\langle\mathbf{p}'|S|\mathbf{p}\rangle = \delta(\mathbf{p}' - \mathbf{p}) - 2\pi i\delta(E' - E)\langle\psi_{p'}^0|T|\psi_p^0\rangle \tag{9.10}$$

where the energy, E', is associated with the momentum state, $\mathbf{p}'$, the energy E with $\mathbf{p}$. $\delta(E' - E)$ rather obviously requires the T matrix to be on the energy shell, although it is true that certain treatments do not rigorously maintain this. If $\mathbf{p}' = \mathbf{p}$, the incident particle is not scattered by the collision. Therefore, the first term in Eq. (9.10) describes no-scatter scatterings, the second term actual scatterings or changes in incident particle momentum.

Now let us appeal to Eq. (8.9) in order to obtain a form of Eq. (9.8a). Thus for the plane wave $|\mathbf{p},\mathbf{r}\rangle = \exp(i\mathbf{p}\cdot\mathbf{r})$,

$$G^+(\mathbf{r},\mathbf{r}';E) = \sum_k |\mathbf{p},\mathbf{r}\rangle[E - E(\mathbf{p}) + i\delta]^{-1}\langle\mathbf{p},\mathbf{r}'|$$

$$= \frac{1}{(2\pi)^3}\int d\mathbf{p}\, e^{i\mathbf{p}\cdot\mathbf{r}}[\kappa^2 - p^2 + i\delta]^{-1} e^{-i\mathbf{p}\cdot\mathbf{r}'} \tag{9.11}$$

where $E=\kappa^2$ for future convenience. As the system is homogeneous, the Greens function depends only on the coordinate difference, $R=\mathbf{r}-\mathbf{r}'$. The integration is straightforward [cf. § 4.12 of Ziman (52)] and yields

$$\mathsf{G}^{0^+}(\mathbf{R}; E)=\frac{e^{i\kappa R}}{4\pi R} \tag{9.12}$$

which means that Eq. (9.8a) may be replaced by

$$|\Psi^+,\mathbf{r}\rangle=|\Phi,\mathbf{r}\rangle+\int \mathsf{G}^{0^+}(\mathbf{r},\mathbf{r}'; E)\mathsf{V}(r')|\Psi^+,\mathbf{r}'\rangle\, d\mathbf{r}'$$

which, by Eq. (8.7b), we may associate with the Lippman-Schwinger equation,

$$|\Psi^+\rangle=|\Phi\rangle+[E-H_0+i\delta]^{-1}\mathsf{V}|\Psi^+\rangle \tag{9.13}$$

The reader will doubtless find it exhilarating to locate a similar expression for $\langle\Psi^-|$ and multiply it into Eq. (8.13) in order to obtain

$$\begin{aligned}\langle\Psi_p^-|\Psi_p^+\rangle&=\langle\Phi_{p'}|\Phi_p\rangle+\{[E-E_i'+i\delta]^{-1}+[E'-E+i\delta]^{-1}\}\langle\Phi_{p'}|\mathsf{V}|\Psi_p^+\rangle\\ &=\delta(p-p')-2\pi i\delta(E-E')\langle\Phi_{p'}|\mathsf{V}|\Psi_p^+\rangle\end{aligned} \tag{9.14}$$

It is probably apparent that the first term, an integral over a pair of wave functions, is, by definition, the delta function. Further, a little rumination tells us that we have gotten Eq. (9.10) back again with the concomitant definition of T. Therefore, from Eqs. (8.9), (8.10), and (8.14), we see that

$$2\pi\delta(E'-E)\frac{f}{m}=\langle\mathbf{p}'|T|\mathbf{p}\rangle \tag{9.15}$$

Now let us substitute $S=1+2\pi iT$ into $S^+S=1$ in order to obtain

$$T+T^+=2\pi i T^+T\Rightarrow$$

$$\langle\mathbf{p}'|T|\mathbf{p}\rangle-\langle\mathbf{p}|T|\mathbf{p}'\rangle^*=+2\pi i\int\frac{d\mathbf{p}''\langle\mathbf{p}''|T|\mathbf{p}'\rangle^*\langle\mathbf{p}''|T|\mathbf{p}\rangle}{(2\pi)^3} \tag{9.16}$$

by elementary matrix multiplication. When we substitute for the T-matrix elements from Eq. (9.15), Eq. (9.16) becomes

$$f-f^*=i\int d\mathbf{p}''\frac{\delta(E-E')f^*(\mathbf{p}'',\mathbf{p}')f(\mathbf{p}'',\mathbf{p})}{2\pi m}$$

$$\begin{aligned}\mathscr{I}[f]&=[f-f^*]/2i\\ &=\int dp''\,dE''\,d\Omega\frac{\delta(E''-E')\delta(E''-E)f^*(\mathbf{p}'',\mathbf{p}')f(\mathbf{p}'',\mathbf{p})}{4\pi}\\ &=p\int d\Omega_{k''}\frac{|f(\mathbf{p}'',\mathbf{p})|^2}{4\pi}=p\int d\Omega\frac{d\sigma}{d\Omega}/4\pi\\ &=p\sigma(p)/4\pi\end{aligned}$$

Therefore,

$$\mathscr{I}[T]\propto-\sigma(\mathbf{p}) \tag{9.17}$$

10 Greens Functions and Projection Operators in Liouville Space

We begin with an obvious extrapolation of Eq. (8.7b),

$$\mathsf{G}=[E-H]^{-1}=[E-H_0-\mathsf{V}]^{-1} \tag{10.1}$$

In order to rewrite this, let us recall the following operator relationship:

$$A^{-1}-B^{-1}=A^{-1}(B-A)B^{-1}=B^{-1}(B-A)A^{-1}$$

We now let $A=[E-H]$, $B=[E-H_0]$ so that

$$[E-H]^{-1}-[E-H_0]^{-1}=[E-H_0]^{-1}\mathsf{V}[E-H]^{-1}$$

$$\mathsf{G}=\mathsf{G}^0[1+\mathsf{VG}]=\mathsf{G}^0[1+\mathsf{VG}^0+\mathsf{VG}^0\mathsf{VG}^0+\cdots] \tag{10.2}$$

The first Born approximation or simply the born approximation [cf. Eq. (9.9)] is obtained by cutting off the series of Eq. (10.2) after the second term, thus

$$\mathsf{G}=\mathsf{G}^0[1+\mathsf{VG}^0] \tag{10.3}$$

Now whereas Eq. (10.3) corresponds to the Born approximation wherein we deal with matrix elements of the potential, V, over states unperturbed by this potential, in the T-matrix approximation the effects of the potential enter the state description as is evident from Eqs. (9.9) and (9.10). We then replace the Born approximation of Eq. (10.3) by the T-matrix, thus

$$\mathsf{G}=\mathsf{G}^0[1+T\mathsf{G}^0] \tag{10.4}$$

(We shall find out somewhat later why we have obtained Eq. (10.4). For now suffice it to say that it is the T version of the Born approximation.)

From Eqs. (9.14), (9.15), and (9.13) it would appear that

$$T|\Phi_p^0\rangle=\mathsf{V}|\Psi_p^+\rangle=\mathsf{V}|\Phi_p\rangle+[E-H_0+i\epsilon]^{-1}\mathsf{V}|\Psi_p^+\rangle=\mathsf{V}|\Phi_p\rangle+\mathsf{G}^0\mathsf{V}T|\Phi_p\rangle$$

This result then tells us that

$$T=\mathsf{V}+\mathsf{VG}^0T=\mathsf{V}\sum_n(\mathsf{G}^0\mathsf{V})^n=\mathsf{V}+T\mathsf{G}^0\mathsf{V}=\mathsf{V}+\mathsf{VGV} \tag{10.5}$$

which, the reader may remark, may be of general interest but apparently has little to do with Liouville space. Such a remark would be unassailable; however, we shall find a use for Eqs. (10.4) and (10.5) in the development of an analogy between the simple space of the Hamiltonian operation (H-space) and the product space of the Liouville operator ($\mathfrak{L}$-space). We turn now to this latter space, and, in doing so, remark that Ben-Reuven (67, 69, 71) should probably be recognized as the principal expositor of the very important $\mathfrak{L}$-space formalism in line broadening theory.

The Liouville operator is particularly applicable to the spectral line broadening problem: The eigenvalues of this operator are not the energies or frequencies of the molecular states but the frequencies of the observed spectral lines. The eigenvectors are the transition operators. We will recall encountering these transition operators in Section 8 as $\mathsf{P}_{if}=|f\rangle\langle i|$, this operator inducing a transition by annihilating the state $|i\rangle$ and creating the state $|f\rangle$. The $|i\rangle$, $|f\rangle$, and so on, are eigenvectors of the Hamiltonian. From the defining equation for $\mathfrak{L}$, Eq. (1.7), we see that

$$\begin{aligned} \mathfrak{L}\mathsf{P}_{fi} &= H|f\rangle\langle i| - |f\rangle\langle i|H^{+} = E_f|f\rangle\langle i| - |f\rangle\langle i|E_i \\ \mathfrak{L}|\mathsf{P}_{fi}\rangle\rangle &= (E_f - E_i)|\mathsf{P}_{fi}\rangle\rangle = \omega_{fi}|\mathsf{P}_{fi}\rangle\rangle \end{aligned} \tag{10.6}$$

where we now designate an eigenvector in $\mathfrak{L}$-space by $|\rangle\rangle$, and wherein we have demonstrated that an operator in H-space is an eigenvector in $\mathfrak{L}$-space.

We could proceed in a quite abstract fashion for some time before we attempted to make contact with the detail which will be of importance to line broadening theory, but there is no obvious reason for our doing so. Therefore, for reasons which the reader will probably infer and which will be discussed at the appropriate time, we consider the polarization of a gaseous collection of molecules. As we have seen, some molecular moment—usually the electric dipole moment is chosen—will interact with the electromagnetic vector potential in order to induce emission or absorption. (The phrase may be subject to unfortunate interpretation, for we are not excluding the third member of the Einstein triad.) If we take the molecules as point particles, the ith component of such a polarization is expressible as

$$M^i(\mathbf{x}) = \sum_{A=1}^{n} \mu_A^i \delta(\mathbf{x} - \mathbf{r}_A) \tag{10.7a}$$

where μ_A^i is the molecular moment of the Ath molecule, $\mathbf{r}_A$ the position vector of that molecule. The reader will agree that the delta function renders the determination of the kth Fourier component of this moment immediate:

$$M^i(\mathbf{k}) = \sum_{A=1}^{n} \mu_A^i e^{-i\mathbf{k}\cdot\mathbf{r}} \tag{10.7b}$$

Therefore,

$$M^i(\mathbf{x}) = V^{-1} \sum_{k} e^{i\mathbf{k}\cdot\mathbf{x}} M^i(\mathbf{k}) \tag{10.7c}$$

At which point we have an H-space operator which it is very useful to use as a vector in $\mathfrak{L}$-space manipulation, the operator basically being the μ_A^i. However, we must do more than merely write it down with a double bracket round it, indeed, it becomes useful only when we write it in terms of some basis set of vectors in $\mathfrak{L}$-space. As a basis set the transition operators of Eq. (10.6) are chosen, and surely

$$|M\rangle\rangle = \sum_{ij} \mu_{A,ij} |\mathsf{P}_{ij}\rangle\rangle \tag{10.8}$$

With which we are very little further ahead, having no idea as to how the coefficients, $\mu_{A,ij}$, are to be evaluated. If, however, we were somewhat more familiar with the inner product in $\mathfrak{L}$-space, such would not be the case.

Perhaps the most expeditious way to determine both $\mu_{A,ij}$ and a specific expression for the inner product is to assume a definition for the former, find the latter and then demonstrate our expression for $\mu_{A,ij}$.

We suppose $\mu_{A,ij}$ to be the matrix element for μ_A in H-space

$$\mu_{ij}=\langle i|\mu e^{i\mathbf{k}\cdot\mathbf{r}_A}|j\rangle \tag{10.9}$$

Ignoring the summation over A in Eq. (10.7b) we may now apply Eq. (1.2) for the inner product in product space in order to obtain

$$\begin{aligned}\langle\langle M|M\rangle\rangle &= \sum_{ijkl}\mu_{ji}^{+}\mu_{kl}\langle\langle \mathsf{P}_{ij}|\mathsf{P}_{kl}\rangle\rangle \\ &= \sum_{ijkl}\mu_{ji}^{+}\mu_{kl}\langle i|k\rangle\langle j|l\rangle \\ &= \sum_{ij}\mu_{ji}^{+}\mu_{ij}=\mathrm{Tr}\{\mu^{+}\mu\}\end{aligned}$$

wherein we have temporarily dropped the exponential of Eq. (10.9) and let $\mu_{ij}=\langle i|\mu|j\rangle$. In general, the inner product in $\mathfrak{L}$-space of two vectors $|A\rangle\rangle$ and $|B\rangle\rangle$, which correspond to operators A and B in H-space, will thus be

$$\langle\langle A|B\rangle\rangle=\mathrm{Tr}\{A^{+}B\} \tag{10.10a}$$

[This was apparently first defined by Primas (221).] In order to determine the μ_{ij} in Eq. (10.8) we multiply Eq. (10.8) through on the left by $\langle\langle \mathsf{P}_{ij}|$. The right side will then be $\mu_{A,ij}$. The left side may be evaluated using Eq. (10.10):

$$\begin{aligned}\langle\langle \mathsf{P}_{ij}|M\rangle\rangle &=\mathrm{Tr}\{\mathsf{P}_{ij}^{+}M\}=\sum_{\alpha\beta}\langle\alpha|\mathsf{P}_{ij}^{+}|\beta\rangle\langle\beta|\mu e^{i\mathbf{k}\cdot\mathbf{r}_A}|\alpha\rangle \\ &=\sum_{\alpha\beta}\langle\alpha|j\rangle\langle i|\beta\rangle\langle\beta|\mu e^{i\mathbf{k}\cdot\mathbf{r}_A}|\alpha\rangle=\langle i|\mu e^{i\mathbf{k}\cdot\mathbf{r}_A}|j\rangle\end{aligned} \tag{10.11}$$

wherein the reader will remark that we have dropped the A summation over molecules in Eq. (10.7b). This may be judiciously reinserted as circumstances warrant.

So far we have defined a matrix element of the operator, $\mathfrak{L}$, by a single-bracketed expression in Eq. (1.7). We may think of this as a matrix element of $\mathfrak{L}$ in H-space and use the subscript notation, $\mathfrak{L}_{im,jk}$, to represent it, thus abiding by the usual convention wherein the single bracket and the subscript notation are interchangeable. However, now it will be useful for us to have a relationship between the matrix element of our operator in $\mathfrak{L}$-space and that of the same operator in H-space. That the relationship is

$$\langle\langle ij|A|kl\rangle\rangle = \sum_{mn} \langle\langle ij|mn\rangle\rangle A_{mn,kl} \qquad (10.10b)$$

it is worthwhile for us to verify with $\mathfrak{L}$.

$$\begin{aligned}\langle\langle ij|\mathfrak{L}|kl\rangle\rangle &= \langle\langle ij|HI^*|kl\rangle\rangle - \langle\langle ij|IH^*|kl\rangle\rangle \\ &= \mathrm{Tr}\{(\mathsf{P}_{ij})^+ HI^*\mathsf{P}_{kl}\} - \mathrm{Tr}\{(\mathsf{P}_{ij})^+ IH^*\mathsf{P}_{kl}\} \\ &= \mathrm{Tr}\{|j\rangle\langle i|HI^*|k\rangle\langle l|\} - \ldots = \sum_m \langle m|j\rangle\langle i|HI^*|k\rangle\langle 1|m\rangle - \ldots \\ &= \sum_m \langle m|j\rangle\langle l|m\rangle H_{ik}\delta_{jl} - \ldots = \sum_m \langle m|j\rangle\langle i|i\rangle\langle j|m\rangle H_{ik}\delta_{jl} - \ldots \\ &= \mathrm{Tr}\{\mathsf{P}_{ij}^+\mathsf{P}_{ij}\} H_{ik}\delta_{jl} - \ldots = \langle\langle ij|ij\rangle\rangle \mathfrak{L}_{ij,kl}\end{aligned}$$

Because μ_A is to be a function only of the internal coordinates of molecule A—such need not be the case, for example, in induced spectra—does not mean that $|\alpha\rangle$ comprises only internal states of the molecules. In full it includes both the internal and translational states so that we write $|i\rangle$ as $|i,\mathbf{p}\rangle$ where $\mathbf{p}$ is the momentum of a molecule in internal state i. From Eq. (10.11) we see that our $\mathfrak{L}$-space vectors must be defined by means of

$$\mathsf{P}_{ij}(k) = \{|i,\mathbf{p}+\mathbf{k}\rangle\langle j,\mathbf{p}|\}_A \qquad (10.12)$$

where we have reintroduced the sum of Eq. (10.7b). At which point we have developed a certain bumbling expertise in the vectors of $\mathfrak{L}$-space. That these vectors obey a pseudo Schrödinger equation has already been demonstrated in Eq. (10.6). Without changing this result we may enlarge somewhat upon it. We consider H_0, which is diagonal in the $|i\rangle$, and $\mathfrak{L}_0$ which is hence diagonal in the $|\mathsf{P}_{if}\rangle\rangle$. Then

$$\begin{aligned}\mathfrak{L}_0|\mathsf{P}_{if}(k)\rangle\rangle &= \left[-\nabla^2|i,\mathbf{p}+\mathbf{k}\rangle\langle j,\mathbf{p}| + |i,\mathbf{p}+\mathbf{k}\rangle\langle j,\mathbf{p}|\nabla^{+^2}\right]/2\mathsf{m} \\ &= \left[\omega_{ij} + k^2/2\mathsf{m} + \mathbf{k}\cdot\mathbf{v}\right]|\mathsf{P}_{if}(k)\rangle\rangle \qquad (10.13)\end{aligned}$$

In Eq. (10.13) ω_{ij} is the frequency separation of the internal states. The term $k^2/2\mathsf{m}$ is the recoil energy of the molecule, a factor often and quite justifiably ignored, and $\mathbf{k}\cdot\mathbf{v}$ is the shift in energy due to the Doppler effect, the interaction between the momentum of the photon and that of the emitting or absorbing molecule. We shall consider this effect *in extenso* in Chapter 7. Now we turn preliminarily to projection vectors.

We have largely referred to P_{ij} as a transition operator, but we have been aware, if subliminally, that its diagonal elements are projection operators. Consider an arbitrary vector, $|A\rangle$, expanded in terms of the $|j\rangle$ of H-space,

$$|A\rangle = \sum_j a_j|j\rangle$$

Operating on $|A\rangle$ with P_{ii} will yield

$$\mathsf{P}_{ii}|A\rangle=\sum_j a_j|i\rangle\langle i|j\rangle=a_i|i\rangle$$

that is, P_{ii} projects the $|i\rangle$ component out of $|A\rangle$. We can phrase this another way: P_{ii} selects the component along the axis of $|i\rangle$ in H-space or the component in the $|i\rangle$ (one-dimensional) subspace of H-space. And we now extrapolate the concept to $\mathfrak{L}$-space.

We define P as the operator which projects a vector in $\mathfrak{L}$-space into a subspace, $\mathfrak{L}_1$. The residual operator, $\mathsf{M}=1-\mathsf{P}$, projects a vector into the complementary subspace $\overline{\mathfrak{L}}_1$, where

$$\mathfrak{L}=\mathfrak{L}_1\oplus\overline{\mathfrak{L}}_1 \tag{10.14}$$

We may construct such a projection operator in $\mathfrak{L}$-space by a simple extrapolation from P_{ii}. For example,

$$\mathsf{P}=\mathfrak{N}|\mathsf{P}_{ij}\rangle\rangle\langle\langle\mathsf{P}_{ij}| \tag{10.15}$$

$\mathfrak{N}$ being a normalization factor with which we shall not concern ourselves in the meantime. At which point we have hopefully increased our fund of not entirely useless information vis-à-vis $\mathfrak{L}$-space.

And encountered a number of precise analogies between operators and operator relations in H-space and operators and operator relations in $\mathfrak{L}$-space. Which should prepare us to accept certain quite important operators and operator relations in $\mathfrak{L}$-space as extrapolations of those in H-space. First and quite importantly, we introduce the $\mathfrak{L}$-space Greens function,

$$\mathfrak{G}(\omega)=(\omega-\mathfrak{L}+\mathrm{i}\epsilon)^{-1}=\frac{1}{2\pi\mathrm{i}}\int_{-\infty}^{\infty}d\epsilon\,\mathsf{G}(\epsilon+\omega)\mathsf{G}^*(\epsilon) \tag{10.16a}$$

$$\mathfrak{G}(t)=-\mathrm{i}\vartheta(t)\exp[-\mathrm{i}\mathfrak{L}t]=-\mathrm{i}\mathsf{G}(t)\mathsf{G}^*(t) \tag{10.16b}$$

That the equality in Eq. (10.16a) holds may be shown as follows:

$$\begin{aligned}(2\pi\mathrm{i})^{-1}\int d\epsilon[\epsilon+\omega-H+\mathrm{i}(\eta-\eta')]^{-1}[\epsilon-H^*-i\eta']^{-1}&\\ =(2\pi\mathrm{i})(2\pi\mathrm{i})^{-1}[\omega-H+H^*+\mathrm{i}\eta]^{-1}&\\ =[\omega-\mathfrak{L}+\mathrm{i}\eta]^{-1}=\mathfrak{G}(\omega)&\end{aligned}$$

when the integral is evaluated by taking the residue at the pole $[\epsilon-H^*-\mathrm{i}\eta']^{-1}$.

The analogy to Eq. (10.4) in $\mathfrak{L}$-space is

$$\mathfrak{G}(\omega)=\mathfrak{G}^0(\omega)[\mathfrak{J}+\mathfrak{T}(\omega)\mathfrak{G}^0(\omega)] \tag{10.17a}$$

where

$$\mathfrak{J}=II^* \tag{10.17b}$$

$$\mathfrak{G}^0(\omega)=[\omega-\mathfrak{L}_0+\mathrm{i}\epsilon]^{-1} \tag{10.17c}$$

$$\mathfrak{L}=\mathfrak{L}_0+\mathfrak{L}' \tag{10.17d}$$

$$\mathfrak{L}_0 = H_0 I - I H_0^+ \tag{10.17e}$$

$$\mathfrak{L}' = \mathsf{V} I - I \mathsf{V}^+ \tag{10.17f}$$

Here $\mathfrak{I}$ is the identity operator in $\mathfrak{L}$-space, I the one in H-space. The Hamiltonian has been broken down in a familiar fashion into perturbed and unperturbed parts, the Liouville following suit in an obvious way. The T-matrix relation Eq. (10.5) becomes

$$\begin{aligned}\mathfrak{T} &= \mathfrak{L}' + \mathfrak{L}'\mathfrak{G}^0\mathfrak{T} = \mathfrak{L}'\sum(\mathfrak{G}^0\mathfrak{L}')^n = \mathfrak{L}'\left[1-(\omega-\mathfrak{L}_0)^{-1}\mathfrak{L}'\right]^{-1} \\ &= \mathfrak{L}' + \mathfrak{T}\mathfrak{G}^0\mathfrak{L}' = \mathfrak{L}' + \mathfrak{L}'\mathfrak{G}\mathfrak{L}' \end{aligned} \tag{10.18}$$

With which we return to the projection operators encountered in and around Eq. (10.15.) If the reader will apply Eqs. (1.2) and (10.15), he may readily show that

$$\mathsf{P}\mathfrak{L}_0\mathsf{M} = \mathsf{M}\mathfrak{L}_0\mathsf{P} = 0 \tag{10.19}$$

that is to say, he may show that there are no matrix elements connecting the subspaces $\mathfrak{L}_0$ and $\mathfrak{L}_1$. Since this is the case, Eq. (10.17c) is reduced by such a projection operator, as we shall now demonstrate.

$$\begin{aligned}\mathsf{P}\mathfrak{G}^0(\omega)\mathsf{P} &= \mathfrak{G}_p(\omega) = \mathsf{P}\left[\omega - \mathfrak{L}_0 + \mathrm{i}\epsilon\right]^{-1}\mathsf{P} = \mathsf{P}\omega^{-1}\sum\left[(\mathfrak{L}_0 + \mathrm{i}\epsilon)/\omega\right]^n\mathsf{P} \\ &= \mathsf{P}\omega^{-1}\left[1 + (\mathfrak{L}_0 + \mathrm{i}\epsilon)/\omega + \left[(\mathfrak{L}_0+\mathrm{i}\epsilon)/\omega\right]\left[(\mathfrak{L}_0+\mathrm{i}\epsilon)/\omega\right] + \dots\right]\mathsf{P}\end{aligned}$$

If there is some convergence question in the expansion, we may expand it in some slightly different fashion and similarly prove what we have in mind. In short, convergence is of no real consequence. First, ω is a c number and hence commutes with and is diagonal in all operators. Therefore,

$$\mathsf{P}\omega^{-1}\mathsf{P} = [\mathsf{P}\omega\mathsf{P}]^{-1} = [\omega\mathsf{P}^2]^{-1} = [\omega\mathsf{P}]^{-1}$$

due to the idempotent quality of the projection operators, that is, $\mathsf{P}^2 = \mathsf{P}$. Next, consider a typical $\mathfrak{L}_0$ product, say $\mathfrak{L}_0^2$.

$$\begin{aligned}&\mathsf{P}\mathfrak{L}_0^2\mathsf{P} = \mathsf{P}\mathfrak{L}_0\mathsf{P}\mathsf{P}\mathfrak{L}_0\mathsf{P} + \mathsf{P}\mathfrak{L}_0\mathsf{M}\mathsf{P}\mathfrak{L}_0\mathsf{P} + \dots = \mathsf{P}\mathfrak{L}_0\mathsf{P}\mathsf{P}\mathfrak{L}_0\mathsf{P} \\ &\Rightarrow \mathsf{P}\mathfrak{L}_0^n\mathsf{P} = \mathsf{P}\mathfrak{L}_0\mathsf{P}\mathsf{P}\mathfrak{L}_0\mathsf{P}\cdots\mathsf{P}\mathfrak{L}_0\mathsf{P}\end{aligned}$$

by Eq. (10.19). Finally, the convergence factor, ϵ, need hardly concern us so that we find

$$\mathsf{R}\mathfrak{G}^0\mathsf{R} = \mathfrak{G}_R(\omega) = \left[\mathsf{R} - \mathfrak{L}_R + \mathrm{i}\epsilon\right]^{-1}, \qquad \mathsf{R} = \mathsf{P} \text{ or } \mathsf{M} \tag{10.20a}$$

$$\mathfrak{L}_R = \mathsf{R}\mathfrak{L}_0\mathsf{R} \tag{10.20b}$$

A definition introduced by Zwanzig and used, as we shall see, by Fano (115) is the connected part of $\mathfrak{T}(\omega)$. This is that part which is obtained by allowing only vectors of the subspace $\mathfrak{L}_1$ to appear as intermediate vectors $\mathfrak{G}^0$ in Eq. (10.18). Thus,

$$\mathfrak{T}_c(\omega) = \mathfrak{L}_1 + \mathfrak{L}_1\mathfrak{G}_Q(\omega)\mathfrak{T}_c(\omega) \tag{10.21}$$

Now since $\mathfrak{G}^0=\mathfrak{G}_P+\mathfrak{G}_M$, Eqs. (10.18) and (10.21) tell us that

$$\mathfrak{T}(\omega)=\mathfrak{T}_c(\omega)+\mathfrak{T}_c(\omega)\mathfrak{G}_P(\omega)\mathfrak{T}(\omega) \tag{10.22}$$

Now we are in a position to obtain an important expression for $\mathsf{P}\mathfrak{G}(\omega)\mathsf{P}$. From Eq. (10.17a) we obtain

$$\mathsf{P}\mathfrak{G}(\omega)\mathsf{P}=\mathfrak{G}_P(\omega)[1+\mathfrak{T}(\omega)\mathfrak{G}_P(\omega)] \tag{10.23}$$

The second term on the right is actually

$$\mathsf{P}\mathfrak{G}_0\mathfrak{T}\mathfrak{G}_0\mathsf{P}=\mathsf{P}\mathfrak{G}_0\mathsf{P}\mathsf{P}\mathfrak{T}\mathsf{P}\mathsf{P}\mathfrak{G}_0\mathsf{P}=\mathsf{P}\mathfrak{G}_0\mathsf{P}\mathfrak{T}\mathsf{P}\mathfrak{G}_0\mathsf{P}$$

either because of the idempotent quality of P or because, sandwiched between two $\mathfrak{G}_P$'s as it is, for all practical purposes $\mathsf{P}\mathfrak{T}\mathsf{P}=\mathfrak{T}$. From Eq. (10.18)

$$\begin{aligned}1+\mathsf{P}\mathfrak{T}\mathsf{P}\mathfrak{G}_P&=1+\mathsf{P}\mathfrak{L}'\mathsf{P}[1+\mathfrak{G}_P\mathsf{P}\mathfrak{L}'\mathsf{P}+\mathfrak{G}_P\mathsf{P}\mathfrak{L}'\mathsf{P}\mathfrak{G}_P\mathsf{P}\mathfrak{L}'\mathsf{P}+\ldots]\mathfrak{G}_P\\&=1+\mathsf{P}\mathfrak{L}'\mathsf{P}\mathfrak{G}_P+\mathsf{P}\mathfrak{L}'\mathsf{P}\mathfrak{G}_P\mathsf{P}\mathfrak{L}'\mathsf{P}\mathfrak{G}_P+\cdots=\sum(\mathsf{P}\mathfrak{L}'\mathsf{P}\mathfrak{G}_P)^n\\&=[1-\mathsf{P}\mathfrak{L}'\mathsf{P}\mathfrak{G}_P]^{-1}=\mathfrak{G}_P^{-1}[\mathfrak{G}_P^{-1}-\mathsf{P}\mathfrak{T}_c\mathsf{P}]^{-1}\end{aligned} \tag{10.24}$$

by Eq. (10.21), and since $[\mathsf{P}\mathfrak{L}'\mathsf{P}\mathfrak{G}_P]^{-1}=\mathfrak{G}_P^{-1}[\mathsf{P}\mathfrak{L}'\mathsf{P}]^{-1}$. When we substitute Eq. (10.24) into Eq. (10.23) we obtain

$$\mathsf{P}\mathfrak{G}\mathsf{P}=[\mathfrak{G}_P^{-1}-\mathsf{P}\mathfrak{T}_c(\omega)\mathsf{P}]^{-1}=[\omega\mathsf{P}-\mathfrak{L}_P-\mathsf{P}\mathfrak{T}_c(\omega)\mathsf{P}]^{-1} \tag{10.25}$$

We have not really restricted the subspace into which P projects.

We shall find that Eq. (10.25) plays a role of considerable consequence in the work of Fano and Ben-Reuven.

We postpone the derivation of the following expression for $\mathfrak{T}$ to what we consider a more appropriate placement in Section 24:

$$\mathfrak{T}(\omega)=\int d\epsilon[(T\Delta_0^*-\Delta_0T^*)+(2\pi\mathrm{i})^{-1}\mathfrak{D}TT^*\mathfrak{D}] \tag{24.12a}$$

where

$$\mathfrak{D}=\mathsf{G}_0I^*-I\mathsf{G}_0^* \tag{24.12b}$$

$$\Delta_0(x)=\delta(x-H_0) \tag{24.12c}$$

We remark that such operators in product space have been termed superoperators, their four-index elements, supermatrix elements [Fiutak (119, 120)]. In the basis in which H_0 is diagonal, one of these supermatrix elements may be written

$$\mathfrak{T}_{ij,kl}(\omega)=T_{ik}(\epsilon+\omega)\delta_{jl}^*-\delta_{ik}T_{jl}^*(\epsilon_i-\omega)+\int\frac{d\epsilon d_{ij}}{2\pi\mathrm{i}}T_{ik}(\epsilon+\omega)T_{jl}^*(\epsilon)\,d_{kl} \tag{10.26a}$$

where

$$d_{ij}\delta_{ii'}\delta_{jj'}=D_{ij,i'j'} \tag{10.26b}$$

$$d_{ij}=[(\epsilon+\omega-\epsilon_i+\mathrm{i}\eta)^{-1}-(\epsilon-\epsilon_j-\mathrm{i}\eta')^{-1}] \tag{10.26c}$$

Equation (10.26b) tells us the diagonal elements of the supermatrix given by Eq. (24.12b).

Now we write down Eq. (9.10) as

$$S_{ik}=\delta_{ik}-2\pi i\delta(\epsilon-\epsilon_i)T_{ik}(\epsilon), \qquad \epsilon_i=\epsilon_k \tag{10.27}$$

which reminds us that the asymptotic behavior of the wave functions is determined by on-energy-shell matrix elements of S and suggests that the asymptotic behavior of other operators is to be determined from on-frequency-shell elements of Eq. (10.26a) where $\omega=\omega_{ij}=\omega_{kl}$. Then

$$d_{ij}(\omega-\omega_{ij})=-2\pi i\delta(\epsilon-\epsilon_j)$$

and the integral of Eq. (10.26a) vanishes unless $\epsilon_j=\epsilon_l$. This means that on-frequency-shell elements of $\mathfrak{T}(\omega)$ contain only on-energy-shell elements of $T(\epsilon+\omega)$ and $T^*(\epsilon)$, which are on the energy shell $\epsilon_i=\epsilon+\omega$ and $\epsilon_j=\epsilon$, respectively. Thus Eq. (10.27) tells us that Eq. (10.26a) becomes

$$[\mathfrak{T}_{ij,kl}(\omega)]_{\omega=\omega_{ij}=\omega_{kl}}=\int\frac{d\epsilon}{2\pi i}[\delta_{ik}\delta_{jl}-S_{ik}S^*_{jl}]$$

$$=\int\frac{d\epsilon}{2\pi i}[\mathfrak{J}-\mathfrak{S}]_{ij,kl} \tag{10.28a}$$

$$\mathfrak{S}=SS^* \tag{10.28b}$$

where the integration of Eq. (10.28a) is carried across the energy shells.

11 Equations for the Three-Particle T Matrix

We now consider the derivation of a particular set of equations for the T matrix in the three-body scattering problem [Fadeev (112)]. We begin with three different, spinless nonrelativistic particles of masses m_1, m_2, and m_3 and having system Hamiltonian,

$$H=K_1+K_2+K_3+V_{23}+V_{31}+V_{12}, \qquad H_0=K_1+K_2+K_3 \tag{11.1a}$$

$$K_i=-\nabla_i^2/2m_i \tag{11.1b}$$

Asymptotic basis functions are chosen corresponding to (1) all three particles free, (2) particles 2 and 3 bound, particle 1 free, (3) 3 and 1 bound, 2 free and (4) 1 and 2 bound, 3 free.

case (1)

$$\Phi_{n_0}=\exp\{i\mathbf{p}_1\cdot\mathbf{r}_1+i\mathbf{p}_2\cdot\mathbf{r}_2+i\mathbf{p}_3\cdot\mathbf{r}_3\} \tag{11.2a}$$

$$E_{n_0}=p_1^2/2m_1+p_2^2/2m_2+p_3^2/2m_3 \tag{11.2b}$$

$\mathbf{p}_i$ is the translational momentum of particle i.

Case (2)

$$\Phi_{n_{23}}=\exp\{i\mathbf{p}_1\cdot\mathbf{r}_1+i\mathbf{P}_{23}\cdot\mathbf{R}_{23}\}\varphi_{23}^{(l)}(\mathbf{r}_{23}) \tag{11.3a}$$

$$\mathbf{R}_{23}=(\mathrm{m}_2\mathbf{r}_2+\mathrm{m}_3\mathbf{r}_3)/(\mathrm{m}_2+\mathrm{m}_3), \qquad \mathbf{r}_{23}=\mathbf{r}_2-\mathbf{r}_3 \tag{11.3b}$$

$$\left[\frac{\nabla_r^2}{2\mathrm{m}_{23}}-\mathrm{v}_{23}(r)\right]\varphi_{23}^{(l)}(\mathbf{r})=\lambda_{23}^{(l)}\varphi_{23}^{(l)}(\mathbf{r}) \tag{11.3c}$$

$$E_{n_{23}}=p_1^2/2\mathrm{m}+P_{23}^2/2(\mathrm{m}_2+\mathrm{m}_3)-\lambda_{23}^{(l)} \tag{11.3d}$$

$$\mathrm{m}_{23}=\mathrm{m}_2\mathrm{m}_3/(\mathrm{m}_2+\mathrm{m}_3), \qquad \lambda_{23}^{(l)}\geq 0, \qquad l=1,2,\ldots \tag{11.3e}$$

Case (3) and Case (4) may be constructed in obvious analogy to Case (2).

If the Ψ_n are eigenfunctions of the total Hamiltonian, we may use Eq. (11.10a) in order to express Ψ_n in terms of the Φ_n of Eqs. (11.2) and (11.3):

$$\Psi_n=-\lim_{\epsilon\to 0+} i\epsilon \mathrm{G}(E_n+i\epsilon)\Phi_n \tag{11.4}$$

$\mathrm{G}_0(z)=(H_0-z)^{-1}$ with H_0 given by Eq. (8.1a) which provides us with the ingredients for the right side of Eq. (10.2) in terms of which we rewrite Eq. (11.4) as follows:

$$\Psi_n=-\lim_{\epsilon\to 0+} i\epsilon G_0(E_n+i\epsilon)\Phi_n-\mathrm{G}_0(E_n+i0)\mathrm{V}\Psi_n$$

For Case (1) Eq. (8.10b) may be used to obtain

$$\Psi_{n_0}=\Phi_{n_0}-\mathrm{G}_0(E_{n_0}+i0)\mathrm{V}\Psi_{n_0} \tag{11.5}$$

Next, by analogy to Eq. (8.11) we know that

$$-\lim_{\epsilon\to 0+} i\epsilon \mathrm{G}_0(E_{n_{23}}+i\epsilon)\Phi_{n_{23}}=0$$

so that

$$\Psi_{n_{23}}=-\mathrm{G}_0(E_{n_{23}}+i0)\mathrm{V}\Psi_{n_{23}} \tag{11.6}$$

The reader will agree that Eq. (11.6) does not, at least as it stands, have a unique solution. One may try another form for the Greens function,

$$\mathrm{G}_{23}(z)=(H_0+\mathrm{V}_{23}-z)^{-1}$$

Fadeev does, but he does so in order to establish the conclusion at which he arrives, namely, such Lippman-Schwinger equations do not lead him to a satisfactory result. He therefore seeks alternative equations, specifically, the $\mathrm{G}(z)$ satisfying a slightly modified Eq. (10.4),

$$\mathrm{G}(z)=\mathrm{G}_0(z)-\mathrm{G}_0(z)T(z)\mathrm{G}_0(z) \tag{11.7a}$$

where

$$T(z)=\mathrm{V}-\mathrm{V}\mathrm{G}_0(z)T(z)\mathrm{G}(z). \tag{11.7b}$$

Using Eqs. (11.7) and (11.1), the following iterated solution may be written out:

$$\begin{aligned} T(z) &= \mathsf{V}_{23}+\mathsf{V}_{31}+\mathsf{V}_{12}-(\mathsf{V}_{23}+\mathsf{V}_{31}+\mathsf{V}_{12})\mathsf{G}_0(z)(\mathsf{V}_{23}+\mathsf{V}_{31}+\mathsf{V}_{12})+\cdots \\ &= \mathsf{V}_{23}-\mathsf{V}_{23}\mathsf{G}_0(z)\mathsf{V}_{23}+\mathsf{V}_{23}\mathsf{G}_0(z)\mathsf{V}_{23}\mathsf{G}_0(z)\mathsf{V}_{23}-\cdots+\mathsf{V}_{31}-\mathsf{V}_{31}\mathsf{G}_0\mathsf{V}_{31} \\ &\quad +\cdots+\mathsf{V}_{12}-\mathsf{V}_{12}\mathsf{G}_0(z)\mathsf{V}_{12}+\cdots(\mathsf{V}_{23}-\mathsf{V}_{23}\mathsf{G}_0(z)\mathsf{V}_{23}+\cdots)\mathsf{G}_0(z)(\mathsf{V}_{31} \\ &\quad -\mathsf{V}_{31}\mathsf{G}_0(z)\mathsf{V}_{31}+\cdots)+\cdots \end{aligned} \tag{11.8}$$

The sum of the 23 chain, for example, is denoted as

$$T_{23}(z)=\mathsf{V}_{23}-\mathsf{V}_{23}\mathsf{G}_0(z)\mathsf{V}_{23}+\cdots \tag{11.9a}$$

T_{23} a solution of

$$T_{23}(z)=\mathsf{V}_{23}-\mathsf{V}_{23}\mathsf{G}_0(z)T_{23}(z) \tag{11.9b}$$

Equation (11.9b) is obtained from Eq. (11.8) with $\mathsf{V}_{31}=\mathsf{V}_{32}=0$. $T_{31}(z)$ and $T_{12}(z)$ are defined analogously. In order to write Eq. (11.8) in terms of these T_{ij} we merely replace V_{ij} by T_{ij}, being careful to drop terms where T_{ij} appears on both sides of G_0, such as $T_{31}\mathsf{G}_0T_{31}$. If we write

$$T(z)=T^{(1)}(z)+T^{(1)}(z)+T^{(1)}(z) \tag{11.10a}$$

where

$$T^{(i)}(z)=\mathsf{V}_{jk}-\mathsf{V}_{jk}\mathsf{G}_0(z)T(z) \tag{11.10b}$$

the $T^{(i)}$ satisfy the following set of equations:

$$\begin{bmatrix} T^{(1)}(z) \\ T^{(2)}(z) \\ T^{(3)}(z) \end{bmatrix} = \begin{bmatrix} T_{23}(z) \\ T_{31}(z) \\ T_{12}(z) \end{bmatrix} - \begin{bmatrix} 0 & T_{23}(z) & T_{23}(z) \\ T_{31}(z) & 0 & T_{31}(z) \\ T_{12}(z) & T_{12}(z) & 0 \end{bmatrix} \mathsf{G}_0(z) \begin{bmatrix} T^{(1)}(z) \\ T^{(2)}(z) \\ T^{(3)}(z) \end{bmatrix} \tag{11.11}$$

There is no reason for us to carry this problem further.

12 Certain Bases for Spectral Line Shape Studies

In Chapter 2 we determined the spectral line shape by determining the probability, usually as the square of an SGC, that, after infinite time, there would be a photon of frequency ω in the field, then equating this probability to the relative intensity at frequency ω in the spectral line. Such an SGC will of course contain a matrix element of that portion of the Hamiltonian which induces the absorbing or radiating transition, the portion of course being that describing the interaction between the emitter and the radiation field, $\boldsymbol{\sigma}\cdot\mathbf{A}$. In practice, one usually deals with a simplified form of this interaction wherein the electric dipole moment—or, rarely, the magnetic dipole or electric quadrupole moment—enters in order to represent the atom, its component parallel to the electromagnetic field being indicated by the inner product, an exponential factor representing the field at the emitter entering either directly from the field or as a consequence of an astutely injected Fourier transform. In any event, an SGC as a function of frequency provides an expression for the spectral line shape.

Now this is appropriate since the SGC relates to an isolated atom as does the natural line shape. Should there happen to be N such atoms, only the integrated line intensity is affected, the shape expression as a whole being multiplied by N. (Such a statement of course requires that N be small enough so that pressure-broadening effects are negligible.) However, when we consider the broadening of spectral lines due to the presence of foreign-gas or like molecules, the SGC begins to lose its appeal. Of course, one could set up a wave function for N molecules, so many of this type and having such and such a state distribution and so on, then determine the desired SGC and hence the line shape. Such a procedure has apparently had but little allure for the curious and for quite good reason. The computation of the power or energy per unit time either emitted or absorbed by an assemblage of molecules is a quite straightforward and highly satisfactory approach to the problem and the more devious and obscure approach through the SGC would be a comparative curiosity.

As we shall see, Baranger and many before and after him have found it convenient to consider the power emitted in, say, a dipole transition. As we shall see in Chapter 6, Ross and others have dealt with the absorption of power from an incident electromagnetic field by a collection of molecules. As the reader knows—or as he can find out—such absorption is related to the susceptibility, and this quantity has been the departure point for a number of spectral line shape treatments. We shall see an example of this in perusing the work of Mead and his collaborators in Chapter 8. Whichever of these essentially similar approaches is used, we can see qualitatively—and we shall see specifically—that the expression for the intensity at a particular frequency will depend on the square of the matrix element of the transition-inducing perturbation and the density matrix. The density matrix will enter for the following reasons.

The probability for the transition from a given state will be proportional, from the most elementary of viewpoints, to the number of molecules present in that state. The most general expression for the intensity at a given frequency may contain contributions from a number of spectral lines, each contribution proportional to the number of molecules in the appropriate initial state. Now the density matrix enters by providing these numbers of molecules. So much for these generalized observations. We conclude this section with some more specific reflections on certain of the directions from which the line broadening problem may be approached.

The existence of a complex dielectric constant, $\epsilon(\omega)=\epsilon'(\omega)+i\epsilon''(\omega)$, is a fact of which we are aware. That the expression for the electric field and hence the Poynting or power vector is multiplied by $\exp[i\epsilon(\omega)x]$ means that the traveling electromagnetic field will be attenuated by the factor

$\exp[-\epsilon''(\omega)x]$. As a consequence, a knowledge of the dielectric constant corresponds to a knowledge of the frequency-dependent power absorption, that is to say, to a knowledge of the line shape. We shall return to this, but first we must consider why there is a relationship between the susceptibility—and hence the dielectric constant—and the Greens functions for the electromagnetic field in the presence of our n absorbing molecules.

Just as there is a free-particle Greens function, G_0, which describes the propagation of a particle in the absence of any interacting particles, so there is a free-field Greens function D^0, which describes the propagation of a field in the absence of other fields. (For "fields" we may write "particles which generate fields.") Likewise, a field Greens function, D, corresponds to the particle Greens function, G. That there is a field equation that is a precise analogy to the particle Eq. (10.4) and that relates D and D^0 should come as no particular surprise.

$$\mathsf{D}=\mathsf{D}^0[1+\Pi\mathsf{D}] \tag{12.1}$$

In Eq. (12.1) Π obviously plays the role of the interparticle interaction, V, and embodies the effects of the medium that has changed our free-field Greens function to a field Greens function.

In order to write down these field Greens functions the reader may consult other sources [cf. Chap. 6 of Abrikosov et al. (4)] or Eq. (8.15) in this book. Consulting that equation will hardly render him totally conversant with the field Greens functions, but it will allow him to draw sufficient analogies so as to arrive at the desired result. First of all, we will replace the particle operator, $\psi(x)$, by the field operator,

$$e^{H\tau}\mathsf{A}_\alpha(\mathbf{r})e^{-H\tau} \tag{12.2}$$

In Eq. (12.2) $\mathsf{A}_\alpha(\mathbf{r})$ is the αth component of the four-vector potential in the Heisenberg representation. Here τ is the imaginary time essential to the temperature Greens functions which we encounter in Chapter 6 and elsewhere. These temperature Greens functions require an averaging which is likewise implied by Eq. (8.15) and which leads to the following form for the field Greens functions:

$$D_{\alpha\beta}(\mathbf{r}_1,\mathbf{r}_2;\tau_1-\tau_2)=-\begin{cases}\mathrm{Tr}\{e^{(F-H)\beta}e^{H(\tau_1-\tau_2)}\mathsf{A}_\alpha(\mathbf{r}_1)e^{-H(\tau_1-\tau_2)}\mathsf{A}_\beta(\mathbf{r}_2)\}, & \tau_1>\tau_2\\ \mathrm{Tr}\{e^{(F-H)\beta}e^{-H(\tau_1-\tau_2)}\mathsf{A}_\alpha(\mathbf{r}_1)e^{H(\tau_1-\tau_2)}\mathsf{A}_\beta(\mathbf{r}_2)\}, & \tau_1<\tau_2\end{cases} \tag{12.3}$$

In Eq. (12.3) F is the free energy. The Greek subscripts refer to four-vector subscripts, the Latin to spatial parts.

It is apparent that the effects of the particles are going to enter through the Hamiltonian appearing in the exponentials. It is also apparent that the Greens function is only dependent on time differences. That, since we shall assume isotropic media, those functions are likewise dependent only on coordinate differences, $\mathbf{r}_1-\mathbf{r}_2$, is perhaps not equally apparent but surely equally true. We shall, however, postpone the application of this isotropy for a moment.

We find it convenient to gauge our vector fields such that

$$\mathbf{E} = -\dot{\mathbf{A}}, \qquad \mathbf{H} = \nabla \times \mathbf{A} \tag{12.4a}$$

which means, of course, that the scalar potential is zero.

We recall that

$$\nabla \times \langle \mathbf{H}(\mathbf{r}, t) \rangle = 4\pi \mathbf{j}^{\text{ext}}(\mathbf{r}, t) + \frac{\partial}{\partial t}\left[\epsilon \langle \mathbf{E}(\mathbf{r}, t) \rangle \right] \tag{12.4b}$$

$$\nabla \times \langle \mathbf{E}(\mathbf{r}, t) \rangle = -\frac{\partial}{\partial t} \langle \mathbf{H}(\mathbf{r}, t) \rangle \tag{12.4c}$$

We take the Fourier transform of Eqs. (12.4). Now when we do this, the time dependence in the field quantity goes into an expression $\exp[\pm i\omega t]$ so that taking the time derivative merely has the effect of bringing down a factor $\pm i\omega$. Therefore, Eqs. (12.4b) and (12.4c) become

$$\nabla \times \langle \mathbf{H}(\mathbf{r}, \omega) \rangle = 4\pi \mathbf{j}^{\text{ext}}(\mathbf{r}, \omega) - i\omega\epsilon(\mathbf{r}, \omega) \langle \mathbf{E}(\mathbf{r}, \omega) \rangle \tag{12.5a}$$

$$\nabla \times \langle \mathbf{E}(\mathbf{r}, \omega) \rangle = i\omega \langle \mathbf{H}(\mathbf{r}, \omega) \rangle \tag{12.5b}$$

The substitution of Eq. (12.4a) into Eq. (12.5a) yields

$$\left[\omega^2 \epsilon(\mathbf{r}, \omega) - \nabla \times \nabla \times \right] \langle \mathbf{A}(\mathbf{r}, \omega) \rangle = -4\pi \mathbf{j}^{\text{ext}}(\mathbf{r}, \omega) \tag{12.6}$$

We recall that the curl of a column vector may be written

$$\nabla \times \mathbf{A} = \begin{pmatrix} 0 & -\dfrac{\partial}{\partial z} & \dfrac{\partial}{\partial y} \\ \dfrac{\partial}{\partial z} & 0 & -\dfrac{\partial}{\partial x} \\ -\dfrac{\partial}{\partial y} & \dfrac{\partial}{\partial x} & 0 \end{pmatrix} \begin{pmatrix} A_x \\ A_y \\ A_z \end{pmatrix} \tag{12.7a}$$

the matrix element rather obviously being

$$(\nabla \times \mathbf{A})_i = \sum_j (\nabla \times)_{ij} A_j \tag{12.7b}$$

Elementary matrix multiplication will then allow us to reexpress the second term on the left of Eq. (12.6) as a vector component

$$\left(\epsilon(\mathbf{r}, \omega)\omega^2 \delta_{il} - \sum_{ml} (\nabla \times)_{im} (\nabla \times)_{ml} \right) \langle \mathsf{A}_l(\mathbf{r}, \omega) \rangle = -4\pi \mathbf{j}_i^{\text{ext}}(\mathbf{r}, \omega) \tag{12.8}$$

Eq. (12.8) is the equation for the average potential $\langle \mathbf{A}^{\text{ext}} \rangle$ in that gauge for which the scalar potential is zero. Mathematically, there is a Greens function which is a solution of this equation with the right side replaced by $4\pi\delta_{ik}$ $\delta(\mathbf{r}-\mathbf{r}')$ which we do not write down.

Following Abrikosov et al. (3) we take the Hamiltonian as consisting of the sum of a portion relating to the absorber or emitter and the radiation (H)

and to what remains in the complete system (H^{ext}). We may then write

$$\langle \mathbf{A}^{\text{ext}}(\mathbf{r},t)\rangle = \langle e^{\mathrm{i}(H+H^{\text{ext}})t}\mathbf{A}(\mathbf{r})e^{-\mathrm{i}(H+H^{\text{ext}})t}\rangle$$
$$= \langle e^{\mathrm{i}H^{\text{ext}}t}\mathbf{A}(\mathbf{r},t)e^{-\mathrm{i}H^{\text{ext}}t}\rangle = \langle S_{\text{ext}}^{-1}(t)\mathbf{A}(\mathbf{r},t)S_{\text{ext}}(t)\rangle \tag{12.9a}$$

where

$$S_{\text{ext}}(t) = \mathrm{T}_{\leftarrow}\exp\left\{-\mathrm{i}\int_{-\infty}^{t} H^{\text{ext}}(t')\,dt'\right\} \tag{12.9b}$$

$$H^{\text{ext}} = -\int \mathbf{j}^{\text{ext}}(\mathbf{r},t)\cdot\mathbf{A}(\mathbf{r})\,d\mathbf{r} \tag{12.9c}$$

When S_{ext} is expanded in powers of H^{ext} and only those terms first order in $\mathbf{j}^{\text{ext}}$ retained, we obtain

$$\langle \mathsf{A}_i^{\text{ext}}(\mathbf{r},t)\rangle = -i\int_{-\infty}^{t} dt'\int d\mathbf{r}'\,\mathbf{j}_k^{\text{ext}}(\mathbf{r}',t)\langle[\mathsf{A}_k(\mathbf{r}',t'),\mathsf{A}_i(\mathbf{r},t)]\rangle$$
$$-\int_{-\infty}^{\infty} dt'\int d\mathbf{r}'\,\mathbb{D}_{ik}^{R}(\mathbf{r},\mathbf{r}';t-t')\mathbf{j}_k^{\text{ext}}(\mathbf{r}',t') \tag{12.9d}$$

where $\mathbb{D}_{ik}^{R}$ is the retarded zero-temperature Greens function. We take the Fourier component with respect to the time difference in order to obtain an expression which need not be written down but from which we conclude that the zero-temperature retarded Greens function is the solution to Eq. (12.8). Therefore, in order to obtain a solution in terms of a Greens function, we need merely go from real to imaginary time for reasons which we shall discuss in Chapter 6. This has the effect of replacing ω by $\mathrm{i}|\omega_n|$, and Eq. (12.8) becomes

$$\left[\epsilon(\mathbf{r},\mathrm{i}|\omega_n|)\omega_n^2\delta_{il} + \sum_{ml}(\nabla\times)_{im}(\nabla\times)_{ml}\right]\mathsf{D}_{lk}(\mathbf{r},\mathbf{r}',\omega) = 4\pi\delta(\mathbf{r}-\mathbf{r}')\delta_{ik} \tag{12.10}$$

It is probably obvious that the absence of matter means a value of one for the dielectric constant and an equation analogous to Eq. (12.10) for D_{ij}^{0}, the free-photon Greens function.

At this point we introduce the isotropy of our medium which means that only the differences, $\mathbf{r}-\mathbf{r}'$, are of importance. We now take the Fourier components with respect to momenta, which has the practical effect of introducing the factors $\exp[\mathrm{i}\mathbf{k}\cdot(\mathbf{r}-\mathbf{r}')]$. If the reader will apply Eq. (12.7a) he will thus obtain

$$\{[\omega^2\epsilon(\mathrm{i}|\omega_n|)+k^2]\delta_{il}+k_ik_l\}\mathsf{D}_{lj}(\mathbf{k},\omega_n) = -4\pi\delta_{ij} \tag{12.11a}$$

Now we recall that the Roman subscripts refer to the spatial components of the four-potential $\mathsf{A}=(\mathbf{A},\varphi)$, D_{ij} the Greens function appropriate to the gauge wherein $\varphi=0$. We next take $\mathsf{D}_{\alpha\beta}^{E}$ as the Greens function in arbitrary gauge, D_{ij}^{E} being the spatial Greens function such that $\mathsf{D}_{ij}^{E}=\omega^2\mathsf{D}_{ij}$, the equation for D^E_{ij} being immediately obtainable from Eq. (12.11a) as

$$\{[\omega^2\epsilon(i|\omega_n|)+k^2]\delta_{il}+k_ik_l\}D^E_{lj}(\mathbf{k},\omega_n)=4\pi\omega_n^2\delta_{ij} \tag{12.11b}$$

Corresponding to the first equation of Eq. (12.4a),

$$\mathbf{E}(\mathbf{r},t)=-\frac{\partial}{\partial t}\mathbf{A}(\mathbf{r},t)-\nabla\varphi(\mathbf{r},t)$$

or

$$E_i(\mathbf{k},\omega)=-\omega A_i(\mathbf{k},\omega)-k_i\varphi(\mathbf{k},\omega) \tag{12.12}$$

for the scalar potential $\varphi\neq 0$. (We have not yet gone to imaginary time.) Therefore, from Eq. (12.3) we see that, corresponding to $D^E_{ij}=\omega^2 D_{ij}$, we now have

$$\mathbb{D}^E_{ij}=\omega^2\mathbb{D}^R_{ij}-\omega k_i\mathbb{D}^R_{0j}-\omega k_j\mathbb{D}^R_{i0}+k_ik_j\mathbb{D}^R_{00} \tag{12.13}$$

wherein the $\mathbb{D}^R_{ij}$ are Greens functions of mixed space (A_i) and time (φ) components, the $\mathbb{D}^R_{00}$ pure time. We may extrapolate Eq. (12.9d) for Eq. (12.13) in an obvious fashion. One now looks pensively at Eq. (12.13) and reasons as follows.

The time vector $\mathbb{D}^R_{00}$ has a factor k_ik_j. Therefore, among other things $\mathbb{D}^E_{ij}$ must depend on at least the units of this factor k_ik_j. Which means that $\mathbb{D}^R_{0j}$, for example, must have a factor k_j, which, with k_i, will distribute the k_ik_j appropriately. Therefore, we suppose

$$\mathbb{D}^R_{i0}=\mathbb{D}^R_{0i}=k_i d \tag{12.14a}$$

and

$$\mathbb{D}^R_{ij}=a\delta_{ij}+bk_ik_j. \tag{12.14b}$$

When we substitute Eqs. (12.13) and (12.14) into Eq. (12.11b), we obtain two equations

$$a[\epsilon(\omega)\omega^2-k^2]=4\pi \tag{12.15a}$$

$$a+\epsilon(\omega)[\omega^2 b+\mathbb{D}^R_{00}-2\omega d]=0, \tag{12.15b}$$

that is to say, two equations in three unknowns, a situation which leaves us free for certain arbitrary selections of constant values. Abrikosov *et al.* (4) remark on three possible choices (1) $b=d=0$: (2) $\varphi=0$, and (3) $d=0$. The latter will be of some interest to us and is

$$D_{ij}=4\pi\delta_{ij}[\epsilon(i|\omega_n|)\omega_n^2+k^2]^{-1}\left[\delta_{ij}-\frac{k_ik_j}{k^2}\right] \tag{12.16a}$$

$$D_{00}=4\pi/\epsilon(i|\omega_n|)k^2 \tag{12.16b}$$

$$D_{i0}=0 \tag{12.16c}$$

where the $D_{\alpha\beta}$ are finite temperature Greens functions.

Which means that we have obtained the dielectric constant in terms of the Greens function. We could carry out a rather involved derivation in order to obtain Eq. (12.1) and hence the Greens function in terms of the polarization operator, Π. There would, however, be no real point in this exercise since the relationship will evolve quite simply and naturally in later chapters. We therefore merely accept Eq. (12.1) as a logical extension of Eq. (12.3).

$$\mathsf{D}_{ij}=\mathsf{D}^0_{ij}+\sum_{lm}\mathsf{D}^0_{il}\Pi_{lm}\mathsf{D}_{mj} \tag{12.17}$$

(We have left out a double integration over the second term on the right, one integral over a position coordinate linking D^0_{il} and Π_{lm}, one over a position coordinate linking Π_{lm} and D_{mj}. By dropping the Dirac delta function on the right of Eq. (12.10) we shall rather irrigorously lay these integrals to rest.) We now multiply Eq. (12.17) through on the left by $[\omega_n^2\delta_{il}+\sum_{ml}(\nabla\times)_{im}(\nabla\times)_{ml}]$, the operator on the left of Eq. (12.10) for $\epsilon=1$, D^0_{ij}, changing the subscripts in the former equation appropriately. As a result we obtain

$$\Pi_{ij}(k,\omega_n)=[\epsilon(\mathrm{i}|\omega_n|)-1]\omega_n^2\delta_{ij}/4\pi \tag{12.18}$$

For $t=0$, $\varphi=0$, this goes over to a relationship of identical appearance save $\epsilon(\mathrm{i}|\omega_n|)\to\epsilon(|\omega_n|)$. We recall writing the dielectric constant as a sum of a real (ϵ') and an imaginary (ϵ'') part. If we suppose negative frequencies to correspond to emission rather than absorption so that ϵ'' will be negative and yield augmentation rather than attenuation of the radiation field, $\epsilon''(\omega)=-\epsilon''(-\omega)$. On the other hand, $\epsilon'(\omega)=\epsilon'(-\omega)$. Therefore, we see that

$$\epsilon'(\omega)=1+\frac{4\pi}{3\omega^2}\mathcal{R}[\Pi_{ii}(\omega)] \tag{12.19a}$$

$$\epsilon''(\omega)=\frac{4\pi}{\omega|\omega|}\mathcal{I}[\Pi_{ii}(\omega)] \tag{12.19b}$$

Van Vleck (44) tells us that the dielectric constant is related to the electric susceptibility as

$$\chi_e=\frac{1}{4\pi}[\epsilon-1]\left(=\frac{\mathbf{P}}{\mathbf{E}}\right) \tag{12.20}$$

and we now have the susceptibility related to the polarization operator. That the imaginary part of this susceptibility will tell us the spectral line shape is certainly apparent and forms the starting point for a number of studies. Finally, we remark on the approach to line broadening which has been used by, among others, Ben-Reuven (71) and which views the absorption process in terms of linear response on the part of the molecular system to the time-dependent constraint imposed by the external field on the molecular moments or polarizations. [DiGiacomo and Feo (132) have considered non-linear response as have Lisitsa and Yakovlenko (184), the latter obtaining essentially the Karplus-Schwinger structure for absorbable power, albeit including an elastic-relaxation parameter.]

We suppose that the constraining electric fields, E^j, are coupled to the intensive fields or polarizations, P^i, which indeed it seems obvious they would be. Then these intensive fields may be expressed as functions of the constraining fields. This means that they may be expanded in terms of the E^j. The linear response idea amounts to retaining only the linear term in this expansion and calling it, straightforwardly enough, the (linear) response function. Then the Fourier transform of this latter function is the (linear) susceptibility.

We conclude this section with a discussion of an expression relating the imaginary part of the diagonal components of the susceptibility tensor to the transform of a moment average. First, let us recall that (1) the βth component of the electric or magnetic moment of an atom, M_β, relates to the αth component of the impressed field, E_α, through the $\alpha\beta$th component of the susceptibility tensor,

$$\langle M_\beta \rangle = \chi_{\beta\alpha}\mathsf{E}_\alpha \exp(\mathrm{i}\omega t) \tag{12.21a}$$

and (2) for $\chi=\chi'-\mathrm{i}\chi''$, the radiant absorption coefficient per unit volume is

$$I(\omega)=4\pi\omega\chi''(\omega)N \tag{12.21b}$$

With this preliminary we turn to a derivation of $\chi''_{\alpha\alpha}(\omega)$, first apparently given by Kubo (166), but for which we follow Huber and Van Vleck (147).

We begin by seeking the change in the component of M_α arising from the interaction $-M_\alpha\mathsf{E}_\alpha\exp(\mathrm{i}\omega t)$ which is first applied at a large negative time, $-t_A$. We suppose there to have been no polarization prior to $-t_A$; we may write

$$\delta M_\alpha = \int_{-t_A}^{t} \mathrm{Tr}\left(\gamma \frac{dM_\alpha(t'')}{dt''}\right) dt'' \tag{12.22a}$$

where

$$\gamma = e^{-H_0/kT}/\mathrm{Tr}\left[e^{-H_0/kT}\right] \tag{12.22b}$$

and

$$\mathrm{i}\dot{M}_\alpha = [H_0, M_\alpha] = [H, M_\alpha] \tag{12.22c}$$

γ is to be evaluated at $t=-t_A$. Now $H=H_0+\mathbf{M}\cdot\mathbf{E}e^{\mathrm{i}\omega t}$, but Eq. (12.22c) holds prior to the application of the field since M_α commutes with $M_\alpha\mathsf{E}_\alpha e^{\mathrm{i}\omega t}$. The change in inner energy corresponds to the work done by the field so that

$$\dot{H} = \dot{M}_\alpha\mathsf{E}_\alpha e^{\mathrm{i}\omega t} \Rightarrow H(t) = \int_{-t_A}^{t} M_\alpha(t')\mathsf{E}_\alpha(t')e^{\mathrm{i}\omega t'}\,dt' \tag{12.22d}$$

since for $t<-t_A$, $\delta M_\alpha=0$. We may therefore combine Eqs. (12.22a) and (12.22d) in order to obtain

$$\delta M_\alpha = i\int_{-t_B}^{t} dt'' \int_{-t_A}^{t} \Theta(t',t'')\mathsf{E}_\alpha e^{\mathrm{i}\omega t'}\,dt' \tag{12.23a}$$

where

$$\Theta(t',t'') = \mathrm{Tr}\left\{\gamma\left[\dot{P}_\alpha(t'), P_\alpha(t'')\right]\right\} \tag{12.23b}$$

Our expression for δM_α need only be correct to the first power in E_α, a precondition which allows us to evaluate $\Theta(t', t'')$ as though the field were not there and to treat H as a constant. $\Theta(t', t'')$ is replaced by $\Theta(t'', t')$ on the basis that, in the unperturbed condition, $\langle A(t')B(t'')\rangle$ is a function only of the argument $t'-t''$, the equilibrium distribution being a stationary one. Said replacement obviously allows integration of Θ over t''. Terms arising from the lower limit of this integration of the form $\langle M_\alpha(-t_B)M_\alpha(t)\rangle$ drop out when we take $-t_B \to -\infty$ if there is no mean polarization in the absence of the field, and we arrive at

$$\delta M_\alpha(t) = \mathrm{i}\int_{-\infty}^{t} \langle M_\alpha(t)M_\alpha(t') - M_\alpha(t')M_\alpha(t)\rangle \mathsf{E}_\alpha e^{\mathrm{i}\omega t'}\,dt' \tag{12.24}$$

We have encountered the unit step function,

$$\vartheta(t-t') = \frac{\mathrm{i}}{2\pi}\int_{-\infty}^{\infty} e^{-\mathrm{i}\omega(t-t')}[\omega + \mathrm{i}\epsilon]^{-1}\,d\omega$$

or sign function of Chapter 2 which we now use in order to rewrite Eq. (12.24) as

$$\delta M_\alpha = \left(-\frac{1}{2\pi}\int_{-\infty}^{\infty}\int_{-\infty}^{\infty} d\omega'\,dt' \exp[-\mathrm{i}(\omega'+\omega)(t-t')] \right.$$
$$\left. \times \frac{\langle M_\alpha(t)M_\alpha(t') - M_\alpha(t')M_\alpha(t)\rangle}{\omega' + \mathrm{i}\epsilon} \right) \mathsf{E}_\alpha e^{\mathrm{i}\omega t} \tag{12.25}$$

Combining Eqs. (12.21a) and (12.25) yields

$$\chi_{\alpha\alpha}(\omega) = \frac{1}{2\pi}\int_{-\infty}^{\infty}\int_{-\infty}^{\infty} d\omega'\,dt e^{\mathrm{i}\omega' t} \frac{\langle M_\alpha(t)M_\alpha(0) - M_\alpha(0)M_\alpha(t)\rangle}{(\omega' - \omega + \mathrm{i}\epsilon)} \tag{12.26}$$

In obtaining Eq. (12.26) we make use of the fact that, for purposes of the ω' integration, ω is a constant. Therefore, a new variable $\omega'_{\text{new}} = \omega' + \omega$ is convenient to introduce. Further, $\chi_{\alpha\alpha}$ is not a function of time. Therefore, t may simply be set equal to zero in Eq. (12.25) in order to obtain Eq. (12.26).

We now make use of the cyclic property of the trace, $\mathrm{Tr}\{ABC\} = \mathrm{Tr}\{BCA\} = \mathrm{Tr}\{CAB\}$, in order to obtain the following:

$$\int_{-\infty}^{\infty} e^{\mathrm{i}\omega t}\langle M_\alpha(0)M_\alpha(t)\rangle\,dt = \mathrm{Tr}\int_{-\infty}^{\infty} e^{-H/kT}M_\alpha e^{\mathrm{i}Ht}M_\alpha e^{-\mathrm{i}Ht}e^{\mathrm{i}\omega t}\,dt$$
$$= \mathrm{Tr}\int_{-\infty}^{\infty} e^{-H/kT}e^{\mathrm{i}H(t-\mathrm{i}/kT)}M_\alpha e^{-\mathrm{i}H(t-\mathrm{i}/kT)}M_\alpha e^{\mathrm{i}\omega t}\,dt$$
$$= e^{-\omega/kT}\int_{-\infty}^{\infty} e^{\mathrm{i}\omega t}\langle M_\alpha(t)M_\alpha(0)\rangle\,dt \tag{12.27}$$

The point of course being that the matrix elements over $\exp(\pm H/kT)$ may be taken out from under the integral sign while leaving a Boltzmann factor for the averaging process.

We take the first factor on the right of Eq. (12.26) and utilize Eq. (12.27) in order to rewrite it as

$$M_\alpha(t)M_\alpha(0)=M_\alpha(t)M_\alpha(0)\left[e^{\omega/2}+e^{-\omega/2}\right]\left[e^{\omega/2}+e^{-\omega/2}\right]^{-1}$$
$$=\left[M_\alpha(t)M_\alpha(0)e^{\omega/2}+M_\alpha(0)M_\alpha(t)e^{\omega/2}\right]\left[e^{\omega/2}+e^{-\omega/2}\right]^{-1}$$

The second factor on the right of Eq. (12.26) may be similarly rewritten, and we find

$$\chi_{\alpha\alpha}(\omega)=\frac{1}{2\pi}\int_{-\infty}^{\infty}\int_{-\infty}^{\infty}\frac{d\omega'\,dt e^{i\omega' t}}{\omega'-\omega+i\epsilon}\tanh\frac{\omega'}{2kT}\langle M_\alpha(0)M_\alpha(t)+M_\alpha(t)M_\alpha(0)\rangle \tag{12.28}$$

When the average in Eq. (12.28) is evaluated in terms of the eigenvectors of H, we see that

$$\langle M_\alpha(0)M_\alpha(t)+M_\alpha(t)M_\alpha(0)\rangle=2\sum_{nn'}e^{-E_n/kT}|\langle n|M|n'\rangle|^2\cos\left[(E_n-E_{n'})t\right] \tag{12.29}$$

Therefore, the bracketed expression in Eq. (12.28) is both a real and even function of t. Since this is so, and since $(\omega+i\epsilon)^{-1}=\mathcal{P}/\omega-i\pi\delta(\omega)$, the real and imaginary parts of the susceptibility follow from Eq. (12.28) as

$$\chi'_{\alpha\alpha}(\omega)=\tfrac{1}{2}\mathcal{P}\int_{-\infty}^{\infty}\int_{-\infty}^{\infty}\frac{d\omega'\,dt\cos\omega' t}{\omega'-\omega}\tanh\frac{\omega'}{2kT}\langle M_\alpha(0)M_\alpha(t)+M_\alpha(t)M_\alpha(0)\rangle \tag{12.30a}$$

$$\chi''_{\alpha\alpha}(\omega)=\tfrac{1}{2}\tanh\frac{\omega}{2kT}\int_{-\infty}^{\infty}dt\cos\omega t\langle M_\alpha(0)M_\alpha(t)+M_\alpha(t)M_\alpha(0)\rangle \tag{12.30b}$$

From Eqs. (12.3) we see that the Kramers-Krönig relation relating the real and imaginary parts of the susceptibility is

$$\chi'_{\alpha\alpha}(\omega)=\frac{\mathcal{P}}{\pi}\int_{-\infty}^{\infty}d\omega'\frac{\chi''_{\alpha\alpha}(\omega)}{(\omega'-\omega)} \tag{12.31}$$

4

The Resolvant Operator and the Density Matrix

By an "isolated spectral line" we mean to describe, as the reader might suppose, a spectral line that is isolated, that is to say, a line in the spectrum of a molecule which is "well separated" from every other line in that spectrum. However, what is "well separated?" If the inquirer will think of a term in which the denominator is the difference between the frequency of the line under consideration and that of its nearest neighbor, the separation idea will clarify itself. When the term is small the line is isolated. Anderson (55) developed the quantum mechanical formulation of interruption or impact broadening after a fashion which was proven extremely valuable by any number of subsequent developments. Breene (9) has reviewed this development together with the important contributory work of Foley (122), and Tsao and Curnutte (269) have done so *in extenso*. We shall not consider this theory in detail, but we do recall the important introduction of the TDO into line broadening theory which was signaled by this work. As Roney (36) has quite justly remarked, the work of Leslie (172), who anticipated Baranger in extending the treatment of the isolated line to include Doppler broadening, has been largely overlooked. Unfortunately perhaps, there is now no reason for us to do otherwise.

With this chapter we continue the development by considering the resolvant formulation of the interruption theory for a group of overlapping spectral lines, a formulation which was largely the work of Baranger (63–65) and to which Kolb and Griem (164), among others, contributed. [Griem and his collaborators have done a great deal of work and published a great many papers in this and related areas which we shall make no effort to reproduce. Many of these references are available in certain of Griem's books (18, 19).] We remark that various analytic properties of the non-Hermitian resolvant operator have been considered by Fonda et al. (120).

Let us suppose that we wish to know the intensity of radiation at a particular frequency due to a particular molecule, this molecule having several, closely spaced spectral lines, each of which contributes at this frequency. Let us suppose further that we know precisely what the contribution from each line would be were all the lines isolated. The fact that the total intensity is not the simple sum of the individual intensities arises from a phenomenon which has come to be called "line coupling." Of course the origin of the effect is to be found in the level coupling induced in the molecule by the perturbing portions of the Hamiltonian. Its treatment we consider in Section 13. In Section 14, we shall consider Baranger's first work in the line broadening arena, in which he studied the interruption problem for an isolated line, this study providing a number of worthwhile insights which are not so obvious in the overlapping study. Sections 15 and 16 are again Baranger dominated, this time with certain ruminations in reduction, some of which will be directly useful to us in later chapters.

As we have gathered by this time, the S matrix is of considerable consequence to spectral line broadening. We may, with various authors, expound aloofly upon it, but any specific calculations are apt to require its detailed evaluation. It is not our intention to discuss much in the way of specific calculations, but it certainly behooves us to have some idea as to how they are to be carried out. We should be at least vaguely aware of how one should evaluate S-matrix elements. This we discuss in Sections 17 and 18, the first devoted to evaluation by truncation of the series, the second to evaluation by the use of the classical path, the elimination of time ordering, and the consequent elimination of the necessity for expansion.

We have fleetingly encountered the density matrix to this point; in future chapters we shall deal with it *in extenso*. That density matrices should enter averages and that dipole averages should enter intensity expressions will hardly come as a surprise. A rather straightforward general introduction to this entity is provided by a treatment of Cooper (104) to which we devote Section 19.

Finally, we recall the Van Vleck-Weisskopf [cf. § 6.19 of Breene (9)] expression for the spectral line shape

$$I \sim \frac{\omega}{\omega_0}\left\{\left[(\omega-\omega_0)^2+\delta^2\right]^{-1}+\left[(\omega+\omega_0)^2+\delta^2\right]^{-1}\right\}$$

an expression which is deeply dependent, for its origin, on the Boltzmann distribution. This expression, we will also recall, differs but trivially from the simple interruption shape save at relatively low frequencies, it being important, for example, in the microwave. Van Vleck's interest in the effect of the Boltzmann distribution on spectral line shape did not wane with the years, an example of his later work occupying us in Section 20 of this chapter.

13 General Formulation of the Baranger Resolvant Formalism

Let us suppose that we wish to begin with the general expression for the intensity in a spectral line arising from the spontaneous emission associated with a dipole transition. (It is really not very important, but one must begin a long journey with the first step, as the author of the Tao Teh King was wont to remark. Incidentally, Ko Hung tells us the Lao-Tse spent sixty years in the womb, emerging therefrom with a lengthy white beard. History does not record the reaction of his maternal ancestor, but she must surely have been somewhat relieved at his departure.) For a single atom the intensity at a frequency, ω, will be the energy of the emitted photon, ω, multiplied by the probability that the radiator will spontaneously emit it. This probability is, excluding certain constants, a product of the absolute square of the dipole matrix element and ω^3, the latter arising from Planck's radiation density function. Therefore, our intensity will be at least proportional to

$$\omega^4 |\langle f|d|i\rangle|^2 \tag{13.1}$$

The assumption is usually made that ω^4 is a constant over the spectral line. The reader may wish to mull this over, but we shall not concern ourselves with the microwave in any event. Our initial level, i, here will not be of zero width for various reasons among which we may remark translational effects and level broadening by collision. Since it follows that ω_{if}, the frequency corresponding to the $i \to f$ transition, may take on a spectrum of values, and since we are interested in the value ω, we select out the intensity corresponding to ω with the Dirac delta function, thus

$$I(\omega) = |\langle f|\mathbf{d}|i\rangle|^2 \delta(\omega - \omega_{ij}) \tag{13.2}$$

We will recall that

$$\delta(\omega - \omega_{ij}) = (2\pi)^{-1} \int_{-\infty}^{\infty} e^{i(\omega - \omega_{if})t}\, dt \tag{13.3}$$

Therefore, the intensity in the spectral line may be written as

$$I(\omega) = (2\pi)^{-1} \int_{-\infty}^{\infty} \Phi(t) e^{i\omega t}\, dt \tag{13.4a}$$

where

$$\begin{aligned}
\Phi(t) &= e^{-i\omega_{if}t} |\langle f|\mathbf{d}(0)|i\rangle|^2 \\
&= \langle f|\mathbf{d}(0)|i\rangle \langle i|e^{-iE_i t}\mathbf{d}(0)e^{-E_f t}|f\rangle \\
&= \langle f|\mathbf{d}(0)|i\rangle \langle i|e^{-iH_0 t}\mathbf{d}(0)e^{iH_0 t}|f\rangle \\
&= \langle f|\mathbf{d}(0)|i\rangle \langle i|\mathbf{d}(t)|f\rangle
\end{aligned} \tag{13.4b}$$

We have thus provided the dipole moment with time dependence, although we reserve the right to modify this time dependence to a true Heisenberg representation by including the perturbing portion of the Hamiltonian. We have here presumed of course that H_0 is diagonal in $|i\rangle$ and $|f\rangle$.

The emitted power and hence the line shape must be real. As a consequence, the integral on the right of Eq. (13.3a) must be real. Taking $\Phi(t)$ as complex ($= \Phi_R + i\Phi_I$), the requirement that $\Phi(t)e^{i\omega t}$ be real leads to

$$\Phi(-t) = \Phi^*(t) \tag{13.5}$$

Anderson (55), Bloom and Margenau (80), and others (269) have devoted considerable effort to the derivation with which we are concerning ourselves, albeit they invoke somewhat more sophistication. In straightforward fashion we now (1) take the absolute square, (2) sum over all final states, (3) average over all initial states, and (4) average over the motion of the perturbers. The result is

$$I(\omega)=(2\pi)^{-1}\int_{-\infty}^{\infty} e^{i\omega t}\Phi(t)\,dt \tag{13.6a}$$

$$\Phi(t)=\sum_{if}\Upsilon_i\left[\langle\psi_i(0)|\mathbf{d}|\psi_f(0)\rangle\langle\psi_f(t)|\mathbf{d}|\psi_i(t)\rangle\right]_{\text{av}} \tag{13.6b}$$

where "av" designates the average over perturber motion. The Υ_i are the elements of the density matrix. The summation is carried over all initial and final states that contribute to the overlapping lines. Φ is what Foley (122) and Anderson (55) called the correlation function. An excellent discussion of the meaning of this function is given by Margenau and Lewis (193), and we shall essentially repeat our earlier definition [§ 2.12 of Breene (9)] at a point where such repetition is more appropriate, that is, after Eq. (13.9) below. In any event, Eq. (13.6a) becomes

$$I(\omega)=\pi^{-1}\Re\left(\int_0^{\infty} e^{i\omega t}\Phi(t)\,dt\right) \tag{13.7}$$

It is at this point that the TDO is introduced. Anderson initiated extensive use of the TDO in broadening work, and, indeed, it is the heart of the whole matter. From this point one ponders, manipulates, and evaluates the collision TDO. This is hardly surprising since by definition this TDO tells us how the wave functions involved react to the collision by relating those at the start of the time period to those at the end, t, as follows,

$$\mathfrak{T}(t)\psi(0)=\psi(t) \tag{13.8}$$

We utilize Eq. (13.8) in order to transform Eq. (13.6b) as follows:

$$\sum_{if}\Upsilon_i\left[\langle\psi_i(0)|\mathbf{d}|\psi_f(0)\rangle\langle(\mathfrak{T}_f(t)\psi_f^*(0))|\mathbf{d}|\mathfrak{T}_i(t)\psi_i(0)\rangle\right]_{\text{av}}$$

$$=\sum_{if}\Upsilon_i\left[\langle\psi_i(0)|\mathbf{d}|\psi_f(0)\rangle\langle\psi_f(0)|\mathfrak{T}_f^*(t)\mathbf{d}\mathfrak{T}_i(t)|\psi_i(0)\rangle\right]_{\text{av}}$$

$$=\sum_{i}\Upsilon_i\left[\langle\psi_i(0)|\mathbf{d}\mathfrak{T}_f^*(t)\mathbf{d}\mathfrak{T}_i(t)|\psi_i(0)\rangle\right]_{\text{av}}$$

$$=\mathrm{Tr}\left\{\Upsilon\left[\mathbf{d}\mathfrak{T}_f^*(t)\mathbf{d}\mathfrak{T}_i(t)\right]\right\}_{\text{av}} \tag{13.9}$$

At this point we diverge from the Baranger-Kolb-Griem (BKG) development for a brief discussion of the correlation function. Now let us assume the classical path. This will mean that a matrix element of the interaction potential is simply the interaction potential, perhaps of the form $\mathsf{V}(r)=\mathsf{V}(t)=f(\rho^2+v^2t^2)$ if we assume straightline trajectories of closest approach, ρ. Then if we suppose the trace over the unperturbed emitter functions, we obtain

$$\langle i|\mathfrak{T}_i|j\rangle = \exp\left[-\mathrm{i}\int_0^t H_0\,dt' - \mathrm{i}\int_0^t \mathrm{V}_i\,dt'\right]_{ij}$$

$$= \exp(-\mathrm{i}\omega_i t)\exp\left[-\mathrm{i}\int_0^t \mathrm{V}_i(t')\,dt'\right]\delta_{ij}$$

As a consequence, Eq. (13.9) becomes

$$\Phi(t) = \sum_{if} \Upsilon_i |\mathbf{d}_{if}|^2 \langle e^{\mathrm{i}(\omega_f-\omega_i)t} e^{-\mathrm{i}\int_0^t(\mathrm{V}_i-\mathrm{V}_f)\,dt'}\rangle_{\mathrm{av}}$$

$$= \mathrm{const}\int_0^\infty dt_0 \exp\left[-\mathrm{i}\int_{t_0}^{t_0+t}(\mathrm{V}_i-\mathrm{V}_f)\,dt'\right]$$

where the integral over t_0 is taken as the averaging process over the perturber coordinates. The correlation function here is at least reminiscent of a zeta function—the sum of a principal part and a delta function—although it does not actually take this form for a simple potential. However, a qualitative consideration of the details of the correlation function tells us that it is of delta function form.

For $t=0$, we may replace the exponential by unity, and the integral is certainly very large. As t increases from zero the following will not generally happen, but let us suppose that it does: the imaginary exponential will oscillate with extreme rapidity, and the correlation function will have effectively zero value for $t \neq 0$. Then we take $\Phi(t) = \delta(t)$. When this is substituted into Eq. (13.2) we will of course obtain

$$I(\omega) = \mathrm{const}\int \delta(t) e^{\mathrm{i}\omega t}\,dt = \mathrm{const}\; e^{\mathrm{i}\omega\cdot 0} = \mathrm{const}$$

so that the intensity will be independent of frequency. Thus rapid loss of correlation between the value of $\Phi(t)$ at $t=0$ and at $t\neq 0$ will lead to broad, ill-defined intensity distributions in the spectral line. As soon, however, as we move away from a delta-function-like correlation function, there is a frequency dependence in the spectral line. As the correlation function broadens as a consequence of a closer relationship between the values of this function for increasing times, so the intensity distribution narrows and tends to become more precisely defined. Although this is all qualitative in the extreme, it does tend to indicate the origin of the term. Now we return to Eq. (13.9).

In doing so, we should reiterate the fact that the trace is over emitter coordinates. Thus the density matrix, usually taken as a Boltzmann distribution over these coordinates, provides us with the emitter average, the "av" with the average over perturber coordinates. The density matrix is included within the "av" symbol only for convenience. In proceeding, let us take the substates of the initial level to be $\gamma, \gamma', \cdots$ and those of the final to be

$\beta, \beta', \cdots$ Then we may rewrite Eq. (13.9) as follows:

$$\begin{aligned}\mathrm{Tr}\{[d\mathfrak{T}_f^* d\mathfrak{T}_i]\}_{\mathrm{av}} &= \sum_{\gamma\gamma'\beta\beta'} \langle\gamma'|\mathbf{d}|\beta\rangle\langle\beta|\mathfrak{T}_f^*|\beta'\rangle\langle\beta'|\mathbf{d}|\gamma\rangle\langle\gamma|\mathfrak{T}_i|\gamma'\rangle \\ &= \sum_{\gamma\gamma'\beta\beta'\sigma} \langle\gamma'|d_\sigma|\beta\rangle\langle\beta'|d_\sigma|\gamma\rangle\langle\beta|\mathfrak{T}_f^*|\beta'\rangle\langle\gamma|\mathfrak{T}_i|\gamma'\rangle\end{aligned} \tag{13.10}$$

in which we have taken the inner product of the dipole moment operator.

The complete TDO, of course, includes operators relating to the isolated molecule as well as those relating to the broadening collision. One may go to another form of the TDO, $\mathfrak{U}$, in the following rather obvious fashion:

$$\mathfrak{T}_i = e^{-iH_i t}\mathfrak{U}_i \tag{13.11}$$

wherein H_i is the Hamiltonian for the isolated molecule in its initial state. The line shape is given by

$$\begin{aligned}I_{if}(\omega) = \Re\Big\{ \sum_{\gamma\gamma'\beta\beta'\sigma} \int_0^\infty dt \langle\langle\gamma\beta|e^{i[\omega-(H_i-H_f)]t}|\gamma'\beta'\rangle\rangle\langle\gamma'|d_\sigma|\beta\rangle\langle\beta'|d_\sigma|\gamma\rangle \\ \cdot\{\langle\beta|\mathfrak{U}_f^*|\beta'\rangle\langle\gamma|\mathfrak{U}_i|\gamma'\rangle\}_{\mathrm{av}}\Big\}\end{aligned} \tag{13.12}$$

since the averaging over perturbers hardly applies to the exponential term. We follow Baranger in using the symbol $\langle\langle|\rangle\rangle$ for the $(\gamma\gamma')$ matrix element over H_i multiplied by the $(\beta\beta')$ matrix element over H_f, although we have previously used this symbol (and shall again in future) for the matrix element in $\mathfrak{L}$-space. One can make the connection, but it really seems much ado about not much, a mote in the eye of the beholder. Let us simply take the symbol as inferentially defined by Eqs. (13.10) and (13.11).

Note that collisional quenching is excluded by the absence of TDO matrix elements between initial and final states as are radiating transitions between the sublevels (no d_σ matrix elements $\gamma\to\gamma'$ or $\beta\to\beta'$). "av" specifies an average over the thermal motion of the perturbers.

We now find an equation for the interaction TDO. To begin with, the general TDO must satisfy the equation,

$$i\dot{\mathfrak{T}} = [H_a + \mathsf{V}]\mathfrak{T} \tag{13.13}$$

where V is the interaction potential. Equation (13.13) has the formal solution,

$$\mathfrak{T} = e^{-i[H_a+\mathsf{V}]t} \tag{13.14}$$

The definition of $\mathfrak{U}$ is obvious from Eqs. (13.11) and (13.14). Substituting Eq. (13.11) into Eq. (13.13) yields

$$e^{-iH_a t} i\dot{\mathfrak{U}} = \mathsf{V}e^{-iH_a t}\mathfrak{U}$$

from which

$$i\dot{\mathfrak{U}} = e^{iH_a t}\mathsf{V}e^{-iH_a t}\mathfrak{U} = \mathsf{V}'\mathfrak{U} \tag{13.15}$$

We iterate this twice in order to obtain

$$\mathcal{U}(t)=1-\mathrm{i}\int_0^t dt'\,\mathsf{V}'(t')-\int_0^t dt'\,\mathsf{V}'(t')\int_0^{t'} dt''\,\mathsf{V}'(t'') \tag{13.16}$$

Next, a change in the TDO product of Eq. (13.12) in the time interval Δt may surely be written

$$\Delta\{\mathcal{U}_i(t,0)\mathcal{U}_f^*(t,0)\}_{\text{av}}=\{\mathcal{U}_i(t+\Delta t,0)\mathcal{U}_f^*(t+\Delta t,0)-\mathcal{U}_i(t,0)\mathcal{U}_f^*(t,0)\}_{\text{av}} \tag{13.17}$$

The notation is slightly different here; we use $\mathcal{U}(t,0)$ for the time interval between 0 and t. Since the TDO's are exponential in form, Eq. (13.13) may be written

$$\Delta\{\mathcal{U}_i(t,0)\mathcal{U}_f^*(t,0)\}_{\text{av}} = \{[\mathcal{U}_i(t+\Delta t,t)\mathcal{U}_f^*(t+\Delta t,t)-1][\mathcal{U}_i(t,0)\mathcal{U}_f^*(t,0)]\}_{\text{av}} \tag{13.18}$$

The impact approximation is brought into play when we say that Δt is so large that there is no correlation between the first and second factors on the right of Eq. (13.14). That is, if there is a collision in the t interval and another in the subsequent interval, the time between them is much longer than their duration. Hence one may average the two separately, thus

$$\Delta\{\mathcal{U}_i(t,0)\mathcal{U}_f^*(t,0)\}=\{\mathcal{U}_i(t+\Delta t,t)\mathcal{U}_f^*(t+\Delta t,t)-1\}\{\mathcal{U}_i(t,0)\mathcal{U}_f^*(t,0)\} \tag{13.19}$$

If $\mathcal{U}(t,0)$ is given by Eq. (13.16) with additional terms for the continued iteration, then

$$\mathcal{U}_\alpha(t+\Delta t,t)=1-\mathrm{i}\int_t^{t+\Delta t} dt'\,\mathsf{V}'(t')-\int_t^{t+\Delta t} dt'\,\mathsf{V}'(t')\int_t^{t'} dt''\,\mathsf{V}'(t'')+\cdots \tag{13.20}$$

where

$$\mathsf{V}'(t)=e^{\mathrm{i}H_\alpha t}\mathsf{V}(t)e^{-\mathrm{i}H_\alpha t} \tag{13.21}$$

It is perhaps appropriate to point out here that the potential $\mathsf{V}(t)$ is classical in the sense that it is based on the classical path, the separation of the emitter and the perturber being precisely defined therein.

The time scale is now shifted so that

$$\mathcal{U}(t+\Delta t,t)=1-\left[\mathrm{i}\int_0^{\Delta t} dt'\,\mathsf{V}'(t')+\int_0^{\Delta t} dt'\,\mathsf{V}'(t')\int_0^{t'} dt''\,\mathsf{V}'(t'')+\cdots\right] \tag{13.22}$$

The evaluation of the first term on the right of Eq. (13.19) now calls for the straightforward multiplication of two equations of the form Eq. (13.22),

$$\{\mathfrak{U}_i(t+\Delta t,t)\mathfrak{U}_f^*(t+\Delta t,t)-1\}=\Big\{-\int_t^{t+\Delta t}dt'[\mathrm{V}_i'(t')-\mathrm{V}_f'(t')]$$
$$-\Big[\int_t^{t+\Delta t}dt'\mathrm{V}_i'(t')\int_0^{t'}dt''\mathrm{V}_i'(t'')+\int_t^{t+\Delta t}dt'\mathrm{V}_f'(t')\int_0^{t'}dt''\mathrm{V}_f'(t'')$$
$$-\int_t^{t+\Delta t}dt'\mathrm{V}_i'(t')\int_0^{t'}dt''\mathrm{V}_f'(t'')\Big]+\cdots\Big\} \tag{13.23}$$

If we now define φ_{if} as follows:

$$\varphi_{if}=\frac{1}{\Delta t}\Big\{-\mathrm{i}\int_t^{t+\Delta t}dt'[\mathrm{V}_i'(t')-\mathrm{V}_f'(t')]-\Big[\int_t^{t+\Delta t}dt'\mathrm{V}_i'(t')\int_0^{t'}dt''\mathrm{V}_i'(t'')$$
$$+\int_t^{t+\Delta t}dt'\mathrm{V}_f'(t')\int_0^{t'}dt''\mathrm{V}_f'(t'')-\int_t^{t+\Delta t}dt'\mathrm{V}_i'(t')\int_0^{t'}dt''\mathrm{V}_f'(t'')\Big]+\cdots\Big\} \tag{13.24}$$

Equation (13.19) may be written as the difference equation

$$\Delta\{\mathfrak{U}_i(t,0)\mathfrak{U}_f^*(t,0)\}=\varphi_{if}\{\mathfrak{U}_i(t,0)\mathfrak{U}_f^*(t,0)\}\Delta t \tag{13.25}$$

We have stipulated that the time between collisions is sufficiently long so that separate averaging may take place; we now specify it short enough so that the difference equation [Eq. (13.25)] may be taken as a differential equation of obvious solution,

$$\{\mathfrak{U}_i(t,0)\mathfrak{U}_f^*(t,0)\}=e^{\mathrm{i}\varphi_{if}t} \tag{13.26}$$

The major portion of the derivation is now completed, one need but substitute Eq. (13.26) into Eq. (13.12). We then write Eq. (13.12) as

$$I_{if}=\mathfrak{R}\Big\{\sum_{\alpha\alpha'\beta'\sigma}\Upsilon\langle\gamma|\mathrm{d}_\sigma|\beta\rangle\langle\gamma'|\mathrm{d}_\sigma^*|\beta\rangle\int_0^\infty dt$$
$$\times\langle\langle\alpha\beta|\exp\{\mathrm{i}[\omega-(H_i-H_f)+\mathrm{i}\varphi_{if}]t\}|\gamma'\beta'\rangle\rangle\Big\} \tag{13.27}$$

where we have reintroduced the density matrix, Υ. Finally,

$$I_{if}(\omega)=\mathfrak{R}\Big\{\sum_{\alpha\alpha'\beta'\sigma}\Upsilon\langle\gamma|\mathrm{d}_\sigma|\beta\rangle\langle\gamma'|\mathrm{d}_\sigma^*|\beta'\rangle$$
$$\times\langle\langle\gamma\beta|[\mathrm{i}\omega-\mathrm{i}(H_i-H_f)+\varphi_{if}]^{-1}|\gamma'\beta'\rangle\rangle\Big\} \tag{13.28}$$

The H_i and H_f are diagonal in the γ and β, but φ_{if} may not be.

In order to rewrite Eq. (13.28) so that the matrix elements will be in the denominator, we must find a set of functions such that $\mathrm{i}(H_i-H_f)-\varphi_{if}$ is diagonal. We suppose the desired functions to be given by

$$\chi_\gamma \chi_\beta = \sum_{\gamma'\beta'} C_{\gamma\gamma'}^{\beta\beta'} \psi_{\gamma'} \psi_{\beta'} \tag{13.29}$$

In rather standard fashion one obtains these coefficients from the equation,

$$\sum_{\gamma'\beta'} C_{\gamma\gamma'}^{\beta\beta'} \langle\langle \gamma''\beta''|\varphi_{if}|\gamma'\beta'\rangle\rangle - (G_{\gamma\beta} + \omega_{if})\delta_{\gamma'\gamma''}\delta_{\beta'\beta''} = 0 \tag{13.30}$$

Let us use the following notation for the inverse situation

$$\psi_\gamma \psi_\beta = \sum_{\gamma'\beta'} D_{\gamma\gamma'}^{\beta\beta'} \chi_{\gamma'} \chi_{\beta'} \tag{13.31}$$

Now one introduces the transformation Eq. (13.31) to put the matrix elements in the denominator and subsequently introduces the transformation Eq. (13.29) in order to evaluate them in terms of available wave functions. As a consequence Eq. (13.28) becomes

$$I_{if}(\omega) = \Re\left\{ \sum_{\gamma\gamma'\beta'\sigma} \Upsilon \langle \gamma|\mathsf{d}_\sigma|\beta\rangle^2 \sum_{\gamma''\beta''} D_{\gamma\gamma''}^{\beta\beta''} D_{\gamma'\gamma''}^{\beta'\beta''} \left[\mathrm{i}\omega - \omega_{\gamma\beta} \right.\right.$$

$$\left.\left. + \sum_{\gamma'''\gamma^{(4)}\beta^{(4)}} C_{\gamma''\gamma'''}^{\beta''\beta'''} C_{\gamma''\gamma^{(4)}}^{\beta''\beta^{(4)}} \langle \gamma''' \langle \beta'''|\varphi_{if}|\beta^{(4)}\rangle|\gamma^{(4)}\rangle \right]^{-1} \right\}$$

$$I_{if}(\omega) = \sum_{\gamma\gamma'\beta'\sigma} \sum_{\gamma''\beta''} \frac{\langle \gamma|\mathsf{d}_\sigma|\beta\rangle^2 \delta_i}{(\omega - \omega_\gamma + \Delta_i)^2 + \delta_i^2} \tag{13.32a}$$

$$\delta_i = \Re\left\{ \sum_{\gamma'''\gamma^{(4)}} C_{\gamma''\gamma'''}^{\beta''\beta'''} C_{\gamma''\gamma^{(4)}}^{\beta''\beta^{(4)}} \langle \gamma'''|\langle \beta'''|\varphi_{if}|\beta^{(4)}\rangle|\gamma^{(4)}\rangle \right\} \tag{13.32b}$$

$$\varphi_{if} = -N \int\int 2\pi\rho v f(v)\, dv\, d\rho \left\{ \mathrm{i} \int_{-\infty}^{\infty} dt [\mathsf{V}_i'(t) - \mathsf{V}_f'(t)] \right.$$

$$+ \left[\int_{-\infty}^{\infty} dt \mathsf{V}_i'(t) \int_{-\infty}^{t} dt' \mathsf{V}_i'(t') \right.$$

$$+ \int_{-\infty}^{\infty} dt \mathsf{V}_f'(t) \int_{-\infty}^{t} dt' \mathsf{V}_f'(t')$$

$$\left.\left. - \int_{-\infty}^{\infty} dt \mathsf{V}_i'(t) \int_{-\infty}^{t} dt' \mathsf{V}_f'(t') \right] + \cdots \right\}_{\text{ang}} \tag{13.32c}$$

To obtain Eq. (13.32c) from Eq. (13.24) we have taken the collision frequency, $(\Delta t)^{-1}$, to be $2\pi\rho v f(v)\, dv\, d\rho$ for collisions having velocities between v and $v + dv$ and optical collision diameters between ρ and $\rho + d\rho$. N is, of course, the perturber density. The integral limits have been extended since nothing occurs outside the original limits anyhow. The average indicated is over the angles associated with the orientations of ρ and v. Let us see how the shape of the group of overlapping lines predicted by the BKG theory differs from a simple superposition of isolated lines.

From Eq. (13.9) we may arrive at the following expression for the intensity:

$$I=\mathfrak{R}\left\{\mathrm{Tr}\left[d^2(\mathrm{i}\omega-\mathrm{i}H_i+\varphi_i)^{-1}\right]\right\} \tag{13.33}$$

where we consider only interactions in the initial state. We require a diagonal representation of $\mathrm{i}H_i+\varphi_i$ in order to get the matrix elements in the denominator. To find these we take the eigenstates of the operator, $-\mathrm{i}H_i+\varphi_i$, to be ψ_i, the eigenvalues $\omega_i-\mathrm{i}\delta_i$ so that

$$(-\mathrm{i}H_i+\varphi_i)\psi_i=(\delta_i-\mathrm{i}\omega_i)\psi_i \tag{13.34}$$

Since $H_i+\varphi_i$ is not an observable, we cannot expect the matrix to be Hermitian or the states orthogonal. However, a set of states, χ_j, may be defined such that

$$\langle\chi_j|\psi_i\rangle=\delta_{ij}\Rightarrow\sum_i|\psi_i\rangle\langle\chi|=1 \tag{13.35}$$

With these functions Eq. (13.33) becomes

$$I\propto\mathfrak{R}\left\{\sum_i\langle\chi_i|\mathrm{d}^2[i\omega-iH_i+\varphi_i]^{-1}|\psi_i\rangle\right\}$$

$$=\sum_i\left[\left\{\frac{\delta_i}{(\omega-\omega_i)^2+\delta_i^2}\right\}\mathfrak{R}[\langle\chi_i|\mathrm{d}^2|\psi_i\rangle]-\left\{\frac{(\omega-\delta_i)}{(\omega-\omega_i)^2+\delta_i^2}\right\}\mathfrak{I}[\langle\chi_i|\mathrm{d}^2|\psi_i\rangle]\right. \tag{13.36}$$

The summation in Eq. (13.36) is over the various overlapping lines, and it is rather obvious that the result is not simply a summation over a number of dispersion shapes equal to the number of lines. Superficially, the second term within the bracket would appear to make the integral over the group of lines logarithmically divergent since, for ω large, the term is simply ω^{-1}. Note, however, that, for large ω, the second term becomes $\omega^{-1}\mathfrak{I}\{\mathrm{Tr}[\mathrm{d}]\}$. Thus since the trace of an Hermitian matrix is real, this is zero.

Equation (13.36) is obviously a sum of dispersion line shapes, one for each of the overlapping spectral lines, plus a set of interference terms. The interference terms arise from the non-Hermitian character of what is often called the resolvant operator, $-\mathrm{i}(H_i-H_f)+\varphi_{if}$. These terms disappear if one separates or coalesces the overlapping lines.

If one separates the overlapping lines sufficiently, one may treat the collision operator, φ_{if}, as a perturbation of H_i-H_f by means of ordinary perturbation theory. As a consequence, the difficulties associated with the non-Hermitian character of the resolvant operator disappear and with them the interference effects. Equally obvious is the reduction to the Anderson treatment of an isolated line which follows.

The disappearance of the interference effects with the coalescence of the overlapping lines into a single degenerate line is analogous (141). Let us suppose that the diagonal elements may essentially be included within the G's of the matrix. When this is done the diagonal elements are real, and the matrix is Hermitian. Once again the interference effects disappear.

In Section 49 we shall consider the Royer idea of the interference between various possible "paths" in the sense of histories that a sequence of physical events may follow. That author makes an interesting point vis-à-vis these paths and overlapping lines (241) which, although equivalent to our above remarks, is well worth repetition.

Let us suppose we have two overlapping spectral lines, one from the transition $a \to b$, the other from $c \to d$. Then at some frequency, ω_s, the physical system follows one of two paths, one of which involves the electronic states and translational momenta associated with a, b, and ω, the other those associated with c, d, and ω. In a manner which we shall discuss in Section 49, these paths interact in order to produce the interference term of Eq. (13.36).

14 The Baranger Result for the Isolated Line

Now Ross (236), among others, is quite interested in establishing contact with a highly successful Baranger theory. However, the details of Ross' work are of no concern to us as yet. Further, it is of intrinsic interest to follow the contact that Baranger made with certain of the interruption theories that preceded him. It therefore behooves us to derive the Baranger result for an isolated spectral line which preceded the BKG result of Section 13. This will allow us to show how the former author established his communication with these earlier results. In reproducing Baranger's work it is necessary to begin with the specific wave functions which he used for the $n+1$ particle system consisting of an emitter and n perturbers.

In writing down the wave function for the system we appeal to a Jablonski sort of idea [Chapter 3 of Breene (9)], that is to say, we hypothesize an $(n+1)$ atomic molecule, the electronic structure of one of whose atomic constituents (the emitter) playing the role of providing electronic structure, the translational structure of the remaining atomic constituents (the n perturbers) playing the part of a sort of continuous rotation-vibration structure. It is supposed that the total interaction potential is a sum of n two-particle potentials, one particle in each being the radiator, the other a perturber. Obviously we are ignoring perturber correlations, but this is seldom a matter for concern. Since the potential is thus additive, the Schrödinger equation is separable, and the wave function is of product form. We shall follow the Baranger notation for the system wave functions,

$$u_i(x_A)\psi^+_{ip_1}(x_1)\psi^+_{ip_2}(x_2)\cdots\psi^+_{ip_n}(x_n) \tag{14.1a}$$

$$u_f(x_A)\psi^+_{fp'_1}(x_1)\psi^+_{fp'_2}(x_2)\cdots\psi^+_{fp'_n}(x_n) \tag{14.1b}$$

and the system energies,

$$E_i+\epsilon_1+\epsilon_2+\cdots+\epsilon_n \tag{14.1c}$$

$$E_f+\epsilon_1'+\epsilon_2'+\cdots+\epsilon_n' \tag{14.1d}$$

From which it follows that the emitted spectral line will have energy

$$(E_i-E_f)+(\epsilon_1-\epsilon_1')+\cdots+(\epsilon_n-\epsilon_n') \tag{14.2}$$

which emphasizes the origin of emitted frequencies greater or less than the unperturbed line frequency in the loss or gain of kinetic energy by the perturbers. We remark the plus superscript on the translational functions, which refers to the fact that Baranger chose the outgoing scattered wave to represent them. The subscript "p" refers to broadener momentum. In proceeding with our derivation a certain amount of repetition of our earlier work will allow us to be somewhat cavalier regarding detail.

Our spectral line shape will still be given by Eq. (13.6a), but our correlation function may now be specifically written out as

$$\Phi(t)=\sum_{\substack{p_1\cdots p_n\\ p_1'\cdots p_n'}} \Upsilon_{p_1}\cdots\Upsilon_{p_n}|\langle\psi_{fp_1'}^+|\psi_{ip_1}^+\rangle|^2\cdots|\langle\psi_{fp_n'}^+|\psi_{ip_n}^+\rangle|^2 \cdot\exp[-\mathrm{i}(\epsilon_1-\epsilon_1'+\cdots+\epsilon_n-\epsilon_n')t] \tag{14.3}$$

First with respect to Eq. (14.3), we have simply dropped the matrix element over the dipole moment—although this of course is where these overlap integrals originate—as a constant having no effect on the spectral line shape. It is important to emphasize why the overlap integrals do not in general disappear.

Let us suppose that we had a scattering potential, V, from which we calculated translational wave functions. Then overlap integrals for $p\neq p'$, both computed using V, would disappear according to elementary quantum mechanical principles. However, here we have two potentials, V_i and V_f, so that $|\psi_{ip}\rangle$ is only fortuitously orthogonal to $|\psi_{fp'}\rangle$. It is equally important to remark, however, that, if $V_i=V_f$, Eq. (14.3) will zero out.

If n independent random variables have the same distribution, the Fourier transform of the distribution of their sum is the nth power of the Fourier transform of a single distribution [cf. p. 188 of Cramer (11)]. Baranger uses this theorem in order to replace Eq. (14.3) by

$$\Phi(t)=[\varphi(t)]^n \tag{14.4a}$$

$$\varphi(t)=\sum_{pp'}\Upsilon_p|\langle\psi_{fp'}^+|\psi_{ip}^+\rangle|^2e^{-\mathrm{i}(\epsilon-\epsilon')t} \tag{14.4b}$$

Let us suppose first that there is simply no perturber present. Then the radiator will have a single set of orthonormal translational states, and only one of the overlap integrals of Eq. (14.4b) will exist, say, $\langle\psi_{fp}^+|\psi_{ip}^+\rangle=1$. This is the same as saying that, insofar as broadening a spectral line is concerned, mostly nothing happens. Obviously this initial assumption will be germane only when the density is "sufficiently low." Although such may be almost correct, it obviously cannot be entirely correct if we really desire any line broadening. Therefore "1" must be corrected. When there is a perturber present, there will be a V_i in the initial state and a V_f in the final state and two sets of orthonormal states, $|\psi_{ip}^{+}\rangle$ and $|\psi_{fp}^{+}\rangle$. Now, however, the states $|\psi_{ip}^{+}\rangle$ will not be orthogonal to the states $|\psi_{fp}^{+}\rangle$ as was the case with no perturber present. Just what their contribution to the overlap integral will be we do not, as yet, know. We shall assume, however, that there is a random probability for the location of the perturber in the configuration space. Then, if the volume containing radiator and perturber becomes infinite, the probability for perturber influence goes to zero so that $\varphi(t)=1$. This can be arranged by supposing that

$$\varphi(t)=1-V^{-1}g(t) \tag{14.5}$$

Thus

$$\begin{aligned}\Phi&=\left[1-V^{-1}g(t)\right]^{n}=1-nV^{-1}g(t)+\frac{1}{2!}n(n-1)V^{-2}g(t)\\&\quad-\frac{1}{3!}n(n-1)(n-2)V^{-3}g^{3}(t)+\cdots\\&\doteq 1-Ng(t)+\frac{1}{2!}N^{2}g^{2}(t)-\frac{1}{3!}N^{3}g^{3}(t)+\cdots\\&=\exp[-Ng(t)]\end{aligned} \tag{14.6}$$

where $N=nV^{-1}$. With slightly less justification we may let n and V go to infinity and replace the approximate equality by an equality.

By elementary matrix multiplication we may rewrite Eq. (14.4b) as

$$\begin{aligned}\varphi&=\sum_{pp'}\Upsilon_p\langle\psi_{ip}^{+}|\psi_{fp'}^{+}\rangle e^{i\epsilon' t}\langle\psi_{fp'}^{+}|\psi_{ip}^{+}\rangle e^{-i\epsilon t}\\&=\sum_{pp'}\Upsilon_p\langle\psi_{ip}^{+}|e^{iH_f t}|\psi_{fp'}^{+}\rangle\langle\psi_{fp'}^{+}|e^{-iH_i t}|\psi_{ip}^{+}\rangle\\&=\sum_{p}\Upsilon_p\langle\psi_{ip}^{+}|e^{iH_f t}e^{-iH_i t}|\psi_{ip}^{+}\rangle\end{aligned} \tag{14.7}$$

where H_i and H_f are the perturber Hamiltonians when the emitter is in state i and state f, respectively, that is,

$$H_j=K+\mathsf{V}_j \tag{14.8}$$

For notational convenience Baranger leaves the averaging to be inferred in Eq. (14.7) and drops the summation and density matrix. We now suppose the wave functions of that equation to be volume normalized, a supposition which allows us to multiply the matrix element by V^{-1} with an eye toward Eq. (14.5) with which we are seeking comparison. The introduction of the integral equation (118),

$$e^{iH_ft}e^{-iH_it}=1-i\int_0^t dt'\,e^{iH_ft'}\Delta V e^{-iH_it'}$$

into Eq. (14.7) allows us to make such a comparison with

$$g(t)=i\int_0^t dt'\,e^{-i\epsilon t'}\langle\psi_{ip}^+|e^{iH_ft'}\Delta V|\psi_{ip}^+\rangle \tag{14.9}$$

where Baranger has removed $\exp[-iH_it]|\psi_{ip}^+\rangle=\exp[-i\epsilon t]|\psi_{jp}^+\rangle$ from the matrix element.

We shall encounter the Lippmann-Schwinger equation (183) with sufficient frequency so that there is no point in dwelling upon it. We now utilize the following form of it:

$$\langle\psi_{ip}^+|=\langle\psi_{fp}^+|+\langle\psi_{ip}^+|\Delta V[\epsilon-H_f-i\eta]^{-1} \tag{14.10}$$

in Eq. (14.9) with the result

$$g(t)=i\int_0^t dt'\,e^{-i\epsilon t'}\langle\psi_{fp}^+|e^{iH_ft'}\Delta V|\psi_{ip}^+\rangle+i\int_0^t dt'\,e^{-i\epsilon t'}\langle\psi_{ip}^+|\Delta V[\epsilon-H_f$$

$$-i\eta]^{-1}e^{iH_ft'}\Delta V|\psi_{ip}^+\rangle=i\int_0^t dt'\,e^{-i\epsilon t'}\langle\psi_{fp}^+|\Delta V|\psi_{ip}^+\rangle+i\int_0^t dt'\int dp'$$

$$\cdot e^{-i\epsilon t'}\langle\psi_{ip}^+|\Delta V|\psi_{fp'}^+\rangle\frac{\langle\psi_{fp}^+|[\epsilon-H_f-i\eta]^{-1}e^{-iH_ft'}|\psi_{fp'}^+\rangle\langle\psi_{fp'}^+|\Delta V|\psi_{ip}^+\rangle}{(2\pi)^3}$$

$$=it\langle\psi_{fp}^+|\Delta V|\psi_{ip}\rangle+\int\frac{dp'}{(2\pi)^3}|\langle\psi_{fp'}^+|\Delta V|\psi_{ip}^+\rangle|^2\frac{1-e^{-i(\epsilon-\epsilon')t}}{(\epsilon-\epsilon')(\epsilon-\epsilon'-i\eta)} \tag{14.11}$$

where the integral over p' enters through matrix multiplication, H_f being diagonal in the $|\psi_{fp'}^+\rangle$ and where integration over time has been carried out.

Equations (14.5) and (14.11) then constitute the general result which could be dealt with, in theory at least, if the wave functions are known. With Baranger, we shall now specialize this result for the interruption approximation, first for the case where $V_f=0$, then for the case $V_f\neq0$. It is definitely less cumbersome to proceed in this fashion. Further, for the final state a lower, perhaps a ground state, such is often practically the case. The emitting atom may well be much smaller in physical size—even for the purist, such a statement has a certain amount of physical meaningfulness—its orbital electrons more tightly bound and interacting negligibly with the perturber as compared to their behavior in the upper state.

Since, for $V_f=0$, our final state translational function will be a plane wave, we replace $\langle\psi_{fp}^+|$ by $\langle p|$. For simplicity, we let $V_i=V$, $H_i=H$. Now Eq. (14.9) becomes

$$g(t)=i\int_0^t dt'\,e^{-i\epsilon t'}\langle\psi_p^+|e^{iKt'}V|\psi_p^+\rangle \tag{14.12}$$

A qualitative argument now leads to an important replacement in Eq. (14.12). It goes about like this:

$\langle\psi_p^+|$ includes the effects of the scatterer (radiator) at the origin. After this scattering, $\langle\psi_p^+|e^{iKt}$ is a description, during some time, t, of the propagation of $\langle\psi_p^+|$ by the free Hamiltonian, K. During this time the scattered wave will recede from the radiator, no new scattered waves being formed, no new collisions occurring. Thus at least near the emitter, $\langle\psi_p^+|e^{iKt}$ will look like $\langle p|e^{i\epsilon t}$. On the other hand, $\mathsf{V}|\psi_p^+\rangle$ vanishes, by the nature of the interaction, save near the radiator. Therefore, we may replace Eq. (14.12) by

$$g(t)=it\langle p|\mathsf{V}|\psi_p^+\rangle \tag{14.13}$$

In short we have eliminated the second term from Eq. (14.11), and Eqs. (14.4a) and (13.2) become

$$\Phi(t)=\exp\left[-iN\langle p|\mathsf{V}|\psi_p^+\rangle t\right]\Rightarrow$$

$$I(\omega)=\frac{\delta}{\pi}\frac{1}{(\omega-\Delta)^2+\delta^2} \tag{14.14a}$$

$$\Delta=N\mathcal{R}\left[\langle p|\mathsf{V}|\psi_p^+\rangle\right] \tag{14.14b}$$

$$\delta=-N\mathcal{I}\left[\langle p|\mathsf{V}|\psi_p^+\rangle\right] \tag{14.14c}$$

certainly an old familiar form.

As we can see from Eq. (9.9), the matrix elements in Eqs. (14.14) correspond to the scattering amplitude with $\mathbf{p}\|\mathbf{p}'$, that is, the forward scattering amplitude, $f(0)$. Therefore, from Eqs. (9.9) and (14.14),

$$\Delta=-\frac{2\pi N}{\mathsf{m}}\mathcal{R}[f(0)]_{\text{av}} \tag{14.15a}$$

$$\delta=\frac{2\pi N}{\mathsf{m}}\mathcal{I}[f(0)]_{\text{av}} \tag{14.15b}$$

where the averaging which was temporarily dropped from Eq. (14.7) has been replaced. Now we turn to the case $\mathsf{V}_f\neq 0$ and Eq. (14.9).

The matrix element in that equation is replaced, through a qualitative argument, by $\langle\psi_{fp}^+|e^{i\epsilon t}\Delta\mathsf{V}|\psi_{ip}^+\rangle$ so that

$$g(t)\doteq it\langle\psi_{fp}^+|\Delta\mathsf{V}|\psi_{ip}^+\rangle \tag{14.16}$$

Perhaps the strongest argument for this is the fact that Baranger's theory has been quite successful. For the purists among us his thesis is the following: $\langle\psi_{ip}^+|e^{iH_ft}$ is obviously the result of propagating the wave scattered from the initial state potential by the potential of the final state. Can we replace it with something? Well, if, during propagation, the scattered parts of $\langle\psi_{ip}^+|$ begin to look like those of $\langle\psi_{fp}^+|$, we could replace $\langle\psi_{ip}^+|e^{iH_ft}$ by $\langle\psi_{fp}^+|e^{i\epsilon t}$. Near the radiator, it would seem quite reasonable to suppose that such would be the case. Again $\Delta\mathsf{V}|\psi_{ip}^+\rangle$ vanishes far from the emitter so that the result follows.

For the two-potential then, Eqs. (14.14) are modified to

$$\Delta = N\mathcal{R}[\langle\psi_{fp}^{+}|\Delta\mathsf{V}|\psi_{ip}^{+}\rangle_{\rm av}] \qquad (14.14b')$$

$$\delta = -N\mathcal{I}[\langle\psi_{fp}^{+}|\Delta\mathsf{V}|\psi_{ip}^{+}\rangle_{\rm av}] \qquad (14.14c')$$

In order to express this result in terms of scattering amplitudes, Baranger appeals to the following Lippmann-Schwinger equations:

$$|\psi_p^{+}\rangle = |p\rangle + [\epsilon - K + i\eta]^{-1}\mathsf{V}|\psi_p^{+}\rangle \qquad (14.17a)$$

$$\langle\psi_p^{+}| = \langle p| + \langle\psi_p^{+}|\mathsf{V}[\epsilon - K - i\eta]^{-1} \qquad (14.17b)$$

$$[\epsilon - K + i]^{-1} = \mathcal{P}[\epsilon - K]^{-1} - i\pi\delta(\epsilon - K) \qquad (14.17c)$$

$$[\epsilon - K - i]^{-1} = \mathcal{P}[\epsilon - K]^{-1} + i\pi\delta(\epsilon - K) \qquad (14.17d)$$

Recalling that $\mathsf{V} = \mathsf{V}_i - \mathsf{V}_f$, we apply Eqs. (14.17) to the matrix element of Eqs. (14.14),

$$\langle\psi_{fp}^{+}|\Delta\mathsf{V}|\psi_{ip}^{+}\rangle = \langle\psi_{fp}^{+}|\mathsf{V}_i|\psi_{ip}^{+}\rangle - \langle\psi_{fp}^{+}|\mathsf{V}_f|\psi_{ip}^{+}\rangle = \langle p|\mathsf{V}_i|\psi_{ik}^{+}\rangle$$
$$- \langle\psi_{fp}^{+}|\mathsf{V}_f|p\rangle + 2\pi i\langle\psi_{fp}^{+}|\mathsf{V}_f\delta(\epsilon - K)\mathsf{V}_i|\psi_{ip}^{+}\rangle \qquad (14.18)$$

Elementary matrix multiplication allows us to write the last term on the right of Eq. (14.18) as

$$(2\pi)^{-3}\int dp'\langle\psi_{fp}^{+}|\mathsf{V}_f|k'\rangle\delta(\epsilon-\epsilon')\langle p'|\mathsf{V}_i|\psi_{ip}^{+}\rangle$$

Integration over a part of this equation is immediate when $dp' = \mathsf{m}p'\,d\epsilon'\,d\Omega$, Ω the solid angle, is introduced.

Once again Eq. (9.9) is utilized in order to obtain the result in terms of scattering amplitudes:

$$\langle\psi_{fp}^{+}|\Delta\mathsf{V}|\psi_{ip}^{+}\rangle = -\frac{2\pi}{\mathsf{m}}[f_i(0) - f_f^*(0)] + iv\int d\Omega\, f_f^*(\vartheta,\varphi)f_i(\vartheta,\varphi), \qquad (14.19)$$

so that, as in Eq. (14.14),

$$\Delta = \left\{-\frac{2\pi N}{\mathsf{m}}\mathcal{R}[f_i(0) - f_f(0)] + \tfrac{1}{2}iNv\int d\Omega[f_f^*(\Omega)f_i(\Omega) - f_f(\Omega)f_i^*(\Omega)]\right\}_{\rm av} \qquad (14.20a)$$

$$\delta = \left\{-\frac{2\pi N}{\mathsf{m}}\mathcal{I}[f_i(0) + f_f(0)] + \tfrac{1}{2}Nv\int d\Omega[f_f^*(\Omega)f_i(\Omega) + f_f(\Omega)f_i^*(\Omega)]\right\}_{\rm av}$$
$$= \left\{\tfrac{1}{2}Nv\int d\Omega|f_i(\Omega) - f_f(\Omega)|^2\right\}_{\rm av} \qquad (14.20b)$$

where the last term on the right of Eq. (14.20a) is obtained by application of the optical theorem. It is with Eq. (14.20a) that Ross, for one, will establish contact.

If, instead of writing the result in terms of the scattering amplitude, we derive it in terms of the S matrix, we may begin with

$$\langle p'|S_i|p\rangle = \langle p'|p\rangle - 2\pi i\delta(\epsilon-\epsilon')\langle p'|V_i|\psi_{ip}^{+}\rangle \tag{14.21a}$$

$$\langle p|S_f^{+}|p'\rangle = \langle p|p'\rangle + 2\pi i\delta(\epsilon-\epsilon')\langle \psi_{fp}^{+}|V_f|p'\rangle \tag{14.21b}$$

The product is apparently

$$(2\pi)^{-3}\int dp'\langle p|S_f^{+}|p'\rangle\langle p'|S_i|p\rangle = \langle p|p\rangle - 2\pi i\delta(\epsilon-\epsilon')\Big[\langle p|V_i|\psi_{ip}^{+}\rangle - \langle\psi_{fp}^{+}|V_f|p\rangle + 2\pi i\int\frac{dp'}{2\pi^3}\langle\psi_{fp}^{+}|V_f|p'\rangle\delta(\epsilon-\epsilon')\langle p'|V_i|\psi_{ip}^{+}\rangle\Big] \tag{14.22}$$

The reader will therefore see that Eqs. (14.20) may be written in terms of the S matrix as

$$\Delta \sim \mathcal{I}\{\langle 1 - S_i S_f^{+}\rangle\} \tag{14.22a}$$

$$\delta \sim \mathcal{R}\{\langle 1 - S_i S_f^{+}\rangle\} \tag{14.22b}$$

which is of interest to the Reck (224) comparison of the classical limit of the quantal interruption theory with the classical theory.

From, say, Eq. (2.121) of Breene (9) we may obtain the classical expression for shift and width as

$$\Delta \sim \mathcal{I}\langle 1 - \exp(i\eta)\rangle \tag{14.23a}$$

$$\delta \sim \mathcal{R}\langle 1 - \exp(i\eta)\rangle \tag{14.23b}$$

where η is the phase shift given by the integral over the broadening potential. When we recall that S is related to $\exp(i2\delta_l)$, the l referring to the lth partial wave, a relationship between the quantal and classical results begins to appear. As Reck demonstrates, however, the quantal limit, wherein S is replaced by a classical S, reduces to the classical result for strictly straight paths and hence relatively high energies. Reck concludes that the classical limit to the quantal treatment is to be preferred.

For certain interesting and detailed calculations which utilize the importance of the classical limit of the quantal interruption result, the reader is referred to the work of Reck and Hood (225, 146). However, let us discuss the interruption result in general.

If not in his first paper, at least in his third, Baranger describes his theory as an "impact" theory. Now over the years a great deal of often qualitative discussion of this subject has appeared in the literature, and certainly the general ideas should be easily grasped. Baranger, however, goes into a reasonable amount of quantitative detail, some of which it will be beneficial for us to reproduce.

15 The Validity of the Interruption Approximation

First, and under the onus of repeating ourselves, let us recall the physical inferences of the interruption or impact approximation.

The intermolecular forces responsible for spectral line broadening may in some fashion be assigned a range or maximum effective intermolecular separation. One of the early methods of doing this was to select the distance at which an approaching perturber has induced a phase shift of unity—or π or 2π or whatever other number the selector is clever enough to justify—in the natural frequency of the emitter. There are various others. This range may be used as the radius of an interaction sphere about the emitter. A broadening collision occurs when a perturber penetrates within this sphere. The time of residence of a perturber within this sphere is the collision time, the time between penetrations of the sphere the intercollision time. As has been discussed in some detail [Breene (9)], if the collision time is much greater than the intercollision time, the statistical theory is valid. If the intercollision time is much greater than the collision time, the interruption theory is valid. (Many authors, as a consequence of an historical injustice, call the resulting shape a "Lorentz" shape. If any name were to be attached to it, it should have been that of Michelson. We call it an interruption shape.) Now we turn to Baranger's ruminations on the subject.

As we have seen, the validity of Eq. (14.13) or Eq. (14.16) depends on t being sufficiently large. If t is sufficiently large, these equations are acceptable and lead to an interruption line shape which Baranger takes as being equivalent to the interruption approximation. If we define τ as a collision time, the condition may be written

$$t \gg \tau \tag{15.1}$$

and what Baranger calls a collision volume—our interaction sphere—may be defined. Now, as we shall see in Section 16, when the interruption approximation is valid, the real exponential factor in Φ will be $\exp[-iNv\sigma t/2]$. Therefore, the important values of t will be of the order $2/Nv\sigma$, and Eq. (15.1) tells us that, for interruption validity

$$2/Nv\sigma \gg \tau \Rightarrow \tfrac{1}{2}v\sigma\tau \ll N^{-1} \tag{15.2}$$

$\frac{1}{2}v\sigma\tau$ has the units volume; indeed, the velocity, cross section and time involved in this expression essentially describe the volume which is presented to the perturber as our interaction sphere. (We recall that distant collisions during which there is really no penetration of the interaction sphere contribute largely to shift, little to broadening [§ 2.7 of Breene (9)]. Particularly since Baranger's definition of interruption validity relates to width, such collisions will hardly concern us here.) This we call U and arrive at the following criterion for interruption theory validity:

$$U \ll N^{-1} \tag{15.3}$$

Baranger, however, presents a considerably more erudite and informative development of very nearly this same result.

We begin with Eq. (14.11) and inquire as to the conditions which must obtain in order to replace it with its first term as in Eq. (14.13). If the difference between the real parts of Eq. (14.11) and (14.11) is small, such a replacement may obviously be made, and Eq. (14.17d) tells us that this difference is approximately

$$(2\pi)^{-3}\int d\mathbf{p}'|\langle p'|\mathsf{V}|\psi_p\rangle|^2\mathcal{P}(\epsilon-\epsilon')^{-2}\sin[(\epsilon-\epsilon')t] \tag{15.4}$$

As to how small this difference should be, it should be smaller than the real part of Eq. (14.13) so that corrections to the width will be small in comparison to the width itself.

Assuming that $\epsilon^{-1} \ll t$, Baranger used the odd-function characteristic of the integrand in

$$(2\pi)^{-3}\int d\mathbf{p}|\langle p''|\mathsf{V}|\psi_p^+\rangle|^2\mathcal{P}(\epsilon-\epsilon')^{-2}\sin[(\epsilon-\epsilon')t]=0$$

$\mathbf{p}''$ a vector whose direction is that of $\mathbf{p}'$, its length that of $\mathbf{p}$, in order to write Eq. (15.4) as

$$(2\pi)^{-3}\int d\mathbf{p}\left(\frac{|\langle p'|\mathsf{V}|\psi_p^+\rangle|-|\langle p''|\mathsf{V}|\psi_p^+\rangle|}{\epsilon-\epsilon'}\right)\left(\frac{|\langle p|\mathsf{V}|\psi_p^+\rangle|+|\langle p''|\mathsf{V}|\psi_p^+\rangle|}{\epsilon-\epsilon'}\cdot\sin[(\epsilon-\epsilon')t]\right) \tag{15.5}$$

According to Schwartz's inequality [cf. Chapter 4 of Beals (6)], Eq. (15.5) is smaller than

$$\left\{(2\pi)^{-3}\int d\mathbf{p}\left(\frac{|\langle p'|\mathsf{V}|\psi_p^+\rangle|-|\langle p''|\mathsf{V}|\psi_p^+\rangle|}{\epsilon-\epsilon'}\right)^2\right\}^{1/2}$$
$$\cdot\left\{(2\pi)^{-3}\int d\mathbf{p}'[|\langle p'|\mathsf{V}|\psi_p^+\rangle|+|\langle p''|\mathsf{V}|\psi_p^+\rangle|]^2\frac{\sin^2[(\epsilon-\epsilon')t]}{(\epsilon-\epsilon')^2}\right\}^{1/2} \tag{15.6}$$

The reader may wish to verify the fact that $(\epsilon-\epsilon')^{-2}\sin^2[(\epsilon-\epsilon')t]$ and $\pi t\delta(\epsilon-\epsilon')$ have the same integral. Baranger uses this fact to replace the former by the latter in Eq. (15.5). The optical theorem tells us that

$$(2\pi)^{-3}\int d\mathbf{p}'|\langle p'|\mathsf{V}|\psi_p^+\rangle|^2\pi\delta(\epsilon-\epsilon')=\tfrac{1}{2}v\sigma t \tag{15.7}$$

so that the second curly bracket in Eq. (15.6) is $4(\frac{1}{2}v\sigma t)$. Therefore, we now have an expression,

$$2\sqrt{\tfrac{1}{2}v\sigma t}\left[(2\pi)^{-1}\int d\mathbf{p}\left(\frac{|\langle p'|\mathsf{V}|\psi_p^+\rangle|-|\langle p''|\mathsf{V}|\psi_p^+\rangle|}{\epsilon-\epsilon'}\right)^2\right]^{1/2} \tag{15.8}$$

which must be smaller than the interruption result in order that the approximation be valid. Thus Eq. (15.8) must be smaller than

$$\mathcal{I}\langle p'|\mathsf{V}|\psi_p^+\rangle = (2\pi t/\mathsf{m})\mathcal{I}[f(0)] = (2\pi t/\mathsf{m})(p/4\pi)\sigma = \tfrac{1}{2}v\sigma t \qquad (15.9)$$

From Eqs. (15.8) and (15.9) then, we see that

$$4(2\pi)^{-3}\int d\mathbf{p}'\left(\frac{|\langle p'|\mathsf{V}|\psi_p^+\rangle| - |\langle p''|\mathsf{V}|\psi_p^+\rangle|}{\epsilon-\epsilon'}\right)^2 \ll \tfrac{1}{2}v\sigma t = N^{-1} \qquad (15.10)$$

since the t values of consequence are in the neighborhood of $t = 2/Nv\sigma$.

Equation (15.10) provides a precise criterion for interruption theory applicability. A comparison of Eqs. (15.10) and (15.3) indicates that our interaction sphere may be defined by the integral of Eq. (15.10), the criterion presented by this equation being a factor of four more stringent than was the earlier case. Practically speaking, of course, "much greater than" will have to be specifically defined if one is to deal with a particular situation. In such a case, keep in mind that the introduction of the Schwartz inequality militates toward the cancellation of the factor of four.

16 Reduction of the Baranger Theory to Previous Theories

First, let us assume the validity of the classical path approximation. This approximation, we will recall [§ 4.25 of Breene (9)], allows us to ignore the quantum induced uncertainty in perturber location and hence the perturber translational wave functions. As Baranger has phrased it, we may replace these translational functions with strongly localized wave packets. Suppose we do this in Eq. (14.7). Then Eq. (14.7) becomes

$$\varphi(t) = \langle \exp[-\mathrm{i}(\Delta\mathsf{V}t)]\rangle_{\mathrm{av}} \qquad (16.1\mathrm{a})$$

where

$$\Delta\mathsf{V} = H_i - H_f = \mathsf{V}_i - \mathsf{V}_f. \qquad (16.1\mathrm{b})$$

Now let us suppose that the broadening potential is of the form r^{-j}. If we assume straight paths, Eq. (16.1b) will become

$$\Delta V = C_i/\left(\rho_i^2 + v^2t^2\right)^{j/2} - C_f/\left(\rho_f^2 + v^2t^2\right)^{j/2} = \Delta\mathsf{V}(t) \qquad (16.2)$$

where ρ_j is the distance of closest approach, v the perturber velocity (emitter assumed stationary), and t the temporal separation of the perturber from its point of closest approach. (The reader will see that different trajectories are precisely the same in principle.)

Then what $\Delta\mathsf{V}t$ tells us is the phase shift after time, t, for $\Delta\mathsf{V}$ not a function of t. This must, for Eq. (16.2), be replaced by an integral, and we obtain the Anderson classical path result.

$$\varphi(t) = \left\langle \exp\left[-\mathrm{i}\int_0^t \Delta\mathsf{V}(t')\,dt'\right]\right\rangle_{\mathrm{av}} \qquad (16.3)$$

Lorentz-Lenz-Weisskopf theories [Chapter 2 of Breene (9)] tell us that the width of the spectral line is the reciprocal of the intercollision time. If we take l as the mean free path, ρ as the optical collision diameter and, as usual, σ as the collision cross section, it is obvious that

$$\delta = \frac{1}{\tau} = \frac{v}{l} = v(N\pi\rho^2) = Nv\sigma \tag{16.4}$$

l^{-1} being taken as $N\pi\rho^2$, σ as $\pi\rho^2$. From Eq. (14.15b) and the optical theorem, $\mathcal{I}[f(0)] = p\sigma/4\pi$, we see that the Baranger result reduces, in the interruption limit, to

$$\delta = \left(\tfrac{1}{2}Nv\sigma\right)_{\text{av}} \tag{16.5}$$

The "av", as we may show, is appropriate to Eq. (16.4) so these equations differ by the factor $\frac{1}{2}$. As Baranger points out, however, this is precisely what would be expected for complete agreement between the two. We will recall that, in the Lorentz-Lenz-Weisskopf formulation, the interaction sphere is precisely defined so that, if it is penetrated, a collision occurs; if not, there is no effect. Under these conditions the quantum cross section in the classical limit is twice the cross section of classical theory as it also contains the diffraction cross section. Thus the σ of Eq. (16.5) is twice the σ of Eq. (16.4), and the agreement is exact.

We will recall that, for a spherically symmetrical potential, the scattering amplitude may be written as

$$f(\vartheta) = \sum_l \frac{(2l+1)(e^{2i\delta_l}-1)P_l(\cos\vartheta)}{2ip} \tag{16.6}$$

For forward scattering, $\vartheta = 0$, $P_l(\cos\vartheta)$ is replaced by unity for all l, and Eqs. (14.15) become

$$\Delta = -\left[\pi N \sum_l \frac{(2l+1)\sin 2\delta_l}{mp}\right]_{\text{av}} \tag{16.7a}$$

$$\delta = \left[\pi N \sum_l \frac{(2l+1)(1-\cos 2\delta_l)}{mp}\right]_{\text{av}} \tag{16.7b}$$

which are precisely Eqs. (11) and (10) of Lindholm (181). Thus Baranger has certainly demonstrated the reduction of his theory to previous theories in the classical path, interruption approximation.

That such a theory reduces to the statistical theory for collision times long compared to intercollision times has already been demonstrated [cf. § 3.16 of Breene (9)]. Although objections might be raised as to the inconsequence of semantics, words must occasionally be used, and it is generally advisable for us to agree as to what meaning a particular word is intended to convey. For example, consider the minor controversy over whether to describe long-collision-time theories as "statistical" or "static."

Almost from the time of its inception by Margenau (188) the broadening induced by slowly moving, long-collision-time broadeners was called "statistical broadening." Since various statistical distributions of broadeners contributed to the various frequencies in the spectral line, this nomenclature seemed eminently satisfactory. Then in the middle years of this century certain authors, peering at the phenomenon through lenses of a different polarization, began to maintain that "static" not "statistical" was the proper description of such slow motion, such almost static behavior. However, as Margenau (192) pointed out, if these theories were truly static, there would be no broadened spectral line, simply a sharp line corresponding to the "static" distribution of perturbers.

17 The Truncated *S*-Matrix Expansion

As we shall have ample occasion to observe, the averaged elements of the S matrix play a role of considerable importance in spectral line broadening theory, the theories of Anderson and Baranger, Kolb and Griem being cases very much in point. The S operator is an exponential, and one would anticipate that it would be dealt with as a truncated expansion. Indeed, such an expansion proves quite satisfactory where distant or weak collisions play the preponderant part. That expansions would converge readily for such collisions is obvious from the fact that the expansion is nothing more than the time-ordered powers of the factor $\int_{\text{coll.}} \mathsf{V}\,dt$. Here V is the collision potential, and the integration is over the time of the collision. For r large and V consequently weak, this will obviously be small. In Section 18 we shall consider one method of evaluating the S matrix without reference to an expansion, a procedure which is often useful when close, strong collisions are important. In this section, however, we consider the case where the distant collisions are of consequence.

One vital area in which distant collisions are important is that of Stark broadening where long-range Coulomb forces dominate (not that van der Waals forces are not Coulomb in origin). Because Griem (18, 19) has covered the area so thoroughly, we shall not concern ourselves with Stark broadening. This omission has its unfortunate aspects among which is our consequent failure to deal with the interesting and extensive work of van Regemorter and his associates (227, 83, 84, 228). However, although the most common interest in distant collision S matrices is involved with Stark broadening, such is not the sole application of these techniques. We follow Sahal-Bréchot (245) in considering an example of such an evaluation.

The correlation function contains a TDO, $\mathfrak{U}(t,0)$, which tells us the effect of the perturbers on the radiator. Let us suppose that, during the time interval $(0, t)$, there are k binary collisions. Thus the interaction, $\hat{\mathsf{V}}$ ($=\exp[\mathrm{i}H_s t]\mathsf{V}\exp[-\mathrm{i}H_s t]$, H_s the Hamiltonian of the isolated radiator), is the sum of k pair potentials so that

$$\mathfrak{U}(t,0)=\mathsf{T}\exp\left(-\mathrm{i}\int_0^t \hat{\mathsf{V}}_1(t')\,dt'\right)\mathsf{T}\exp\left(-\mathrm{i}\int_0^t \hat{\mathsf{V}}_2(t')\,dt'\right)\cdots\mathsf{T}\exp\left(-\mathrm{i}\int_0^t \hat{\mathsf{V}}_k(t')\,dt'\right), \tag{17.1}$$

if the interruption approximation is acceptable so that what was a single TOO in front of the product may be written as a TOO in front of each exponential.

If the duration of each collision is small compared to t, we may—and Mme. Sahal does—replace $\mathfrak{U}(t,0)$ by the S matrix, $\mathfrak{U}(\infty,-\infty)$,

$$\mathfrak{U}(\infty,-\infty)=\mathsf{T}\exp\left(-\mathrm{i}\int_{-\infty}^{\infty}\hat{\mathsf{V}}_1(t)\,dt\right)\cdots\mathsf{T}\exp\left(-\mathrm{i}\int_{-\infty}^{\infty}\hat{\mathsf{V}}_k(t)\,dt\right)=S_1S_2\cdots S_k$$

so that, since the matrix element of the dipole moment,

$$\langle f|\mathbf{d}(t)|i\rangle=\mathfrak{U}_{ff}^{-1}(t,0)\mathfrak{U}_{ii}(t,0)\langle f|\mathbf{d}|i\rangle e^{-\mathrm{i}\omega_{if}t}$$

between initial and final states is desired, we have

$$\left[\mathfrak{U}_{ff}^{-1}(t,0)\mathfrak{U}_{ii}(t,0)\right]_{\mathrm{av}}=\left[S_{1_{ff}}^{-1}S_{1_{ii}}S_{2_{ff}}^{-1}S_{2_{ii}}\cdots S_{k_{ff}}^{-1}S_{k_{ii}}\right]_{\mathrm{av}} \tag{17.2}$$

to calculate.

From Eq. (2.54a) of Breene (9) we take the Lindholm expression for the probability of there being k collisions during the time interval $(0,t)$ as

$$P=(\nu t)^k e^{-\nu t}/k! \tag{17.3a}$$

Here ν is the total number of collisions per second. Now if the perturbers are treated classically so that the collision diameter is ρ, the velocity, v, the density, N, and the velocity distribution, $f(v)$, we will agree that

$$\nu=N\int vf(v)\,dv\,2\pi\rho\,d\rho \tag{17.3b}$$

From Eq. (2.54b) of Breene (9) we obtain the probability of there being n_q collisions corresponding to the interaction $\hat{\mathsf{V}}_q(t)$ as

$$P_q=k!\prod_{n_q}\left(\frac{\nu_q}{\nu}\right)^n/n_q! \tag{17.3c}$$

where $\nu_q=Nv_qf(v_q)\,dv\,2\pi\rho_q\,d\rho$. Therefore, Eq. (17.2) becomes, with the help of Eqs. (17.3),

$$\begin{aligned}\left[\mathfrak{U}_{ff}^{-1}(t,0)\mathfrak{U}_{ii}(t,0)\right]&=(\nu t)^k e^{-\nu t}\sum_{n_q}k!\prod_{n_q}(\nu_q/\nu)^n\left(S_{q_{ff}}^{-1}S_{q_{ii}}\right)^n/k!n_q!\\&=e^{-\nu t}\prod_{n_q}\exp\left[\nu_q tS_{q_{ff}}^{-1}S_{q_{ii}}\right]=\exp\left[t\sum_{n_q}\nu_q\left(S_{q_{ff}}^{-1}S_{q_{ii}}-1\right)\right]\end{aligned} \tag{17.4}$$

since $\nu = \Sigma \nu_q$. We have not yet carried out the averaging which will, under the classical path assumption, be over the optical collision diameter and velocity, so that

$$[\mathcal{U}_{ff}(t,0)\mathcal{U}_{ii}(t,0)]_{\text{av}} = \exp[-(i\Delta+\delta)t]$$

$$= \exp\left\{-tN\int vf(v)\,dv\int 2\pi\rho\,d\rho\left[1 - S_{ff}^{-1}(\rho,v)S_{ii}(\rho,v)\right]\right\} \quad (17.5a)$$

where the averaging process can quite straightforwardly be taken into the exponential, there being no quantum commutation conditions of the sort which sometimes render such procedures a problem.

We will probably recall that the angular momentum of a perturber on a straightline path and having an optical collision diameter, ρ, in collision with a stationary emitter is $\mathsf{m}\rho v$, m being the perturber mass. The quantum mechanical angular momentum is of course $\sqrt{l(l+1)}$. (Remember that $\hbar = 1$.) Therefore

$$\mathsf{m}^2\rho^2v^2 = l(l+1) \Rightarrow 2\rho\,d\rho = (2l+1)/\mathsf{m}^2v^2 \quad (17.5b)$$

and Eq. (17.5a) becomes, for this semiquantization,

$$\exp\left\{-tN\int vf(v)\,dv\,\pi\sum_l (2l+1)\frac{\left[1 - S_{ff}^{-1}(l,v)S_{ii}(l,v)\right]}{\mathsf{m}^2v^2}\right\} \quad (17.5c)$$

where the summation over l has replaced the integration over ρ. Sahal-Bréchot is now prepared to deal with the magnetic degeneracy.

The writings of and about our old friend, Samuel Johnson, are a source of endless delight, although listening to him must have been a dreadful experience. In anticipation of the conversational proclivities of Dr. Johnson, Leonhard Euler was able to rattle off *The Aeneid* from one end to another from memory and would apparently do so whenever he succeeded in entrapping an audience. Doubtless there is much with which to be impressed in such a performance, although it sounds like the sort of thing that would have driven even Publius Vergilius Maro out for a stroll about his snail farm. With all due gratefulness, therefore, for the two centuries which separate us from Euler's histrionic propensities, we once again make use of his angles.

In this case in order to relate the collision direction to the atomic axes, we presume familiarity with the three-dimensional pure rotation group (cf. 34, 46, 48). It is this group that is going to tell us how to go about developing such a relationship. One may draw intricate diagrams with angles indicated, but let us rather consider the affair qualitatively.

We are given a wave function for the atom, say $|Jm\rangle$, where m is the projection of J on the z axis of the reference frame to which this wave function is referred. Now this atomic frame may of course be oriented any way at all; indeed, in the absence of an impressed field we should expect these orientations to be purely random. In order to orient the atomic frame parallel to the collision frame, a frame defined mainly by the perturber path here, we simply apply a three-dimensional rotation operator to the wave function. As may be seen from, say, Section 11.5 of Wigner (48), such a rotation may be expressed in terms of the space-dependent set and the appropriate irreducible representation as

$$\mathsf{P}_R|nJM\rangle=\sum_{M'}|nJM'\rangle D^{(J)}_{MM'}$$

Here P_R is the rotation operator, $|nJM\rangle$ the state vector, and $D^{(J)}_{MM'}$ the MM' element in the Jth order matrix corresponding to the Jth irreducible representation of the three-dimensional pure rotation group. We specify the rotation via the set, $\{\alpha\beta\gamma\}$, of Euler angles. Then Eq. (15.27) of Wigner may be used to obtain any desired $D^{(J)}_{MM'}(\{\alpha\beta\gamma\})$. Finally, because the radiator orientations are random, there will be equal a priori probability for any particular $(\alpha\beta\gamma)$, so we integrate over these angles. We therefore obtain

$$\begin{aligned}\langle J_fm_f|[\mathbf{d}(t)]_{\text{av}}|J_im_i\rangle=&\sum_{\substack{m_1m_2m_i'\\ m_1'm_2'm_f'}}\big[e^{-i\omega_{if}t}\langle J_fm_f'|\mathcal{U}^{-1}(t,0)|J_fm_1'\rangle\\ &\cdot\langle J_fm_1|\mathbf{d}|J_im_2\rangle\langle J_im_2'|\mathcal{U}(t,0)|J_im_i'\rangle(8\pi^2)^{-1}\\ &\times\int_0^{2\pi}\int_0^{\pi}\int_0^{2\pi}d\alpha\,d\gamma\sin\beta\,d\beta D^{J_f*}_{m_f'm_f}D^{J_f}_{m_1'm_1}D^{J_i*}_{m_2'm_2}D^{J_i}_{m_i'm_i}\end{aligned}\tag{17.6}$$

Let us remark in Eq. (17.6) that the wave functions in the matrix element over **d** are not rotated; hence there are no corresponding D's under the integral sign. The reader will have realized that this is because **d** operates solely on internal radiator coordinates and hence the wave functions in its matrix element need hardly be reoriented. Next we note, say, the sum over both m_1 and m_1'. The sum over m_1 arises from simple matrix multiplication before $|J_fm_1\rangle$ is rotated, the sum over m_1' from the rotation of $|J_fm_1\rangle$.

We may use Eq. (15.27) of Wigner in order to evaluate the angular integral in Eq. (17.6). In the α and γ integrations we have

$$e^{-im_i\alpha}e^{im_1'\alpha}e^{-im_2'\alpha}e^{im_i'\alpha}$$

and

$$e^{-im_f\gamma}e^{im_1\gamma}e^{-im_2\gamma}e^{im_i\gamma}$$

respectively. Since we must have $-m_f'+m_1'-m_2'+m_i'=0$ and $-m_f+m_1-m_2+m_i=0$ in order that the integrations do not zero out, we choose $m_f'=m_1'$, $m_i'=m_2'$, and so on. We shall assume that the reader is familiar with the Wigner-Eckart theorem, which may be written as

$$\langle nJM|T_q^{(k)}|n'J'M'\rangle=(-)^{J'-M'}\langle J'kM'q|JM\rangle\langle nJ\|T^{(k)}\|n'J'\rangle/\sqrt{2J+1}\tag{17.7a}$$

$$=(-)^{J-M}\begin{pmatrix}J&k&J'\\-M&q&M'\end{pmatrix}\langle nJ\|T^{(k)}\|n'J'\rangle\tag{17.7b}$$

Here $T^{(k)}$ is a tensor operator of rank k, $T_q^{(k)}$ the qth component thereof. $\langle|\rangle$ is the vector coupling coefficient (VCC), the matrix the related $3J$ coefficient. We now substitute from Eq. (17.7b) for the matrix elements over $\mathbf{d}$ on both sides of Eq. (17.6). We are left with an equation having one Wigner coefficient on each side. We multiply through by the coefficient on the left side and apply the orthogonality relation for the Wigner coefficient in order to obtain

$$\langle J_f|[\mathbf{d}(t)]_{\mathrm{av}}|J_i\rangle = \langle J_f\|\mathbf{d}\|J_i\rangle \sum_{\substack{m_i'm_f'm_i\\ m_fM}} (-)^{-2J_f+m_i+m_i'} \begin{pmatrix} J_f & 1 & J_i \\ m_f' & M & -m_i' \end{pmatrix}$$

$$\cdot \begin{pmatrix} J_f & 1 & J_i \\ m_f & m & -m_i \end{pmatrix} e^{-\mathrm{i}\omega_{if}t} \langle J_i m_i'|\mathfrak{U}(t,0)|J_i m_i\rangle\langle J_f m_f|\mathfrak{U}^{-1}(t,0)|J_f m_f'\rangle \tag{17.8}$$

where M runs over the components of $\mathbf{d}$.

At this point we are prepared to repeat the steps beginning with Eq. (17.2). We obtain

$$\mathrm{i}\Delta+\delta = \sum_{n_q} \nu_q \sum_{\substack{m_i'm_f'm_i\\ m_fM}} (-)^{2J_f+m_i+m_i'} \begin{pmatrix} J_f & 1 & J_i \\ m_f' & M & -m_i' \end{pmatrix}\begin{pmatrix} J_f & 1 & J_i \\ m_f & M & -m_i \end{pmatrix}$$

$$\cdot\left[1-\langle J_i m_i'|S|J_i m_i\rangle\langle J_f m_f|S^{-1}|J_f m_f'\rangle\right] \tag{17.9}$$

Next we turn, with Mme. Sahal, to the $T\,(=1-S)$ rather than the S matrix. She covers a great many points, not all of which we would have any real reason for describing. Instead, we shall attempt to give a general idea of the thrust of the development. To begin with, the reader should be able to see that either the elastic ($i=j$) collision cross section or the inelastic ($i\to j$) collision cross section, both specified by σ_{ij}, will be given by

$$\sigma_{ij} = \int 2\pi\rho\, d\rho (2J_i+1)^{-1} \sum_{m_i m_j} |\langle J_i m_i|T(\rho,v)|J_j m_j\rangle|^2 \tag{17.10a}$$

in the semiclassical case, by

$$\sigma_{ij} = (\pi/\mathrm{m}^2v^2)\sum_l (2l+1)(2J_i+1)^{-1}\sum_{m_i m_j} |\langle J_i m_i|T_1|J_j m_j\rangle|^2 \tag{17.10b}$$

in the quantal case. We say that the reader should see for the following reasons. (1) The square of the T-matrix element gives us the probability of the $J_i\to J_j$ transition, this being summed over the final states, m_j, and averaged, via the factor $(2J_i+1)$, over the initial states. (2) When this probability is in turn averaged over collision diameter, it seems logical that a cross section should emerge.

It is obvious that the matrix elements of S in terms of T will take the form

$$\langle J_i m_i'|S|J_i m_i\rangle = \delta_{m_i m_i'} - \langle J_i m_i'|T|J_i m_i\rangle$$

and similarly for J_f. Thus Eq. (17.9) becomes

$$(i\Delta+\delta)=\sum_{n_q} v_q \Bigg[(2J_i+1)^{-1}\sum_{m_i}\langle J_i m_i|T|J_i m_i\rangle+(2J_f+1)^{-1}\sum_{m_f}\langle J_f m_f|T|J_f m_f\rangle$$

$$-\sum_{\substack{m_i' m_f' m_i \\ m_f M}} (-)^{2J_f+m_i+m_i'}\begin{pmatrix} J_f & 1 & J_i \\ m_f' & M & -m_i' \end{pmatrix}$$

$$\cdot\begin{pmatrix} J_f & 1 & J_i \\ m_f & M & -m_i \end{pmatrix}\langle J_i m_i'|T|J_i m_i\rangle\langle J_i m_i|T^*|J_f m_f'\rangle \Bigg] \qquad (17.11)$$

The relationship

$$2\Re\{\langle J_i m_i|T|J_i m_i\rangle\}=\sum_{m_i'}|\langle J_i m_i|T|J_i m_i'\rangle|^2+\sum_{\substack{m_j \\ j\neq i}}|\langle J_i m_i|T|J_j m_j\rangle|^2 \qquad (17.12)$$

may be used in conjunction with Eqs. (17.10) in order to show that the linewidth may be displayed as a sum of elastic and inelastic cross sections and an interference term as, of course, we will anticipate from the work of Baranger among others. Therefore, if a calculation is to be carried out, a large number of inelastic cross sections—"en pratique une dizaine suffit," says Sahal-Bréchot—and two elastic ones must be evaluated. The utilization of the Born approximation in such calculation is rejected as overestimating the true cross sections by factors varying "de 1"—which would not seem too serious—"à 10"—which would. Mme. Sahal remarks on various scattering formalisms—distorted waves, coupled equations, and so on—which, if not familiar to the reader, should hardly be rendered familiar by us. She chooses to use what she calls the semiclassical theory (SCT), remarking that its principal defect lies in the necessity for choosing a lower limit for the collision diameter, that is, a cutoff. These remarks relate to the inelastic collisions. For elastic collisions the results might be anticipated to be unsatisfactory due to the importance of collisions having small diameters, but such has apparently not been the case (83, 84).

Our purpose here is most assuredly not to compare minutely the results of using this method of computing cross sections with, say, that of Vainshtein (270, 271) or of anyone else. Although it may seem rather roundabout, we are simply looking at an example of the evaluation of the averaged S matrix by means of a truncated expansion. Now we turn to the approximations on which the SCT is based.

1 The perturber is presumed to be an electron or an ion. Then the broadening interaction is the electrostatic one,

$$\mathsf{V}=\mathsf{Z}_p e\left[(|\mathbf{r}-\mathbf{r}'|)^{-1}-r^{-1}\right] \tag{17.13}$$

where $r(xyr\vartheta\varphi)$ refers to the perturber of charge Z_p, $r'(x'y'r'\vartheta'\varphi')$ to the atomic electron referred to the atomic nucleus. (Obviously we could substitute for this approximation, given some other broadening mechanism for which distant collisions predominate.)

2 The perturber is a classical particle, its coordinates as functions of time being well specified, its trajectory a straight line if the radiator is neutral, an hyperbola if an ion. (The classical path criterion, $\mathsf{m}_p v\rho \ll 1$, is appealed to.)

3 The T matrix is expanded through second order, this perhaps being the most important of the approximations from our point of view. Surely

$$T=\mathrm{i}\int_{-\infty}^{\infty}\hat{\mathsf{V}}(t)\,dt+\int_{-\infty}^{\infty}\hat{\mathsf{V}}(t)\,dt\int_{-\infty}^{\infty}\hat{\mathsf{V}}(t')\,dt'\Rightarrow$$

$$T_{ij}=\mathrm{i}\int_{-\infty}^{\infty}e^{\mathrm{i}\omega_{ij}t}\mathsf{V}_{ij}(t)\,dt+\sum_k\int_{-\infty}^{\infty}\mathsf{V}_{ik}(t)e^{\mathrm{i}\omega_{ik}t}\,dt\int_{-\infty}^{t}\mathsf{V}_{kj}(t')e^{\mathrm{i}\omega_{kj}t'}\,dt' \tag{17.14}$$

We have already remarked on the validity condition for Eq. (17.14): weak collisions. That is, the validity condition applies to those collisions for which the first term on the right of Eq. (17.14) is "much less" than one; such collisions are those for which $\mathsf{V}(t)$ remains small during the time of collision.

We shall presume familiarity with the Legendre expansion of the inverse of the separation of two particles, both of which are referred to a common origin. Here the two particles are the atomic electron and the perturber, each referred to the nucleus, the inverse of the separation being expanded in terms of $r_<$ and $r_>$ where $r_<$ refers to the coordinate of the particle having the lesser of the two separations, $r_>$ to that having the greater. In any event, we find

$$\mathsf{V}=4\pi\mathsf{Z}_p e^2\sum_{\lambda=0}^{\infty}\sum_{\mu=-\lambda}^{\lambda}(2\lambda+1)^{-1}\mathsf{Y}_{\lambda\mu}(\vartheta,\varphi)Y^*_{\lambda\mu}(\vartheta',\varphi')\frac{r_<^{\lambda}}{r_>^{\lambda+1}}-\frac{\mathsf{Z}_p e^2}{r} \tag{17.15}$$

At this point Sahal-Bréchot makes the approximation which restricts the treatment to (more or less) distant collisions. We suppose the perturber coordinate always to correspond to $r_>$, the atomic electron to $r_<$. And at least the angular portions of the V_{ij} matrix element may be evaluated. Surely

$$\mathsf{V}_{ij}(t)=4\pi\mathsf{Z}_p e^2\sum_{\lambda=1,2}\sum_{\mu=-\lambda}^{\lambda}(2\lambda+1)^{-1}\mathsf{Y}_{\lambda\mu}(\vartheta,\varphi)r^{-(\lambda+1)}\langle i|M_{\lambda\mu}|j\rangle \tag{17.16a}$$

where

$$\langle i|M_{\lambda\mu}|j\rangle=\langle n_il_iL_iSJ_im_i|\mathsf{Y}^*_{\lambda\mu}(\vartheta',\varphi')r'^{\lambda}|n_jl_jL_jSJ_jm_j\rangle \qquad (17.16b)$$

In order to evaluate Eq. (17.16b) let us begin a necessary diversion by writing down Wigner's Eq. (17.16b) as

$$D^{(j_1)}(\alpha\beta\gamma)_{m_1'm_1}D^{(j_2)}(\alpha\beta\gamma)_{m_2'm_2}=\sum(j_1m_1'j_2m_2'|j_1j_2jm_1'+m_2') \cdot D^{(j)}(\alpha\beta\gamma)_{m_1'+m_2'm_1+m_2}(j_1j_2jm_1+m_2|j_1m_1j_2m_2) \qquad (17.17)$$

wherein the angular arguments must be the same. The reader, after referring to Wigner's Eq. (15.27), will probably agree that

$$D^{(j)}(\alpha\beta\gamma)_{m0}=(-)^n[4\pi/(2j+1)]\mathsf{Y}_{jm}(\beta\alpha)$$
$$D^{(j)}(\alpha\beta\gamma)_{0m}=[4\pi/(2j+1)]\mathsf{Y}_{jm}(\beta\alpha) \qquad (17.18)$$

From Eqs. (17.17) and (17.18) we see that

$$\mathsf{Y}_{l_1m_1}(\vartheta,\varphi)\mathsf{Y}_{l_2m_2}(\vartheta,\varphi)=\sum_{lm}[(2l_1+1)(2l_2+1)(2l+1)/4\pi]^{1/2}\begin{pmatrix} l_1 & l_2 & l\\ m_1 & m_2 & m\end{pmatrix} \cdot\mathsf{Y}^*_{lm}(\vartheta,\varphi)\begin{pmatrix} l_1 & l_2 & l\\ 0 & 0 & 0\end{pmatrix} \qquad (17.19)$$

Before proceeding, we will simplify Eq. (17.16b) considerably. For a complete treatment one would couple the angular momentum of the active electron, l_i, to that of the atomic core, L_i, then couple this to the spin, S, in order to obtain J_i. Instead, we replace $|n_il_iL_iSJ_im_i\rangle$ by $|l_im_i\rangle$ and consider only the latter. Therefore, $\langle l_im_i|\sim\mathsf{Y}_{l_i-m_i}$, and the first two factors on the right of Eq. (17.16b) are, by Eq. (17.19),

$$\mathsf{Y}_{l_i-m_i}\mathsf{Y}_{\lambda-\mu}=\sum_{lm}[(2l_i+1)(2\lambda+1)(2l+1)/4\pi]^{1/2} \times\begin{pmatrix} l_i & \lambda & l\\ -m_i & -\mu & m\end{pmatrix}\mathsf{Y}^*_{lm}\begin{pmatrix} l_i & \lambda & 1\\ 0 & 0 & 0\end{pmatrix} \qquad (17.20)$$

This leads to

$$\langle i|M_{\lambda\mu}|j\rangle=\sum_{lm}[(2l_i+1)(2\lambda+1)(2l+1)/4\pi]^{1/2}\begin{pmatrix} l_i & \lambda & l\\ -m_i & -\mu & m\end{pmatrix} \cdot\langle n_ilm|r'^{\lambda}|n_jl_jm_j\rangle\cdot\begin{pmatrix} l_i & \lambda & l\\ 0 & 0 & 0\end{pmatrix}$$
$$=\sum_{l_jm_j}\left(\frac{(2l_i+1)(2\lambda+1)(2l+1)}{4\pi}\right)^{1/2}\begin{pmatrix} l_i & \lambda & l_j\\ -m_i & -\mu & m_j\end{pmatrix} \cdot\begin{pmatrix} l_i & \lambda & l_j\\ 0 & 0 & 0\end{pmatrix}\langle n_il_j|r'^{\lambda}|n_jl_j\rangle \qquad (17.21)$$

Because Sahal-Bréchot expresses her result in terms of the Racah coefficient, we indicate how Eq. (17.21) may be so expressed. Suppose we have three electrons, each of course possessed of a particular angular momentum, and we wish to know the result of coupling the three momenta. We could use Wigner coefficients in order to couple two of the momenta, then use Wigner coefficients again in order to couple this resultant to the momentum of the third electron. It will be obvious that there are three ways in which this may be accomplished so that we will have three different vectors, all three of which, however, have the same angular momentum and the same z component of angular momentum associated with them. Therefore, we may recouple these three states in order to obtain the proper eigenvector for the description of the three coupled momenta yielding a particular resultant. The development is quite straightforward and need hardly be reproduced. Suffice it to say that the coefficients in this recoupling have been christened $6J$ symbols—for the good and sufficient reason that they have six J's in them—and arise naturally in writing all this down. They are defined by

$$\begin{pmatrix} J & \mu & m \\ M & j_3 & j \end{pmatrix}\begin{pmatrix} j & \kappa & \lambda \\ m & j_1 & j_2 \end{pmatrix} = \sum_{j'} (-)^{2j}(2j+1)\begin{Bmatrix} J & j_2 & j' \\ j_1 & j_3 & j \end{Bmatrix} \cdot \begin{pmatrix} J & \lambda & m \\ M & j_2 & j' \end{pmatrix} \cdot \begin{pmatrix} j' & \kappa & M \\ m & j_1 & j_3 \end{pmatrix} \tag{17.22a}$$

where the curly bracketed symbol is the $6J$ symbol. The Racah or W coefficient is related to the $6J$ symbol by

$$W(j_1 j_2 l_2 l_1;\, j_3 l_4) = (-)^{j_1+j_2+l_1+l_2}\begin{Bmatrix} j_1 & j_2 & j_3 \\ l_1 & l_2 & l_3 \end{Bmatrix} \tag{17.22b}$$

We could now use Eqs. (17.21) in order to rewrite Eq. (17.20) in terms of Racah coefficients, which is the form in which it was presented by Sahal-Bréchot. That author then continues the evaluation for a number of special cases, all of importance but somewhat beyond the scope of what we can hope to essay. [That line broadening theory must on occasion appeal to even higher-order symbols is illustrated by the work of Rebane (223). Rebane uses a basically interruption approach to study spectral lines having hyperfine structure and encounters the necessity for the $12J$ symbol. For the higher-order J symbols, the reader may consult Edmonds (13) and Yutsis et al. (50).] Now we turn our attention to an example of the treatment of the S matrix without appeal to its truncated expansion.

18 A Semiclassical Treatment of the S Matrix

Dillon, Smith, Cooper, and Mizushima (DSCM) (110) developed a semiclassical method of treating the S operator in its entirety. It seems reasonable to begin by defining what we are discussing:

$$S=\mathsf{T}\exp\left[-\mathrm{i}\int_{-\infty}^{\infty}\hat{\mathsf{V}}_1(t)\,dt\right] \tag{18.1a}$$

$$\hat{\mathsf{V}}_1(t)=\exp(\mathrm{i}H_0^s t)\mathsf{V}_1\exp(-\mathrm{i}H_0^s t) \tag{18.1b}$$

Let us rewrite Eq. (18.1a) slightly,

$$S=\mathsf{T}\left\{1-\mathrm{i}\int_{-\infty}^{\infty}\hat{\mathsf{V}}_1(t)\,dt\left[1-\mathrm{i}\frac{1}{2!}\int_{-\infty}^{t}\hat{\mathsf{V}}_i(t')\,dt'\right.\right.$$

$$\left.\left.-\frac{1}{3!}\int_{-\infty}^{t}\hat{\mathsf{V}}_i(t')\,dt'\int_{-\infty}^{t}\hat{\mathsf{V}}_i(t'')\,dt''+\cdots\right]\right\}$$

$$=1-\mathrm{i}\int_{-\infty}^{\infty}\hat{\mathsf{V}}_1(t)\,dt\,\mathcal{U}(t,-\infty)$$

where the numerical coefficient problem which the reader may encounter [e.g., $(2!)^{-1}$ is hardly needed in the second term of an exponential expansion] is overcome by a precise reversal of the procedure leading to the expansion. Thus, the S-matrix elements are

$$\langle i|S|j\rangle=\langle i|j\rangle-\mathrm{i}\sum_k\int_{-\infty}^{\infty}e^{\mathrm{i}\omega_{ik}t}\langle i|\mathsf{V}_1|k\rangle\langle k|\mathcal{U}(t,-\infty)|j\rangle\,dt \tag{18.2}$$

DSCM now go about eliminating the exponential in Eq. (18.2) as follows. (1) For intermediate states such that $\omega_{ik}\tau_c\gg 1$, τ_c the collision time, $\exp(\mathrm{i}\omega_{ik}t)$ will obviously oscillate with sufficient rapidity so as to zero out the integral. (2) For $\omega_{ik}\tau_c\sim 1$, the interaction is assumed to be weak, that is to say, $\langle i|\mathsf{V}_1|k\rangle\tau_c\ll 1$, and these terms are dropped. The authors feel that, although such an approximation is usually unjustifiable, it is often useful. (3) For $\omega_{ik}\tau_c\ll 1$, the exponential may be replaced by unity, and the exponential is eliminated. This state of affairs is formally indicated by the following projection operator:

$$\begin{aligned}\langle i|\mathsf{PV}_1|k\rangle&=\langle i|\mathsf{V}_1|k\rangle &&\text{for } \omega_{ik}\tau_c\ll 1\\ \langle i|\mathsf{PV}_1|k\rangle&=0 &&\text{for } \omega_{ik}\tau_c\gtrsim 1\end{aligned} \tag{18.3}$$

The potentials in use are defined as follows: V is the interaction between the radiator and the perturbers. V_0 is that part of the interaction which does not operate on radiator states. Finally, $\mathsf{V}_1=\mathsf{V}-\mathsf{V}_0$. Equation (18.3) allows us to rewrite Eq. (18.1a) as

$$S=\mathsf{T}\exp\left[-\mathrm{i}\int_{-\infty}^{\infty}\mathsf{PV}_1\,dt\right] \tag{18.4}$$

From this point these authors are going to: (1) assume the classical path and (2) eliminate time ordering by eliminating the time dependence of the operators. Intuitively at least, we can see that one should be able to do something with Eq. (18.4) after such steps have been taken. [Although our interests do not include Stark broadening per se, we might remark the study by Godfrey, Vidal, Smith, and Cooper (136) of the specific effect on certain Stark-broadened H profiles of the inclusion and noninclusion of time ordering.]

We will recall the requirements for the classical path: the perturber will reside in the interaction sphere for a period of time equal to the time of collision. Quantum mechanically, both emitter and perturber may be described by a wave packet which diffuses or extends its limits in space with time. It should be kept in mind, however, that, the heavier the particle, the slower the diffusion. Now if, at the moment of interaction sphere penetration —the commencement of collision—the separation of the collision pair is much greater than the extension of either wave packet, this separation may be described classically. (Here, as always in such situations, the vague "much greater than" enters. Is ten times greater than what "much greater than" means? If ten times yields agreement with experiment, it probably is, but only for that particular experiment.) If, during the time of perturber residence in the interaction sphere, the spatial diffusion of the wave packets is "much less than" sphere radius, the classical path description is often considered acceptable. In practice of course some have ended by equating diffusion and radius, and we have grist for an angels and pins sort of discussion. (Ibn-Roshd or Averroes felt that the intellect must be "transparent like a crystal which permits nothing to pass but the image of objects." The reader may wish to exercise his crystal on this one.)

DSCM base their choice of a potential on a review article by Buckingham (89) wherein the geometries are as given in Fig. 18.1, the potentials functions of $r(t)$, $\chi(t)$, and the molecular moments. Azimuthal symmetry is obvious from the figure. We shall simply accept the following form for the potential:

$$V_1 = \eta_0 V_0(r) + \mathsf{V}_1(r)\eta_k \bar{P}_k(\cos\chi(t)) \tag{18.5}$$

In Eq. (18.5) $\mathsf{V}_0(r)$ and $\mathsf{V}_1(r)$ are scalar quantities associated with the vibrational and rotational states of the molecule while η_0 and η_k are functions of time-independent moments which are scalar quantities vis-à-vis rotational states but generally operators on vibrational states. The bar on P_k, for example, denotes an operator on radiator states.

We have here interjected what is, with no more explanation than we shall extend, a paragraph or two of gibberish purporting to relate to a potential. Such behavior is only defensible in that: (1) we have no intention of discussing intermolecular potentials in detail. The subject is of course important, but we cannot hope to cover it in any detail. This despite the fact that many authors feel spectral line shape to be an excellent avenue of approach to these potentials. (2) There is no reason to suppose that Eq. (18.5) is not an excellent approximation to whatever the "true" potential may be. However, even should it prove to be very poor indeed, the methods that we shall describe and by means of which we shall apply it to S-matrix calculation will retain their value.

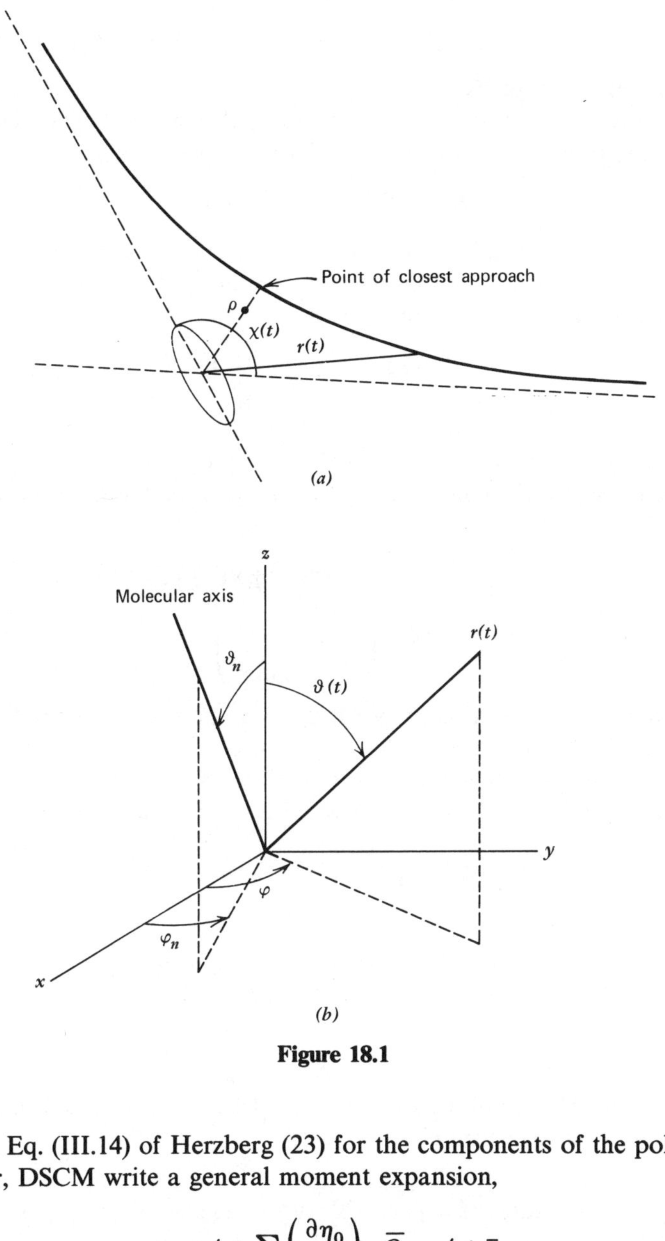

Figure 18.1

From Eq. (III.14) of Herzberg (23) for the components of the polarizability tensor, DSCM write a general moment expansion,

$$\eta_0=\eta_0'+\sum\left(\frac{\partial\eta_0}{\partial Q}\right)_{eq}\bar{Q}_r=\eta_0'+\bar{\eta}_0 \tag{18.6}$$

where η_0' is the scalar part, $\bar{\eta}_0$ the operator, and Q_r the normal vibrational coordinate. The subscript on the derivative means that it is to be evaluated at the equilibrium nuclear configuration of the molecule. The operator PV_1 is thus

$$\mathsf{PV}_1(r)=\mathsf{V}_0(r)\mathsf{P}\eta_0+\mathsf{V}_1(r)\mathsf{P}(\eta_k'+\eta_k)\mathsf{P}[\cos\chi(t)] \tag{18.7}$$

while the scalar part of V_0 is $\mathsf{V}_0(r)\eta_0'$.

Since there is azimuthal symmetry, we may write a Lagrangian for the classical scattering problem in plane polar coordinates as

$$L=\tfrac{1}{2}\mathsf{m}(\dot{r}^2+r^2\dot{\vartheta}^2)-\mathsf{V}(r)$$

where we suppose the radiator of infinite mass, the perturber of mass m, and the reader not in need of a diagram. We remark that the forces are central [$\mathsf{V}(r)$]. One of the Lagrange equations is

$$\mathsf{p}=\frac{\partial L}{\partial\dot{\vartheta}}=\mathsf{m}r^2\dot{\vartheta}\Rightarrow\dot{p}_\vartheta=\frac{d}{dt}\left(\frac{\partial L}{\partial\dot{\vartheta}}\right)=\frac{\partial L}{\partial\vartheta}=0$$
$$\mathsf{m}r^2\dot{\vartheta}=l \tag{18.8a}$$

where l is of course the constant angular momentum. On the basis of energy conservation,

$$E=\tfrac{1}{2}\mathsf{m}(\dot{r}^2+r^2\dot{\vartheta}^2)+\mathsf{V}(r)$$
$$\dot{r}=\frac{2}{\mathsf{m}}\left(E-\mathsf{V}-\frac{1^2}{2\mathsf{m}r^2}\right)^{1/2} \tag{18.8b}$$

and we call m the reduced mass of the collision pair where the radiator is no longer considered of infinite mass. Therefore, an integral over t may be changed to one over r by using Eq. (18.8b) in $dt=\dot{r}^{-1}dr$ if we suppose the classical path and central forces as DSCM do. We hence obtain

$$\int_{-\infty}^{\infty}\mathsf{PV}_1(r)\,dt=2\left(\int_{\rho_0}^{\infty}dr\frac{\mathsf{V}_0(r)}{\dot{r}}\right)\mathsf{P}\bar{\eta}_0$$
$$+\mathsf{P}\left[\int_{\rho_0}^{\infty}\mathsf{V}_1(r)\frac{\{\bar{P}_k[\cos\chi(t)]+\bar{P}_k[\cos\chi(-t)]\}}{\dot{r}}\right]dr(\eta_k'+\bar{\eta}_k) \tag{18.9}$$

where ρ_0 is the distance of closest approach. Using Fig. 18.1 and the addition theorem for spherical harmonics, we see that

$$\bar{P}_n[\cos\chi(t)]=4\pi(2k+1)^{-1}\sum_{m=-k}^{k}\mathsf{Y}_{km}^*[\vartheta(t),\varphi_c]\bar{\mathsf{Y}}_{km}(\vartheta_n,\varphi_n) \tag{18.10a}$$

$$\bar{P}_n[\cos\chi(t)]=4\pi(2k+1)^{-1}\sum_{m=-k}^{k}\mathsf{Y}_{km}^*[\vartheta_c-\vartheta(t),\varphi_c]\bar{\mathsf{Y}}_{km}(\vartheta_n,\varphi_n) \tag{18.10b}$$

At which point the following argument is utilized to eliminate the time dependence of the operators and hence the necessity for time ordering: Since $\dot{r}$ is, at r_0, going through its zero in passing from, say, negative to positive values, $\mathsf{V}_1(r)\dot{r}^{-1}$ will obviously peak there, changing rapidly in that vicinity. Therefore, since most of the important phenomena occur near r_0, for small n, $\bar{P}_n[\cos\chi(\pm t)]$ varies slowly compared to $\mathsf{V}_1(r)\dot{r}^{-1}$. Hence it is assumed that Eq. (18.9) may simply be evaluated for $\vartheta(t)=\vartheta_c$, $\chi=\chi^0$. From the orbit equation, Eq. (18.8a),

$$\vartheta(t)-\vartheta_c=\int_{r_0}^{r(t)} dr\,\frac{1}{\mathsf{m}r^2\dot{r}}$$

since the direction of initial approach was chosen parallel to the z axis $[\vartheta(\infty)=0]$, ϑ_c and hence χ^0 may be evaluated.

$$P_n(\cos\chi^0)=4\pi(2k+1)^{-1}\sum_m \mathsf{Y}^*_{km}(\vartheta_c,\varphi_c)\mathsf{Y}_{km}(\vartheta_n,\varphi_n) \tag{18.11}$$

The time dependence has thus disappeared from Eq. (18.11), and Eq. (18.4) has become

$$S=\exp\left[-\mathrm{i}K_0\mathsf{P}\bar{\eta}_0-\mathrm{i}K_1\mathsf{P}(\bar{\eta}'_k+\bar{\eta}_k)P_k(\cos\chi^0)\right] \tag{18.12a}$$

where

$$K_i=2\int_{r_0}^{\infty} dr\,\frac{\mathsf{V}_i(r)}{\dot{r}} \tag{18.12b}$$

The approximation will surely be most accurate for close or strong collisions where $\vartheta(t)$ varies slowly as compared to $\mathsf{V}_1(r)\dot{r}^{-1}$; however, DSCM state that, by comparison with the straight-path approximation—which should be quite good for distant collisions—a discrepancy of less than 25% is found for the worst case, that of weak or distant collisions.

For rotational lines $\bar{\eta}_0$ and $\bar{\eta}_k$ vanish, and S may be expanded as

$$S=\left[\exp\{-\mathrm{i}K_1\eta'_k\mathsf{P}P_k(\cos\chi^0)\}\right]=\sum_\alpha A_\alpha^{(k)}\mathsf{P}P_\alpha(\cos\chi^0) \tag{18.13}$$

By differentiating with respect to x $(=\cos\chi^0)$ DSCM are able to obtain a recursion relation for the $A_\alpha^{(k)}$, which we do not reproduce and which they do not pursue further. Since the laborer in this particular vineyard will usually be interested only in first- and second-order Legendre polynomials in the exponential for S, these cases are dealt with *in extenso* for the rotational lines under consideration. Here the orthogonality relation—multiply Eq. (18.13) through by $P_\alpha(x)$ and integrate—

$$A_\alpha^{(k)}=\tfrac{1}{2}(2\alpha+1)\int_{-1}^{1} P_\alpha(x)\exp\left[-\mathrm{i}K_1\eta'_kP_k(x)\right]dx \tag{18.14}$$

is utilized. For a first-order polynomial we obviously have

$$S=\exp\left[-\mathrm{i}K_1\eta'_1\mathsf{P}P_1(x)\right]=\sum_{\alpha=0}^{\infty} A_\alpha^{(1)}\mathsf{P}P_\alpha(x) \tag{18.15}$$

If the reader will keep the familiar relation, $xP'_\alpha(x)=\alpha P_\alpha(x)+P'_{\alpha-1}(x)$, in mind, he will be able to integrate Eq. (18.14) by parts in order to obtain

$$A_\alpha^{(1)}=A_{\alpha-2}^{(1)}\left[\frac{(2\alpha+1)}{(2\alpha-3)}\right]-\mathrm{i}A_{\alpha-1}^{(1)}\left[\frac{(2\alpha+1)}{K_1\eta'_1}\right] \tag{18.16a}$$

wherein we see that

$$A_0^{(1)}=\sin\frac{(K_1\eta'_1)}{K_1\eta'_1} \tag{18.16b}$$

$$A_1^{(1)}=3\mathrm{i}\frac{[\cos(K_1\eta'_1)-\sin(K_1\eta'_1)/K_1\eta'_1]}{K_1\eta'_1} \tag{18.16c}$$

We shall not repeat the analogous derivation for $P_2(x)$ in S. Although these recursion relations are obviously unstable for small values of $|K_1\eta'_1|$, such a region will have $|P_k(x)|\leq 1$, so that a power series expansion for S will converge rapidly. Such an expansion is obtained by DSCM for a second-order Legendre polynomial in the S operator.

We remark that the apparently infinite series on the right of Eq. (18.13) is truncated for finite values of the rotational quantum number. The vibrational situation is purported to be in complete analogy to the rotational one which we have considered above, and indeed there is no obvious reason why it should not be.

19 Enter the Density Matrix

As an introduction to the density matrix we consider the work of Cooper (104). What that author did was the following.

(1) The probability that our absorbing system—we deal here with foreign-gas broadening and hence distinguishability of emitter and perturbers—will be in a particular final state may be expressed in terms of the density matrix. This means that (2) the power absorbed from the radiation field and hence the spectral line shape may be expressed in terms of that quantity. (3) The splitting of the density matrix into the product of radiator and thermal bath (perturbers) density matrices together with the interruption approximation allows a determination of the radiator density matrix as a sum of two components: (a) a part corresponding to the preponderant interaction between collisions, that with the radiation field, and (b) a part corresponding to the preponderant interaction during collisions, that with the perturbers. (4) The interruption approximation, the assumption of frequencies near line center and the restriction to isolated spectral lines then allows the development of a difference equation, a differential equation and, finally, the solution of the latter. This leaves us with the details to be discussed.

In order to follow the development of Cooper, it behooves us to apportion out the various parcels of the Hamiltonian (H) more or less as did he, designating a particle-field part (H_F) and a particle part (H_c) which is in turn broken down into a radiator portion (H_0), a perturber portion (H_p) and a radiator-perturber interaction (V) which requires further explanation. We suppose the particle portion of the wave function—the only portion to which we shall devote any attention—to be a product, $|\varphi\rangle = |\psi\rangle|\chi\rangle$, of that for the radiator ($|\psi\rangle$) and that for the perturbers ($|\chi\rangle$). The interaction portion of the Hamiltonian (V) we shall only really encounter after it has been integrated over perturber coordinates. Of course a TDO ($\mathfrak{T}$) will correspond to the total Hamiltonian, equations and so on being familiar. Another TDO ($\mathfrak{T}_c$) can obviously be taken as corresponding to the absence of the electromagnetic field when $H \to H_c$. Therefore, if we are dealing with absorption, the field will be present in the initial state and the development of the wave function will be described by

$$|\Psi_i(t)\rangle = \mathfrak{T}(t, t_0)|\Psi_i(t_0)\rangle$$

while in the field free final state

$$|\Psi_f(t)\rangle = \mathfrak{T}_c(t, t_0)|\Psi_f(t_0)\rangle = |\varphi_f\rangle$$

Therefore, the probability for the final state at time, t, will be the absolute square of the overlap integral, $\langle\Psi_f|\Psi_i\rangle$, multiplied by the probability, p_i, for the existence of the initial state summed over initial states (the average):

$$\begin{aligned}
P_f(t) &= \sum_i p_i |\langle\Psi_f(t)|\Psi_i(t)\rangle|^2 \\
&= \sum_i \langle\Psi_f(t_0)|\mathfrak{T}_c^{-1}(t, t_0)\mathfrak{T}(t, t_0)|\Psi_i(t_0)\rangle p_i \langle\Psi_i(t_0)| \\
&\quad \cdot \mathfrak{T}^{-1}(t, t_0)\mathfrak{T}_c(t, t_0)|\Psi_f(t_0)\rangle \\
&= \sum_i \langle\varphi_f|\mathfrak{T}(t, t_0)\Upsilon(t_0)\mathfrak{T}^{-1}(t, t_0)|\varphi_f\rangle
\end{aligned}$$

by Eq. (6.2). When we recall the fashion in which a TDO transforms an operator, this becomes

$$P_f(t) = \langle\varphi_f|\Upsilon(t)|\varphi_f\rangle \tag{19.1a}$$

and, as we are well aware, the diagonal elements of the density matrix tell us the probability for the existence of the states to which they refer. To which we add the familiar

$$i\dot{\Upsilon}(t) = [H, \Upsilon(t)] \tag{19.1b}$$

Associating an energy, E_f, with each $|\varphi_f\rangle$, we see that the total probable energy of the system may be written as

$$E(t) = \sum_f E_f \langle\varphi_f|\Upsilon|\varphi_f\rangle$$

from which a rate of change of energy (power) immediately follows:

$$\dot{E}(t)=\sum_f E_f\langle\varphi_f|\dot{\Upsilon}|\varphi_f\rangle=\mathrm{i}\sum_f E_f\langle\varphi_f|H_f\Upsilon-\Upsilon H_f|\varphi_f\rangle$$

$$=-\mathrm{i}\sum_{if} E_f[\langle\varphi_f|H_f|\varphi_i\rangle\langle\varphi_i|\Upsilon|\varphi_f\rangle-\langle\varphi_f|\Upsilon|\varphi_i\rangle\langle\varphi_i|H_f|\varphi_f\rangle]$$

where now the H_f commutator has replaced that of Eq. (19.1b) since we are only interested in power absorbed from the field. Since, in general, $E_i\neq 0$ and $E_f=E_i+\omega$, ω the frequency absorbed from the field, this should be written

$$\begin{aligned}\dot{E}(t)=&-\mathrm{i}\{E_f[\langle\varphi_f|H_f|\varphi_i\rangle\langle\varphi_i|\Upsilon|\varphi_f\rangle-\langle\varphi_f|\Upsilon|\varphi_i\rangle\langle\varphi_i|H_f|\varphi_f\rangle]\\&+E_i[\langle\varphi_i|H_f|\varphi_f\rangle\langle\varphi_f|\Upsilon|\varphi_i\rangle-\langle\varphi_i|\Upsilon|\varphi_f\rangle\langle\varphi_f|H_f|\varphi_i\rangle]\}\\=&-\mathrm{i}\{(E_f-E_i)[\langle\varphi_f|H_f|\varphi_i\rangle\langle\varphi_i|\Upsilon|\varphi_f\rangle-\langle\varphi_f|\Upsilon|\varphi_i\rangle\langle\varphi_i|H_f|\varphi_f\rangle]\}\\=&\,2\omega\mathcal{I}[\langle\varphi_f(t)|H_f|\varphi_i(t)\rangle\langle\varphi_i(t)|\Upsilon|\varphi_f(t)\rangle]\end{aligned}\tag{19.2a}$$

because the matrices of H_f and Υ are Hermitian so that $\langle\varphi_f|H_f|\varphi_i\rangle=[\langle\varphi_i|H_f|\varphi_i\rangle]^*$. From Eq. (19.2a) Cooper defines a transition probability per unit time, $\dot{P}(t)$, such that $\omega=E_f-E_i$, as

$$\dot{P}(t)=2\mathcal{I}\left\{\sum_{if}\langle\varphi_f(t)|H|\varphi_i(t)\rangle\langle\varphi_i(t)|\Upsilon|\varphi_f(t)\rangle\right\}\tag{19.2b}$$

To this point the wave functions have referred to radiator plus perturbers. Now the product form for the wave function and the dilute approximation for the density matrix, $\Upsilon=\Upsilon_S\Upsilon_B$, where Υ_S refers to the radiator, Υ_B to the perturbers, are introduced. Equation (19.2b) becomes

$$\begin{aligned}\dot{P}(t)&=2\mathcal{I}\left\{\sum_{if}\langle\chi_f|\langle\psi_f|H_f|\psi_i\rangle|\chi_i\rangle\langle\chi_i|\langle\psi_i|\Upsilon_S\Upsilon_B|\psi_f\rangle|\chi_f\rangle\right\}\\&=2\mathcal{I}\left\{\sum_{if}\langle\chi_f|\langle\psi_f|H_f|\psi_i\rangle\Upsilon_B\langle\psi_i|\Upsilon_S|\psi_f\rangle|\chi_f\rangle\right\}\\&=2\mathcal{I}\left\{\sum_{if}\langle\langle\psi_f(t)|H_f|\psi_i(t)\rangle\langle\psi_i(t)|\Upsilon_S|\psi_f(t)\rangle\rangle_{\mathrm{av}}\right\}\end{aligned}\tag{19.3a}$$

where the average is obviously over perturber density matrix and states, and where now

$$\mathrm{i}|\dot{\psi}(t)\rangle=[H_0+\mathsf{V}]|\psi(t)\rangle\tag{19.3b}$$

We will recall that the radiator-perturber potential has already been integrated over $|\chi\rangle$ in order to obtain V.

The introduction of the interruption approximation is used by Cooper in order to replace the eigenvectors of Eq. (19.3a) by the eigenvectors, $|i\rangle$, of the unperturbed radiator Hamiltonian, H_0, the basis for the replacement being that, most of the time, the radiator exists as a free, noninteracting atom for which $\mathsf{V}=0$ in Eq. (19.3b). Equation (19.3a) becomes

$$\dot{P}(t)=2\mathcal{I}\left\{\sum_{if}\langle H_{f_{fi}}\Upsilon_{S_{if}}\rangle_{\text{av}}\right\} \tag{19.4}$$

and we remark that, since only off-diagonal elements of H_F exist, only off-diagonal elements of Υ_S will be required.

The next step is to obtain the density matrix, which Cooper effects as follows. Strong collisions are well separated in time by the admissibility of the interruption approximation. Then the radiator is either (1) between such collisions when the particle-field interaction (H_F) dominates and the von Neumann equation may be taken for the density matrix or (2) undergoing such a collision at which time an expression for Υ_A may be obtained ignoring H_F. To a reasonable level of approximation, then, it may be supposed that the density matrix is a sum of these two solutions. We begin following through this development by writing the density matrix and the TDO in the particular form of the interaction representation favored by Cooper,

$$\Upsilon_S'(t)=\exp[\mathrm{i}H_0(t-t_0)]\Upsilon_S(t)\exp[-\mathrm{i}H_0(t-t_0)] \tag{19.5a}$$

$$\mathfrak{U}_S(t,t_0)=\exp[\mathrm{i}H_0(t-t_0)]\mathfrak{T}_S(t,t_0) \tag{19.5b}$$

where

$$\mathfrak{T}(t,t_0)=\mathfrak{T}_S(t,t_0)\mathfrak{T}_B(t,t_0) \tag{19.5c}$$

Again we have perforce supposed the radiator-perturbers interaction ignorable in writing the product form for the TDO. The operators H_F and V are also written in the interaction representation. In this representation the von Neumann equation will take the form,

$$\dot{\Upsilon}_S'=\mathrm{i}[\Upsilon_S'(\mathsf{V}'+H_F')-(\mathsf{V}'+H_F')\Upsilon_S'] \tag{19.6}$$

which becomes

$$\dot{\Upsilon}_S'=\mathrm{i}[\Upsilon_S'H_F'-H_F'\Upsilon_S'] \tag{19.7}$$

between the strong collisions. In general, of course,

$$\mathrm{i}\dot{\mathfrak{T}}_S(t,t_0)=[H_0+H_F+\mathsf{V}]\mathfrak{T}_S(t,t_0) \tag{19.8a}$$

so that

$$\mathrm{i}\dot{\mathfrak{U}}_S(t,t_0)=[\mathsf{V}'+H_F']\mathfrak{U}_S(t,t_0) \tag{19.8b}$$

whose iterative solution will surely be

$$\mathfrak{U}_S(t,t_0)=\exp[\mathrm{i}H_0(t'-t_0)]\mathfrak{U}(t-t',0)\exp[-iH_0(t'-t_0)]\mathfrak{U}_S(t',t_0) \tag{19.9a}$$

where

$$\mathfrak{U}(t-t',0)=\left[1-\mathrm{i}\int_0^{t-t'}\mathsf{V}'(t_1)\,dt_1+\mathrm{i}^2\int_0^{t-t'}\mathsf{V}'(t_1)\,dt_1\int_0^{t_1}\mathsf{V}'(t_2)\,dt_2+\cdots\right] \tag{19.9b}$$

$$\mathsf{V}'(t)=\exp[\mathrm{i}H_0t]\mathsf{V}(t)\exp[-\mathrm{i}H_0t] \tag{19.9c}$$

for a strong collision commencing at time t'. If this collision is completed at time, $t'+\tau$, we see that

$$\Upsilon_S'(t'+\tau)=\mathcal{U}_S(t'+\tau,t_0)\Upsilon_S'(t_0)\mathcal{U}_S^{-1}(t'+\tau,t_0)$$
$$=\exp[\mathrm{i}H_0(t'-t_0)]\mathcal{U}(\tau,0)\Upsilon_S(t')\mathcal{U}^{-1}(\tau,0)\exp[-\mathrm{i}H_0(t'-t_0)] \tag{19.10}$$

by Eqs. (19.9). We shall often encounter the argument that if a collision lasts for a period, say, $t'\le t\le t'+\tau$, and if integration includes the entire period, then the extension of the temporal limits to $-\infty$ and $+\infty$ does not affect the result (the temporal extension argument). Since we will be dealing with a V describing a single collision, such, for Eq. (19.9b), will obviously be the case. However, the limit extension is important here because it allows us to go from the TDO to the S matrix as $\mathcal{U}(+\infty,-\infty)\equiv S$. Using this fact in conjunction with Eqs. (19.5a) and (19.10) yields

$$\Upsilon_S(t'+\tau)=\exp[-\mathrm{i}H_0\tau]S\Upsilon_S(t')S^{-1}\exp[\mathrm{i}H_0\tau]$$
$$\Upsilon_S(t'+\tau)_{if}=\exp[-\mathrm{i}\omega_{if}\tau]\sum_{jk}S_{ij}\Upsilon_S(t')_{jk}S_{kf}^{-1} \tag{19.11}$$

where the matrix elements are taken over the eigenvectors of H_0.

Having obtained the during-collision solution as Eq. (19.11), we simply add the between-collision solution from Eq. (19.7) in order to obtain the Cooper approximation to the density matrix

$$\Upsilon_S'(t'+\tau)\doteq\mathrm{i}\int_{t'}^{t'+\tau}[\Upsilon_S'(t')H_F'(t)-H_F'(t)\Upsilon_S'(t')]\,dt$$
$$+\exp[\mathrm{i}H_0(t'-t_0)]S\Upsilon_S(t')S^{-1}\exp[-\mathrm{i}H_0(t'-t_0)] \tag{19.12}$$

If the unperturbed Hamiltonian for the radiator is considerably larger than that for its interaction either with the field or the perturbers, the radiator density matrix will of course be expressible as

$$\Upsilon_S=e^{-H_0/kT}/Z_s,\qquad Z_s=\mathrm{Tr}\{e^{-H_0/kT}\} \tag{19.13}$$

Thus only the diagonal elements of Υ_S will exist. Between collisions when the first term of Eq. (19.12) holds, $\mathrm{V}\ll H_0$ and may be ignored. We may suppose $H_F<H_0$ but large enough so that it will induce off-diagonal elements of Υ_S in a perturbative fashion. Therefore, the matrix element of Eq. (19.7) will be

$$\dot{\Upsilon}_{S_{if}}'=\mathrm{i}\sum_j\left[\Upsilon_{S_{ij}}'H_{jf}'-H_{ij}'\Upsilon_S'\right]_{jf}\doteq\mathrm{i}\left[\Upsilon_{S_{ii}}-\Upsilon_{S_{ff}}\right]H_{if}' \tag{19.14}$$

since $\Upsilon_{S_{ii}}'=\Upsilon_{S_{ii}}$ and so on. Now we will recall that the vector field potential entering H_F will introduce a factor $\exp(\mathrm{i}\omega t)$ so that H_{if}' may be written $H_{F_{if}}'=\hat{H}_{F_{if}}'\exp(\mathrm{i}\omega t)$, $H_{fi}=\hat{H}_{fi}\exp(-\mathrm{i}\omega t)$. Keeping in mind that

$$\langle i|e^{\mathrm{i}H_0t}\Upsilon_S e^{-\mathrm{i}H_0t}|f\rangle=e^{-\mathrm{i}\omega_{fi}t}\Upsilon_{S_{if}}$$

we integrate both sides of Eq. (19.14) between t and $t+\Delta t$, using Eq. (19.5a) and an analogous equation for H_F' in order to obtain

$$\Upsilon_S(t+\Delta t)_{if} = C_{if}\exp[\mathrm{i}\omega(t+\Delta t)]\{1-\exp[-\mathrm{i}(\omega-\omega_{fi})\Delta t\}/\mathrm{i}(\omega-\omega_{fi}) + \Upsilon_S(t)\exp(\mathrm{i}\omega_{fi}\Delta t) \tag{19.15a}$$

where

$$C_{if} = \mathrm{i}\hat{H}_{F_{if}}(\Upsilon_{S_{ii}} - \Upsilon_{S_{ff}}) \tag{19.15b}$$

Equation (19.15a) may now be used for the immediate obtention of

$$\Delta(H_{F_{ji}}\Upsilon_{S_{if}})_{\mathrm{rad}} = H_{F_{ji}}(t+\Delta t)\Upsilon_{S_{ij}}(t+\Delta t) - H_{F_{ji}}(t)\Upsilon_{S_{ij}}(t)$$
$$= \hat{H}_{F_{ji}}\{[C_{if}/\mathrm{i}(\omega-\omega_{fi})] - \Upsilon_{S_{if}}(t)\exp(-\mathrm{i}\omega t)\}\{1-\exp[-\mathrm{i}(\omega-\omega_{fi})t]\} \tag{19.16a}$$

In like manner, the second term in Eq. (19.12) tells us that

$$\Delta(H_{F_{fi}}\Upsilon_{S_{if}})_{\mathrm{coll}} = \hat{H}_{F_{fi}}e^{-\mathrm{i}\omega t}\left\{e^{-(\omega_{fi}-\omega)\tau}\sum_{jk} S_{ij}\Upsilon_{S_{jk}}(t)S_{kf}^{-1} - \Upsilon_{S_{if}}(t)\right\} \tag{19.16b}$$

where it is supposed that Δt is larger than but comparable to the collision time τ.

Taking the collision frequency for a collision having a particular set of collision parameters as $d\nu$, the probability that such a collision will occur during Δt is obviously $d\nu\Delta t$. Further, if T is the time between strong collisions, $(1-\Delta t/T)$ will surely be the probability that an atom has not undergone a strong collision during Δt. The former probability will naturally be associated with the collisional change in $H_F\Upsilon_S$ [Eq. (19.16b)], the latter with the radiational [Eq. (19.16a)], so that combining these equations and probabilities will yield a total change in that quantity

$$\Delta(H_{F_{fi}}\Upsilon_{S_{if}})_{\mathrm{tot}} = \Delta(H_{F_{fi}}\Upsilon_{S_{if}})_{\mathrm{rad}}(1-\Delta t/T) + \int \Delta(H_{F_{fi}}\Upsilon_{S_{if}})_{\mathrm{coll}}\, d\nu\,\Delta t \tag{19.17}$$

We apparently have the ingredients for a difference equation that, after a fashion of many years standing, is to be metamorphosed into a differential equation. Anderson more or less originated this technique using arguments involving the interruption approximation, and the author of the paper under consideration did not deviate from this precedent. First we say (1) $\Delta t \sim \tau$, the collision time. Now we demand small collision times, $\tau \ll T$, and (2) frequencies near line center, $(\omega_{fi}-\omega)\Delta t \ll 1$. Condition (2) replaces the curly bracketed term on the right of Eq. (19.16a) by $\mathrm{i}(\omega-\omega_{fi})\Delta t$, and we have the factor Δt multiplying the entire right side. Further, condition (1) cancels the factor $\Delta t/T$ in Eq. (19.17), the right side of this latter equation now being multiplied by Δt when Eqs. (19.16) are substituted into it. Remarking that condition (2) cancels the exponential preceding the summation sign in Eq. (19.16b), we obtain the desired differential equation:

$$\frac{d}{dt}(H_{F_{fi}}\Upsilon_{S_{if}})=\frac{\Delta}{\Delta t}(H_{F_{fi}}\Upsilon_{S_{if}})=\mathrm{i}\hat{H}_{F_{fi}}e^{-\mathrm{i}\omega t}\Upsilon_{S_{if}}(t)(\omega_{fi}-\omega)+C_{if}\hat{H}_{F_{fi}}$$
$$+\hat{H}_{F_{fi}}e^{-\mathrm{i}\omega t}\int\left\{\sum_{jk}S_{ij}\Upsilon_{S_{jk}}(t)S_{kf}^{-1}-\Upsilon_{S_{if}}(t)\right\}_{\text{ang av}}d\nu \tag{19.18}$$

(For purposes of indexing we call the procedure the differential interruption transformation.)

If a steady state exists, the ensemble average, to which $(H_{F_{fi}}\Upsilon_{S_{if}})$ surely corresponds, will hardly vary with time, so that the left side of Eq. (19.18) will be zero. Now at least one way that such a result may be arranged is by supposing that

$$\begin{aligned}\Upsilon_{S_{if}}&=\hat{\Upsilon}_{if}e^{\mathrm{i}\omega t}\\ \Upsilon_{S_{fi}}&=\hat{\Upsilon}^{*}_{if}e^{-\mathrm{i}\omega t}\end{aligned}, \qquad E_f>E_i \tag{19.19}$$

so that the $\exp(-\mathrm{i}\omega t)$ present in H_F is canceled by the time factor of Eq. (19.19), the resulting $\hat{H}_{F_{fi}}\hat{\Upsilon}_{if}$ then being time independent. The diagonal elements of Υ_S are time independent.

Next, the zeroing out of the left side of Eq. (19.18) will mean that

$$\frac{d}{dt}(H_{F_{fi}}\Upsilon_{S_{if}})=-\mathrm{i}\omega H_{F_{fi}}\Upsilon_{S_{if}}+H_{F_{fi}}\left[\frac{d}{dt}(\Upsilon_{S_{if}})_{\text{rad}}+\frac{d}{dt}(\Upsilon_{S_{if}})_{\text{coll}}\right]=0 \tag{19.20}$$

By comparing the right sides of Eqs. (19.18) and (19.20) we may sort out the following:

$$(H_{F_{fi}})\frac{d}{dt}(\Upsilon_{S_{if}})_{\text{coll}}=\hat{H}_{F_{fi}}\int\left\{\left(\sum_{jk>j}S_{ij}\hat{\Upsilon}_{jk}S_{kf}^{-1}-\hat{\Upsilon}_{if}\right)\right.$$
$$\left.+e^{-\mathrm{i}\omega t}\sum_{j}S_{ij}\Upsilon_{S_{jj}}S_{jf}^{-1}+e^{-2\mathrm{i}\omega t}\sum_{jk<j}S_{ij}\hat{\Upsilon}^{*}_{jk}S_{kf}^{-1}\right\}_{\text{ang av}}d\nu \tag{19.21}$$

where $j>k\Rightarrow E_j>E_k$.

Time averaging will, due to the presence of the exponential factors, render the second and third terms within the curly bracket of Eq. (19.21) negligible compared to the first, and the random nature of the collisions implies the calculation of such an average. We therefore retain only the first term—for which $E_k>E_j$—and write the result as

$$(H_{F_{fi}})\frac{d}{dt}(\Upsilon_{S_{if}})_{\text{coll}}=-\hat{H}_{F_{fi}}\hat{\Upsilon}_{if}\Gamma_{if}+\hat{H}_{F_{if}}\sum_{\substack{j<k\\ j\neq i,\,k\neq f}}\Upsilon_{jk}\vartheta_{jk}^{if} \tag{19.22a}$$

where

$$\Gamma_{if}=\int\{1-S_{ii}S_{ff}^{-1}\}_{\text{ang av}}\,dv \tag{19.22b}$$

and

$$\hat{\Upsilon}_{jk}\vartheta_{jk}^{if}=\int\{S_{ij}\hat{\Upsilon}_{jk}S_{kf}^{-1}\}_{\text{ang av}}\,dv \tag{19.22c}$$

We simply pick out the term involving $(\Upsilon_{S_{if}})_{\text{rad}}$ from Eq. (19.18) and substitute it together with Eq. (19.22a) into Eq. (19.20) in order to obtain

$$\hat{H}_{F_{fi}}\hat{\Upsilon}_{if}[i(\omega-\omega_{fi})+\Gamma_{if}]=\mathrm{i}\hat{H}_{F_{fi}}\hat{H}_{F_{if}}(\Upsilon_{S_{ii}}-\Upsilon_{S_{ff}})+\hat{H}_{F_{fi}}\sum_{\substack{j<k\\ j\neq i,\,k\neq f}}\hat{\Upsilon}_{jk}\vartheta_{jk}^{if} \tag{19.23}$$

The solution of the set of equations implied by Eq. (19.23) is rendered somewhat less tedious by supposing the spectral line corresponding to it well separated from those represented by jk. Then from, say, Eq. (19.15a) we see that $\hat{\Upsilon}_{if}\gg\hat{\Upsilon}_{jk}$, and the last term on the right of Eq. (19.23) may be dropped. Obviously then

$$\hat{H}_{F_{fi}}\hat{\Upsilon}_{if}=\mathrm{i}\hat{H}_{F_{fi}}\hat{H}_{F_{if}}(\Upsilon_{S_{ii}}-\Upsilon_{S_{ff}})/[\mathrm{i}(\omega-\omega_{fi})+\Gamma_{if}] \tag{19.24}$$

so that, from Eq. (19.4)

$$I(\omega)=2|H_{F_{fi}}|^2(\Upsilon_{S_{ii}}-\Upsilon_{S_{ff}})\frac{\Delta}{(\omega-\omega_{if}+\Delta)^2+\delta^2} \tag{19.25a}$$

$$\Gamma_{if}=\Delta+\mathrm{i}\delta \tag{19.25b}$$

With this we conclude our density matrix manipulations under the aegis of the assumptions that (1) the density matrix is separable, (2) the interruption approximation is valid, and (3) the spectral line is isolated.

20 The Boltzmann Operator at Low Frequencies

Since the density matrix is in essence the matrix of the Boltzmann operator, our first serious brush with the density matrix may logically be followed by certain ruminations on the Boltzmann operator. Here we enter the arena of Van Vleck and various of his collaborators, an arena that we shall not penetrate far, it being mostly concerned with the solid state. However, certain of the ideas that emerge are important to the gaseous state, and we shall touch briefly on these.

As we have pointed out, Huber and Van Vleck (147) have given a derivation for the imaginary part of the susceptibility, our Eq. (12.10b), which depends on the existence of the Boltzmann operator. For our purposes, the susceptibility is the spectral line shape. A distinguishing feature of this result is the hyperbolic tangent in front of the integral sign,

$$\tanh(\omega/2kT)=\left[e^{\omega/2kT}-e^{-\omega/2kT}\right]/\left[e^{\omega/2kT}+e^{-\omega/2kT}\right]$$

For ω very small, that is, for the microwave or lesser frequencies of the spectrum, this factor may not be negligible. For ω large, on the other hand, the negative exponentials zero out, and the tanh goes to unity. Therefore, we would expect the Boltzmann operator to loom important, say, in the microwave, a statement which could possibly be modified by extreme values of the temperature.

So what do Van Vleck and Huber do with Eq. (12.10b)? Basically, they consider the special case of a two-level system, one level corresponding to spin up, the other to spin down, and inquire as to the shape of the spectral line radiated by such a system under various circumstances. Such a specific physical situation is probably not of much interest to us; at least we shall use that thought as an excuse for not following their treatment in detail. Instead, we discuss what we hope is informative and certainly is a qualitative description of their development. We begin by recalling that

$$\begin{aligned}\langle i|M_\alpha(t)|j\rangle &= \langle i|e^{iHt}M_\alpha(0)e^{-iHt}|j\rangle\\ &= \sum_k \langle i|e^{iHt}|k\rangle\langle k|M_\alpha(0)|1\rangle\langle 1|e^{-iHt}|j\rangle\\ &= e^{i\omega_0 t}\langle i|M_\alpha(0)|j\rangle, \qquad \omega_0=E_i-E_j\end{aligned}$$

if H is diagonal in $|i\rangle$ (as we suppose). Therefore.

$$\begin{aligned}\langle M_\alpha(0)M_\alpha(t)+M_\alpha(t)M_\alpha(0)\rangle &= \mathrm{Tr}\left\{e^{-H/kT}\left[M_\alpha(0)M_\alpha(t)+M_\alpha(t)M_\alpha(0)\right]\right\}\\ &= 2|M_\alpha|^2_{\mathrm{av}}\cos(\omega_0 t) \qquad (20.1\mathrm{a})\end{aligned}$$

where

$$|M|^2_{\mathrm{av}}=\left[e^{-E_i/kT}+e^{-E_j/kT}\right]|\langle i|M_\alpha(0)|j\rangle|^2 \qquad (20.1\mathrm{b})$$

To this point the fact that collisions occur and, presumably, cause a broadening of the spectral line has not entered our considerations. We now introduce collisions by introducing a damping factor, $\exp[-|t|/\tau]$ into Eq. (12.10b). This is quite straightforward based on the following argument.

The fact of collisions will introduce a mean lifetime, τ_+, in the upper state and hence into the wave functions, a mean lifetime, τ_-, into the lower state. Without concerning ourselves overmuch with how these will combine, we simply presume that they will produce τ. This allows us to write the imaginary part of the susceptibility as

$$\chi''_{\alpha\alpha}(\omega) = \text{const} \int_{-\infty}^{\infty} \cos(\omega_0 t)\cos(\omega t) e^{-|t|/\tau} dt \tanh(\omega/2kT)$$

$$= \text{const} \tanh(\omega/2kT) \left[\frac{1/\tau}{(\omega-\omega_0)^2+(1/\tau)^2} + \frac{1/\tau}{(\omega+\omega_0)^2+(1/\tau)^2} \right] \tag{20.2}$$

Huber and Van Vleck arrived at the same line shape as Kronig (165), Van Vleck and Weisskopf (274), Fröhlich (125), Karplus and Schwinger (159), Garstens (129), and so on. As Huber and Van Vleck point out, although Van Vleck (272) and Karplus and Schwinger (159) used the factor, $(\omega/\omega_0)\tanh(\omega_0/2kT)$, in place of $\tanh(\omega/2kT)$ above, the latter is undoubtedly correct. The correct factor had been known from its use in the fluctuation dissipation theorem which had been adapted to the quantum mechanical case by Callen and Welton (94) in connection with the study of Johnson noise in circuits.

5

Liouville Operators and Liouville Spaces

In a paper enlighteningly entitled "Pressure Broadening as a Prototype of Relaxation," Fano (115) introduced the space of the Liouville operator to spectral line broadening theory. This work is immensely valuable for the concepts and techniques that it ushered in, but it is only just to say that it served mainly as the harbinger of contact which was to be established with experiment by Ben-Reuven (67–71), Smith and Hooper (248), Fiutak (119–121), Czuchaj (105–107), and others, Fano's efforts concluding—or perhaps his interest flagging—with an expression of awesome unwieldiness, the justly renowned Eq. (55). In a series of papers culminating in a masterly review, Ben-Reuven reduced, as we shall see, the ravening lion of Eq. (55) to a very compact lamb. (See also Fiutak and Czuchaj and remark that the Ben-Reuven result should perhaps be modified somewhat for higher-than-binary collisions.) We could, and with considerable justification, simply reproduce Ben-Reuven's refined development, but, in doing so, we would lose touch with certain, perhaps not as sophisticated but surely educational, facets of Fano's development. Therefore, we shall first consider essentially the early and middle portions of Fano's work and conclude, albeit not immediately, with that of Ben-Reuven.

Between the efforts of these two authors we shall sandwich the work of Royer whose approach, although couched in the language of Liouville space, is sufficiently unique to demand presentation. Here too a density expansion presents certain interesting features.

It has been not unusual for writers on spectral line broadening to discuss many perturbers and deal with one, that is, with the binary collision. We may recall that one of the earliest studies of the ternary collision was that of Margenau (191) for the rather special case of three linear vibrators with mirror potentials placed at the vertices of an equilateral triangle. We conclude our studies in the present chapter with the work of Czuchaj on the ternary collision, his work being carried out within the framework of the Liouville formulation.

Although we shall have occasion to mention two of the so-called "unified" line broadening theories in Sections 50, neither of which could be described as Liouvillian, several of them are Liouvillian theories and might consequently be considered as falling within the purview of this chapter. Although Lee (177) might accuse us of triviality, we may conceive of a unified theory as one that reduces to the interruption and statistical results in the appropriate limits. Such theories are of course not new, but perhaps the first Stark-directed attempt using the Liouvillian formulation of such a theory was made by Smith, Cooper, and Vidal (249). It was followed by the work of Voslamber (276, 277), Lee (177), and Bottcher (81) among others. We shall not consider these theories in detail, not only because of their Stark orientation, but also because of our consideration of much the same general formulations in what is to follow.

21 The Liouville Approach with Separable Density Matrix

With Fano, let us begin our considerations with Eq. (13.7) slightly rewritten as

$$I(\omega)=\pi^{-1}\mathcal{R}\left\{\int_0^{\omega} e^{i\omega t}\mathrm{Tr}[\mathsf{d}\mathsf{d}(t)\Upsilon]\,dt\right\} \tag{21.1}$$

Transformation of the time-dependent dipole moment to the Heisenberg representation yields an expression,

$$\mathrm{Tr}[\Upsilon\mathsf{d}\mathsf{d}(t)]=\mathrm{Tr}\{\mathsf{d}e^{-iHt}(\Upsilon\mathsf{d})e^{iHt}\} \tag{21.2}$$

wherein the exponentials may quite properly be regarded as a TDO transforming the operator $\Upsilon\mathsf{d}$. Fano (113) had earlier discussed the value of regarding the elements of Υ as vector components in a product space that is familiar. When Υ is so constituted, the rotational transformation effect of the TDO's is emphasized as a consequence. As we have seen in Section 6, the density matrix, Υ, will satisfy an equation of motion involving the Liouville operator,

$$\dot{\Upsilon}=-i\mathfrak{L}\Upsilon \tag{21.3}$$

From a development analogous to that following Eq. (8.15) we may rewrite Eq. (21.2) as

$$\mathrm{Tr}\{\mathsf{d}\mathsf{d}(t)\Upsilon\}=\mathrm{Tr}\{\mathsf{d}e^{-i\mathfrak{L}t}(\Upsilon\mathsf{d})\} \tag{21.4}$$

When Eq. (21.4) is substituted into Eq. (21.1), we obtain

$$I(\omega)=-\pi^{-1}\mathcal{I}\{\mathrm{Tr}[\mathsf{d}(\omega-\mathfrak{L})^{-1}(\Upsilon\mathsf{d})]\} \tag{21.5}$$

One must, of course, bear in mind the operator nature of $\mathfrak{L}$, but such considerations should by now be familiar to us. Fano remarks that "$\cdots(\omega-\mathfrak{L})^{-1}$ plays the role of the resolvant operator$\cdots$"

Now apparently Eq. (21.5) contains all one needs to know about spectral line broadening when its description is cast in Zwanzig's Liouville operator form (293). Certainly we seem to have introduced no approximations save that of the dipole. Before reproducing whatever portions of Fano's paper it is our intention to reproduce, it might be well for us to ask just what it was that Fano did with this equation.

First of all, he began with the approximation that the radiating molecule, which he called the "system," could be distinguished from the remaining molecules (or whatever they might happen to be), which he called the "thermal bath." The Hamiltonian can then be written as

$$H = H_0^{(s)} + H_0^{(b)} + \mathsf{V} \tag{21.6a}$$

V referring to the interaction between system and bath, the Liouville then expressible as

$$\mathfrak{L} = \mathfrak{L}_0 + \mathfrak{L}_1 \tag{21.6b}$$

$$\mathfrak{L}_0 = H_0^{(s)} I^* - I H_0^{(s)*} + H_0^{(b)} I^* - I H_0^{(b)*} \tag{21.6c}$$

$$\mathfrak{L}_1 = \mathsf{V} I^* - I \mathsf{V}^* \tag{21.6d}$$

As Ben-Reuven (67) was to point out later, the system need not be restricted to a single radiating molecule; indeed, it must not be when we are dealing with induced spectra or self broadening, for example.

In the midst of certain operator manipulations that follow, Fano then introduced an approximation to the effect that the density matrix is effectively independent of the system-bath interaction so that

$$\Upsilon = \Upsilon^{(s)} \Upsilon^{(b)} \tag{21.7}$$

Such a density matrix assumption is rather important to this and several subsequent treatments [cf. Smith and Hooper (248)] which utilize it, although Ben-Reuven (71) showed that its introduction is often unnecessary.

The operator manipulations bring to light an operator which has been called the "effective interaction tetradic" (EIT) and is of importance, not because of its resounding title, but because it embodies the influence of the bath on the system and is hence the quantity of principal interest to line broadening considerations. This EIT is then expanded in powers of the gas density, the remainder of the development being a matter of operator manipulations and contours in one or more complex planes. We now return to Eq. (21.5) and insert Eq. (21.6b) into it. We let

$$(\omega - \mathfrak{L}_0 - \mathfrak{L}_1)^{-1} = \mathfrak{A} \Rightarrow \mathfrak{A}^{-1} = \omega - \mathfrak{L}_0 - \mathfrak{L}_1$$

Multiplication through on the left by $(\omega - \mathfrak{L}_0)^{-1}$, then by $[1 - (\omega - \mathfrak{L}_0)^{-1}\mathfrak{L}_1]$ and on the right by $\mathfrak{A}$ leads to an expression for $\mathfrak{A}$ which may be modified as follows:

$$\mathfrak{A}=\left[1-(\omega-\mathfrak{L}_0)^{-1}\mathfrak{L}_1\right]^{-1}(\omega-\mathfrak{L}_0)^{-1}=\left[1-(\omega-\mathfrak{L}_0)^{-1}\mathfrak{L}_1\right]^{-1}(\omega-\mathfrak{L}_0)^{-1}$$
$$+(\omega-\mathfrak{L}_0)^{-1}\mathfrak{L}_1\left[1-(\omega-\mathfrak{L}_0)^{-1}\mathfrak{L}_1\right]^{-1}(\omega-\mathfrak{L}_0)^{-1}$$
$$-(\omega-\mathfrak{L}_0)^{-1}\mathfrak{L}_1\left[1-(\omega-\mathfrak{L}_0)^{-1}\mathfrak{L}_1\right]^{-1}(\omega-\mathfrak{L}_0)^{-1}$$
$$=\frac{1}{\omega-\mathfrak{L}_0}\left[1+\mathfrak{M}(\omega)\frac{1}{\omega-\mathfrak{L}_0}\right] \tag{21.8a}$$

where

$$\mathfrak{M}(\omega)=\mathfrak{L}_1\frac{1}{1-(\omega-\mathfrak{L}_0)^{-1}\mathfrak{L}_1} \tag{21.8b}$$

It may likewise be shown that

$$\mathfrak{M}(\omega)=\mathfrak{L}_1\left[1-(\omega-\mathfrak{L}_0)^{-1}\mathfrak{L}_1\right]^{-1}$$
$$=\mathfrak{L}_1+\mathfrak{L}_1(\omega-\mathfrak{L}_0-\mathfrak{L}_1)^{-1}\mathfrak{L}_1=\mathfrak{L}_1\sum_{n=0}^{\infty}\left[\frac{1}{\omega-\mathfrak{L}_0}\mathfrak{L}_1\right]^n \tag{21.8c}$$

The following simplification of the density matrix is assumed:

$$\Upsilon=\Upsilon^{(s)}\Upsilon^{(b)}, \qquad \Upsilon^{(b)}_{\alpha\beta}=f_\alpha\delta_{\alpha\beta} \tag{21.9}$$

We take the eigenvalues of $H_0^{(b)}$, the unperturbed bath Hamiltonian, as $\epsilon_\alpha, \epsilon_\beta, \ldots,$ those of $H_0^{(s)}$, the unperturbed system Hamiltonian as $\epsilon_\mu, \epsilon_\nu, \ldots$ Then the matrix elements of $\mathfrak{L}_0$ will be given by

$$\langle\mu\alpha,\nu\beta|\mathfrak{L}_0|\mu'\alpha',\nu'\beta'\rangle=\langle\mu\nu|\mathfrak{L}_0^{(s)}|\mu'\nu'\rangle\delta_{\alpha\alpha'}\delta_{\beta\beta'}+\langle\alpha\beta|\mathfrak{L}_0^{(b)}|\alpha'\beta'\rangle\delta_{\mu\mu'}\delta_{\nu\nu'}$$
$$=\langle\mu|H|\mu'\rangle\delta_{\alpha\alpha'}\delta_{\beta\beta'}\delta_{\nu\nu'}-\langle\nu'|H^*|\nu\rangle\delta_{\alpha\alpha'}\delta_{\beta\beta'}\delta_{\mu\mu'}+\cdots$$
$$=\left[(\epsilon_\mu-\epsilon_\nu)+(\epsilon_\alpha-\epsilon_\beta)\right]\delta_{\alpha\alpha'}\delta_{\beta\beta'}\delta_{\mu\mu'}\delta_{\nu\nu'}$$
$$=(\omega_{\mu\nu}+\omega_{\alpha\beta})\prod_i\delta_{ii} \tag{21.10}$$

It is apparent that the form of the bath density matrix given by Eq. (21.9) will render the $\mathfrak{L}_0^{(b)}$ contribution zero by Eq. (21.1) when the factors $(\omega-\mathfrak{L}_0)^{-1}$ on the left and right in Eq. (21.8a) are considered. Therefore, in the portion of the trace operation relating to bath variables in Eq. (21.5), $\mathfrak{A}$ reduces to $\mathfrak{M}(\omega)$ so that we let

$$\langle\mathfrak{M}(\omega)\rangle=\mathrm{Tr}_b\left[\mathfrak{M}(\omega)\Upsilon^{(b)}\right] \tag{21.11a}$$

with

$$\langle\mu\nu|\langle\mathfrak{M}(\omega)\rangle|\mu'\nu'\rangle=\sum_{\alpha\alpha'}\mathfrak{M}_{\mu\alpha,\nu\alpha;\mu'\alpha',\nu'\alpha'}f_{\alpha'}. \tag{21.11b}$$

Equation (21.5) now becomes

$$I(\omega)=\pi^{-1}\mathcal{I}\Big(\mathrm{Tr}_s\Big\{\mathsf{d}\big(\omega-\mathfrak{L}_0^{(s)}\big)^{-1}\Big[1+\langle\mathfrak{M}(\omega)\rangle\big(\omega-\mathfrak{L}_0^{(s)}\big)^{-1}\Big]\Upsilon^{(s)}\mathsf{d}\Big\}\Big) \tag{21.12}$$

We remark that the trace is now over the system variables, and we emphasize that the $\langle\mathfrak{M}(\omega)\rangle$ embodies the entire perturbing effect of the bath on the system.

It is essential to the Fano development to recast Eq. (21.12) in a form originally developed by Zwanzig. Let us begin by remarking that Eq. (21.12) is interestingly reminiscent of a somewhat rewritten Eq. (10.23) without projection operators. Retracing our steps somewhat leads us to Eq. (21.8a), which is even more reminiscent of Eq. (10.23). From the definition of an inner product in $\mathfrak{L}$-space, Eq. (10.10), we see further that Eq. (21.12) or, more basically, Eq. (21.5) is such an inner product. One need not, however, decide which are the two operators in these equations which correspond to the two $\mathfrak{L}$-space vectors whose inner products have been taken.

Thus, since $\mathfrak{M}(\omega)$ of Eq. (21.8a) corresponds precisely to the $\mathfrak{T}(\omega)$ of Eq. (10.18), we see that the appropriate insertion of projection operators on both sides of Eq. (21.8a) yields Eq. (10.23). Therefore, we may use Eq. (10.25) to rewrite Eq. (21.12) as

$$I(\omega)=\pi^{-1}\mathcal{I}\Big(\mathrm{Tr}\big\{\mathsf{d}\big[\omega\mathsf{P}-\mathfrak{L}_P-\mathsf{P}\mathfrak{M}_c(\omega)\mathsf{P}\big]^{-1}\Upsilon^{(b)}\Upsilon^{(s)}\mathsf{d}\big\}\Big) \tag{21.13a}$$

where we have replaced $\mathfrak{T}_c$ by $\mathfrak{M}_c$ in order to conform to Fano's notation and where $\mathfrak{T}_c$ is given by Eq. (10.21). The reader will see that Eq. (10.21) may be rewritten slightly as

$$\mathfrak{M}_c=\big[1-\mathfrak{L}_1(1-\mathsf{P})(\omega-\mathfrak{L}_0)^{-1}\big]^{-1}\mathfrak{L}_1=\mathfrak{L}_1\sum_{n=0}^{\infty}\big[(1-\mathsf{P})(\omega-\mathfrak{L}_0)^{-1}\mathfrak{L}_1\big]^n \tag{21.13b}$$

the difference lying in our having written $(1-\mathsf{P})\mathfrak{G}^0=\mathsf{M}\mathfrak{G}^0$ for $\mathsf{M}\mathfrak{G}^0\mathsf{M}$, our earlier arguments demonstrating the complete legitimacy of the procedure. In order to carry this further we must needs define a specific projection operator. Fano used

$$\mathsf{P}\big[A\Upsilon^{(b)}\big]=\Upsilon^{(b)}\mathrm{Tr}_b\{A\Upsilon^{(b)}\} \tag{21.14}$$

which we now apply.

For simplicity's sake, let us treat each term in the denominator of Eq. (21.13a) as if it were not. Then

$$\begin{aligned}\mathrm{Tr}\{\mathsf{d}\omega\mathsf{P}\Upsilon^{(s)}\Upsilon^{(b)}\mathsf{d}\}&=\mathrm{Tr}\{\mathsf{d}\omega\Upsilon^{(b)}\Upsilon^{(s)}\mathsf{d}\,\mathrm{Tr}_b[\Upsilon^{(b)}]\}\\&=\mathrm{Tr}\{\mathsf{d}\omega\Upsilon^{(b)}\Upsilon^{(s)}\mathsf{d}\}=\mathrm{Tr}_s\{\mathsf{d}\omega\Upsilon^{(s)}\mathsf{d}\}\end{aligned} \tag{21.15a}$$

since neither ω nor d is a function of bath coordinates. Similarly, we find that

$$\begin{aligned}\operatorname{Tr}\{\mathsf{dP}(\mathfrak{L}_0^{(b)}+\mathfrak{L}_0^{(s)})\mathsf{Pd}\Upsilon^{(b)}\Upsilon^{(s)}\}&=\operatorname{Tr}\{\mathsf{dP}(\mathfrak{L}_0^{(b)}+\mathfrak{L}_0^{(s)})\mathsf{d}\Upsilon^{(b)}\Upsilon^{(s)}\}\\&=\operatorname{Tr}_s\{\mathsf{d}\mathfrak{L}_0^{(s)}\mathsf{d}\Upsilon^{(s)}\}\end{aligned}\tag{21.15b}$$

In this fashion then we arrive at the form originally attributed to Zwanzig and utilized by Fano,

$$I(\omega)=\pi^{-1}\mathcal{I}\left[\operatorname{Tr}_s\left\{\mathsf{d}\left[\omega-\mathfrak{L}_0^{(s)}-\langle\mathfrak{M}_c(\omega)\rangle\right]^{-1}\Upsilon^{(s)}\mathsf{d}\right\}\right]\tag{21.16a}$$

where

$$\langle\mathfrak{M}_c(\omega)\rangle=\operatorname{Tr}_b\{\mathfrak{M}_c(\omega)\Upsilon^{(b)}\}.\tag{21.16b}$$

For our future reference it is important to remark that we could have proceeded in a somewhat different fashion from Eq. (21.13a) wherein the separable approximation for Υ has as yet no real significance. To this we shall return somewhat later.

As Fano remarks, Eq. (21.16a) corresponds to the BKG result, Eq. (13.28), except that the interruption approximation, which was introduced subsequent to Eq. (13.18), has not been utilized, nor has the classical path approximation which, in the BKG work, makes itself felt in Eq. (13.31). In Eq. (21.16a) $\mathfrak{L}_0^{(s)}+\mathfrak{M}_c(\omega)$ corresponds to the BKG resolvant operator, one of the attributes characterizing the EIT [or the relaxation matrix (67)]. There are several other simplifications, of which we mention a few.

For no system-bath interaction, $\mathfrak{L}_1=0$, and, by Eq. (21.13b), $\langle\mathfrak{M}_c\rangle$ vanishes. As a consequence

$$\begin{aligned}I(\omega)&=-\pi^{-1}\mathcal{I}\left\{\sum_{\mu\nu}\mathsf{d}_{\mu\nu}^*(\omega-\omega_{\mu\nu'})^{-1}(\Upsilon^{(s)}\mathsf{d})_{\mu\nu}\right\}\\&\xrightarrow[I(\omega)\to0+]{}\sum_{\mu\nu}\mathsf{d}_{\mu\nu}^*\delta(\omega-\omega_{\mu\nu})(\Upsilon^{(s)}\mathsf{d})_{\mu\nu}\end{aligned}\tag{21.17}$$

as one would surely expect.

As was the case with the resolvant remarked on preceding Eq. (13.29), if a transformation, S, can be found such that

$$\left(S\left[\mathfrak{L}_0^{(s)}+\langle\mathfrak{M}_c(\omega)\rangle\right]S^{-1}\right)_{jk}=\omega_j\delta_{jk}\tag{21.18}$$

a series of spectral lines having for their width the imaginary part of the eigenvalue, ω_j, shift the real part, will result. The fact that the operator which is being diagonalized $(\mathfrak{L}_0+\langle\mathfrak{M}_c\rangle)$ is not Hermitian leads, as usual, to these complex eigenvalues. This diagonalization may be carried out only if $\mathfrak{L}_0+\langle\mathfrak{M}_c\rangle$ commutes with its Hermitian conjugate.

If the separation of the unperturbed spectral lines is of considerably greater magnitude than the broadening interaction, that is to say, if

$$|\langle\mathfrak{M}_c\rangle_{\mu\nu,\mu'\nu'}/(\omega_{\mu\nu}-\omega_{\mu'\nu'})|\ll1,\qquad \mu\nu\neq\mu'\nu'\tag{21.19}$$

the resolvant may be diagonalized approximately. Letting $\omega - \mathfrak{L}_0^{(s)} = \mathfrak{A}$,

$$\langle \mu\nu | [\mathfrak{A} - \langle \mathfrak{M}_c \rangle]^{-1} | \mu'\nu' \rangle = \left\langle \mu\nu \middle| \mathfrak{A}^{-1} \left[1 + \frac{\langle \mathfrak{M}_c \rangle}{\mathfrak{A}} + \cdots \right] \middle| \mu'\nu' \right\rangle$$

$$\doteq \left\langle \mu\nu \middle| \frac{1}{\mathfrak{A}} \left[1 + \frac{\langle \mathfrak{M}_c \rangle}{\mathfrak{A}} \right] \middle| \mu'\nu' \right\rangle \tag{21.20}$$

In Eq. (21.20), $\mathfrak{A}$ is "diagonal" so that the matrix element of a denominator is the matrix element in the denominator. Therefore, in Eq. (21.6a), we replace a matrix element of the denominator by matrix elements in the denominator, and we find the spectral line intensity distribution for an isolated spectral line arising from a transition between levels i and j as given by

$$I(\omega) = \frac{1}{\pi} \mathrm{Tr}_s \left\{ \frac{\delta(\mathbf{d}^* \Upsilon^{(s)} \mathbf{d})_{ij}}{(\omega - \omega_{ij} - \Delta)^2 + \delta^2} \right\} \tag{21.21a}$$

$$\delta = \mathfrak{I}\{\langle \mathfrak{M}_c(\omega) \rangle\} \tag{21.21b}$$

$$\Delta = \mathfrak{R}\{\langle \mathfrak{M}_c(\omega) \rangle\} \tag{21.21c}$$

The result is rather an important one. [We remark that the off-diagonal elements of $\langle \mathfrak{M}_c(\omega) \rangle$ become important for overlapping spectral lines. Cherkasov (103) has derived certain useful formulas for the evaluation of these off-diagonal elements of this operator, which he calls the "relaxation operator."]

22 An Alternative Excursion Into Liouville Space

At the point which we have reached in the Fano development of Section 21, it is obvious that, in following Fano further, we would next address ourselves to the evaluation of $\mathfrak{M}_c$. However, currently it is of considerably more value to continue with the work of Ben-Reuven, and this we shall do in Section 24. Now although the line of future development will probably be that delineated by the latter author, the Liouville space work of Royer (239, 240) contains so much that is uniquely informative that its presentation is essential.

Where Fano simply dealt with the Liouville per se in Eq. (21.4) and subsequent, Royer dealt with a nonunitary TDO, $\mathfrak{U}(t)$, which acts in the Liouville space of the system (radiator) and which describes the motion of this system in contact with the bath. Of course such a TDO must contain the Liouvillian in some form, as does Eq. (21.4). Because the perturbation expansions of $\mathfrak{U}(t)$ do not converge uniformly and hence cannot be truncated, Royer reexpressed $\mathfrak{U}(t)$ as a function of certain well behaved functions, $\mathfrak{L}_i$, whose expansions apparently converge, at least for the first few terms. Two forms for the function of $\mathfrak{L}_i$ were considered, the ordered exponential,

$$\mathfrak{U}(t) = \mathsf{T}\exp\left[\mathrm{i}\int_0^t dt'\,\mathfrak{L}_i(t')\right]$$

and the resolvant,

$$\mathrm{i}\,\mathfrak{U}(\omega) = [\omega - \mathfrak{L}_2(\omega)]^{-1}$$

We shall consider only the ordered exponential in detail, the principles involved in the resolvant study being sufficiently similar to allow of our merely mentioning it. Several interesting techniques arise during the course of this work, among which we may remark the imaginary time and the cumulant graphing.

Royer begins with the familiar expansion for the power radiated, say Eqs. (13.6), with the correlation function given by Eq. (21.2). As had Fano, Royer considers a single radiator, the system, immersed in a bath of perturbers. The immediate introduction of a TDO, however, is more reminiscent of Anderson–Baranger than of Fano. This TDO governs the development of the system with time and may, of course, be readily defined with reference to Eqs. (12.6) and (21.2) as

$$\mathfrak{U}(t) = (\mathrm{Tr}_s\Upsilon)^{-1}\mathrm{Tr}_B\,\Upsilon e^{\mathrm{i}tL}, \qquad \mathfrak{U}(0) = 1 \Rightarrow \tag{22.1a}$$

$$\Phi(t) = \mathrm{Tr}_s\{\mathsf{d}\sigma\,\mathfrak{U}(t)\mathsf{d}\} \tag{22.1b}$$

$$\sigma = \mathrm{Tr}_s\,\Upsilon, \qquad \mathrm{Tr}_s\,\sigma = 1 \tag{22.1c}$$

[Here $\mathfrak{L}$ is the Liouville operator for which, following Kubo (167) and as had Mizushima, Robert and Galatry (206), Royer uses the notation $H^{\times}$.] The notation for the Liouville is essentially that of Eqs. (21.6). A somewhat different manner of writing the potential, albeit one we shall encounter again,

$$\mathsf{V} = \mathsf{V}_{SB} + \mathsf{V}_B = \sum_{j=1}^{n} \mathsf{V}_{sj} + \sum_{i<j} \mathsf{V}_{ij} \tag{22.2}$$

where the interaction between the emitter and perturbers is segregated from the interaction between perturbers, is used by Royer. The system ($\Upsilon_s = e^{-\beta H_s}/Z_s$, $Z_s = \mathrm{Tr}_s\{e^{-\beta H_s}\}$) and bath density matrices are about as one would expect. A bath operator (BO),

$$\begin{aligned}\mathfrak{B}(t) &= \Upsilon_s^{-1}(\mathrm{Tr}_B\{\Upsilon e^{\mathrm{i}t\mathfrak{L}}\})e^{-\mathrm{i}t\mathfrak{L}_s}\\ &= (Z_S Z_B/Z)\,\mathrm{Tr}_B\{\Upsilon_B(e^{\beta H_0}e^{-\beta H})(e^{\mathrm{i}t\mathfrak{L}}e^{-\mathrm{i}t\mathfrak{L}_0})\}\end{aligned} \tag{22.3}$$

which rather obviously embodies all the effects of the bath on the radiator, is introduced. The term on the extreme right of Eq. (22.3) evolves as follows: first, $\mathfrak{L}_0 = \mathfrak{L}_B + \mathfrak{L}_S$. Since H_B commutes with H_S and with operators pertaining to the radiator alone, $\mathfrak{L}_B d = H_B\mathsf{d} - \mathsf{d}H_B = 0$, and we simply insert $e^{-\mathrm{i}t\mathfrak{L}_B}$ in order to obtain $e^{-\mathrm{i}t\mathfrak{L}_0}$.

Equation (22.3) yields the following form for the TDO:

$$\mathfrak{U}(t)=\mathfrak{B}^{-1}(0)\mathfrak{B}(t)e^{it\mathfrak{L}_s} \tag{22.4}$$

Royer's task is now apparent: the obtention of this TDO, which he accomplishes, first, as an expansion and, second, in resolvant form as Fano did. We begin the first variation with what Royer calls, and we must agree is, a trivial expansion,

$$\dot{\mathfrak{U}}(t)=\mathfrak{U}(t)\left[\mathfrak{U}(t)^{-1}\dot{\mathfrak{U}}(t)\right]\Rightarrow$$

$$\dot{\mathfrak{U}}(t)=\mathrm{i}\,\mathfrak{U}(t)[\mathfrak{L}_s+\mathfrak{R}(t)], \qquad \mathfrak{U}(0)=1 \tag{22.5a}$$

$$\begin{aligned}\mathfrak{R}(t)&=-\mathrm{i}\,\mathfrak{U}(t)^{-1}\dot{\mathfrak{U}}(t)-\mathfrak{L}_s\\ &=-\mathrm{i}\exp[-\mathrm{i}t\mathfrak{L}_s]\mathfrak{B}(t)^{-1}\dot{\mathfrak{B}}(t)\exp[\mathrm{i}t\mathfrak{L}_s]\end{aligned} \tag{22.5b}$$

by Eq. (22.4). The reader may readily verify that

$$\dot{\mathfrak{B}}(t)=\mathrm{i}\mathfrak{B}(t)\hat{\mathfrak{R}}(t) \tag{22.6a}$$

where

$$\hat{\mathfrak{R}}(t)=\exp[\mathrm{i}t\mathfrak{L}_s]\mathfrak{R}(t)\exp[-\mathrm{i}t\mathfrak{L}_s] \tag{22.6b}$$

Equation (22.6a) has the formal solution

$$\mathfrak{U}(t)=\mathsf{T}_{\rightarrow}\exp\left\{\mathrm{i}\int_0^t dt'[\mathfrak{L}_s+\mathfrak{R}(t')]\right\} \tag{22.7}$$

If we suppose $\mathfrak{R}(t)$ to tend, as $t\rightarrow\infty$, to a superposition of a constant, $\overline{\mathfrak{R}}$, plus an oscillatory term $\tilde{\mathfrak{R}}(t)$, then

$$\mathfrak{R}(t)=\overline{\mathfrak{R}}+\tilde{\mathfrak{R}}(t) \tag{22.8a}$$

where

$$\overline{\mathfrak{R}}=\lim_{T\rightarrow\infty}T^{-1}\int_0^T\mathfrak{R}(t)\,dt \tag{22.8b}$$

$$\tilde{\mathfrak{R}}(t)=\mathfrak{R}(t)-\overline{\mathfrak{R}} \tag{22.8c}$$

By this separation of the $\mathfrak{R}(t)$ in Eq. (22.5a) we are in a position to divide the operator, $\mathfrak{L}_s+\mathfrak{R}(t)$, into a time-dependent operator,

$$\mathfrak{L}=\mathfrak{L}_s+\overline{\mathfrak{R}} \tag{22.9a}$$

and a perturbation, $\mathfrak{R}(t)$, where now we suppose

$$\tilde{\mathfrak{R}}_t(t)\equiv e^{it\mathfrak{L}}\tilde{\mathfrak{R}}(t)e^{-\mathrm{i}t\mathfrak{L}} \tag{22.9b}$$

Combining Eqs. (22.7) and (22.9b) yields

$$\mathfrak{U}(t) = \mathsf{T}_{\rightarrow}\exp\left[\mathrm{i}\int_0^t dt'\tilde{\mathfrak{R}}_t(t')\right]e^{\mathrm{i}t\mathfrak{L}}$$

$$= e^{\mathrm{i}t\mathfrak{L}} + \mathrm{i}\int_0^t dt'\, e^{\mathrm{i}t'\mathfrak{L}}\tilde{\mathfrak{R}}(t')e^{\mathrm{i}(t-t')\mathfrak{L}}$$

$$+\mathrm{i}^2\int_0^t dt' \int_0^{t'} dt''\, e^{\mathrm{i}t''\mathfrak{L}}\tilde{\mathfrak{R}}(t'')e^{\mathrm{i}(t'-t'')\mathfrak{L}}\tilde{\mathfrak{R}}(t')e^{\mathrm{i}(t-t')\mathfrak{L}} + \cdots \tag{22.10}$$

Royer considers this TDO as the sum of an unperturbed TDO, $\exp[\mathrm{i}t\mathfrak{L}]$, governing the "average" motion of the radiator in the bath plus the various corrections due to the "interaction" $\mathfrak{R}(t)$. This is obviously a self-consistent and informative point of view. Now if we take τ_c as the collision time, it will be true that, when $t > \tau_c$, $\mathfrak{R}(t)$ may be considered as assuming its asymptotic value $\mathfrak{R}$. That this excludes emitter-perturber bound states Royer points out, and we remark. Therefore, $\mathfrak{U}(t)$ becomes

$$\mathfrak{U}(\infty) = \mathsf{T}_{\rightarrow}\exp\left[\mathrm{i}\int_0^{\infty} dt'\,\tilde{\mathfrak{R}}_t(t')\right]e^{\mathrm{i}t\mathfrak{L}} \equiv \mathfrak{A}e^{\mathrm{i}t\mathfrak{L}} \tag{22.11}$$

where $\mathfrak{A}$ is definable from the equation. Since $\mathfrak{A}$ and $\mathfrak{L}$ are time independent, the time integration of Eq. (22.1b) which leads to the $I(\omega)$ may surely be carried out in order to obtain the interruption result,

$$I_{\mathrm{int}}(\omega) = -\mathrm{i}\,\mathrm{Tr}_s\left[\mathsf{d}\sigma\mathfrak{A}(\omega-\mathfrak{L})^{-1}\mathsf{d}\right] \tag{22.12}$$

As usual, this expression cannot be expected to hold in the line wings. $\mathfrak{A}$ is a non-Hermitian operator which tends to render the spectral line asymmetric, its principal effect in the adiabatic case being the introduction of said asymmetry (263).

If we define

$$\mathfrak{N}(t) = \left\{\mathsf{T}_{\leftarrow}\exp\left[-\mathrm{i}\int_t^{\infty} dt'\,\tilde{\mathfrak{R}}_t(t')\right] - 1\right\}e^{\mathrm{i}t\mathfrak{L}}, \qquad \mathfrak{L} = \mathfrak{L}_s + \overline{\mathfrak{R}} \tag{22.13}$$

it follows immediately from Eq. (22.10) that

$$\mathfrak{U}(t) = \mathfrak{A}e^{\mathrm{i}t\mathfrak{L}} + \mathfrak{A}\mathfrak{N}(t) \Rightarrow \tag{22.14a}$$

$$\mathfrak{U}(\omega) = -\mathrm{i}\mathfrak{A}(\omega-\mathfrak{L})^{-1} + \mathfrak{A}\mathfrak{N}(\omega) \tag{22.14b}$$

As an alternate to this perturbation treatment Royer considers the bath effect as a frequency-dependent operator, $\mathfrak{C}(\omega)$, added to $\mathfrak{L}_s$ in the unperturbed resolvant, $-\mathrm{i}(\omega - \mathrm{i}0 - \mathfrak{L}_s)^{-1}$. As, we may remark, had Fano. (With the infinitesimal, i0, we presume ourselves familiar.) If we take

$$\mathrm{i}\mathfrak{U}(\omega) = \left[\omega - \mathfrak{L}_s - \mathfrak{C}(\omega)\right]^{-1} \tag{22.15}$$

appropriate operator manipulation leads to

$$\begin{aligned}\mathfrak{C}(\omega)&=\omega-\mathfrak{L}_s-[\mathrm{i}\mathfrak{U}(\omega)]^{-1}\\&=\{1-[\mathrm{i}(\omega-\mathfrak{L}_s)\mathfrak{U}(\omega)]^{-1}\}(\omega-\mathfrak{L}_s)\\&=\left\{1-\left[1+\langle\mathfrak{M}(\omega)\rangle(\omega-\mathfrak{L}_s)^{-1}\right]^{-1}\right\}(\omega-\mathfrak{L}_s)\\&=\left[1+\langle\mathfrak{M}(\omega)\rangle(\omega-\mathfrak{L}_s)^{-1}\right]^{-1}\langle\mathfrak{M}(\omega)\rangle\end{aligned}\tag{22.16a}$$

where

$$\begin{aligned}\langle\mathfrak{M}(\omega)\rangle&=(\omega-\mathfrak{L}_s)\left[\mathrm{i}\mathfrak{U}(\omega)-(\omega-\mathfrak{L}_s)^{-1}\right](\omega-\mathfrak{L}_s)\\&=\mathrm{i}(\omega-\mathfrak{L}_s)\int_0^\infty dt\left[\mathfrak{B}(0)^{-1}\mathfrak{B}(t)-1\right]e^{\mathrm{i}t(\omega-\mathfrak{L}_s)}(\omega-\mathfrak{L}_s)\end{aligned}\tag{22.16b}$$

Equation (22.14a) is an equation that was obtained by Fano where $\mathfrak{C}(\omega)\equiv\langle\mathfrak{M}_c(\omega)\rangle$ and where we have obtained $\mathfrak{M}_c$ in Eq. (21.13b), except that Royer imposed no restrictions on the density matrix in its obtention.

Equation (22.16b) may be used in order to obtain the perturbation expansion of $\langle\mathfrak{M}(\omega)\rangle$ and hence that of $\mathfrak{C}(\omega)$. We shall consider neither this nor the approach to the correlation function through the expansion of $\mathfrak{C}(\omega)$.

Royer found that he could considerably simplify certain manipulations of $\mathfrak{B}(t)$ by taking advantage of the exponential nature of the density matrix in order to consider it an imaginary time TDO. Thus a density matrix, $\exp[-\beta H]$, becomes a TDO, $\exp[-\mathrm{i}tH]$, when t is the imaginary time, $t=-\mathrm{i}\beta$. Taking

$$\mathsf{V}(-\mathrm{i}b)=e^{bH_0}\mathsf{V}e^{-bH_0}=e^{\mathrm{i}(-\mathrm{i}b)H_0}\mathsf{V}e^{-\mathrm{i}(-\mathrm{i}b)H_0}\tag{22.17a}$$

$$\mathsf{V}(t)=e^{\mathrm{i}tH_0}\mathsf{V}e^{-\mathrm{i}tH_0}\tag{22.17b}$$

Eq. (22.3) becomes

$$\begin{aligned}\mathfrak{B}(t)&=\frac{Z_sZ_B}{Z}\mathrm{Tr}_B\left\{\Upsilon_B\mathsf{T}_{\leftarrow b}\exp\left[-\int_0^\beta db\mathsf{V}(-\mathrm{i}b)\right]\mathsf{T}_{t\rightarrow}\exp\left[\mathrm{i}\int_0^t dt'\mathsf{V}(t')\right]\right\}\\&=Z_sZ_B\mathrm{Tr}_B\left\{\Upsilon_B\mathsf{T}_{\leftarrow b,t\rightarrow}\exp\left[-\int_0^\beta db\mathsf{V}(-\mathrm{i}b)+\mathrm{i}\int_0^t dt\mathsf{V}(t')\right]\right\}\Big/Z\end{aligned}\tag{22.17c}$$

where the TDO, $\mathsf{T}_{\leftarrow b,t\rightarrow}$, orders imaginary time b such that it increases from right to left, time t so that it increases from left to right. $\mathfrak{B}(t)$ may be even more summarily exhibited by adopting the complex time, $z=t+\mathrm{i}b$, and defining

$$\mathfrak{W}(z)\equiv\mathfrak{V}(z)=\mathfrak{V}(t)\text{ on the real time axis}\tag{22.17d}$$

$$\equiv\mathsf{V}(z)=\mathsf{V}(t+\mathrm{i}b)\text{ off the real time axis}\tag{22.17e}$$

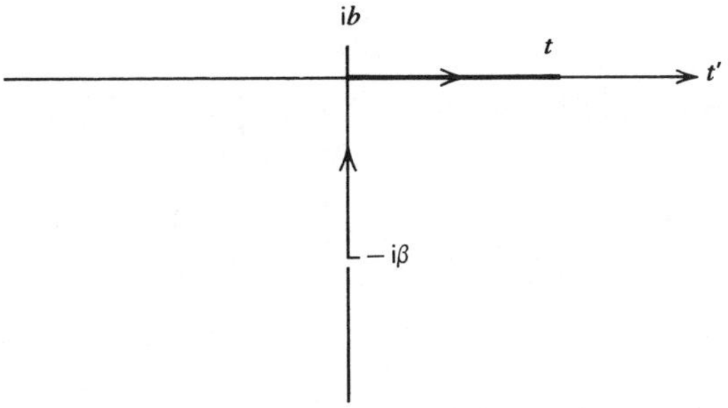

Figure 22.1 After Royer (238).

thus yielding

$$\mathfrak{B}(t) = Z_s Z_B \mathrm{Tr}_B \left\{ \Upsilon_B \mathsf{T}_{z\rightarrow} \exp\left[\mathrm{i} \int_{-\mathrm{i}\beta}^{t} dz\, \mathfrak{W}(z) \right] \right\} \tag{22.17f}$$

The integral of Eq. (22.17f) is taken along the contour given in Fig. 22.1.

The definition of the brac operator,

$$\langle \cdots \rangle \equiv \mathrm{Tr}_B[\Upsilon_B \mathsf{T}_{z\rightarrow 0}(\cdots)], \qquad \langle 1 \rangle = 1, \tag{22.18}$$

is generally familiar save perhaps for the introduction of the TOO. Eq. (22.18) allows the display of the BO in the form,

$$\mathfrak{B}(t) = (Z_s Z_B / Z) \left\langle \exp\left[\mathrm{i} \int_{-\mathrm{i}\beta}^{t} dz\, \mathfrak{W}(z) \right] \right\rangle, \tag{22.19}$$

which provides an informative description of the physical processes under consideration: At an imaginary initial time, $z = -\mathrm{i}\beta$, the radiator and perturbers are uncorrelated, the density matrix for the particle system being a product of one-particle density matrices. As imaginary time increases, $z \rightarrow 0$, and the correlations contained in the $t = 0$ density matrix and describing the particle interactions are established. These of course govern the behavior of $\mathfrak{B}(t)$ during the subsequent progress out the real time axis. That $\mathfrak{B}(t) \rightarrow 0$ as $t \rightarrow \infty$ may be seen from the integral over $\mathsf{V}(t')$ in Eq. (22.17c). The phase, which is given by this integral or, rather, its spectrum of values becomes more and more broad, the average over its exponential tending to vanish.

Using Eq. (23.25b) we may reexpress Eq. (22.19) in terms of cumulants as follows:

$$\begin{aligned} B(t) &= Z_s Z_B \mathsf{T}_{z\rightarrow} \exp\left[\mathrm{i} \int_{-\mathrm{i}\beta}^{t} dz\, \hat{\mathfrak{R}}(z) \right] \Big/ Z \\ &= \left\{ Z_s Z_B \mathsf{T}_{z\rightarrow} \exp\left[\mathrm{i} \int_{-\mathrm{i}\beta}^{0} dz\, \hat{\mathfrak{R}}(z) \right] \Big/ Z \right\} \mathsf{T}_{t\rightarrow} \exp\left[\mathrm{i} \int_{0}^{t} dt'\, \hat{\mathfrak{R}}(t') \right] \\ &= B(0) \mathsf{T}_{z\rightarrow} \exp\left[\mathrm{i} \int_{0}^{t} dt'\, \hat{\mathfrak{R}}(t') \right] \end{aligned} \tag{22.20a}$$

where

$$\hat{R}(z)=\left\langle \exp\left[\mathrm{i}\int_{-\mathrm{i}\beta}^{z} dz'\,\mathfrak{W}(z')\right]W(z)\right\rangle_{c\to}=e^{\mathrm{i}t\mathfrak{L}_s}\mathfrak{R}(t)e^{-\mathrm{i}t\mathfrak{L}_s}$$

$$=e^{it\mathfrak{L}_s}\left\langle \exp\left[\mathrm{i}\int_{-\mathrm{i}\beta}^{t} dz'\,\mathfrak{W}(z'-t)\right]V\right\rangle_{c\to} e^{-\mathrm{i}t\mathfrak{L}_s}$$

by Eq. (23.5b) and

$$\left\langle \prod_i \mathfrak{W}(z_i)\right\rangle=\left\langle e^{\mathrm{i}t\mathfrak{L}_0}\left(\prod_i \mathfrak{W}(z_i-t)e^{-\mathrm{i}t\mathfrak{L}_0}\right\rangle=e^{\mathrm{i}t\mathfrak{L}_s}\left\langle \prod_i \mathfrak{W}(z_i-t)\right\rangle e^{-\mathrm{i}t\mathfrak{L}_s}$$

which follows because H_s and H_B commute with Υ_B so that $\exp[\pm \mathrm{i}t\mathfrak{L}_B]$ cancel with the cyclic property of the trace.

The expansion of $R(t)$ and hence of the BO, in powers of the density is given by Eqs. (23.32) and the following. We shall return to this in a moment, but first we will find it beneficial to write out the following expansion of Eq. (23.29):

$$\mathfrak{B}(t)=\sum_{m=0}^{\infty}\left(\frac{1}{m!}\right)\sum_{j_1}{}'\sum_{j_2}{}'\cdots\sum_{j_m}{}'\int_{-\mathrm{i}\beta}^{t}dz_1\int_{-\mathrm{i}\beta}^{t}dz_2\cdots$$
$$\cdot\int_{-\mathrm{i}\beta}^{t}dz_m\langle\dot{\Gamma}'_{j_1}(z_1)\cdots\dot{\Gamma}'_{j_m}(z_m)\rangle \tag{22.21}$$

In Eq. (22.21) $\Sigma'_j\Gamma'_j$ represents the sum over all irreducible graphs, each $\dot{\Gamma}'_j$ corresponding to a collision of the radiator with the bath,

$$\left\langle \prod_{i=1}^{m}\dot{\Gamma}'_{j_i}(z_i)\right\rangle$$

corresponding to a series of m collisions. The integrations and summations then effectively (1) integrate the series of collisions over all possible sets of collision times, (2) sum the series over all possible sets of m, and (3) sum the result over all m. $B(t)$ thus represents the sum over all possible histories of the radiator during the time interval $-\mathrm{i}\beta$ to t.

As is usually presumed, low densities correspond to binary collisions. (This is a very useful rule-of-thumb sort of statement, but, as we shall see, it cannot be defended *à l'outrance*.) In which case only the binary irreducible graphs, f_{sj}, are retained, and, the collisions well separated and hence time disentangled, $\langle\Pi_j\dot{f}_{s_j}(z_j)\rangle=\Pi_j\langle\dot{f}_{s_j}(z_j)\rangle$. Thus

$$\mathfrak{B}(t)=\sum_{m=0}^{\infty}(m!)^{-1}\int_{-\mathrm{i}\beta}^{t}dz_1\cdots\int_{-\mathrm{i}\beta}^{t}dz_m\mathsf{T}_{z\to}\langle\dot{f}_{s_1}(z_1)\rangle\cdots\langle\dot{f}_{s_m}(z_m)\rangle$$
$$=\mathsf{T}_{z\to}\exp\left[\int_{-\mathrm{i}\beta}^{t}dz\langle\dot{f}_{s_1}(z)\rangle\right] \tag{22.22}$$

These collisions, in addition to being binary, are independent so that, although overlapping binary collisions are included, they enter independently.

The next level of approximation includes binary plus ternary collisions. This means that the $\mathfrak{B}(t)$ expansion will include $\hat{\mathfrak{R}}^{(1)}+\hat{\mathfrak{R}}^{(2)}$, $\hat{\mathfrak{R}}^{(1)}$ corresponding to the $\langle \dot{f}_{s_1} \rangle$ (first) term on the right of Eq. (22.32), $\hat{\mathfrak{R}}^{(2)}$ to the second term. It originally appeared (237) that the series for R converged in all powers as $t \to \infty$, but it was later found (240), in agreement with certain other authors, that such was really the case only for $\hat{\mathfrak{R}}^{(1)}$ and $\hat{\mathfrak{R}}^{(2)}$. Convergent or otherwise, the reader will probably agree that the Royer expansion can be quite helpful in illuminating the details of the physical processes comprehended by the collision.

At this point we are in a position to consider the intensity distribution in the spectral line to the first order in the density. In doing so it behooves us to take into specific consideration the fact that the radiator coordinates on which the TDO, $\mathfrak{U}(t)$, acts relate to internal (electronic for the radiator an atom) and translational motions. Thus the correlation function may be written

$$\Phi(t) = \mathrm{Tr}_e\left[\mathsf{d}\Upsilon_e \mathfrak{B}_e(t) e^{it\mathfrak{L}_e}\mathsf{d}\right] \tag{22.23a}$$

where

$$\mathfrak{B}_e(t) = \mathrm{Tr}_t\left[\Upsilon_t \mathfrak{B}(t)\right] \tag{22.23b}$$

Υ_e and Υ_t being the density matrices relating to internal and translational coordinates, respectively. Given the anticipated linear resolution of the radiator (system) Hamiltonian, $H_s = K_s + H_e$, into translational (K_s) and internal parts (H_e), the reader will encounter no difficulty in writing down the various density matrices. The equation analogous to Eq. (22.20a) will be

$$\mathfrak{B}_e(t) = \mathfrak{B}_e(0)\mathsf{T}_{\rightarrow}\exp\left[\mathrm{i}\int_0^t dt'\,\hat{\mathfrak{R}}_e(t')\right], \tag{22.23c}$$

where $\mathfrak{R}_e(t)$ has its bath averages replaced by bath-translational averages. We do not detail arrival at

$$\Phi(t) = \mathrm{Tr}_e\{\mathsf{d}[\mathrm{Tr}_t \mathrm{Tr}_B \Upsilon]\,\mathfrak{U}_e(t)\mathsf{d}\} \tag{22.24a}$$

where

$$\mathfrak{U}_e(t) = [\mathrm{Tr}_t \mathrm{Tr}_B \Upsilon]^{-1} \mathrm{Tr}_t \mathrm{Tr}_B \Upsilon e^{it\mathfrak{L}} = \mathfrak{B}_e(0)^{-1}\mathfrak{B}_e(t) e^{it\mathfrak{L}_e} \tag{22.24b}$$

To this point we have simply separated the electronic and translational motions. If we now replace all appropriate quantities in Eq. (22.14a) by their approximation to first order in the density, our TDO becomes

$$\mathfrak{U}^{(1)}(t) = \mathfrak{A}_e^{(1)} e^{it\mathfrak{L}_e^{(1)}} - \mathrm{i}\int_t^\infty dt'\, e^{it'\mathfrak{L}_{es}} \tilde{R}^{(1)} e^{-it'\mathfrak{L}_{es}} e^{it\mathfrak{L}_{es}} \tag{22.25}$$

where the subscript "e" indicates general replacement of $\mathfrak{R}$ by $\mathfrak{R}_e$, the superscript "(1)" the first order in the density. We emphasize here that $\mathfrak{L}_{es}$ is the Liouvillian relating to the internal coordinates of the radiator while $\mathfrak{L}_e$ is defined by Eq. (22.13), now as $\mathfrak{L}_e = \mathfrak{L}_{es} + \overline{\mathfrak{R}}$. The second term in Eq. (22.25) may be written as

$$-(\mathfrak{A}_e^{(1)} - 1)e^{\mathrm{i}t\mathfrak{L}_{es}} + i\int_0^t dt'\, e^{\mathrm{i}t'\mathfrak{L}_{es}}\tilde{R}^{(1)}(t')e^{\mathrm{i}(t'-t)\mathfrak{L}_{es}}$$

$$= \mathfrak{U}_1(t) - \left[\mathfrak{A}_e^{(1)} - 1 + \mathfrak{U}_1(0)\right]e^{\mathrm{i}t\mathfrak{L}_{es}} - i\int_0^t dt'\, e^{\mathrm{i}t'\mathfrak{L}_e}\tilde{\mathfrak{R}}^{(1)}e^{\mathrm{i}(t-t')\mathfrak{L}_e} \tag{22.26}$$

using Eqs. (22.11) and (22.8c). What we have obviously done is set

$$\mathfrak{U}_1(t) = \mathrm{i}\int_0^t dt'\, e^{\mathrm{i}t'\mathfrak{L}_{es}}\mathfrak{R}_e^{(1)}(t)e^{-\mathrm{i}t'\mathfrak{L}_{es}}e^{\mathrm{i}t\mathfrak{L}_{es}} + \mathfrak{U}_1(0)e^{\mathrm{i}t\mathfrak{L}_{es}} \tag{22.27}$$

We evaluate the integral of Eq. (22.27) by first remarking that, by Eq. (22.20a), it may be written as

$$\mathrm{i}\int_0^t dt'\, \hat{\mathfrak{R}}_e^{(1)}(t')e^{\mathrm{i}t\mathfrak{L}_{es}} = N\int_0^t dt' \langle \dot{f}_{s_1}(t')\rangle e^{\mathrm{i}t\mathfrak{L}_{es}}$$

$$= N\langle f_{s_1}(t) - f_{s_1}(0)\rangle e^{\mathrm{i}t\mathfrak{L}_{es}} \tag{22.28}$$

by Eq. (23.33) and where we have absorbed V into the average which we suppose over bath and translational radiator coordinates. Now since, by Eq. (23.2b),

$$f_{s_1}(0) = \exp\left[\mathrm{i}\int_{-\mathrm{i}\beta}^{0} dz\, \mathfrak{W}_{s_1}(z)\right] - 1,$$

it follows that

$$\mathfrak{U}_1(t) = N\,\mathrm{Tr}_B\mathrm{Tr}_t\left\{\left[\Upsilon\left(f_{s_1}(t) + 1\right)\right]e^{\mathrm{i}t\mathfrak{L}_{es}}\right\} \tag{22.29a}$$

$$\mathfrak{U}_1(0) = \exp\left[\mathrm{i}\int_{-\mathrm{i}\beta}^{0} dz\, \mathfrak{W}_{s_1}(z)\right] \tag{22.29b}$$

The Fourier transform of the *sp* (single perturber) correlation function follows:

$$I_{sp}(\omega) = \mathrm{Tr}_e\left[\mathsf{d}\Upsilon_e\mathfrak{U}_1(\omega)\mathsf{d}\right] \tag{22.30}$$

Strictly speaking, the intensity distribution is the real part of Eq. (22.30).

The interruption contribution to the correlation function is given by the first term in Eq. (22.25), its Fourier transform being

$$I_{\mathrm{int}}(\omega) = \mathrm{Tr}_e\left(\mathsf{d}\,\mathrm{Tr}_t\mathrm{Tr}_B\left\{\Upsilon\mathfrak{A}_e^{(1)}\left[\mathrm{i}\left(\omega - \mathfrak{L}_e - \overline{\mathfrak{R}}_e^{(1)}\right)\right]^{-1}\mathsf{d}\right\}\right). \tag{22.31}$$

This result we have already obtained as Eq. (22.12). We are left with the second and third terms on the right of Eq. (22.26) as yet untreated. Taking the Fourier transform of the second term is immediate. In the fourth term we essentially take two transforms, one over t' and one over $(t-t')$. Following the notation of Royer we obtain

$$K(\omega)=\mathrm{Tr}_e\Big(\mathsf{d}\Upsilon_e\Big\{\mathsf{i}\big[\mathfrak{A}_e^{(1)}-1+\mathfrak{U}_1(0)\big](\omega-\mathfrak{L}_{es}-\mathsf{i}0)^{-1}$$
$$+\mathsf{i}(\omega-\mathfrak{L}_{es}-\mathsf{i}0)^{-1}\overline{\mathfrak{R}}_e^{(1)}(\omega-\mathfrak{L}_{es}-\mathsf{i}0)^{-1}\Big\}\mathsf{d}\Big). \tag{22.32}$$

The intensity in our spectral line is given by

$$I(\omega)=\mathrm{const}\,\mathfrak{R}\big[I_{\mathrm{int}}(\omega)+I_{sp}(\omega)+K(\omega)\big]. \tag{22.33}$$

The correction, $K(\omega)$, rather obviously diverges at the resonance frequencies of the radiator. The sum $I_{sp}+K$, however, whose Fourier transform is given by the second term in Eq. (22.25), is regular at these resonance frequencies, the transform vanishing for large t. Thus $K(\omega)$ eliminates the singularities in the single-perturber spectrum near the resonance frequencies. In the line wing, on the other hand, where $|(\omega-\mathfrak{L}_{es})^{-1}\overline{R}_e^{(1)}|\ll 1$, we see that the bracketed factor in Eq. (22.21) becomes

$$\mathfrak{R}\Big[\mathsf{i}\big(\omega-\mathfrak{L}_{es}-\overline{\mathfrak{R}}_e^{(1)}\big)\Big]^{-1}=\mathfrak{R}\left[\mathsf{i}(\omega-\mathfrak{L}_{es})^{-1}\sum_n\left(\frac{\mathfrak{R}_e^{(1)}}{(\omega-\mathfrak{L}_{es})}\right)^n\right]$$
$$=\frac{\mathsf{i}}{(\omega-\mathfrak{L}_{es})}\,\frac{\mathfrak{I}\big[\overline{\mathfrak{R}}_e^{(1)}\big]}{(\omega-\mathfrak{L}_{es})}$$

The reader may wish thus to show that $\mathfrak{R}[I_{\mathrm{int}}]=-\mathfrak{R}[K]$ which means that, in the wings of the line the interruption contribution cancels that of the correction, and the shape is governed by the one-perturber spectrum. This of course corresponds to the well-known importance of strong individual collisions in the line wing. This essentially concludes our detailed study of the Royer work.

Now what has been done is the following: the correlation function, $\Phi(t)$, the real part of whose Fourier transform is the spectral line intensity, is expressed in terms of a TDO, $\mathfrak{U}(t)$, which operates in the space of the radiator. Royer was aware that density expansions of $\mathfrak{U}(t)$ would not converge so that he replaced the TDO, first by a time-dependent effective Liouvillian,

$$\mathfrak{U}(t)=\mathsf{T}_{\rightarrow}\exp\left(\mathsf{i}\int_0^t dt'\,\mathfrak{L}_1(t')\right)$$

and, secondly, by a frequency-dependent effective Liouvillian,

$$\mathfrak{U}(\omega)=\int_0^{\omega}dt\,e^{-\mathsf{i}\omega t}\,\mathfrak{U}(t)=-\mathsf{i}\big[\omega-\mathfrak{L}_2(\omega)\big]^{-1}$$

where we have seen that $\mathfrak{L}_1$ involves $\mathfrak{R}(t)$ [Eq. (22.8a)], $\mathfrak{L}_2$ involves $\mathfrak{C}(\omega)$ [Eq. (22.15)]. We have followed through the study of $\mathfrak{L}_1$ in order to arrive at a really rather informative Eq. (22.33); we do not detail the analogous study of $\mathfrak{L}_2$.

The work is of course critically dependent on the density expansion of Section 23, which Royer originally felt was convergent and hence truncatable but which he later realized was probably not. Even so, he felt that approximating $\mathfrak{R}(t)$ by $N\mathfrak{R}^{(1)}$ or $N\mathfrak{R}^{(1)}+N^2\mathfrak{R}^{(2)}$ was still justifiable. It is worth our while to quote precisely what he did say about this:

> It appears that recollision cycles cause the higher-order terms to grow indefinitely with time.... Thus, the discussion (239)...only shows that the density expansion of $\mathfrak{R}(t)$ is free of the 'worst' divergences present in the expansion of $\mathfrak{U}(t)$, namely, those arising from successions of uncorrelated collisions of the radiator with clusters of perturbers. Even though the terms $\mathfrak{R}^{(k)}(t)$, $k\geq 3$, grow with time, it is still believed that $\mathfrak{R}(t)$ itself exists in the limit $t\to\infty$, so that one is still justified in approximating $\mathfrak{R}(t)$ by $N\mathfrak{R}^{(1)}(t)$ or $N\mathfrak{R}^{(1)}+N^2\mathfrak{R}^{(2)}$, which also exists as $t\to\infty$. In order to get approximations of higher order in the density, resummation procedures have been proposed, but these still do not seem to eliminate all divergences.

One probably is so justified. In any event, agreement with experiment is the final arbiter in this discussion, and Royer's work provides us with considerable education in methods and techniques. We now turn to the density expansion that is the subject of Royer's controversy with himself.

23 A Density Expansion of the BO (Graphs and Cumulants)

In his original presentation of the Liouville formalism Fano developed a density expansion for the EIT, an expansion of some complexity, of which he utilized only the first term. Somewhat later, Reck, Takebe, and Mead (216) used mostly dimensional arguments in order to develop a density expansion in their line broadening treatment. Not that the consideration of density expansions either began or ended with these authors; the remark was merely intended to indicate a level of interest. In this section we shall discuss the rather different approach of Royer to such an expansion.

This author performed his density expansion using methods more or less peculiar to those *de rigueur* in the equilibrium theory of classical fluids (25, 258), beginning with the expression for V and $\mathfrak{W}(z)$ from Section 22,

$$\mathfrak{W}=\sum_{j=1}^{n}\mathfrak{W}_{s_j}+\sum_{i<j}\mathfrak{W}_{ij}=\sum_{\mu<\nu}\mathfrak{W}_{\mu\nu} \tag{23.1}$$

where the Greek subscripts run over system and n perturbers, $s,1,2,\ldots,n$. "s" is taken as <1 for notational convenience. The function, $F(t)$, appearing in the BO, may be expressed in terms of an $f_{\mu\nu}$ as follows:

$$F(t)\equiv\exp\left(\mathrm{i}\int_{-\mathrm{i}\beta}^{t}dz\,\mathfrak{W}(z)\right)=\prod_{\mu<\nu}(f_{\mu\nu}+1) \tag{23.2a}$$

where

$$f_{\mu\nu} \equiv \exp\left(\mathrm{i} \int_{-\mathrm{i}\beta}^{t} dz\, \mathfrak{W}_{\mu\nu}(z) \right) - 1. \qquad (23.2\mathrm{b})$$

Now we saw in the last section that $\mathfrak{B}(t)$ is very simply related to the right side of Eq. (23.2a) which fact has the consequence that

$$\mathsf{B}(t) = Z_s Z_B \langle F(t) \rangle / Z \qquad (23.2\mathrm{c})$$

At this point the graphical method peculiar to this particular field of the art is introduced. We represent the system (radiator) by an empty circle, the n perturbers by shaded circles, and $f_{\mu\nu}$ by a line connecting two of these circles. Taking our example from the pictorial lexicon of Royer, we define Fig. 23.1 as $f_{s1} f_{12} f_{45} f_{46} f_{56}$, one of the terms which the reader will see arising from the right side of Eq. (23.2a). It surely follows that

$$F(t) = \text{sum of all distinct labeled graphs} \qquad (23.2\mathrm{a}')$$

Now we will agree that the radiator, s, will only be affected by particles "connected," by the $f_{s\nu}$, to it. The 456 triangle of Fig. 23.1, for example, should really be irrelevant to our problem. Therefore, such unconnected graphs should certainly be removed from consideration. As a first step toward accomplishing this we may write our $F(t)$ as follows:

$$F(t) = \Gamma(s)\Xi^{(s)} + \sum_{j=1}^{n} \Gamma_{\text{conn}}(s, j)\Xi^{(s,j)} + \sum_{i<j} \left(\sum \Gamma_{\text{conn}}(s, i, j) \right) \Xi^{(s,i,j)} + \cdots \qquad (23.3)$$

In Eq. (23.3) $\Gamma(s)$ is the graph consisting of the radiator while $\Xi^{(s)}$ is the sum of all graphs not containing the radiator. $\Gamma_{\text{conn}}(s, j)$ is the sum of all connected graphs (o——•) containing the radiator and the particle j while $\Xi^{(s,j)}$ is the sum of all graphs not containing the radiator and particle j. Finally, $\sum \Gamma_{\text{conn}}(s, i, j \cdots)$ is the sum of all connected graphs containing the radiator, particle i, particle j, and so on, while $\Xi^{(s,i,j,\cdots)}$ is the sum of all graphs not containing the radiator, particle i, particle j, and so on. This may

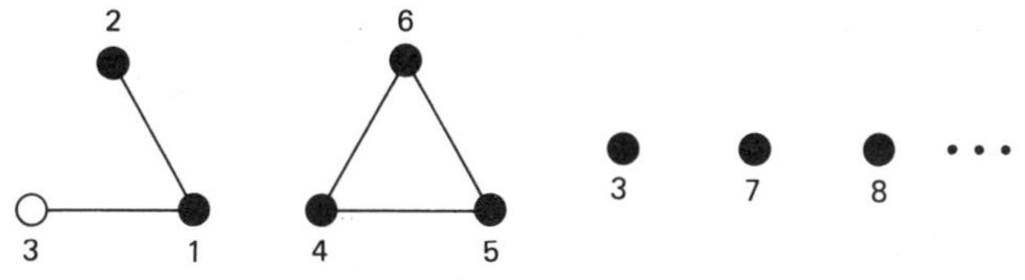

Figure 23.1

be written as

$$F(t)=\sum_{m=0}^{n}\sum_{j_1<}\sum_{j_2<}\cdots\sum_{<j_m}\Big(\sum\Gamma_{\text{conn}}(s,j_1,\cdots,j_m)\Big)\prod_{\mu<\nu}^{(s,j_1,\cdots,j_m)}(f_{\mu\nu}+1) \tag{23.4}$$

where the superscript, $(s, j_1, j_2, \cdots, j_m)$, on the product symbol means that the product is to exclude the emitter and particle j_1 through j_m, $\Sigma\Gamma_{\text{conn}}(s, j_1,\cdots, j_m)$ is the sum of all distinct labeled graphs of vertices $s, j_1,\cdots, j_m$. Royer's example is an appropriate one,

$$\sum\Gamma_{\text{conn}}(s,i,j)=f_{si}f_{ij}+f_{sj}f_{ij}+f_{si}f_{sj}+f_{si}f_{sj}f_{ij} \tag{23.5}$$

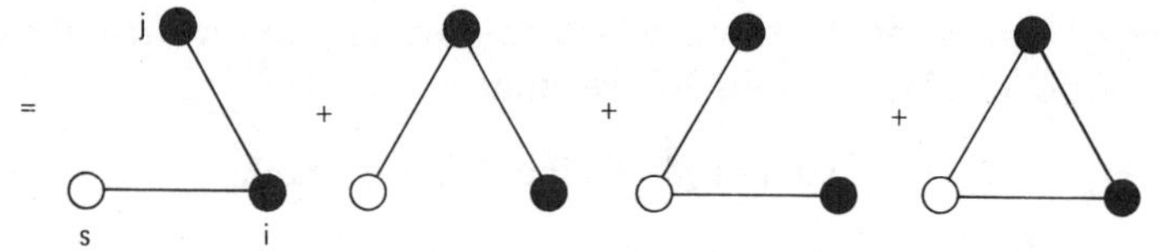

(23.5′)

We shall eventually be interested in the average, $\langle F(t)\rangle$, meaning the average of each m term in Eq. (23.4). Since the quantities $\Gamma^{(s,j_1,\cdots,j_m)}(f_{\mu\nu}+1)$ and $\Sigma\Gamma_{\text{conn}}(s, j_1,\cdots, j_m)$ depend on mutually exclusive sets of atoms, the average of their product will be equal to the product of their averages. If then all perturbers are identical,

$$\langle F(t)\rangle=\sum_{m=0}^{n}\binom{n}{m}\Big\langle\sum\Gamma_{\text{conn}}(s,1,\cdots,m)\Big\rangle\langle F(t)\rangle^{(s,\{m\})} \tag{23.6a}$$

$$\langle F(t)\rangle^{(s,\{m\})}\equiv\Big\langle\prod_{\mu<\nu}^{(s,1,\cdots,m)}(f_{\mu\nu}+1)\Big\rangle \tag{23.6b}$$

In Eq. (23.6a)

$$\binom{n}{m}$$

is of course the binomial coefficient which replaces the sums over $j_1<j_2<\cdots<j_m$. From Eq. (23.2a) it is apparent that $\langle F(t)\rangle^{(s,\{m\})}$ equals $\langle F(t)\rangle$ with the coordinates of the radiator and m perturbers deleted, which fact provides a method for determining the quantity designated Eq. (23.6b).

$F(t)$ is related to $\mathfrak{B}(t)$ through Eq. (23.2c) and hence to the right side of Eq. (22.3). Now, however, the radiator coordinates have been deleted so that, on the adapted right side of Eq. (22.3), $H_0\to H_B$, $H\to H_B+V_B=H_B^{(T)}$, where V_B is the second term on the right of Eq. (22.2). We might say the case is similar for the Liouville operators in Eq. (22.3). Therefore, we obtain

$$\langle F(t)\rangle^{(s,\{m\})}=\mathrm{Tr}_B\Big[\Upsilon_B e^{\beta H_B}e^{-\beta H_B^{(T)}}e^{it\mathfrak{L}_B^{(T)}}e^{-it\mathfrak{L}_B}\Big]_{(n-m)}$$

The Liouvillians in this expression vanish since nothing to the right of them depends on bath coordinates. Therefore,

$$\langle F(t)\rangle^{(s,\{m\})} = \mathrm{Tr}_B\left(e^{-\beta H_B} e^{\beta H_B} e^{-\beta H_B^{(T)}}\right)/Z_B$$
$$= \mathrm{Tr}_B\left(e^{-\beta H_B^{(t)}}\right)/Z_B = Z_B^{(T)}/Z_B \tag{23.7}$$

If $Z_1 = \mathrm{Tr}_1[\exp(-\beta H_1)]$ is the partition function for a single perturber,

$$[Z_B]_{(n-m)} = Z_1^{n-m} \tag{23.8}$$

Now we know (probably) that (1) the partition function for n identical particles is the product of n one-particle partition functions and that (2) the one-particle partition functions will be of exponential form. Consequently,

$$\frac{\partial}{\partial n}\ln\left(Z_B^{(T)}\right)_n = -\beta\mu \tag{23.9}$$

Therefore, we suppose that

$$\left[Z_B^{(T)}\right]_{n-m} = \left[Z_B^{(T)}\right]_n e^{m\beta\mu} \tag{23.10}$$

which puts us in a position to evaluate Eqs. (23.7), (23.6a), and hence $\mathfrak{B}(t)$. In doing so we let n go to infinity, which allows us to use Sterlings formula for certain of the asymptotic factorials in

$$\binom{n}{m}.$$

At the same time, the factor V/V is introduced and the containing volume, V, allowed to go to infinity. The result is

$$\mathfrak{B}(t) = Z_s Z_B^{(T)} \sum_{m=0}^{\infty} \frac{N^m}{m!} \lim_{V\to\infty} \langle V^m \sum \Gamma_{\mathrm{conn}}(s,1,2,\cdots,m)\rangle/Z \tag{23.11a}$$

where

$$N \equiv N_B Z_1 e^{\beta\mu} = N_B Z_1 / Z_1^{(T)} \tag{23.11b}$$

and where N_B is the perturber density, $Z_1^{(T)} \equiv e^{-\beta\mu} = [Z_B^{(T)}]_{(n)}/[Z_B^{(T)}]_{(n-1)}$ the inverse fugacity for the perturber gas, Z_1^{-1} the fugacity for the noninteracting perturber gas.

As $\Gamma_{\mathrm{conn}}(s,1,2,\cdots,m)$ is of order V^{-m} so that the averaged expression in Eq. (23.11a) does not diverge as $V\to\infty$. That this is so is a consequence of the following: $\Gamma_{\mathrm{conn}}(s,1)$ will involve a matrix element, V_{ij}, which is, among other things, over the translational states of the radiator and the perturber. Now each of these states involves a normalization factor $V^{-1/2}$ so that the matrix element introduces the factor V^{-1}. The argument extrapolates straightforwardly for higher values of m. Finally, we emphasize that this expansion is of a quantity which Royer calls the "reduced density," not of the perturber density.

Equation (23.11a) may be rewritten, as the reader has probably inferred, as an exponential. We approach the problem of doing so by dropping the

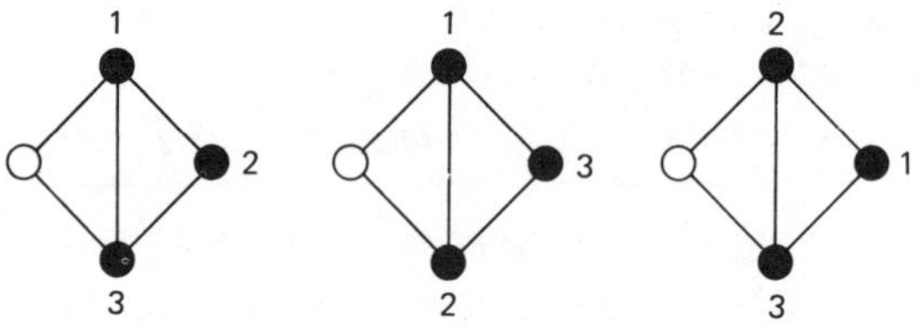

Figure 23.2

multiplicative constant, $Z_s Z_B^{(T)}/Z$, as playing no particular role, and absorbing the factor $N^m V^m$ into what is now termed the "weighted graph," $\Gamma_{\text{conn}}(s, 1, \cdots, m)$. With which we ruminate for a while on graphs, labeled or otherwise.

Let us consider the three-perturber graph of Fig. 23.2. The graphs in that figure are topologically different from, say,

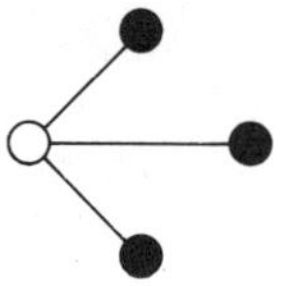 or 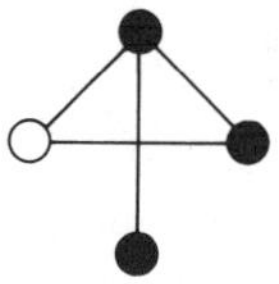 or etc.,

and so on as we cannot simply twist "bonds" about in order to obtain one from the other. There are thus three of the topologically indistinct in Fig. 23.2, that is to say, the topologically distinct 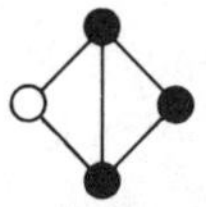 will enter with a factor of 3 in the summation of Eq. (23.6). The multiplicative factor for such a topologically distinct diagram is $m!/\sigma(\Gamma)$ where m is the number of shaded circles and $\sigma(\Gamma)$ is what we shall call the symmetry number. Let us see why this is so.

First, of course, there are, in the case of the three shaded circles of Fig. 23.2, 3! ways in which the numbers 1, 2, 3 may be attached to the circles. Of these six ways, three are eliminated by the symmetry number, $\sigma(\Gamma)$.

The symmetry number, $\sigma(\Gamma)$, is the number of permutations, to include the identity, of the shaded circles which do not change the bonds. As an example, consider Fig. 23.3. This figure would essentially belong to the symmetry point group C_{2v}, although all the covering operations of this group do not contribute to $\sigma(\Gamma)$. The dashed line in Fig. 23.3 corresponds to a twofold symmetry axis, that is, an axis about which a 180° rotation is a covering operation, an axis which is established by the intersection of two reflection planes, one in the plane of the paper. Of these covering operations, remark that one may rotate about the axis in order to obtain Fig. 23.3 from Fig. 23.2, and, since the bonds are unaffected, the rotation yields an acceptable contribution to $\sigma(\Gamma)$. At the same time, a reflection through the plane perpendicular to the plane of the paper is a covering operation, but it simply repeats the effect of our earlier rotation and is thus not to be counted in $\sigma(\Gamma)$.

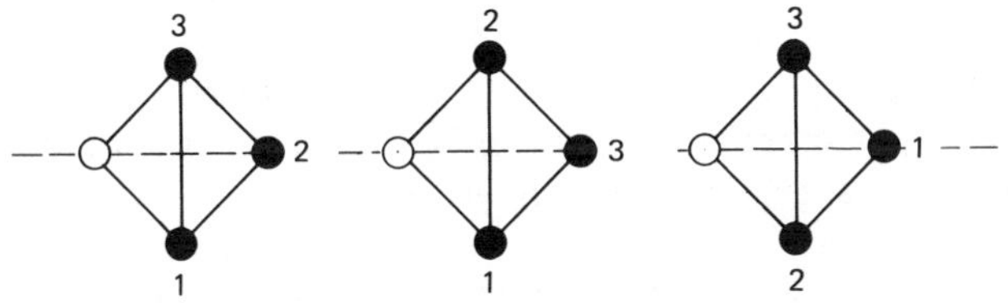

Figure 23.3

The identity operation, however, must be included so that, for this case, $\sigma(\Gamma)=2$. Further, when all the nuclei of Fig. 23.2 are reflected through the symmetry plane, Fig. 23.3 results. Thus, Fig. 23.2 includes, in essence, Fig. 23.3, the total number of nuclear arrangements being reduced by the number of symmetry elements.

We have therefore discovered that

$$\sum \Gamma_{\text{conn}}^{m(\text{labeled})} = \frac{m!\Gamma_{\text{conn}}^{m}}{\sigma(\Gamma_{\text{conn}}^{m})} \tag{23.12a}$$

which surely leads to

$$\sum_{m=0}^{\infty} (m!)^{-1} \sum \Gamma_{\text{conn}}^{m(\text{labeled})} = \frac{\sum \Gamma_{\text{conn}}}{\sigma(\Gamma_{\text{conn}})} \tag{23.12b}$$

where, in Eq. (23.12b), the sum on the right is over all topologically distinct, unlabeled graphs of one unshaded circle and varying numbers of shaded circles. Thus we have arrived at the notion of irreducible graphs, or very nearly so.

An "articulation circle" in a connected graph is a circle whose removal yields two or more disconnected parts, that is,

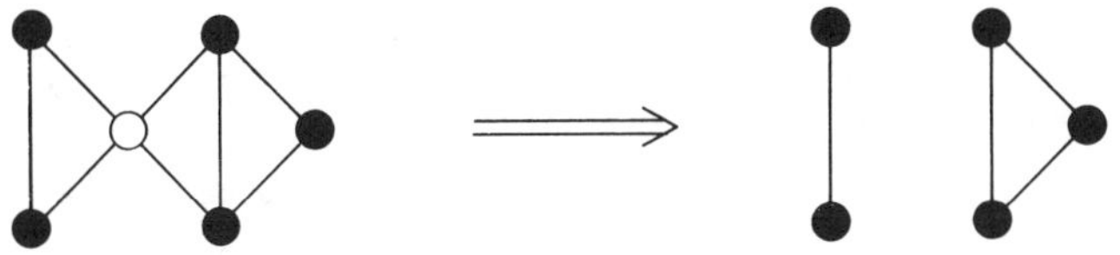

with the removal of the unshaded circle so that the graph has an articulation circle. The graphs of Fig. 23.3 contain no articulation circles and hence are "one-irreducible."

A product is next defined as

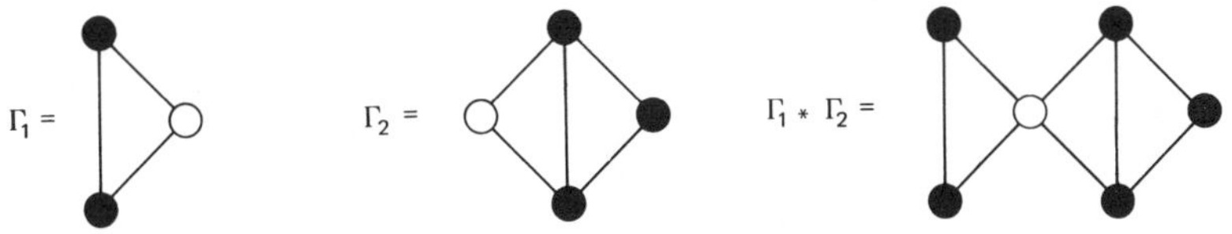

We see that any graph can be formed by one-irreducible graphs attached to articulation circles. We remark, however, that attachment at a shaded circle is written as a product without an asterisk. A graph is "irreducible" if its unshaded circle is not an articulation circle.

The $*$ powers of a graph are defined as $\Gamma *^{k} = \Gamma * \Gamma * \cdots * \Gamma$, k times, the zeroth power being defined for convenience as the unshaded circle alone. From which it follows that

$$\exp^*\Gamma = \sum_{m=0}^{\infty} \frac{\Gamma^{*m}}{m!}$$

and we shall now show that

$$\sum \frac{\Gamma_{\text{conn}}}{\sigma(\Gamma_{\text{conn}})} = \exp^* \sum{}' \frac{\Gamma_{\text{irr}}}{\sigma(\Gamma_{\text{irr}})} \tag{23.13}$$

where the sum on the right is over all distinct irreducible graphs of one unshaded and one ore more shaded circles, the prime excluding the no-shaded-circle case.

Any graph, Γ, of one unshaded and an arbitrary number of shaded circles may, by definition, be written as

$$\Gamma = \prod_j{}' *(\Gamma_{\text{irr}}^{(j)})^{*^{p}} \tag{23.14a}$$

where the prime excludes the case of no shaded circles, the product being over all distinct irreducible graphs of one unshaded and one or more shaded circles. The p_j is simply an exponent that excludes undesirable diagrams by taking the value zero or brings in desired diagrams as many times as required.

The symmetry number of the product is obtainable from the product of the subdiagram symmetry numbers as follows:

$$\sigma(\Gamma) = \prod_j{}' p_j!(\sigma_j)^{p_j}, \qquad \sigma_j \equiv \sigma(\Gamma_{\text{irr}}^{(j)}). \tag{23.14b}$$

Let us work this out for the Γ^{*2} of Fig. 23.4, Γ^{*3} of Fig. 23.5. To begin with we have found that $\sigma(\Gamma) = 2$ for the Γ of the same figure. Therefore, $\sigma(\Gamma)$ will have a factor $4(=2\cdot 2)$ from the rotations (reflections) within the subdiagrams, Fig. 23.3, taken together or individually. But there will be additional symmetry due to the twofold axis (A-A) through the white circle and separating the subdiagrams. A rotation about this axis (corresponding of

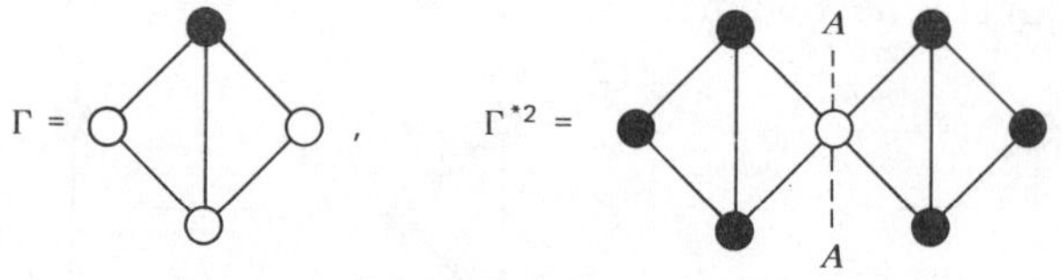

Figure 23.4

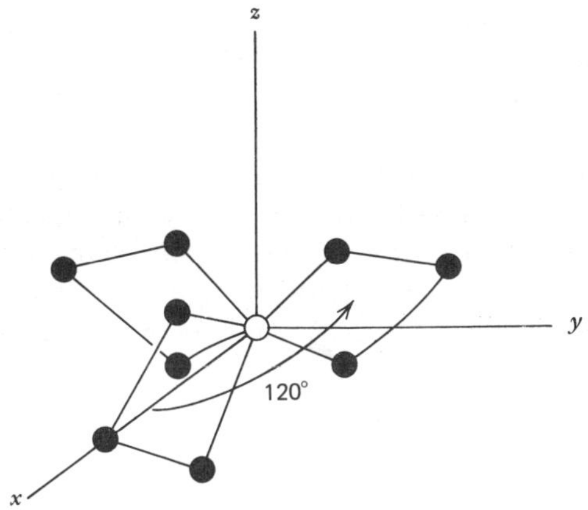

Figure 23.5

course to a reflection through its plane) will be one element, the identity operation another. Therefore, a factor of two will be introduced.

For the Γ^{*3} of Fig. 23.5, we will again have $\sigma_j = 2$. Insofar as permutations which retain the numbers within the three subdiagrams are concerned we should expect $8(=2\cdot2\cdot2)$, but let us investigate the permutations between subdiagrams. The overall figure belongs to symmetry point group C_{3v}, that is, the z axis is a threefold axis, the intersection of three symmetry planes. The permutations arising from this latter fact are obtainable as: (1) three reflections through the three symmetry planes; (2) two rotations, one through 120°, the other through 240°, about the axis; and (3) the identity operation, for a total of six. Therefore, $p_j! = 6$, $p_j = 3$ in Eq. (23.14b) as would appear appropriate.

We may now use Eq. (23.14) in Eq. (23.12) in order to obtain Eq. (23.13) as

$$\sum \Gamma_{\text{conn}}/\sigma(\Gamma_{\text{conn}}) = \prod_j{}' (p_j!)^{-1}(\Gamma_{\text{irr}}^{(j)}/\sigma_j)^{*p_j} = \exp^*\left(\sum_j{}' \frac{\Gamma_{\text{irr}}^{(j)}}{\sigma_j}\right)$$

We now return to Eq. (23.11a) and use Eq. (23.13) in order to rewrite it as

$$\begin{aligned}\mathfrak{B}(t) &= \lim_{V\to\infty} \left\langle \exp\left(\sum{}' \frac{\Gamma_{\text{irr}}}{\sigma(\Gamma_{\text{irr}})}\right)\right\rangle \\ &= \lim_{V\to\infty} \left\langle \sum{}' \Gamma'_{\text{conn}} \right\rangle = \lim_{V\to\infty} \left\langle \exp\left[\sum{}' \Gamma'_{\text{irr}}\right]\right\rangle \qquad (23.15\text{a})\end{aligned}$$

where

$$\Gamma' \equiv \Gamma/\sigma(\Gamma) \qquad (23.15\text{b})$$

Now we encounter the cumulants, an encounter which can no longer be postponed. Not that the subject should be postponed because of any inherent incomprehensibility, the development is reasonably straightforward albeit the notation becomes somewhat alarming. These cumulants came into being as convenient representations of the time-ordered operator products. These products in turn appear in a rather round about expression for the time-ordered exponential, all this time ordering having originated, we will recall, in the desire to express a series of products of an operator, which does not commute with itself for different times, as an exponential. If we can present the general idea of these cumulants without concerning ourselves overmuch with detail, the rote formulas may simply be written down.

If $G(t)$ is an operator, or even if it is not, we will surely agree that

$$G = \exp[\ln G] \tag{23.16a}$$

$$\ln G = \ln[1 + (G-1)] = -\sum_{n=1}^{\infty} n^{-1}(1-G)^n \tag{23.16b}$$

[That Colin Maclaurin, born at Kilmodan in 1698, may be credited with our ability to come to such an agreement has been discussed on many occasions and by many authors. That his equally important contribution to the distillation of (singe malt) Scots whiskey has been almost completely overlooked is one more illustration of the regrettable aspect of overspecialization.]

Now we let $G = \exp[\Sigma_i \lambda_i X_i]$ and use Eq. (23.16b) in order to write

$$\begin{aligned}
\langle \exp \sum_i \lambda_i X_i \rangle &= \langle \exp[\ln G] \rangle = \langle \sum_{n=1} \frac{(-)^{n+1}}{n} (e^{\Sigma \lambda_i X_i} - 1)^n \rangle \\
&= \sum_{n=1}^{\infty} \frac{(-)^{n+1}}{n} \Big[\lambda_{11} \langle X_1(t_1) \rangle + \lambda_{21} \langle X_2(t_1) \rangle \\
&\quad + \cdots + \tfrac{1}{2} \langle [\lambda_{11} X_1(t_1) + \lambda_{21} X_2(t_1) + \cdots] \\
&\quad \times [\lambda_{12} X_1(t_2) + \lambda_{22} X_2(t_2) + \cdots] \rangle + \cdots \Big]^n \\
&= \sum_{n=1}^{\infty} \frac{(-)^{n+1}}{n} \Big[\lambda_{11} \langle X_1(t_1) \rangle + \lambda_{21} \langle X_2(t_1) \rangle \\
&\quad + \cdots + \tfrac{1}{2} \lambda_{11} \lambda_{12} \langle X_1(t_1) X_1(t_2) \rangle + \tfrac{1}{2} \lambda_{11} \lambda_{22} \langle X_1(t_1) X_2(t_2) \rangle + \cdots \Big]^n \\
&= \lambda_1 \langle X_1 \rangle + \lambda_2 \langle X_2 \rangle + \tfrac{1}{2} \lambda_1^2 \langle X_1^2 \rangle \\
&\quad + \tfrac{1}{2} \lambda_1 \lambda_2 \langle X_1 X_2 \rangle - \tfrac{1}{2} \lambda_1^2 \langle X_1 \rangle^2 - \lambda_2^2 \langle X_2 \rangle^2 + \cdots
\end{aligned} \tag{23.17}$$

where we have let $\lambda_{11} \to \lambda_1$ and so on, with the time ordering henceforth being understood. It is a convenience to write this expansion in terms of ordinary cumulants so that

$$\begin{aligned}
\langle e^{\Sigma \lambda_i X_i} \rangle &= \exp\left[\sum_{i=1}^{\infty} \sum_{j_1 \le} \sum_{j_2 \le} \cdots \sum_{\le j_i} \lambda_{j_1} \lambda_{j_2} \cdots \lambda_{j_i} \langle X_{j_1} X_{j_2} \cdots X_{j_i} \rangle_c \right] \\
&= \exp(\langle e^{\Sigma \lambda_i X_i} - 1 \rangle_c)
\end{aligned} \tag{23.18a}$$

where

$$\langle X_1 X_2 \cdots X_i \rangle_c = \text{Sym} \sum_{m=1}^{i} (-)^m (m-1) \sum_{P(m)} \prod_{j=1}^{m} \langle \prod_{k \in K_j} X_k \rangle \quad (23.18b)$$

In Eq. (23.18b) $P(m)$ is a partition of the set $\{1,2,\dots,i\}$ into m subsets, K_j, $j=1,2,\dots,m$ while "Sym" denotes symmetrization with respect to (WRT) the K_k's and WRT the different factors $\langle \Pi_k K_k \rangle$.

From Eq. (23.21) the following simple examples may be worked out and then checked against Eqs. (23.22):

$$\langle X \rangle_c = \langle X \rangle \quad (23.19a)$$

$$\langle X^2 \rangle_c = \langle X^2 \rangle - \langle X \rangle^2 \quad (23.19b)$$

$$\langle X^3 \rangle_c = \langle X^3 \rangle - \tfrac{1}{2} 3 [\langle X^2 \rangle \langle X \rangle + \langle X \rangle \langle X^2 \rangle] + 2 \langle X \rangle^3 \quad (23.19c)$$

$$\langle X_1 X_2 \rangle_c = \tfrac{1}{2} [\langle X_1 X_2 \rangle - \langle X_1 \rangle \langle X_2 \rangle + \langle X_2 X_1 \rangle - \langle X_2 \rangle \langle X_1 \rangle] \quad (23.19d)$$

We now proceed from the ordinary to the $c^{\rightarrow}$ cumulants and, in so doing, introduce the notion of time-ordered differentiation.

First, we remark that we would expect Eq. (23.16a) to hold only if G is the time-ordered exponential of some operator—for reasons given in the last paragraph—which we take as $F(t)$. Something like

$$G(t) = G(0) \mathsf{T}_{\rightarrow} \exp\left[\int_0^t dt' F_{\rightarrow}(t') \right] = \mathsf{T}_{\leftarrow} \exp\left[\int_0^t dt' F_{\leftarrow}(t') \right] G(0) \quad (23.20)$$

where we have indicated time ordering on the operator F by the arrow subscript in order to provide for the likely eventuality of F itself being a product of operators.

We differentiate Eq. (23.17) WRT time with the result

$$\dot{G}(t) = G(t) F_{\rightarrow}(t) = F_{\leftarrow}(t) G(t) \Rightarrow$$

$$F_{\rightarrow}(t) = G^{-1}(t) \dot{G}(t) = [1 + (G-1)]^{-1} \dot{G} = \sum_{n=1}^{\infty} (-)^{n-1} (G-1)^{n-1} \dot{G} \quad (23.21)$$

which, using the fact from Eq. (22.16b) that

$$\frac{d}{dt} \ln G = \sum_{n=1}^{\infty} (-)^{n-1} n^{-1} \frac{d}{dt} (G-1)^n$$

tells us

$$F_{\rightarrow}(t) = G^{-1}(t) \dot{G}(t) = \left(\frac{d}{dt} \right)_{\rightarrow} \ln G(t) \quad (23.22)$$

We have introduced the time-ordered differentiation symbol that says differentiate each member of the product and place it to the right of the product, thus:

$$\left(\frac{d}{dt} \right)_{\rightarrow} \mathrm{ABC} = \mathrm{BC\dot{A}} + \mathrm{AC\dot{B}} + \mathrm{AB\dot{C}} \quad (23.23)$$

Equations (23.22) and (22.23) are indispensable to the definition of the $c^{\rightarrow}$ cumulants whose derivation is begun by specificizing our brac operator thus

$$\langle\ \rangle \equiv \{\mathsf{T}_{\rightarrow}\}_{\text{av}} \tag{23.24}$$

wherein the av referred to is a statistical average. The $c^{\rightarrow}$ cumulants are now defined in a manner analogous to the ordinary cumulants

$$\left\langle \exp\left[\int^{t} dt' \sum_{j} \lambda_j(t')X_j(t')\right]\right\rangle = \mathsf{T}_{t_k^{\rightarrow}} \exp\left[\int^{t} dt_1 \int^{t_1} dt_2 \cdots \int^{t_{k-1}} dt_k \right.$$

$$\left. \cdot \sum_{j_1 \le} \sum_{j_2 \le} \cdots \sum_{\le j_k} \lambda_{j_1}(t_1)\lambda_{j_2}(t_2)\cdots\lambda_{j_k}(t_k)\langle X_{j_k}(t_k)\cdots X_{j_1}(t_1)\rangle_{c^{\rightarrow}}\right] \tag{23.25a}$$

where $T_{t_k^{\rightarrow}}$ orders the cumulants such that t_k increases from left to right.

Thus, the $c^{\rightarrow}$ cumulants are the coefficients of $\Pi_i[dt_i\lambda_{j_i}(t_i)]$ in the expansion of the time-ordered derivative,

$$\left(\frac{d}{dt}\right)_{\rightarrow} \ln\left\langle \exp\left[\int^{t} dt' \sum_{j} \lambda_j(t)X_j(t)\right]\right\rangle$$

A compact form of Eq. (23.25) is obtained by setting $\Sigma_j \lambda_j(t)X_j(t) = X(t)$,

$$\left\langle \exp\left[\int^{t} dt' X(t')\right]\right\rangle = \mathsf{T}_t \exp\left\{\int^{t} dt' \left\langle \exp\left[\int^{t'} dt'' X(t'')\right] X(t')\right\rangle_{c^{\rightarrow}}\right\}$$

$$= \mathsf{T}_{c^{\rightarrow}} \exp\left\{\left\langle \exp\left[\int^{t} dt' X(t')\right] - 1\right\rangle_{c^{\rightarrow}}\right\} \tag{23.25b}$$

The symbol $\mathsf{T}_{c^{\rightarrow}}$ orders the cumulants such that the largest set of time arguments in each increase from left to right.

Through Eqs. (23.22) we may relate the $c^{\rightarrow}$ to the ordinary cumulants as follows:

$$\left\langle \exp\left[\int^{t} dt' X(t')\right] X(t)\right\rangle_{c} = \left(\frac{d}{dt}\right)_{\rightarrow} \left\langle \exp\left[\int^{t} dt' X(t')\right] - 1\right\rangle_{c} \tag{23.25c}$$

Sufficient rumination on Eq. (23.25c) tells us that we may use Eq. (23.18b) to determine $c^{\rightarrow}$ cumulants if we (1) replace X_i by $X_i(t_i)$ and (2) assure that, in each product

$$\prod_{j} \left\langle \prod_{k \in K_j} X_k \right\rangle$$

the factor $\langle \Pi_k X_k \rangle$ containing the largest time argument is placed to the right of all other factors. The two following examples correspond to $t_1 > t_2 > t_3$, $X_i(t_i) = X_i$:

$$\langle X_1 X_2 \rangle_{c^{\rightarrow}} = \langle X_2 X_1 \rangle - \langle X_2 \rangle \langle X_1 \rangle \tag{23.26a}$$

$$\langle X_3 X_2 X_1 \rangle_{c^{\rightarrow}} = \langle X_3 X_2 X_1 \rangle - \langle X_3 X_2 \rangle \langle X_1 \rangle - \langle X_3 \rangle \langle X_2 X_1 \rangle$$
$$- \langle X_2 \rangle \langle X_3 X_1 \rangle + \langle X_3 \rangle \langle X_2 \rangle \langle X_1 \rangle + \langle X_2 \rangle \langle X_3 \rangle \langle X_1 \rangle \tag{23.26b}$$

where we remark the application of the hitherto unique $\mathsf{T}_{c\rightarrow}$, in the continuing appearance of X_1 on the right and then turn to the fundamental theorem of cumulants. In order to do so, however, we must first define brac independence and brac connectedness.

Given two sets of operators $\{X_i\colon i\in I\}$ and $\{X_j\colon j\in J\}$. These two sets are said to be brac independent if the brac average of a product of two operators factors into a product of the brac average of the operators from each set, that is,

$$\left\langle \prod_{l\in L} X_l \right\rangle = \left\langle \prod_{i\in I\cap L} X_i \right\rangle \left\langle \prod_{j\in J\cap L} X_j \right\rangle$$

when $L\subset I\cup J$.

If a set of operators cannot be divided into two or more brac independent sets, the set is said to be brac connected.

The reader might wish to mull over the connection between brac independence and the breakup of the density matrix into product form by the dilute approximation. The following theorem is due to Kubo (167).

Fundamental Cumulant Theorem If the X_i's in a cumulant are not brac connected, the cumulant vanishes.

Proof In order to prove this, we choose two sets $\{X_i(t)\colon i\in Z\}$ and $\{Y_j(t)\colon j\in Z\}$ that are statistically independent and whose element combinations are expressible as

$$X\equiv\int_0^a dt'\sum_i \lambda_i(t')X_i(t')$$

$$Y\equiv\int_a^t dt'\sum_j \kappa_j(t')Y_j(t'), \qquad t>a$$

where $\lambda_i(t)$ and $\kappa_j(t)$ are arbitrary scalar functions. We may use the statistical independence of the two sets of Eq. (23.25b) in order to write

$$\langle e^{X+Y}\rangle = \langle e^X\rangle\langle e^Y\rangle = \left[\mathsf{T}_{c\rightarrow}\exp(\langle e^X-1\rangle_{c\rightarrow})\right]\left[\mathsf{T}_{c\rightarrow}\exp(\langle e^Y-1\rangle_{c\rightarrow})\right]$$
$$=\mathsf{T}_{c\rightarrow}\exp\left[\langle e^X-1\rangle_{c\rightarrow}+\langle e^Y-1\rangle_{c\rightarrow}\right] \tag{23.27a}$$

Again using Eq. (23.25b) but not the statistical independence we see that

$$\langle e^{X+Y}\rangle = \mathsf{T}_{c\rightarrow}\exp(\langle e^{X+Y}-1\rangle_{c\rightarrow})$$
$$=\mathsf{T}_{c\rightarrow}\exp\left[\langle e^X-1\rangle_{c\rightarrow}+\langle e^Y-1\rangle_{c\rightarrow}+\langle (e^X-1)(e^Y-1)\rangle_{c\rightarrow}\right] \tag{23.27b}$$

Since Eq. (23.27a) must equal Eq. (23.27b),

$$\langle (e^X-1)(e^Y-1)\rangle_{c\rightarrow}=0$$

in the latter equation. The functions $\lambda_i(t)$ and $\kappa_i(t)$ have been introduced in such a way that they can be made arbitrary. This implies the disappearance of all $c^{\rightarrow}$ cumulants containing both $X_i(t_i)$'s and $Y_j(t_j)$'s. ■

This theorem has important ramifications, perhaps the most important from our point of view is the uniform convergence of the cumulant series, which allows us to view it as a perturbation expansion in powers of X which may legitimately be truncated. To see that such may indeed be the case, we first consider the situation where $X(t)$ is never correlated with itself at different times, that is to say, where $X(t_k)$ is statistically independent of $X(t_{k-1})$ is statistically independent of $X(t_{k-2})$ and so on. The fundamental cumulant theorem leads to the following result:

$$\left\langle \exp\left[\int^t dt' X(t')\right]\right\rangle = 1 + \sum_{k=1}^{\infty} \int^t dt_1 \int^{t_1} dt_2 \cdots \int^{t_{k-1}} dt_k \langle X(t_k) X(t_{k-1}) \cdots X(t_1)\rangle$$

$$\Rightarrow 1 + \sum_{k=1}^{\infty} \int^t dt_1 \cdots \int^{t_{k-1}} dt_k \langle X(t_k)\rangle \langle X(t_{k-1})\rangle \cdots \langle X(t_1)\rangle$$

$$= \mathsf{T}_{\rightarrow} \exp\left[\int^t dt' \langle X(t)\rangle\right] = \mathsf{T}_{\rightarrow} \exp\left[\int^t dt' \langle X(t')\rangle_{c\rightarrow}\right] \quad (23.28a)$$

that is, only the first-order cumulant is retained. Pairwise correlations analogously lead to

$$\left\langle \exp\left[\int^t dt' X(t')\right]\right\rangle \Rightarrow \mathsf{T}_{\rightarrow} \exp\left[\int^t dt' \langle X(t')\rangle_{c\rightarrow} + \int^t dt' \int^{t'} dt'' \langle X(t'') X(t')\rangle_{c\rightarrow}\right]$$
$$(23.28b)$$

and so on.

Following Eqs. (23.15) we diverged from what was reputedly the development of a BO expansion in order to cogitate on cumulants. We now return to the BO expansion.

We begin by applying Eq. (23.18a) to Eq. (23.15a);

$$B(t) = \exp\left[\lim_{V\to\infty} \left(\langle e^{\Sigma'\Gamma_{\text{irr}}} - 1\rangle_c\right)\right]$$

$$= \mathsf{T}_{z\rightarrow} \exp\left[\int_{-i\beta}^{t} dz \left(\frac{d}{dz}\right)_{\rightarrow} \lim_{V\to\infty} \left\langle \exp\left[\sum \Gamma'_{\text{irr}}(z)\right]\right\rangle_{c(\Gamma_{\text{irr}})}\right] \quad (23.29)$$

where we have again encountered the time-ordered derivative.

The subscript $c(\Gamma_{\text{irr}})$ refers to cumulants constructed on the irreducible graphs, really the only sort one could construct in the present case. The following example of such a cumulant, given by Royer, appears as appropriate for illustration as any:

$$\langle f_{s1} f_{s2} f_{12} f_{s3}\rangle_{c\{\Gamma_{\text{irr}}\}} = \langle f_{s1} f_{s2} f_{12} f_{s3}\rangle$$
$$- \tfrac{1}{2}\left(\langle f_{s1} f_{s2} f_{12}\rangle \langle f_{s3}\rangle + \langle f_{s3}\rangle \langle f_{s1} f_{s2} f_{12}\rangle\right) \quad (23.30)$$

Or

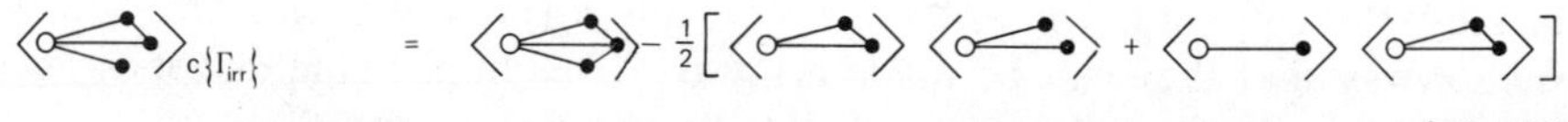

(23.30′)

Equation (23.29) is to be compared to Eq. (23.18b), such a comparison leading to

$$\hat{\mathfrak{R}}(t) = -\mathrm{i}\left(\frac{d}{dt}\right)_{\to} \lim_{V\to\infty} \left\langle \exp\left[\sum{}' \Gamma_{\mathrm{irr}}(t)\right]\right\rangle_{c\{\Gamma_{\mathrm{irr}}\}}$$

$$= -\mathrm{i}\left(\frac{d}{dt}\right)_{\to} \lim_{V\to\infty} \langle \sum \Gamma'_{\mathrm{conn}}(t)\rangle_{c\{\Gamma_{\mathrm{irr}}\}} \tag{23.31}$$

Comparing Eqs. (23.11a) and (23.31) we see that the latter may be written

$$\hat{\mathfrak{R}}(t) = -\mathrm{i}\left(\frac{d}{dt}\right)_{\to} \sum_{m=2}^{\infty} \frac{N^m}{m!} \lim_{V\to\infty} \langle V^k \sum \Gamma_{\mathrm{conn}}(s,1,\cdots,m)\rangle_{c\{\Gamma_{\mathrm{irr}}\}}$$

$$= -\mathrm{i}N \lim_{V\to\infty} \langle V\dot{f}_{s1}\rangle + \frac{N^2}{2\mathrm{i}}\left(\frac{d}{dt}\left[\lim_{V\to\infty} \langle V^2(2f_{s1}f_{12} + f_{s1}f_{s2}f_{12} + f_{s1}f_{s2})\rangle\right]\right.$$

$$\left. -2\lim_{V\to\infty} \langle Vf_{s1}\rangle \lim_{V\to\infty} \langle V\dot{f}_{s2}\rangle\right) + \cdots \tag{23.32}$$

If we designate the approximation of $\mathfrak{R}(t)$ to first order in the density by $\mathfrak{R}^{(1)}(t)$, the latter may surely be obtained from the first term in Eq. (23.32) as modified by Eq. (22.18b) to

$$\hat{\mathfrak{R}}^{(1)}(t) = -\mathrm{i}N \lim_{V\to\infty} \langle V\dot{f}_{s1}(t)\rangle = Ne^{\mathrm{i}t\mathfrak{L}_s} \lim_{V\to\infty} \langle Vm^t_{s1}\rangle e^{-\mathrm{i}t\mathfrak{L}_s} \Rightarrow$$

$$\mathfrak{R}^{(1)} = N \lim_{V\to\infty} \langle Vm^t_{s1}\rangle \tag{23.33}$$

where we have followed the terminology of Royer vis-à-vis m^t_{1s}, which author in turn chose the symbolism to agree with Fano's two-particle collision matrix, pointing out that it may be considered a Liouville space T matrix for the scattering of particles s and 1. This two-particle t matrix is of course

$$m^t_{s1} = -\mathrm{i}e^{-\mathrm{i}t\mathfrak{L}_0}\dot{f}_{s1}(t)e^{\mathrm{i}t\mathfrak{L}_0}$$

$$= \mathsf{T}_{\to}\exp\left[\mathrm{i}\int_{-\mathrm{i}\beta}^{t} dz\, \mathfrak{W}_{s1}(z)\right]\mathfrak{V}_{s1}(t) \Rightarrow \mathsf{T}_{\to}\exp\left[\mathrm{i}\int_{-\mathrm{i}\beta-t}^{0} dz\, \mathfrak{W}_{s1}(z)\right]\mathfrak{V}_{s1}(0) \tag{23.34}$$

where the time scale has been shifted so as to bring the last interaction, $\mathfrak{V}_{s1}$, to time zero, thus changing the lower limit of integration as indicated. Now it is essential that we go to the limit, $t=\infty$, and this may be effected with the following argument.

The time of collision, t_c, will be finite or at least we shall assume it to be so. (That it will be finite is of course true. However, that it will be of short enough duration and sufficiently well defined to allow us to delineate a t_c is obviously an approximation.) For times outside the range t_c, V and hence $\mathfrak{V}$ will be zero. Therefore, if the collision period defined by t_c does not start at a $\mathcal{R}[z] > -t_c$, the $\mathfrak{V}_{s1}(0)$ in Eq. (23.34) will be zero out m^t_{s1}. We may therefore let $t\to-\infty$ with impunity since only that part of the interaction corresponding to $\mathcal{R}[z] \geq -t_c$ will survive. We are thus led to

$$m^t_{s1} \overset{t>t_c}{\to} m^{\infty}_{s1} = \mathsf{T}_{\to}\exp\left[\mathrm{i}\int_{-\infty}^{0} dt\, \mathfrak{V}(t)\right]\mathfrak{V} \tag{23.34}$$

Because $\mathcal{R}[z]$ becomes so much larger than $\mathcal{I}[z]$ when we let $t \to -\infty$ we begin by taking $z \doteq \mathcal{R}[z]$ and end by realizing that such is a quite reasonable result. The imaginary time relates to the correlation of the two particles as illustrated by the entanglement of their coordinates in the density matrix. Now all collisions occur within a period, t_c, of time zero. Therefore, at $t = -\infty$ the two particles were far apart, totally uncorrelated, and the result follows. Equation (23.33) now becomes

$$\mathfrak{R}^{(1)}(\infty) = N \lim_{V \to \infty} \frac{V}{Z_1} \mathrm{Tr}_1\left[e^{-\beta H_1} m_{s1}^{\infty} \right] = N \frac{V}{Z_1} \mathrm{Tr}_1\left[e^{-\beta H} m_{s1}^{\infty} \right] \quad (23.35)$$

The reader will probably recall that Z_1 is proportional to the volume. If he does not, he might look at the matter this way: Z_1 is the sum of the diagonal matrix elements over, among other things, translational states. Because this is a continuum, it may be replaced by an integral multiplied by a volume.

Equation (23.25) is the result first obtained by Baranger, later obtained by Fano and used by us in Section 22. We now consider the work of Ben-Reuven which has carried us so far toward a complete theory of spectral line broadening in the language of the Liouville operator and the Liouville space.

24 The Generalized Liouville Approach

Let us begin with Eq. (21.5) in which we recognize the $\mathfrak{L}$-space Greens function [Eq. (10.16a)], $\mathfrak{G} = [\omega - \mathfrak{L}]^{-1}$. A further inspection of Eq. (21.5) (keeping in mind the lessons learned in Section 10) tells us the following: (1) The dipole moment operators, will, perhaps after a certain amount of recasting, correspond to vectors in $\mathfrak{L}$-space. (2) By Eq. (10.10), the trace can be taken to correspond to an inner product in $\mathfrak{L}$-space. Therefore, Eq. (21.5) can be written in a form which will look something like

$$\langle\langle A | \mathfrak{G} | B \rangle\rangle$$

Our point of departure for a specific determination of the $\mathfrak{L}$-space vectors will be the electric dipole moments of Eq. (21.5). For more generality, these may be replaced by the dipole moment—or some other moment—due to the n molecules of the entire emitting or absorbing gas as in Eqs. (10.7). The jth component of this H-space moment operator is, according to Eq. (10.8),

$$M^i(-k) = \sum_{jl} \langle j | \mathsf{d}^i e^{i\mathbf{k}\cdot\mathbf{R}} | 1 \rangle \mathsf{P}_{jl}(k) \quad (24.1a)$$

where

$$\mathsf{P}_{jl}(k) = \sum_A \mathsf{P}_{jl}^A = \sum_{A=1}^{n} \{ | j, \mathbf{p}+\mathbf{k} \rangle \langle 1, \mathbf{p} | \}_A \quad (24.1b)$$

as in Eq. (10.12), the sum over P_{jl}^A running over the n molecules of the assemblage. This is perhaps an appropriate place for us to comment on the Hamiltonian of Eq. (21.6a) and the dipole moment of Eq. (24.1a).

Let us suppose that we are considering foreign-gas broadening so that the system is a single atom, the bath some number of atoms of a species differing from that of the radiator. Then, as in Eq. (21.6a), the Hamiltonian will be that for an isolated radiator plus that for the certain number of isolated perturbers plus the interaction between these various atoms. The dipole moment in Eq. (24.1a) is taken as that for the isolated radiator. In short, we are inferring that all the atoms maintain their elementary atomic identity during the broadening collisions. Which, as Royer (241) pointed out, may be acceptable for distant collisions but is of doubtful validity for close collisions and frequencies in the line wing. In what, by virtue of the Born-Oppenheimer approximations involved, that author called an adiabatic theory, he chose as Hamiltonian in place of Eq. (21.6a)

$$H = K_{\mathrm{el}} + K_{\mathrm{nucl}} + \mathsf{V}$$

where K_{el} is the kinetic operator for the collection of orbital electrons associated with all involved atoms, K_{nucl} that for all nuclei, and V the potential of interaction between all particles, electrons and nuclei. We now return to Eqs. (24.1).

If we choose the operators of those equations to form our $\mathfrak{L}$-space vectors, we will obviously not be using the vectors spanning the entirety of $\mathfrak{L}$-space. Rather, we will be using those spanning the subspace defined by Eq. (24.1b), what Ben-Reuven called the subspace of the single-molecule excitation modes, the SME subspace. The operator P_{jl}^{A} is a sum of single-molecule excitation operators so that molecules are excited one at a time, and the nomenclature appears appropriate. The point of course is that we wish our Greens function in this subspace, and we may arrange this by replacing $\mathfrak{G}$ by $\mathsf{P}\mathfrak{G}\mathsf{P}$, the P a projection operator projecting into this subspace. It therefore remains for us to select a set of basis vectors appropriate to the subspace.

This may be accomplished by choosing the $\mathsf{P}_{jl}(k)$ as kets or column vectors, $|\rangle\rangle$, a modification of them as bras or row vectors, $\langle\langle|$, as follows:

$$\overline{\mathsf{P}}_{jl}(k) = n^{-1}\Upsilon_l^{-1}\mathsf{P}_{jl}(k) \tag{24.2a}$$

where

$$\Upsilon_1 = \mathrm{Tr}\{|l\rangle\langle l|\Upsilon\} \tag{24.2b}$$

Ben-Reuven at this point has applied what he calls the statistical random phase approximation (SRPA) to the density matrix which simply amounts to diagonalizing it, thus

$$\langle i|\Upsilon_{\mathrm{SRPA}}|j\rangle = \Upsilon_i\delta_{ij} \tag{24.2c}$$

Such an approximation leads to an orthogonal set of basis vectors,

$$\begin{aligned}\langle\langle\overline{\mathsf{P}}_{jl}(\mathbf{k})|\mathsf{P}_{km}(\mathbf{k})\rangle\rangle &= \mathrm{Tr}\{\Upsilon_{\mathrm{SRPA}}\overline{\mathsf{P}}_{jl}^{+}(\mathbf{k})\mathsf{P}_{km}(\mathbf{k})\} \\ &= \delta_{jk}\delta_{1m}, \qquad (\mathbf{k}\neq 0).\end{aligned} \tag{24.3}$$

[For $\mathbf{k}=0$, we must include P_{jj} and $\langle\langle\mathsf{P}_{jj}(0)|\mathsf{P}_{kk}(0)\rangle\rangle = \mathrm{N}\delta_{jk}$.]

Therefore, through the agency of Eqs. (24.1) and (24.2), we have included the operators d and Υ which are to be found in Eq. (21.5), although the reader should remark that the volume factor of Eqs. (10.7c) must be included. [This is of no practical importance to line shape considerations in that V^{-1} and an n to cancel the normalizing one of Eq. (24.2a) multiply the entire line shape expression. Thus the absolute intensities everywhere in the spectral line vary linearly with the emitter density as is to be expected.] The spectral line now looks like this

$$I_{\alpha\beta}(\mathbf{k},\omega)=N\sum_{ijkl}\Upsilon_j(\mathsf{d}_{ij}^{\alpha})^*\mathsf{d}_{kl}^{\beta}\langle\langle\overline{\mathsf{P}}_{ij}(\mathbf{k})|\mathsf{P}_k\mathfrak{G}\mathsf{P}_k|\mathsf{P}_{kl}(\mathbf{k})\rangle\rangle$$

to which we apply Eq. (10.25) in order to obtain

$$I_{\alpha\beta}(\mathbf{k},\omega)=N\sum_{ijkl}\Upsilon_j(\mathsf{d}_{ij}^{\alpha})^*\mathsf{d}_{kl}^{\beta}\langle\langle\mathsf{P}_{ij}(\mathbf{k})|[\omega-\Omega(\mathbf{k})-\Sigma_c(\mathbf{k},\omega)]^{-1}|\mathsf{P}_{kl}(k)\rangle\rangle \tag{24.4a}$$

where

$$\Omega(\mathbf{k})=\mathsf{P}_k\mathfrak{L}_0\mathsf{P}_k \tag{24.4b}$$

$$\Sigma_c(\mathbf{k},\omega)=\mathsf{P}_k\mathfrak{T}_c(\omega)\mathsf{P}_k. \tag{24.4c}$$

Our projection operator, P_k, projects into the subspace $\mathfrak{L}_1(k)$, the residual operator into the space $\overline{\mathfrak{L}}_1(k)$ where of course

$$\mathfrak{L}_1(k)\oplus\overline{\mathfrak{L}}_1(k)=\mathfrak{L}(k) \tag{24.5}$$

The next problem is that of evaluating $\mathfrak{T}$. In order to do so we begin by substituting Eq. (10.16a) for $\mathfrak{G}$ into Eq. (10.18) in order to obtain

$$\mathfrak{T}(\omega)=(\mathsf{V}I^*-I\mathsf{V}^*)\left\{II^*-(\mathrm{i}/2\pi)\int d\epsilon\mathsf{G}(\epsilon+\omega)\mathsf{G}^*(\epsilon)(\mathsf{V}I^*-I\mathsf{V}^*)\right\} \tag{24.6}$$

The first step is to rewrite $(\mathsf{V}I^*-I\mathsf{V}^*)II^*$ as an expression within our integral over ϵ, the idea of course being the consolidation of Eq. (24.6) under the integral sign. In order to do so we consider the following:

$$\int_{-\infty}^{\infty}d\epsilon\mathsf{V}\mathsf{G}^*(\epsilon)=\mathsf{V}\int_{-\infty}^{\infty}d\epsilon[\epsilon-H^*-\mathrm{i}\eta]^{-1} \tag{24.7}$$

which, since the integral has a pole at $\epsilon=H^*+\mathrm{i}\eta$, may be evaluated by a contour in the complex ϵ plane as in Fig. 24.1. The reader will probably wish to demonstrate that the integral around the half circle is zero. Therefore, the integral of Eq. (24.7) is equal to the integral along the contour about the simple pole or isolated singular point, $\epsilon=H^*+\mathrm{i}\eta$. This integral in turn is equal to $2\pi\mathrm{i}$ multiplied by the residue at the pole, that is to say, the coefficient of the inverse linear term in the Laurent expansion,

$$\begin{aligned} f(\epsilon)&=\sum_{n=0}^{\infty}a_n(\epsilon-\epsilon_0)^n+b_1(\epsilon-\epsilon_0)^{-1}+b_2(\epsilon-\epsilon_0)^{-2}+\cdots \\ &=b_1(\epsilon-\epsilon_0)^{-1}=(\epsilon-\epsilon_0)^{-1} \end{aligned} \tag{24.8}$$

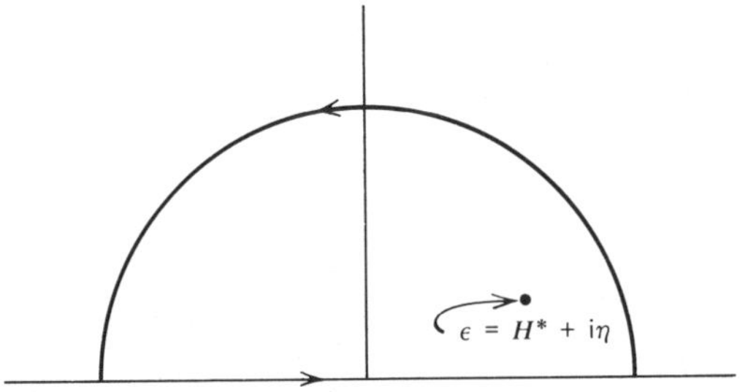

Figure 24.1

Therefore, the residue is unity, the value of the integral $2\pi\mathrm{i}$. Hence we may write

$$\mathsf{V}=(2\pi\mathrm{i})^{-1}\int d\epsilon\,\mathsf{V}\mathsf{G}^*(\epsilon). \tag{24.9a}$$

Now for terms in Eq. (13.6) including V and V^*, say $\mathsf{V}I^*\mathsf{G}\mathsf{G}^*I\mathsf{V}^*$, we utilize relations such as

$$\mathsf{V}\mathsf{G}=\mathsf{V}\mathsf{G}^0+\mathsf{V}\mathsf{G}^0T\mathsf{G}^0=\mathsf{V}\mathsf{G}^0+\left[T\mathsf{G}^0-\mathsf{V}\mathsf{G}^0\right]=T\mathsf{G}^0 \tag{24.9b}$$

from the combination of Eqs. (10.4) and (10.5).

At this point we introduce the fact that the asterisked operators commute. That such is the case may be demonstrated as follows: given an operator B in H-space to which corresponds the vector $|B\rangle\rangle$ in $\mathfrak{L}$-space, then to B^+ corresponds $\langle\langle B|$ as we are aware. It follows that the matrix elements of B^+ in H-space are related to those of B^*—the complex conjugate of B—as $B_{ij}^+=B_{ji}^*$. Therefore, one of the bilinear forms which are operators in $\mathfrak{L}$-space has the matrix element

$$(AX)_{ij}=\sum_{kl}A_{ij,kl}X_{kl}=\sum B_{ik}C_{jl}^*X_{kl}=(BXC^+)_{ij}$$

From this result the product of a pair of bilinear forms may be worked out as

$$(AB^*)(CD^*)X=ACB^*D^*X=ACXD^+B^+$$

which demonstrates the commutation of the asterisked and unasterisked operators. This fact, together with Eqs. (24.9), allows us to rewrite Eq. (24.6) as

$$\begin{aligned}T(\omega)=(2\pi\mathrm{i})^{-1}\int d\epsilon\{&\mathsf{V}[I+\mathsf{G}(\epsilon+\omega)\mathsf{V}]\mathsf{G}^*(\epsilon)\\&+\mathsf{G}(\epsilon+\omega)\mathsf{V}^*[I^*+\mathsf{G}^*(\epsilon)\mathsf{V}^*]-T(\epsilon+\omega)\mathsf{G}^0(\epsilon+\omega)\mathsf{G}^{0*}(\epsilon)T^*(\epsilon)\\&-\mathsf{G}^0(\epsilon+\omega)T(\epsilon+\omega)T^*(\epsilon)\mathsf{G}^{0*}(\epsilon)\}\end{aligned} \tag{24.10}$$

To the square brackets in the first two terms of this equation we apply Eqs.

(10.4) and (10.5) in order to obtain, for example,

$$(2\pi \mathrm{i})^{-1}\int d\epsilon\, T(\epsilon+\omega)\mathsf{G}^*(\epsilon)=(2\pi \mathrm{i})^{-1}\int d\epsilon\, T\mathsf{G}^{0*}[1+T^*\mathsf{G}^{0*}]$$

$$=(2\pi \mathrm{i})^{-1}\int d\epsilon\, T(\epsilon+\omega)\delta(\epsilon-H^*)$$

$$+(2\pi \mathrm{i})^{-1}\int d\epsilon\, \mathsf{G}^{0*}(\epsilon)T(\epsilon+\omega)T^*(\epsilon)\mathsf{G}^{0*}(\epsilon) \tag{24.11}$$

by the arguments involved in evaluating Eq. (24.7). One may use results analogous to Eq. (24.11) together with by now familiar Greens functions relations in order to obtain the following:

$$\mathfrak{T}(\omega)=\int d\epsilon\big[(T\Delta_0^*-\Delta_0 T^*)+(2\pi \mathrm{i})^{-1}\mathfrak{D}TT^*\mathfrak{D}\big] \tag{24.12a}$$

where

$$\mathfrak{D}=\mathsf{G}^0 I^*-I\mathsf{G}^{0*} \tag{24.12b}$$

$$\Delta_0(x)=\delta(x-H_0) \tag{24.12c}$$

The asterisked operators in Eq. (24.12a) have $\epsilon+\omega$ as arguments, the un-asterisked ϵ. These arguments have been suppressed in order to emphasize the immensely impressive compactness of this result. It might be of some interest to write down Fano's analogous expression, his Eq. (55), but it would take the greater part of a page to do so. We recall that we have obtained the supermatrix elements of Eqs. (24.12) as Eqs. (10.28). From a quite general viewpoint then, Eqs. (24.4) and (10.28) tell us the spectral line shape. As Ben-Reuven remarks, the line shape problem has been reduced to one of calculating the proper "self-frequency matrix," $\Sigma_c(\mathbf{k},\omega)$.

Ben-Reuven has considered several approximations to this general result among which we remark:

1 The binary collision approximation (BCA). We will recall from Eqs. (10.2) and (10.4) that the first Born approximation consists in replacing T by V, the second of (roughly) including V^2, and so on. V is almost universally taken as a sum of pairwise interactions between the molecules of the gas. The first Born approximation means that the supermatrix element of Eq. (24.4c) will yield a contribution to the self energy supermatrix proportional to the gas density, the second Born approximation a contribution proportional to N^2, and so on. This occurs in the first case when the result of summing over the pairwise potentials, n, combines with the factor, V^{-1}, introduced by the pairwise translational functions which constitute an obvious portion of the matrix elements over V. An extrapolation of this argument leads to the general result. Ben-Reuven concludes that the BCA

holds for $Nd^3 \ll 1$, a the range of the potential, a condition which usually holds for gas densities below about 10^{-2} amagats.

2 The BCA interruption approximation which, for an isolated spectral line, $\alpha \rightarrow \beta$, in Eq. (24.4a), leads to the classic Michelson-Lorentz shape.

3 The BCA interruption approximation is treated for the case of a group of overlapping lines using a modified S matrix which does not vanish off the energy shell. In this treatment the classical path approximation (CPA) is utilized.

4 The far line wings are considered using the approximations of (3), and an exponential decay with frequency is obtained.

5 The antithesis of the interruption approximation, the statistical approximation, is treated by considering molecular position as a good quantum number, kinetic energy as a perturbation. It is in connection with this approximation that spectral line satellites arise, a subject to which we will return in a later chapter.

The two-body form of Eq. (24.12) is actually what Fano obtained, although, as we shall see, this amounts to about the same thing. Because of the intrinsic importance of the Fano result, however, it behooves us to follow Ben-Reuven's obtention of the precise two body t-matrix. The term in the expansion of the self frequency matrix which corresponds to the BCA, $N\Sigma^{(1)}(\mathbf{k}, \omega)$, is obtained by considering one collision pair at a time and is, of course, linearly proportional to the density. Of considerable importance to this is the separation of the center-of-mass motion of the collision pair from their relative motion. We begin, with Ben-Reuven, by defining the two-molecule t-matrix by

$$T(\epsilon) \rightarrow T^{(1,2)}(\epsilon) = 8\pi^3 \Delta_p t(\epsilon)/V \tag{24.13a}$$

where

$$\Delta_p = \frac{VI_p}{8\pi^3} \tag{24.13b}$$

I_p being the unit operator on the degrees of freedom of the motion of the center-of-mass of the molecular pair. The translational motion of the pair may be separated by writing the eigenvectors as follows:

$$|ab\rangle = |\alpha \mathbf{p}_a, \beta \mathbf{p}_b\rangle = |\alpha\beta, \mathbf{p}_{ab}\rangle |\mathbf{p}_a + \mathbf{p}_b\rangle \tag{24.14a}$$

where

$$\frac{\mathbf{p}_{ab}}{\mathsf{m}} = \frac{\mathbf{p}_a}{\mathsf{m}_1} - \frac{\mathbf{p}_b}{\mathsf{m}_2} \tag{24.14b}$$

$$\mathsf{m} = \frac{\mathsf{m}_1 \mathsf{m}_2}{(\mathsf{m}_1 + \mathsf{m}_2)} \tag{24.14c}$$

In Eqs. (24.14), $\mathbf{p}_{ab}$ represents the relative momentum, m the usual reduced mass. We use Eq. (24.14a) in order to take the following matrix element:

$$\langle ab|\Delta_p t(\epsilon)|cd\rangle = t_{ab,cd}(\epsilon)V(2\pi)^{-3}\langle \mathbf{p}_a+\mathbf{p}_b|I_p|\mathbf{p}_c+\mathbf{p}_d\rangle$$

$$= t_{ab,cd}(\epsilon)(V/V)(2\pi)^{-3}\int_0^{\infty} e^{i(\mathbf{p}_a+\mathbf{p}_b-\mathbf{p}_c-\mathbf{p}_d)\cdot\mathbf{r}}\,d\mathbf{r}$$

$$= t_{ab,cd}(\epsilon)\delta(\mathbf{p}_a+\mathbf{p}_b-\mathbf{p}_c-\mathbf{p}_d) \tag{24.15a}$$

where

$$t_{ab,cd}(\epsilon) = \langle \alpha\beta,\mathbf{p}_{ab}|t(\epsilon)|\gamma\delta,\mathbf{p}_{cd}\rangle \tag{24.15b}$$

The following definition may be generated in analogy to Eqs. (24.15):

$$\langle ab|\Delta_p|cd\rangle = \Delta_{ab,cd}\delta(\mathbf{p}_a+\mathbf{p}_b-\mathbf{p}_c-\mathbf{p}_d) \tag{24.16a}$$

where

$$\Delta_{ab,cd} = \langle \alpha\beta,\mathbf{p}_{ab}|\gamma\delta,\mathbf{p}_{cd}\rangle \tag{24.16b}$$

and will prove useful in what follows.

From Eqs. (10.26), (24.15), and (24.16) we may now obtain

$$\mathrm{t}_{agbg',chdh'}(\omega) = t_{ag,ch}(\epsilon_{bg'}+\omega)\Delta^*_{bg',dh'} - \Delta_{ag,ch}t^*_{bg',dh'}(\epsilon_{ag}-\omega)$$

$$+(2\pi\mathrm{i})^{-1}\int d\epsilon\, d_{ag,bg'}t_{ag,ch}(\epsilon+\omega)t^*_{bg',dh'}(\epsilon)\,d_{ch,dh'} \tag{24.17a}$$

where

$$\epsilon_{ab} = \epsilon_\alpha + \epsilon_\beta + \mathrm{p}^2_{ab}/2\mathrm{m} \tag{24.17b}$$

$$d_{ag,bh} = (\epsilon+\omega-\epsilon_{ag}+i\eta)^{-1} - (\epsilon-\epsilon_{bh}-\mathrm{i}\eta')^{-1} \tag{24.17c}$$

In Eqs. (24.17), ϵ_{ab} is the pair energy in the center-of-mass coordinates.

We therefore now need only evaluate the matrix elements of the self frequency matrix in the BCA, the first term in the density expansion, $N\Sigma^{(1)}(k,\omega)$. In order to do this we use Eqs. (24.13) in Eqs. (24.4) for the t-matrix, Eq. (10.10b) telling us the matrix elements of t that will surive. The result is

$$N\Sigma^{(1)}_{ab,cd} = 8\pi^3 N\Upsilon_b^{-1}\sum_{gh}\Upsilon^{(2)}_{bg}\left[\mathrm{t}_{agbg,chdh}(\omega) + \mathrm{t}_{agbg,hchd}(\omega)\right] \tag{24.18}$$

where $\Upsilon^{(2)}_{bg}$ is the diagonal two-body density matrix element depending only on the relative velocity, $\mathbf{p}_{bg}/\mathrm{m}$.

It is obvious but important that the first term on the right of Eq. (24.18) relates to direct scattering of our absorber or emitter by a perturber, a term that would enter either in foreign-gas or self broadening. In the second term, however, we encounter a situation in which, in essence, the excitation is exchanged during the collision. This term is chiefly of interest in resonance or self broadening, although it could come into play for an accidental resonance between molecules of different species.

Referring to Eq. (24.17a) in conjunction with Eq. (24.18), we see that the supermatrix elements of t divide themselves into elements linear to t or t^* and terms, tt^*, bilinear in this operator. The linear terms are sufficiently simple for us to write them down

$$\left[\Sigma^{(1)}_{ab,cd}(\mathbf{k},\omega)\right]_{\text{linear}}=8\pi^3\Upsilon_b^{-1}\Big\{\sum_g \Upsilon_{bg}\left[\delta_{bd}t_{ag,bg}(\epsilon_{bg}+\omega)-\delta_{ac}t^*_{bg,dg}(\epsilon_{ag}+\omega)\right]$$

$$+\Upsilon_{bd}t_{ad,bc}(\epsilon_{bd}+\omega)-\Upsilon_{bc}t^*_{bc,ad}(\epsilon_{ac}-\omega)\Big\} \qquad (24.19)$$

That Σ is not generally Hermitian follows from the fact that $\mathfrak{T}$ is not. We may, therefore, divide the diagonal elements of the self energy matrix into an Hermitian and an anti-Hermitian part, thus

$$\Sigma=\Sigma'+i\Sigma'' \qquad (24.20)$$

It would follow that Σ relates to the shift of the resonance frequency by collisions, the Σ'' to its damping, a situation which we have encountered frequently enough. On Σ'' Ben-Reuven imposes three restrictions which he terms "impact conditions" and which, *in toto*, he feels constitute the impact or interruption approximation. These conditions are:

1 The time of collision is much greater than the damping time,

$$\tau\ll\left[\Sigma''_{ab,ab}(\mathbf{k},\omega)\right]^{-1}. \qquad (24.21a)$$

2 In the case of an isolated spectral line the effects of translational motion are removable.

$$\Sigma^{(1)}_{ab,cd}=\Sigma^{(1)}_{\alpha\beta,\gamma\delta}\delta_{p_b p_d} \qquad (24.21)$$

3 The dependence of the resonance frequency matrix, $\Omega(\mathbf{k})$, on the Doppler effect is small enough so that the dependence may be ignored, that is,

$$\Sigma''_{ab,ab}\gg\langle|\mathbf{k}\cdot\mathbf{p}_b/\mathrm{m}_1|\rangle \qquad (24.21c)$$

As we have remarked, when these three conditions are met, Ben-Reuven obtains the basic interruption shape. All of which, we recall, relates to the diagonal elements of Σ.

The off-diagonal elements, $N\Sigma^{(1)}_{ab,cd}$, $ab\neq cd$, basically relate to the mixing of distinct resonance modes by collisions. Very illuminatingly, Ben-Reuven considers them as off-diagonal perturbations to Ω, the supermatrix of resonance frequencies. It would seem reasonable to assume a certain familiarity with elementary perturbation theory, and we will recall the manner in which an off-diagonal, Hermitian perturbation tends to push apart the two levels between which it acts. An anti-Hermitian perturbation accordingly tends to pull the unperturbed levels together. Thus we see that Σ'' tends to pull the resonance levels or frequencies together and mix them. Ben-Reuven mentions three effects which these off-diagonals have on spectral lines. The Doppler narrowing and transition from resonant to nonresonant line shape we shall have occasion to remark upon at a more appropriate point.

As we have seen, strongly overlapping spectral lines cannot be adequately represented by the simple superposition of isolated profiles. Such lines call into play the off-diagonal elements of $\Sigma''_{ab,cd}$ with the concomitant mixing and variation from the isolated shapes. That the mixing be of consequence requires the fulfillment of the following condition:

$$|\omega_{ab}-\omega_{cd}|\lesssim\tau^{-1} \tag{24.22}$$

where again τ is the time of collision, a quantity which we now consider in somewhat more detail. There will be a range of energies, $\Delta\epsilon$, over which $t(\epsilon)$ in Eq. (24.17a) changes "appreciably," corresponding to which will be a range of frequencies, $\Delta\omega$, over which Eq. (24.17a) itself changes appreciably. ("Appreciably" will obviously depend on the specifics of a given problem.) By means of this range a collision time is defined, $\tau=|\Delta\epsilon|^{-1}$, and Eq. (24.22) may have been rendered more meaningful.

Should the reader actually avail himself of this formalism in order to attack a particular problem, he will, of course, wish to consider various other factors affecting the importance of various portions of Eq. (24.17a). Since $t(\epsilon)$ is a function which varies smoothly with the energy, Δ will strongly influence the linear terms, d the bilinear terms. From Eq. (24.17c) and various of its antecedents, we see that d is expressible in terms of principal parts and delta functions, both of which will furnish valuable assistance, for example, in determining whether or not to include the bilinear terms.

25 The Ternary Collision Within the Liouville Framework

We will first remark that Ben-Reuven has pointed out the results of Albers and Oppenheim (53, 54) which indicate the possibility of certain logarithmic divergences in the higher-than-binary terms of a $\Sigma(\omega)$ density expansion. Czuchaj, on the other hand, has studied what we might call a generalized method of including the ternary-collision term which is of course quadratic in the density. Now everyone must perforce begin with a line shape expression involving $[\omega-\mathfrak{L}]^{-1}$. From this point, however, he follows the specifics of Fiutak's derivation (121) which differs from what we have considered in the form of the Liouville operators and hence yields an expression for what amounts to a self-frequency matrix that is somewhat more amenable to immediate density expansion. Of course we must start with a Liouville operator which amounts to Eqs. (21.6), say, $\mathfrak{L}=\mathfrak{L}_{0s}-\mathfrak{L}_{0b}+\mathfrak{B}$, $\mathfrak{B}$ the interaction portion. However, this may be reexpressed as the equivalent

$$\mathfrak{L}=\mathfrak{L}'+\mathfrak{W}, \tag{25.1a}$$

where

$$\mathfrak{L}'=\mathfrak{L}_{0s}+\mathfrak{L}_{0b}+\mathsf{P}\mathfrak{B}\mathsf{P}+(1-\mathsf{P})\mathfrak{B}(1-\mathsf{P}) \tag{25.1b}$$

$$\mathfrak{W}=\mathsf{P}\mathfrak{B}(1-\mathsf{P})+(1-\mathsf{P})\mathfrak{B}\mathsf{P} \tag{25.1c}$$

P remains a projection operator. A series of operator manipulations involving Eqs. (25.1) transform the intensity expression involving $[\omega-\mathfrak{L}]^{-1}$ to

$$I_{if}(\omega)=-\pi^{-1}\mathcal{G}\left[\frac{1}{\omega-\omega_{if}-\langle\mathfrak{B}\rangle-\Phi_{if}(\omega)}\right] \tag{25.2a}$$

$$\mathfrak{B}=(X_{if},\mathfrak{B}X_{if}) \tag{25.2b}$$

$$\Phi_{if}(\omega)=(X_{if},\mathfrak{B}M)\left[\omega-\mathfrak{L}_{0s}-\mathfrak{L}_{0b}-\mathsf{M}\mathfrak{B}\mathsf{M}\right]^{-1}(M\mathfrak{B}X_{if}) \tag{25.2c}$$

$$X_{if}=\chi_{if}\sqrt{\Upsilon} \tag{25.2d}$$

X_{if} are eigenvectors in Liouville space which we recall to be defined in terms of dipole moment operators in H-space. M is the residual operator, $1-\mathsf{P}$, with which Czuchaj sets out to compute what he calls the relaxation matrix, $\Phi_{if}(\omega)$, to second order in the gas density, Φ_{if} to be compared to the EIT of Section 24 and preceding sections. In order to do this, he writes

$$\mathfrak{B}=\sum_{k=1}^{n}\left(\mathfrak{v}_k+\sum_{l>k=1}^{n}\mathfrak{v}_{kl}\right) \tag{25.3}$$

where $\mathfrak{v}_k$ represents the interaction between the radiator and the kth perturber, $\mathfrak{v}_{kl}$ that between the kth and lth perturbers. For a binary collision then $\mathfrak{B}$ will be of the form $\mathfrak{v}_1$, for a ternary $\mathfrak{v}_1+\mathfrak{v}_2+\mathfrak{v}_{12}$. We substitute Eq. (25.3) into Eq. (25.2c). There will be n terms wherein $\mathfrak{B}$ is replaced by, say, $\mathfrak{v}_i$ on both sides of the denominator and $n(n-1)$ additional terms where $\mathfrak{B}$ is replaced by $(\mathfrak{v}_i+\mathfrak{v}_j+\mathfrak{v}_{ij})$ on both sides. Therefore, the binary collision term, say, Φ'_{if}, will be Eq. (25.2c) with $\mathfrak{B}$ replaced by $\mathfrak{v}_1$, the whole thing multiplied by n. The ternary term, say Φ''_{if}, will be Eq. (25.2c) with $\mathfrak{B}$ replaced by $(\mathfrak{v}_1+\mathfrak{v}_2+\mathfrak{v}_{12})$, the result multiplied by $n(n-1)$. There is no obvious reason for writing the result down. Czuchaj now carries out several pages of operator manipulations, which we do not reproduce and arrives at

$$\begin{aligned}\Phi_{if}(\omega)=\;&n\left(X_{if},\mathfrak{v}_1\left[\omega-\mathfrak{L}_{0s}-\mathfrak{l}_{01}-\mathsf{M}_1\mathfrak{v}_1\right]^{-1}\mathsf{M}_1\mathfrak{v}_1X_{if}\right)\\&+\tfrac{1}{2}n(n-1)\left(X_{if}^{(2)},(\mathfrak{v}_1+\mathfrak{v}_2+\mathfrak{v}_{12})\left[\omega-\mathfrak{L}_{0s}-\mathfrak{l}_{01}-\mathfrak{l}_{02}\right.\right.\\&\left.\left.-(1-\mathsf{P}_1\mathsf{P}_2)(\mathfrak{v}_1+\mathfrak{v}_2+\mathfrak{v}_{12})\right]^{-1}(1-\mathsf{P}_1\mathsf{P}_2)(\mathfrak{v}_1+\mathfrak{v}_2+\mathfrak{v}_{12})X_{if}^{(2)}\right)\\&-\tfrac{1}{2}n(n-1)\left(X_{if},\mathfrak{v}_1\left[\omega-\mathfrak{L}_{0s}-\mathfrak{l}_{01}\mathsf{M}_1\mathfrak{v}_1\right]^{-1}\mathsf{M}_1\mathfrak{v}_1X_{if}\right)\end{aligned} \tag{25.4}$$

In Eq. (25.4) the subscripts "1" and "2" refer to perturbers "1" and "2." $\mathfrak{l}_{0i}$ is the Liouville operator corresponding to the free translational and internal motions of perturber "i." $X_{if}^{(2)}$ is the Liouville space vector, which is somewhat changed from X_{if} by our consideration of two perturbers. If we now let $\mathfrak{L}_0=\mathfrak{L}_{0s}+\mathfrak{l}_{01}$, the first term in Eq. (25.4) may be written as

$$\Phi_{if}^{(1)}(\omega)=-in\left(X_{if},\mathfrak{v}_1\int_0^{\infty}e^{i\omega t}\exp\left[-i(\mathfrak{L}_0+\mathsf{M}_1\mathfrak{v}_1)t\right]dt\,\mathfrak{v}_1X_{12}\right) \tag{25.5}$$

a portion of which may be written as a TDO in $\mathfrak{L}$-space,

$$e^{-\mathrm{i}[\mathfrak{L}_0+M_1\mathfrak{v}_1]t}=S_1(t,0)e^{-\mathrm{i}\mathfrak{L}_0 t} \tag{25.6}$$

By differentiating Eq. (25.6) we may obtain an equation,

$$\mathrm{i}\dot{S}_1(t,0)=S_1(t,0)e^{-\mathrm{i}\mathfrak{L}_0 t}\mathsf{M}_1\mathfrak{v}_1 e^{\mathrm{i}\mathfrak{L}_0 t} \tag{25.7a}$$

of iterated solution [remembering that $S_1(0,0)=1$]

$$S_1(t,0)=1-\mathrm{i}\mathsf{M}_1\int_0^t e^{-\mathrm{i}\mathfrak{L}_0 t'}\mathfrak{v}_1 e^{\mathrm{i}\mathfrak{L}_0 t'}\,dt'$$

$$+\mathrm{i}^2\mathsf{M}_1\int_0^t\int_0^{t'} e^{-\mathrm{i}\mathfrak{L}_0 t''}\mathfrak{v}_1 e^{\mathrm{i}\mathfrak{L}_0 t''}\,dt''\,\mathsf{M}_1 e^{-\mathrm{i}\mathfrak{L}_0 t'}\mathfrak{v}_1 e^{\mathrm{i}\mathfrak{L}_0 t'}\,dt'+\cdots \tag{25.7b}$$

Equation (25.7b) may be used to replace the appropriate terms in Eq. (25.5), and we will obtain an expansion

$$\Phi_{if}^{(1)}(\omega)=\Phi_0^{(1)}(\omega)+\Phi_1^{(1)}(\omega)+\Phi_2^{(1)}(\omega)+\cdots \tag{25.8}$$

the first term of which we consider as an example. The residual operator, M_1, is of course equivalent to $1-\mathsf{P}_1$, $\mathsf{P}_1=|X_{ij}\rangle\rangle\langle\langle X_{ij}|$. Therefore, $\Phi_0^{(1)}(\omega)$ becomes

$$\Phi_0^{(1)}(\omega)=-\mathrm{i}n\left(X_{if},\mathfrak{v}_1\int_0^{\infty}e^{\mathrm{i}\omega t}e^{-\mathrm{i}\mathfrak{L}_0 t}\,dt\,\mathfrak{v}_1 X_{if}\right)$$

$$+\mathrm{i}n\left(X_{if},\mathfrak{v}_1\int_0^{\infty}e^{\mathrm{i}\omega t}e^{-\mathrm{i}\mathfrak{L}_0 t}\,dt\,X_{if}\right)(X_{if},\mathfrak{v}_1 X_{if}) \tag{25.9}$$

The second term in Eq. (25.9) will be proportional to V^{-2} on which basis this and future terms involving the projection operator will be dropped. (Such would not necessarily be the case for n^2 terms.)

We now return to H-space. Equations (10.10a), (25.2d), and (1.4) allow us to rewrite Eq. (25.9) as

$$\Phi_0^{(1)}(\omega)=-\mathrm{i}n\left[\mathrm{Tr}_1\left(\sqrt{\Upsilon_1}\,\chi_{if}^{+}\int_0^{\infty}e^{\mathrm{i}\omega t}e^{\mathrm{i}H_0 t}\chi_{if}\sqrt{\Upsilon_i}\,\mathsf{v}_1 e^{-\mathrm{i}H_0 t}\mathsf{v}_1\,dt\right)\right.$$

$$\left.+\mathrm{Tr}_1\left(\sqrt{\Upsilon_1}\,\chi_{if}^{+}\mathsf{v}_1\int_0^{\infty}e^{\mathrm{i}\omega t}e^{\mathrm{i}H_0 t}\mathsf{v}_1\chi_{if}\sqrt{\Upsilon_1}\,e^{-\mathrm{i}H_0 t}\,dt\right)\right] \tag{25.10}$$

We now take the frequency near line center so that $\omega\doteq\omega_{if}=E_i-E_f$, and, recalling that

$$e^{\mathrm{i}H_0 t}|\chi_{if}\rangle=e^{\mathrm{i}(E_f+E_0)t}|\chi_{if}\rangle e^{\mathrm{i}Kt},$$

$$\langle\chi_{if}|e^{-\mathrm{i}H_0 t}=e^{-\mathrm{i}(E_i+E_0)t}|\chi_{if}\rangle e^{-\mathrm{i}Kt},$$

we see that Eq. (25.10) may be rewritten as

$$\Phi_0^{(1)}(\omega) = -\mathrm{i}n\mathrm{Tr}_1\Big\{\Upsilon_1\int_0^\infty e^{\mathrm{i}(E_i+E_0)t}\mathrm{Tr}_s\big[\chi_{if}^+\chi_{if}e^{\mathrm{i}Kt}\mathrm{v}_1(\mathbf{R}_1)e^{-\mathrm{i}Kt}e^{-\mathrm{i}(H_s+h_1)t}$$

$$\cdot\mathrm{v}_1(\mathbf{R}_1)\big]\,dt\Big\} - \mathrm{i}n\mathrm{Tr}_1\Big\{\Upsilon_1\int_0^\infty e^{-\mathrm{i}(E_f+E_0)t}\mathrm{Tr}_s\big[\chi_{if}^+\mathrm{v}_1(\mathbf{R}_1)$$

$$\cdot e^{\mathrm{i}(H_s+h_1)t}e^{\mathrm{i}Kt}\mathrm{v}_1(\mathbf{R}_1)e^{-\mathrm{i}Kt}\chi_{if}\big]\,dt\Big\} \tag{25.11}$$

In Eq. (25.11) Tr_1 is over the coordinates of the perturber, Tr_s over those of the radiator. h_1 is the Hamiltonian having to do with the internal coordinates of the perturber, E_{01} the ground state energy of that particle. We consider the first term in Eq. (25.11) as an example.

In doing so we will be taking a trace over the radiator coordinates so that we may choose a perturber translational state, say $|p\rangle$, and internal state corresponding to E_{0p}. Next, suppose that the radiator-perturber separation is $\mathbf{R}_1$ as indicated. Then the initial interaction potential will be $\mathrm{v}_1(\mathbf{R}_1)$, the interaction potential at some later time, t, $\exp[\mathrm{i}Kt]\mathrm{v}_1(\mathbf{R}_1)\exp[-\mathrm{i}Kt]=\mathrm{v}_1(\mathbf{R}_{1t})$, propagation surely being effected by the translational portion of the Hamiltonian, K. (Actually, this will be K_1 if we suppose the radiator at rest.) We now consider the temporal integral in, say, the first term of Eq. (25.11). We follow Czuchaj in taking, as an example, the *ik* matrix element in the trace. Then the integral may be integrated by parts as follows:

$$\lim_{\epsilon\to 0}\int_0^\infty e^{-\epsilon t}e^{\mathrm{i}(\omega_{ik}+\omega_{0k1})t}\langle im_i,0|\mathrm{v}_1(\mathbf{R}_{1t})|km_k,k_1m_{k1}\rangle\,dt$$

$$=\mathrm{i}\langle im_i,0|\mathrm{v}_1(\mathbf{R}_1)|km_k,k_1m_{k1}\rangle/(\omega_{ik}+\omega_{0k1})$$

$$-\lim_{\epsilon\to 0+}\int_0^\infty \frac{e^{-\epsilon t}e^{\mathrm{i}(\omega_{ik}+\omega_{0k1})t}}{\mathrm{i}(\omega_{ik}+\omega_{0k1})}\frac{d}{dt}\langle im_i,0|\mathrm{v}_1(\mathbf{R}_{1t})|km_k,k_1m_{k1}\rangle\,dt \tag{25.12}$$

where $\mathrm{v}_1(\mathbf{R}_{1t})\to\mathrm{v}_1(\mathbf{R}_1)$ in the first term since this term is evaluated at the lower limit, $t=0$, and $\mathrm{v}_1(\mathbf{R}_{1t})=\mathrm{v}_1(\mathbf{R}_1+\mathbf{v}t)$.

If the classical path and $\mathrm{v}_1(\mathbf{R}_1)=\mathbf{d}\cdot\mathbf{d}_1R^{-3}$ are assumed, we see that

$$\langle im_i,0|\mathrm{v}_1(\mathbf{R}_{1t})|km_k,k_1m_{k1}\rangle=\frac{C}{[\rho^2+v^2t^2]^{3/2}} \tag{25.13}$$

if we take t as the time to point of closest approach. In order to compare the first and second terms in Eq. (25.12) we may take the ratio of the time derivative of the matrix element of Eq. (25.13) and the matrix element itself. This ratio will be

$$3t/\big[(\rho/v)^2+t^2\big]\doteq 1/t$$

The meaningful time here will be the collision time, τ, so that an argument involving collision times allows Czuchaj to neglect the second term in Eq. (25.12) as compared to the first. And proceed to obtain the following result for Eq. (25.11):

$$\Phi_0^{(1)} = n\,\mathrm{Tr}_1\{\Upsilon_1[\mathsf{V}_1^{(i)}(R_1) - \mathsf{V}_1^{(f)}(R_1)]\} \tag{25.14}$$

With Eq. (25.14) we have completed our example of Czuchaj's treatment, that author continuing in this vein through seventh order in $\mathbb{S}_1(t,0)$ in order to obtain for Eq. (25.8)

$$\Phi_{if}^{(1)}(\omega) = n\,\mathrm{Tr}_1\{\Upsilon\mathsf{V}_1^{(if)}(R_1)\} - \mathrm{i}n\int_0^{\infty} e^{\mathrm{i}(\omega-\omega_{if})t}\,\mathrm{Tr}_1\{\Upsilon_1\mathsf{V}_1^{(if)}(R_1)\mathbb{S}_1(t,0)$$
$$\cdot\mathsf{V}_1^{(if)}(R_{1t})\}\,dt \tag{25.15a}$$

where

$$\mathbb{S}_1(t,0) = \exp\left[-\mathrm{i}\int_0^{\infty}\mathsf{V}_1^{(if)}(R_{1t'})\,dt'\right] \tag{25.25b}$$

$$\mathsf{V}_1^{(if)}(R_1) = \mathsf{V}_1^{(i)}(R_1) - \mathsf{V}_1^{(f)}(R_1) \tag{25.15c}$$

With this Czuchaj proceeds to obtain a similiar expression for $\Phi_{if}^{(2)}(\omega)$ which, as the reader will most assuredly have surmised, is quite lengthy and will not be displayed in its entirety. Let us, however, write down the first and simplest term in this expression,

$$\tfrac{1}{2}n(n-1)\,\mathrm{Tr}_{12}\{\Upsilon_1\Upsilon_2[\mathsf{V}_1^{(if)}(R_1) + \mathsf{V}_2^{(i)}(R_2) + E_3^{(if)}(R_1, R_2, R_{12})] \tag{25.16a}$$

where

$$E_3^{(if)} = C_9^{(if)}[1 + 3\cos\vartheta_1\cos\vartheta_2\cos\vartheta_3]/R_1^3R_2^3R_{12}^3 \tag{25.16b}$$

There are three angular coordinates because angles referring to the momentum of the radiator plus two perturbers enter. Classically, the trace will mean averaging over the translational motion of the perturbers, that is,

$$\mathrm{Tr}_{12}\{\Upsilon_1\Upsilon_2\cdots\} = \int_{-\infty}^{\infty} d\mathbf{p}_1\int_{-\infty}^{\infty} d\mathbf{p}_2\int_{-\infty}^{\infty} d\mathbf{R}_1\int_{-\infty}^{\infty} d\mathbf{R}_2\{\Upsilon_1(p_1)\Upsilon_2(p_2)\cdots\}$$

$E_3^{(if)}$ is critical to the ternary effects. Czuchaj found that this term is negligible for only one of the two perturbers in close proximity to the radiator. Further, when both perturbers are moving with the same velocity, $E_3^{(if)}$ is greatest when the two particles follow parallel paths, the distance between them remaining constant as a result. The consequences of this are important, for the Czuchaj calculations indicate that one may restrict the averaging to such trajectories with the obvious attendant simplification.

We have seen the compound projection operator, $(1-\mathsf{P}_1\mathsf{P}_2)$, enter Eq. (25.4). Czuchaj finds that this hitherto unencountered operator eliminates certain line-center divergences which had appeared in the earlier Fano work.

Finally, while the author found the interaction between the two perturbers of no importance compared to that between radiator and perturbers for, say, noble-gas-broadened alkalis, he felt that, in the case of long-range forces, such as those of resonance broadening, such would be anything but the case.

6

The Feynman Diagram Approach to Spectral Line Broadening

As we have seen in Chapter 2, Low first applied the Feynman diagram to natural line shape in 1951. There were rather casual encounters between line shape theory and the diagrammatic technique—the work of Mizushima, Robert, and Galatry which we have encountered in Chapter 2 furnishes an example—until the middle 1960s when, as is so often the case, a minor epidemic broke out, Ross treating the problem in 1964 (37), Bezzerides in 1966 (8), and Zaidi in 1967 (51).

We have become familiar with the fact that the distribution of intensities within the spectral line is proportional to the imaginary part of the susceptibility. That this susceptibility may in turn by expressed as a combination of one- and two-particle Greens functions we shall discover. This means that the drawing of Feynman diagrams will be somehow arrangeable, although Feynman himself might feel inclined to disavow some of the artistic endeavors which are associated with his name as a consequence. The principles involved appear to be straightforward enough, although the practice offers perhaps insurmountable difficulties. That such difficulties should arise should come as no surprise, however, since very nearly all possible physical phenomena—the scope of such a statement is vast indeed—are embraced. If nothing else, this approach certainly portrays for us just what physical processes are to be anticipated, and it portrays them with a clarity which no other approach can equal.

In perusing this chapter the reader might keep a weather eye out for what has been forecast by Eq. (8.20). Since the Liouvillian and the Greens function are related by $\mathsf{G}=\exp(-\mathrm{i}\mathfrak{L}t)$, and since we know that the susceptibility is given by $\exp(-\mathrm{i}\mathfrak{L}t)$, we should expect to find the susceptibility expressed by Greens functions. [Note that we do not say that the susceptibility is given solely by $\exp(-\mathrm{i}\mathfrak{L}t)$. We must put in $\exp(\mathrm{i}\omega t)$, Fourier transform it, average it and what not, but, basically, it is given by $\exp(-\mathrm{i}\mathfrak{L}t)$.] Basically, we shall find the susceptibility given by Greens functions in what is to follow. We may immediately put the Liouvillian in the denominator by Fourier transformation. We may immediately put the Hamiltonian of the Greens function in the denominator by presuming a sufficiently simple Greens function. We end by putting everything of consequence in the denominator via the Greens function [cf. Eq. (31.25)] when we simplify the diagrammatic approach sufficiently to obtain a legible frequency-dependent susceptibility. The point of course is that, although often obscured by detail, the connection between the Liouvillian and the diagrammatic is always present. [The analogous correspondence between the correlation function and the two-particle Greens function was emphasized by Yakimets (287).]

26 The Susceptibility and the Current Commutator

Ross begins by taking the entire collection of atoms involved in the absorption as described by H_0. Then the energy of this assemblage will indeed be given by H_0, H_0 being averaged by means of a density function, thus, $\langle H_0 \rangle$. Now the time rate of change of this quantity H_0, if the interaction with the incident electromagnetic field is the only perturbation, will surely be the rate at which energy is absorbed by the assemblage, $d\langle H_0 \rangle / dt$.

A useful form of the particle-field interaction is

$$H_1 = -\int dr[\mathbf{j}(r), \mathbf{A}(r)] \tag{26.1}$$

where the A^2 term in the particle-field portion of the Hamiltonian is, as usual, being neglected. We are thus in a position to write out an expression for the net rate of absorption of energy per unit volume by our collection of atoms,

$$\begin{aligned} P(t) &= V^{-1}\frac{d\langle H_0 \rangle}{dt} = \mathrm{i}V^{-1}\langle [H, H_0] \rangle = -\mathrm{i}V^{-1}\langle [H_0, H] \rangle \\ &= -\mathrm{i}V^{-1}\langle [H_0, H_1] \rangle = -\mathrm{i}V^{-1}\langle [H, H_1] \rangle \\ &= -V^{-1}\frac{d\langle H_1 \rangle}{dt} = V^{-1}\int dr\left(\mathbf{A}(\mathbf{r}), \frac{d\langle \mathbf{j}(\mathbf{r}) \rangle}{dt}\right) \end{aligned} \tag{26.2}$$

wherein we remark that the explicit dependence of $\mathbf{j}$ on $\mathbf{A}$ is being neglected.

Now we shall transform to the time-dependent Heisenberg operator by means of

$$\begin{aligned} \mathbf{A}(\mathbf{r}) &= \exp\left(-\mathrm{i}\int H_1\,dt'\right)\mathbf{A}(\mathbf{r}, t)\exp\left(\mathrm{i}\int H_1\,dt'\right), \\ \mathbf{j}(\mathbf{r}) &= \exp\left(-\mathrm{i}\int H_1\,dt'\right)\mathbf{j}(\mathbf{r}, t)\exp\left(\mathrm{i}\int H_1\,dt'\right) \end{aligned} \tag{26.3}$$

where $\mathbf{A}(\mathbf{r}, t)$ and $\mathbf{j}(\mathbf{r}, t)$ are in the interaction representation, that is, $\mathbf{A}(\mathbf{r}, t) = \exp[-\mathrm{i}H_0 t]\mathbf{A}\exp[\mathrm{i}H_0 t]$ and so on. In Eq. (26.3), $\exp[-\mathrm{i}\int H_1\,dt]$ is a form of the S matrix. Ross took the operators to first approximation so that

$$\mathbf{A}(r) = (1+S_1)\mathbf{A}(\mathbf{r}, t)(1-S_1) = \mathbf{A}(\mathbf{r}, t_0) + S_1\mathbf{A}(\mathbf{r}, t) - \mathbf{A}(\mathbf{r}, t)S_1 \tag{26.4}$$

where

$$S_1 = -\mathrm{i}\int H_1(t')\,dt'$$

and similarly for $\mathbf{j}(\mathbf{r}, t)$. Keeping in mind that $\dot{\mathbf{j}}(\mathbf{r}, t) = 0$, we may substitute Eqs. (26.3) into Eq. (26.2) and retain only first-order terms in order to obtain

$$P = V^{-1}\int_{-\infty}^{\infty} d\mathbf{r}\,d\mathbf{r}'\,dt'\,\mathbf{A}(\mathbf{r}, t)\cdot \mathrm{i}\frac{d}{dt}\langle[\mathbf{j}(\mathbf{r}, t), \mathbf{j}(\mathbf{r}', t')]_R\rangle\cdot \mathbf{A}(\mathbf{r}', t') \quad (26.5)$$

where $[\;]_R$ is the retarded commutator which simply means it will vanish for $t < t'$. This t requirement should be familiar to us by this time and arises from the necessity for time ordering the S matrix.

From Eq. (2.5b) we may write a three-dimensional vector potential as

$$\mathbf{A}(\mathbf{r}, t) = \tfrac{1}{2}\sqrt{\frac{\omega}{V}}\sum_k e\left[e^{\mathrm{i}\mathbf{k}\cdot\mathbf{r} - \mathrm{i}\omega t}\mathbf{A}(\mathbf{k}, \omega) + e^{-\mathrm{i}\mathbf{k}\cdot\mathbf{r} + \mathrm{i}\omega t}\mathbf{A}^+(\mathbf{k}, \omega)\right] \quad (26.6)$$

where e is the polarization of the incident radiation, perpendicular to its propagation $\mathbf{k}$.

We substitute for $\mathbf{A}(\mathbf{r}, t)$ and $\mathbf{A}(\mathbf{r}', t)$ in Eq. (26.5) and integrate the result over t with the intention of obtaining an average power. Several points arise for consideration.

First of all, we will have terms involving (1) A^2 and A^{+2} and including factors $\exp[\mathrm{i}\omega(t + t')]$ and (2) $|\mathsf{A}|^2$ and including factors $\exp[\mathrm{i}\omega(t - t')]$. The terms (1) will oscillate very rapidly during temporal integration as compared to (2), and, on this basis, (1) are dropped. Next we transform from t and t' to $(t - t')$ and t which allows us to carry out the integration over t and eliminate the derivative in Eq. (26.5). Finally, we have a summation over $\mathbf{k}$ which we replace by $V\int d\mathbf{k}''$, and, since we desire P for a particular $\mathbf{k}$, integrate Eq. (26.6) over $\mathbf{k}''$ with a factor $\delta(\mathbf{k}'' - \mathbf{k}')$. The net result is to eliminate the summation over $\mathbf{k}$ and the factor $V^{-1/2}$ in Eq. (26.6). As a consequence of all these machinations we will obtain an expression dependent on only $(\mathbf{r} - \mathbf{r}')$ as the retarded current commutator depends only on the difference in position coordinate in a homogeneous medium. We therefore transform from $\mathbf{r}$ and $\mathbf{r}'$ to $\mathbf{r}$ and $\mathbf{r} - \mathbf{r}'$. As $\int d\mathbf{r} = V$, the $\mathbf{r}$ integration and the V in the denominator of Eq. (26.5) are eliminated, and we obtain

$$I(\mathbf{k}, \omega) = \tfrac{1}{2}\omega^3\left(I[\mathbf{e}\cdot\chi(\mathbf{k}, \omega)\cdot\mathbf{e}]\right)|\mathsf{A}(\mathbf{k}, \omega)|^2 \quad (26.7\text{a})$$

$$\chi(\mathbf{k}, \omega) = \mathrm{i}\omega^{-2}\int d(t - t')\,d(\mathbf{r} - \mathbf{r}')\exp\{-\mathrm{i}[\mathbf{k}\cdot(\mathbf{r} - \mathbf{r}') - \omega(t - t')]\}$$
$$\cdot\langle[\mathbf{j}(\mathbf{r}, t), \mathbf{j}(\mathbf{r}', t')]_R\rangle \quad (26.7\text{b})$$

We remark the bilinear form, corresponding to an $\mathfrak{L}$-space operator, which has been encountered in the susceptibility. The next step in the Ross development is to obtain an expression for the current commutator in Eqs. (26.7b) in terms of the particle-field operators.

We now appeal to the current for the Dirac electron,

$$j_\mu = \tfrac{1}{2} ie\left[\psi^+(r,t)\delta_\mu, \psi(r,t)\right] \tag{26.8a}$$

$\gamma_\mu \to (\mathbf{p}_\mu - e\mathbf{A}_\mu)/\mathsf{m}$ for the nonrelativistic case where now p_μ is the particle momentum, A_μ the vector potential for out incident radiation. Eq. (26.8a) then becomes

$$\begin{aligned} j_\alpha(\mathbf{r},t) &= \tfrac{1}{2}\mathrm{i}e_\alpha\left[\psi^+(\mathbf{r},t)(\mathbf{p}-e_\alpha\mathbf{A}), \psi(\mathbf{r},t)\right]\Big/\mathsf{m}_\alpha \\ &= \tfrac{1}{2}e_\alpha\left[\psi^+(\mathbf{r},t)\nabla\psi(\mathbf{r},t) - (\nabla\psi^+(\mathbf{r},t))\psi(\mathbf{r},t)\right]\Big/\mathsf{m}_\alpha \\ &\quad - \tfrac{1}{2}e_\alpha^2\mathbf{A}(\mathbf{r},t)\left[\psi^+(\mathbf{r},t),\psi(\mathbf{r},t)\right]\Big/\mathsf{m}_\alpha \end{aligned} \tag{26.8b}$$

where the subscript α relates to the species of electron or ion. We proceed [cf. Källén (27)] from j_μ as $\psi^+\gamma_\mu\psi$ to the commutator of Eq. (26.8a) in order to eliminate the zero-point charge. We reverse this procedure vis-à-vis the second term on the right of Eq. (26.8a)

$$\begin{aligned} j_\alpha(\mathbf{r},t) &= \tfrac{1}{2}e_\alpha\left[\psi^+(\mathbf{r},t)\nabla\psi(\mathbf{r},t) - (\nabla\psi^+(\mathbf{r},t))\psi(\mathbf{r},t)\right]\Big/\mathsf{m}_\alpha \\ &\quad - \mathrm{i}e_\alpha^2\mathsf{A}(\mathbf{r},t)\psi_2^+(\mathbf{r},t)\psi_\alpha(\mathbf{r},t)/\mathsf{m}_\alpha \end{aligned} \tag{26.9}$$

Next, with Ross, we obtain an expression for the current density using the hydrogen atom as a prototype. We approach this by determining a charge "density" at a point $\mathbf{r}'$ due to a point electron of mass m at $\mathbf{r}_0$ and a point proton or nucleus of mass M at a point $\mathbf{R}_0$. Since the integral over this charge density will certainly equal the charge at the point, the Dirac delta function will come into play,

$$\begin{aligned} \rho(\mathbf{r}',t) &= -e\left[\delta(\mathbf{r}'-\mathbf{r}_0) - \delta(\mathbf{r}'-\mathbf{R}_0)\right] \\ &= -e\left[\delta(\mathbf{r}'-\mathbf{R}-[\mathsf{M}/(\mathsf{M}+\mathsf{m})]\mathbf{r}) - \delta(\mathbf{r}'-\mathbf{R}+[\mathsf{m}/(\mathsf{M}+\mathsf{m})]\mathbf{r})\right] \end{aligned}$$

where $\mathbf{R}$ is the atomic center-of-mass location, $\mathbf{r}$ the location of the electron with respect to this center of mass. We now carry out a Taylor expansion through first order:

$$\rho(\mathbf{r}',t) = e\left[\mathbf{r}\cdot\left(\nabla_r\delta\left(\mathbf{r}'-\mathbf{R}-\frac{\mathsf{M}}{\mathsf{M}+\mathsf{m}}\mathbf{r}\right)\right)_{r=0} - \mathbf{r}\cdot\left(\nabla_r\delta\left(\mathbf{r}'-\mathbf{R}+\frac{\mathsf{m}}{\mathsf{M}+\mathsf{m}}\mathbf{r}\right)\right)_{\mathbf{r}=0}\right]. \tag{26.10}$$

Now we know [cf. § 11 of Schiff (38)] that $(d/dx)\delta(x) = -(1/x)\delta(x)$. From this relation we may obtain an expression for $\nabla_r\delta$ in terms of $\nabla_{r'}\delta$ which we evaluate at $r=0$ with the result

$$\rho(\mathbf{r}',t) = e\mathbf{r}\cdot\nabla_{r'}\delta(\mathbf{r}'-\mathbf{R}) \tag{26.11}$$

Taking the current as the velocities of these point charges yields

$$\begin{aligned} \mathbf{j}(\mathbf{r}',t) &= -e\left[\delta(\mathbf{r}'-\mathbf{r}_0)\dot{\mathbf{r}}_0 - \delta(\mathbf{r}'-\mathbf{R}_0)\dot{\mathbf{R}}_0\right] = -e\left[\delta(\mathbf{r}'-\mathbf{R}-[\mathsf{M}/(\mathsf{M}+\mathsf{m})]\mathbf{r})\right] \\ &\quad \cdot\left[(\dot{\mathbf{R}}+[\mathsf{M}/(\mathsf{M}+\mathsf{m})]\dot{\mathbf{r}}) - \delta(\mathbf{r}'-\mathbf{R}+[\mathsf{m}/(\mathsf{M}+\mathsf{m})]\mathbf{r})(\dot{\mathbf{R}}-[\mathsf{m}/(\mathsf{M}+\mathsf{m})]\dot{\mathbf{r}})\right] \end{aligned}$$

which we also expand and cut off, using the delta function derivative relation, in order to obtain

$$\mathbf{j}(\mathbf{r}',t) = -e\left[\delta(\mathbf{r}'-\mathbf{R})\dot{\mathbf{r}} - \mathbf{r}\cdot\nabla_{r'}\delta(\mathbf{r}'-\mathbf{R})\dot{\mathbf{R}} - \frac{\mathsf{M}^2-\mathsf{m}^2}{(\mathsf{M}+\mathsf{m})^2}\mathbf{r}\cdot\nabla_{r'}\delta(\mathbf{r}'-\mathbf{R})\dot{\mathbf{r}}\right] \tag{26.12}$$

the last term of which Ross neglects as of quadrupole order. A little manipulation allows us to write Eq. (26.12) as

$$\mathbf{j}(\mathbf{r}',t) = -e\left(\frac{d}{dt}\left[\delta(\mathbf{r}'-\mathbf{R})\mathbf{r}\right] + \mathbf{r}\nabla_{r'}\cdot\delta(\mathbf{r}'-\mathbf{R})\dot{\mathbf{R}} - \nabla_{r'}\cdot\mathbf{r}\delta(\mathbf{r}'-\mathbf{R})\dot{\mathbf{R}}\right) \tag{26.13}$$

the last term directly dependent on the commutation of $\mathbf{r}\cdot\nabla_{r'}$.

The first term in Eq. (26.13) involves the dipole moment, $\mathbf{d}=e\mathbf{r}$, if we suppose the center of mass corresponds with the nucleus, and let us rather carefully transform the operator $\delta(\mathbf{r}'-\mathbf{R})\mathbf{d}$ to second-quantized form. We recall the familiar relationship between a Schrödinger operator, **d**, and a second-quantized operator, **d**,

$$\begin{aligned}\hat{\mathbf{d}} &= \sum_{\alpha\beta}\int d\mathbf{r}\,\mathbf{d}\chi_\alpha^*(\mathbf{r})\chi_\alpha(\mathbf{r})\psi_\alpha^+(\mathbf{R},t)\psi_\beta(\mathbf{R},t)\\ &= \sum_{\alpha\beta}\psi_\alpha^+(\mathbf{R},t)\mathbf{d}_{\alpha\beta}\psi_\beta(\mathbf{R},t)\end{aligned} \tag{26.14}$$

In Eq. (26.14) $\chi_\beta(\mathbf{r})$ is a Schrödinger-like eigenvector, $\psi_\alpha^+(\mathbf{R},t)$ a creation operator in the interaction representation which creates a particle in state α at point **R** and time t. $\psi_\beta(\mathbf{R},t)$ is an analogous annihilation operator. Thus, we are creating our atom at the location of its center of mass. The function $\delta(\mathbf{r}'-\mathbf{R})$ will then have the effect of replacing the center-of-mass coordinate, **R**, by the coordinate of the point at which the current is desired. Thus, the first term in Eq. (26.13) will have the form

$$\frac{d}{dt}\left[\sum_{\alpha\beta}\psi_\alpha^+(\mathbf{r},t)\mathbf{d}_{\alpha\beta}\psi_\beta(\mathbf{r},t)\right] \tag{26.15}$$

We carry the evaluation of Eq. (26.13) no further but, instead, turn to a more useful evaluation of Eq. (26.12). Now we will agree that the velocity of the orbital electron with respect to the center of mass will be

$$\dot{\mathbf{r}} = \mathbf{p}/\mathsf{m}' + e\mathbf{A}(\mathbf{R})/\mathsf{m}' \tag{26.16}$$

where m' is the reduced mass, **p** the electronic momentum relative to the center of mass. Next we apply the procedure which yielded Eq. (26.15) from the first term in Eq. (26.13) in order to obtain for Eq. (26.12), neglecting the last term,

$$\mathbf{j}(\mathbf{r},t)=\sum_{\alpha\beta}\psi_\alpha^+(\mathbf{r},t)(-e\mathbf{p}/\mathrm{m}')_{\alpha\beta}\psi_\beta(\mathbf{r},t)+e^2\mathbf{A}(\mathbf{r},t)\sum_\alpha\frac{\psi_\alpha^+(\mathbf{r},t)\psi_\alpha(\mathbf{r},t)}{\mathrm{m}'}$$
$$+\mathrm{i}\nabla\cdot\sum_{\alpha\beta}\frac{\psi_\alpha^+(\mathbf{r},t)\,\mathbf{d}_{\alpha\beta}\nabla\psi_\beta(\mathbf{r},t)}{\mathrm{m}_\alpha} \tag{26.17}$$

Remark that $\dot{\mathbf{R}}$ in Eq. (26.12) relates to the translational motion of the atom.

As we have noted following Eq. (4.1c),

$$(-e\mathbf{p}/\mathrm{m}')_{\alpha\beta}=\mathrm{i}\omega_{\alpha\beta}\mathbf{d}_{\alpha\beta}\equiv(\dot{\mathbf{d}})_{\alpha\beta} \tag{26.18}$$

so that Eq. (26.17) may be written in the form favored by Ross,

$$\mathbf{j}(\mathbf{r},t)=\sum_{\alpha\beta}\psi_\alpha^+(\mathbf{r},t)\mathrm{i}\omega_{\alpha\beta}\mathbf{d}_{\alpha\beta}\psi_\beta(\mathbf{r},t)-\sum_\alpha\frac{n_\alpha e^2\psi_\alpha^+(\mathbf{r},t)\psi_\alpha(\mathbf{r},t)\mathbf{A}(r,t)}{\mathrm{m}_e}$$
$$+\sum_{\alpha\beta}\nabla\cdot\left[\psi_\alpha^+(r,t)\mathrm{i}\mathrm{m}_\alpha^{-1}\mathbf{d}_{\alpha\beta}\nabla\psi_\beta(r,t)\right] \tag{26.19}$$

Let us remark that Eq. (26.19) has been (approximately) extended from the hydrogen to the general atomic case by the factor n_α for the total number of electrons in the atom. In Eq. (26.19) **d** is, of course, the dipole moment for the n_α-electron atom, m_e the reduced mass of the electron, m_α the reduced mass of the atom. As we have seen in the course of the development, the first two terms in Eq. (26.19) represent the oscillation of the atomic dipoles within a given volume, the second term occurring because the momentum of the electrons does not correspond to their velocities [Eq. (26.16)]. The third term arises due to the motion of these dipoles or atoms in and out of this volume. Ross ignores these second and third terms as having negligible effect on "optical" lines so that he obtains for the commutator

$$\langle[\mathbf{j}(\mathbf{r},t),\mathbf{j}(\mathbf{r}',t')]\rangle=-\sum_{\alpha\beta\gamma\delta}\omega_{\alpha\beta}\omega_{\gamma\delta}\mathbf{d}_{\alpha\beta}\mathbf{d}_{\gamma\delta}\langle[\psi_\alpha^+(\mathbf{r},t)\psi_\beta(\mathbf{r},t),\psi_\gamma^+(\mathbf{r}',t')$$
$$\cdot\psi_\delta(\mathbf{r}',t')]\rangle \tag{26.20a}$$

where

$$\langle\cdots\rangle=\mathrm{Tr}\left[e^{-\beta(H-\mu N)}\cdots\right]/\mathrm{Tr}\left[e^{-\beta(H-\mu N)}\right] \tag{26.20b}$$

$$H-\mu N=\int d\mathbf{r}\sum_\alpha\psi_\alpha^+(\mathbf{r})\left[-\nabla^2/2\mathrm{m}_\alpha+E_\alpha\right]\psi_\alpha(\mathbf{r})$$
$$+\tfrac{1}{2}\int d\mathbf{r}\int d\mathbf{r}'\sum_{\alpha\beta\gamma\delta}\psi_\alpha^+(\mathbf{r})\psi_\beta^+(\mathbf{r}')V_{\alpha\beta\gamma\delta}(\mathbf{r}-\mathbf{r}')\psi_\delta(\mathbf{r}')\psi_\gamma(\mathbf{r})$$
$$-\int d\mathbf{r}\sum_\alpha\mu_\alpha\psi_\alpha^+(\mathbf{r})\psi_\alpha(\mathbf{r}) \tag{26.20c}$$

We remark that the electromagnetic field does not enter into the Hamiltonian used in the averaging ensemble as indeed it will not save for radiation densities so high as to compete with the "temperature" in distributing atoms over electronic states. In Eq. (26.20c), E_α and μ_α are the atomic internal energy and chemical potential, respectively, $V_{\alpha\beta\gamma\delta}$ the interatomic potential.

Therefore, the details of the spectral line contour for the absorption of radiation by an isotropic distribution of molecules between which a pair potential operates is given by Eqs. (26.7) and (26.2) subject to the approximations (1) the dipole approximation [Eq. (26.10)] and (2) point absorbers. The main thrust of the Ross development now is (1) the expression of χ in terms of Greens functions and (2) the application of many-body theory to the result.

27 Of One- and Two-Particle Propagators and Their Diagrams

We can certainly agree that, if we add a particle at point, $\mathbf{r}_1$, and time, t_1, to an n-particle system in its ground state ($T=0$), the probability of finding this particle at point $\mathbf{r}_2$, and time t_2, will be given by Eq. (8.16a) as

$$iG^+(\mathbf{r}_1\mathbf{r}_2; t_2 - t_1) = \langle \Psi_0 | T_\leftarrow \{\psi(\mathbf{r}_2 t_2) \psi^+(\mathbf{r}_1 t_1)\} | \Psi_0 \rangle \qquad (27.1)$$

where $T_\leftarrow$ is the TOO ($t_2 > t_1$), Ψ_0 is the system ground state, ψ the particle operator, all in second-quantized form. G is the Greens function; the operators are in the Heisenberg representation. The poles of G^+ correspond to the energies of the $n+1$ particle system less the energy of the n particle ground state.

That G^+ is a solution to

$$\left[-\frac{\mathbf{p}^2}{2m} + i\frac{\partial}{\partial t} - V \right] G^+(\mathbf{p}, t-t') = \delta(t-t') \qquad (27.2a)$$

where

$$G^+(\mathbf{p}, t-t') = G^0(\mathbf{p}, t-t') + \int_{-\infty}^{\infty} dt'' G^0(\mathbf{p}, t-t') V G^+(\mathbf{p}, t''-t') \qquad (27.2b)$$

we have seen in Eq. (8.4). In Eq. (27.2b) G^0 is the free-particle propagator.

By repeated iteration—substitute the right side of Eq. (27.3b) for G^+ under the integral sign—we easily obtain

$$G^+(\mathbf{p}, t-t') = G^0(\mathbf{p}, t-t') + \int_{-\infty}^{\infty} dt'' G^0(\mathbf{p}, t-t') V G^0(\mathbf{p}, t-t')$$
$$+ \int_{-\infty}^{\infty} \int dt' dt''' G^0 V G^0 V G^0 + \ldots \qquad (27.3a)$$

or

$$\Uparrow = \uparrow + \uparrow^{\nearrow} + \uparrow^{\uparrow} + \ldots = \uparrow [1 + \uparrow(\_\_\lrcorner) + \uparrow^2(\_\_\lrcorner)^2 + \ldots]$$

$$= \frac{\uparrow}{1 - \_\_\lrcorner} \qquad = \frac{1}{\uparrow^{-1} - \_\_\lrcorner} \qquad (27.3b)$$

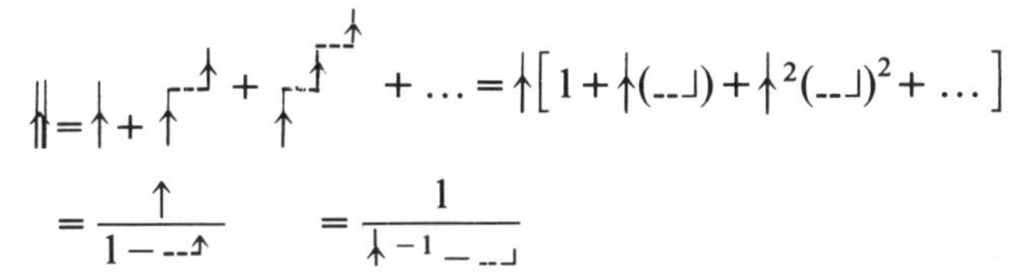

where $\uparrow = \mathrm{i}G^0$, $--- = \mathrm{i}V$, $\Uparrow = \mathrm{i}G^+$ and vertex $(--\lrcorner) = \int$. (We continue to reserve $\sim$ for the photon.)

G^+ corresponds to a particle moving forward in time, G^- to a particle moving backward in time (hole). Thus $G^{0-}(=\mathrm{i}\vartheta(t',t)\cdot\exp[-\mathrm{i}\epsilon_k(t-t')]$ for $t \neq t'$, $=i$ for $t=t'$) becomes

$$G^{0-} = [\omega - \epsilon_k - \mathrm{i}\delta]^{-1} \tag{27.4}$$

for $\epsilon_k < \epsilon_F$, where ϵ_F is the energy of the Fermi surface, that is, the containing energy of the Fermi sea. From G^{0-} follows G^-, the one-hole propagator.

Our expression for the S matrix for the particle-field interaction (Section 5) holds equally well for the particle-particle interaction of interest here when we replace $\mathbf{p}\cdot\mathbf{A}$ with the interparticle V. With a bit of artistic ingenuity then we may pictorially develop the S-matrix expansion for the present case which should prove to be of something more than aesthetic interest if it can be related to the one-particle propagator. That it can be so related we shall use Eq. (27.1) to continue to demonstrate [cf. Appendix E of Mattuck (29)] in a quite nonrigorous fashion. Since the operators are in the Heisenberg representation (we shall not change notation as we change representation) Eq. (27.1) becomes

$$\psi(\mathbf{r}_2t_2)\psi^+(\mathbf{r}_1t_1) = {}^{\mathrm{i}H_1t_2}e^{\mathrm{i}H_0t_2}\psi(\mathbf{r}_2)e^{-\mathrm{i}H_0t_2}e^{-\mathrm{i}H_1(t_2-t_1)}e^{\mathrm{i}H_0t_1}\psi^+(\mathbf{r}_1)$$
$$\cdot e^{-\mathrm{i}H_0t_1}e^{-\mathrm{i}H_1t_1} = e^{\mathrm{i}H_1t_2}\psi(\mathbf{r}_2t_2)\mathcal{U}(t_2t_1)\psi^+(\mathbf{r}_1t_1)e^{-\mathrm{i}H_1t_1} \tag{27.5}$$

where the $\mathcal{U}$ is a TDO, the ψ now in the interaction representation. Now $|\Psi_0\rangle$ in Eq. (27.1) is the state vector for the interacting particles. We replace $|\Psi_0\rangle$ as follows:

$$|\Psi_0\rangle = e^{\mathrm{i}H_1T}|\Phi_0\rangle/N$$

where N is simply a normalization factor. We then obtain

$$\mathrm{i}G(\mathbf{r}_2\mathbf{r}_1;t_2-t_1) = \lim_{T_1T_2\to\infty}[\langle\Phi_0|\mathcal{U}(T_2t_2)\psi(\mathbf{r}_2t_2)\mathcal{U}(t_2t_1)\psi^+(\mathbf{r}_1t_1)\mathcal{U}(t_1T_1)|\Phi_0\rangle]/$$
$$\cdot\langle\Phi_0|\mathcal{U}(T_2T_1)|\Phi_0\rangle = \langle\Phi_0|\mathrm{T}\{S\psi(\mathbf{r}_2t_2)\psi^+(\mathbf{r}_1t_1)\}|\Phi_0\rangle/\langle\Phi_0|S|\Phi_0\rangle \tag{27.6}$$

In this case S_1 is sufficiently instructive to be written down [cf § 12.1 of Kirzhnits (28) but not his Hamiltonian]. This logically demands a change of notation. $\mathbf{r}_1$ refers to position "1," $\mathbf{r}_2$ to position "2," both referring, if you will, to particle "1." Now we are going to consider a binary interaction hence calling forth a particle "2." Therefore, let $\psi(1t_1)$ refer to particle "1" at time t_1. Therefore,

$$G_1 = \frac{-\int dt'\langle\Phi_0|V(12)\mathrm{T}\{\psi^+(1t')\psi^+(2t')\psi(2t')\psi(1t')\psi(1t_\alpha)\psi^+(1t_1)\}|\Phi_0\rangle}{\langle\Phi_0|S|\Phi_0\rangle} \tag{27.7}$$

Now we recall Wick's theorem. Since we suppose Φ_0 to be the vacuum state, any annihilation operator found on the right of the operator product zeroes out the matrix element. Thus all normal products are eliminated from the Wick series leaving only all possible full contractions. (This requires a slightly different representation of contractions, for they now include external lines.) In considering Eq. (27.7) we first consider only the numerator, the only ground rule for forming contractions being that an annihilation operator must be contracted with a creation operator. We thus obtain

$$\psi^+(1t')\psi^+(2t')\psi(2t')\psi(1t')\psi(1t_2)\psi^+(1t_1) \qquad \text{(a)}$$

$$\psi^+(1t')\psi^+(2t')\psi(2t')\psi(1t')\psi(1t_2)\psi^+(1t_1) \qquad \text{(b)}$$

$$\psi^+(1t')\psi^+(2t')\psi(2t')\psi(1t')\psi(1t_2)\psi^+(1t_1) \qquad \text{(c)}$$

$$\psi^+(1t')\psi^+(2t')\psi(2t')\psi(1t')\psi(1t_2)\psi^+(1t_1) \qquad \text{(d)}$$

$$\psi^+(1t')\psi^+(2t')\psi(2t')\psi(1t')\psi(1t_2)\psi^+(1t_1) \qquad \text{(e)}$$

$$\psi^+(1t')\psi^+(2t')\psi(2t')\psi(1t')\psi(1t_2)\psi^+(1t_1) \qquad \text{(f)}$$

The diagrams corresponding are given in Figs. 27.1 and 27.2. Again we are applying what amounts to the Boson ladder method. We see therefore that a two-rung ladder tells us the second approximation (Fig. 27.2) where (1)

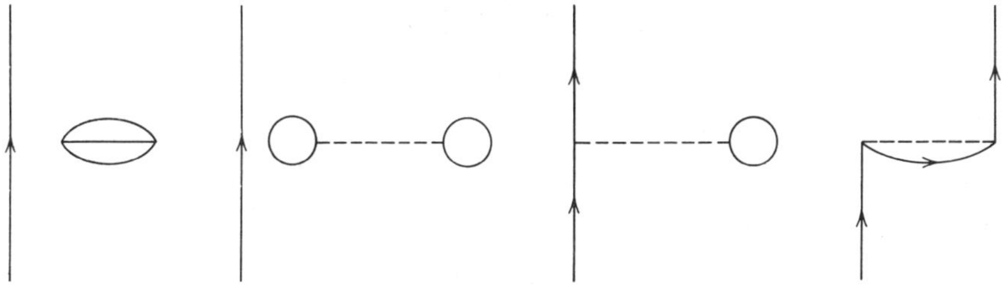

(a) *(b)* *(c), (e)* *(d), (f)*

Figure 27.1

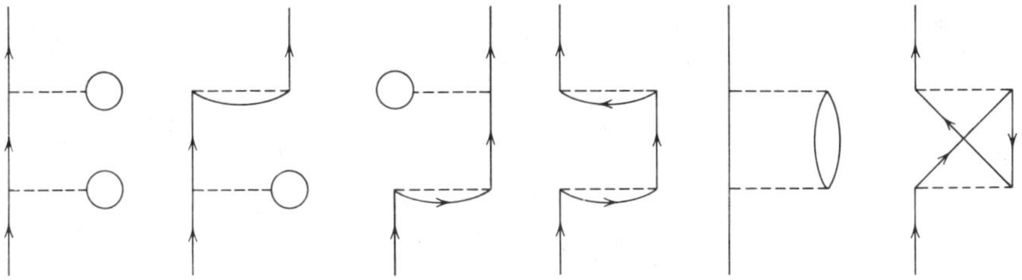

Figure 27.2

anomalous or momentum nonconserving graphs, that is,

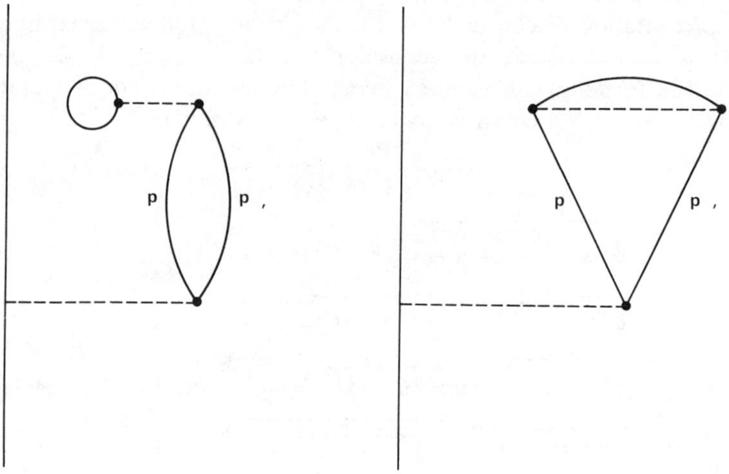

etc.,

and so on have been dropped since they require $\mathbf{p}_{\text{hole}} = \mathbf{p}_{\text{particle}}$ which is impossible, and (2) the full $G(=G^{\dagger}+G^{-})$ is associated with each directed line à la Feynman, so that, for example,

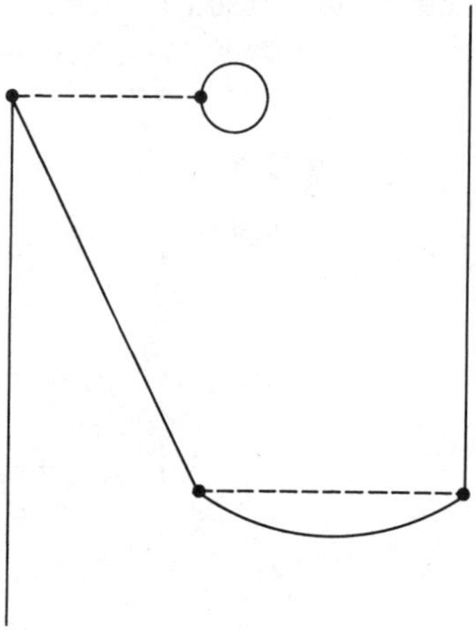

is already included. This puts us in a position to write out Eq. (27.7) or, better yet, Eq. (27.6),

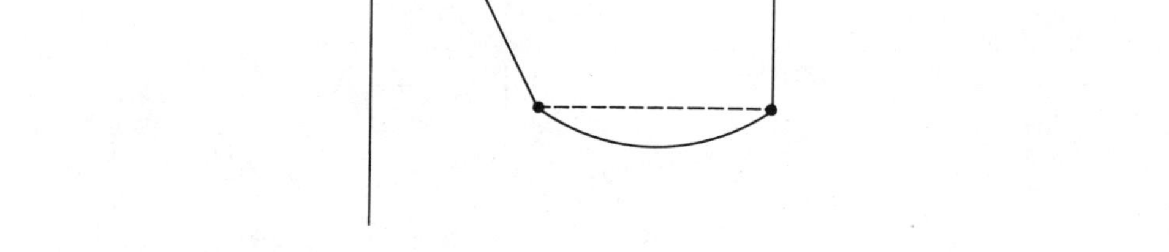

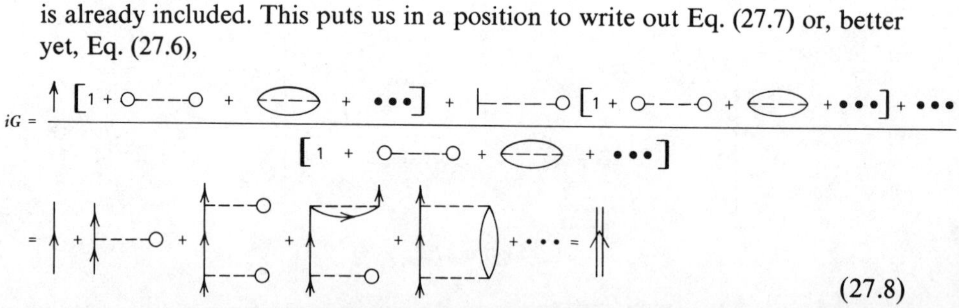

(27.8)

which amounts to a repetition of the linked-diagram theorem of Section 5. (We shall not attempt listings of general rules for algebraic diagram translations but will content ourselves with explaining those whose translation becomes necessary.)

We will recall that the "self-energy part" of a diagram is that portion containing no external lines so that the SEP's appearing in Eq. (27.8) are

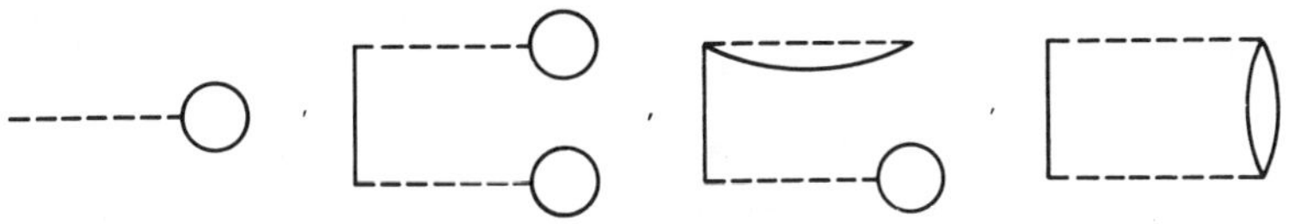

(27.9)

The proper or irreducible SEP's have no diagram components connected by single internal lines, so that the first and fourth figures in Eq. (27.9) are proper, the second and third improper. These improper SEP's may surely be written as products of proper SEP's so that Eq. (27.9) may be written as

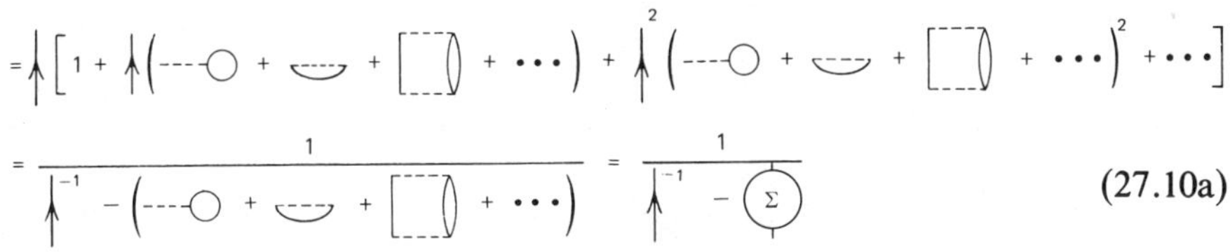

(27.10a)

where, of course,  is the sum of all proper SEP's, the irreducible self-energy.

Equation (27.10a) is the form of Dyson's equation which holds for no external potential. The more general form is

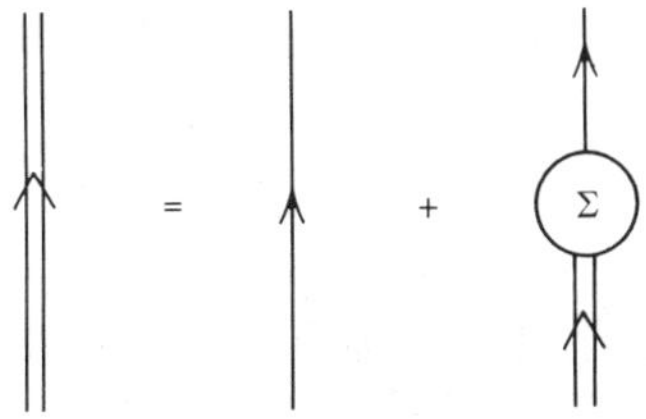

(27.10b)

or

$$iG(\mathbf{p}, t_2 - t_1) = iG^0(\mathbf{p}, t_2 - t_1) + \int\int dt'\, dt''\, iG^0(\mathbf{p}, t_2 - t_1)(-i)$$
$$\cdot \Sigma(\mathbf{r}, t'' - t')\, iG(\mathbf{p}, t' - t_1) \qquad (27.10c)$$

where anomalous diagrams must now be included. It is sometimes convenient to break down the definition of Σ into clothed skeletons, that is, into sums of all SEP's which may be associated with internal lines, thus

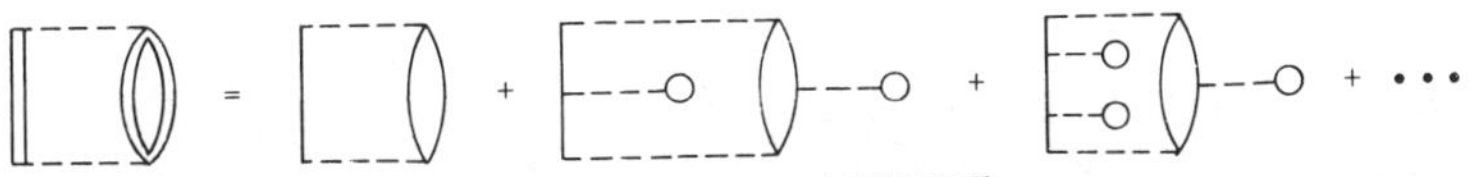

(27.11a)

(27.11b)

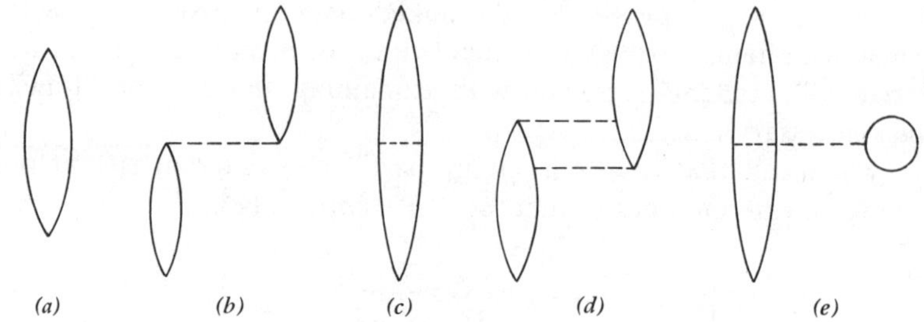

Figure 27.3

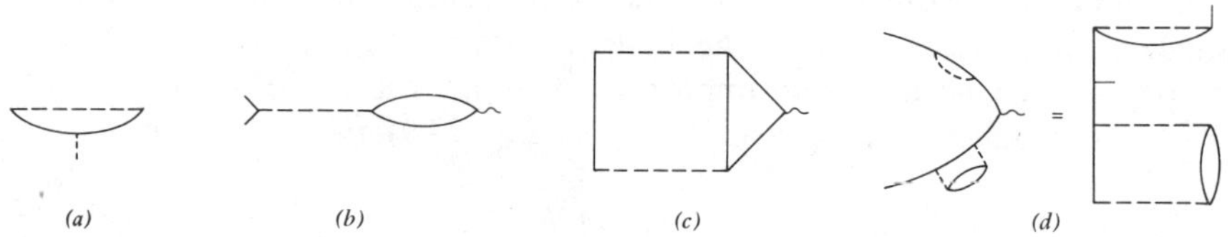

Figure 27.4

This may—or, then again, it may not—be an appropriate point at which to introduce the "polarization part" of the diagram: Any element having no external lines and which may be inserted into an interaction line or a photon line. Examples are given in Fig. 27.3*a*. We have encountered, for example, Delbruck scattering wherein Fig. 27.3*a* is simply inserted into a free-photon propagator.

Figures 27.3*a*, 27.3*c*, and 27.3*d* are irreducible or proper polarization parts since the cutting of a single interaction line does not split up the diagrams. In like manner, Figs. 27.3*b* and *e* are improper.

If we define a "vertex" as the juncture of (1) an interaction, (2) a particle line, and (3) a hole line—as a vertex actually—then a "vertex part" is any diagram component to which may be joined such elements as in Fig. 27.4. We remark that no single particle or interaction line in Figs. 27.4*a* or 27.4*c* may be cut so that the entire diagram is disconnected. Such vertices are called proper or irreducible vertex parts. Figures 27.4*b* and 27.4*c* are improper of course. Which leads to the following definition:

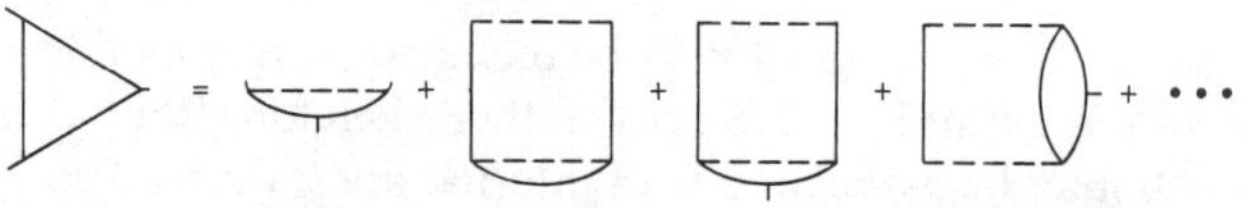

(27.12)

for the sum of all PVP's.

We have now more or less evaluated the one-particle propagator, Eqs. (27.1) and (27.6), or, more precisely, we have indicated how we might evaluate it for $T=0$. However we are really interested in the $T\neq 0$ case, and, in order to make the transition, we merely (1) associate a statistical weight factor with each directed line, (2) replace real time by imaginary time, and (3) subtract the chemical potential from the energy [cf § 14 of Mattuck (29), § 21 of Kirzhnits (28), and § 11 of Abrikosov et al. (3)]. Now we obtain

$$\mathsf{G}(\mathbf{p}, \tau_2 - \tau_1) = -\langle \mathsf{T}_{\leftarrow}\{\psi(\mathbf{p}\tau_2)\psi^+(\mathbf{p}\tau_1)\}\rangle \tag{27.13}$$

where $\langle \cdots \rangle$ includes the grand canonical ensemble, and an operator in the Heisenberg representation is

$$O(\tau) = e^{(H-\mu N)\tau} O e^{-(H-\mu N)\tau}, \qquad \tau \text{ real} \tag{27.14}$$

so that an imaginary time, t-iτ, has apparently been introduced. The point will probably be familiar, but we make it anyway. From the averaging process we will obtain a Boltzmann factor, $\exp[-\beta H]$, where $\beta \sim (kT)^{-1}$ say. It is convenient to behave as if $\beta = \tau = -\mathrm{i}t$ so that we will instead obtain the factor $\exp[\mathrm{i}Ht]$. Since β is real, t is imaginary, the imaginary time. That the identical diagrammatic treatment may now be built on the Bloch equation as the $T = 0$ situation was built on the Schrödinger equation may be demonstrated. There is, however, an important difference. The imaginary-time free-particle propagator is [cf. § 14.3 of Mattuck (29)]

$$\mathsf{G}^0(\mathbf{p}, \tau_2 - \tau_1) = -\left[\vartheta_{\tau_2 - \tau_1} f_p^+ - \vartheta_{\tau_1 - \tau_2} f_p^-\right] e^{-(\epsilon_k - \mu)(\tau_2 - \tau_1)} \tag{27.15}$$

In Eq. (27.15) we see that the energy has been shifted by the chemical potential, μ, and ϑ factors relating to energy by the statistical factors f_p. The result of these f factors is to allow a hole and a particle to be present with the same p which has the effect of reintroducing the anomalous graphs which we eliminated from Fig. 27.2. (We have not emphasized the fact, but we do require momentum conservation at a vertex.) With which we turn to the two-particle propagator.

Let us suppose that one particle is introduced into the system at $(\mathbf{r}_1 t_1)$, another at $(\mathbf{r}_3 t_3)$. Then we call $\mathsf{G}^2(\mathbf{r}_4 t_4 \cdots \mathbf{r}_1 t_1)$ the probability that there will be one particle at $(\mathbf{r}_2 t_2)$ and another at $(\mathbf{r}_4 t_4)$ where $t_3 > t_4 > t_1 > t_2$. Then our diagrammatic experiences of the last few paragraphs suggest that we can write a two-particle propagator as

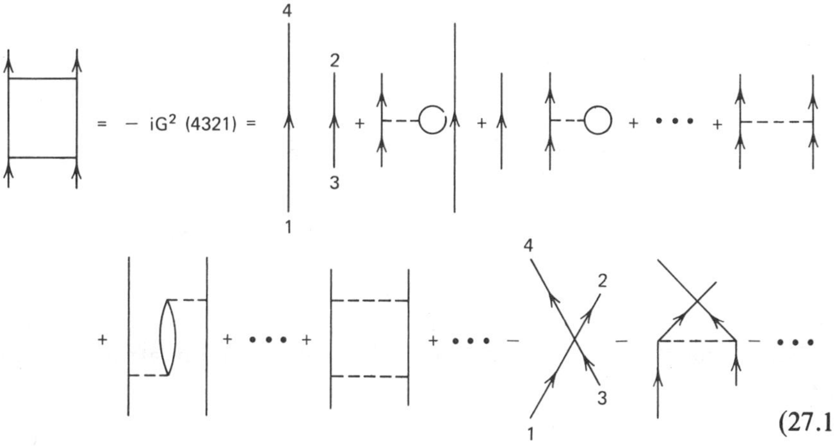

(27.16a)

One may initially find the cross terms with negative sign something of a surprise, but a little rumination will allay any uneasiness. To this point we have only dealt with two external lines corresponding to one real particle. (The indistinguishability principle somewhat obscures this straightforward concept, but we ignore it for the moment.) Virtual particles came and went with no concern over sign. However, now we are dealing with two real particles. In the diagrams having positive (repulsive) sign we have a situation where the particles interact but maintain their individuality. In the diagrams having negative (attractive) sign, we have a situation wherein the particles exchange individuality. Such a sign correspondence is quite familiar from atomic structure [cf. Slater (41)]. In order to show why the sign differences really occur, however, consider the zeroth-order term in the expansion for $G^2(4,3,2,1)$ analogous to Eq. (27.6),

$$\begin{aligned} -iG^2(4,3,2,1) &= -\langle 0|T\{\psi(4)\psi^+(3)\psi(2)\psi^+(1)\}|0\rangle \\ &\quad -\langle 0|T\{\psi(4)\psi^+(3)\psi(2)\psi^+(1)\}|0\rangle . \end{aligned}$$

Now

$$iG^0(12) = \begin{matrix} \langle 0|\psi(x_1)\psi^+(x_2)|0\rangle & t_1 > t_2 \\ -\langle 0|\psi^+(x_2)\psi(x_1)|0\rangle & t_1 < t_2 \end{matrix}$$

so that

$$\begin{aligned} -iG^2 &= -[iG^0(4,1)][-iG^0(2,3)] - [-iG^0(4,3)][-iG^0(2,1)] \\ &= [iG^0(4,1)]\cdot[iG^0(2,3)] - [iG^0(4,3)][iG^0(2,1)] \end{aligned}$$

since $G^0(2,3) = -G^0(3,2)$ and $G^0(1,2)$ has a minus sign for $t_1 < t_2$.

We have thus a particle-particle propagator in Eq. (27.16). However, we could have a particle-hole propagator,

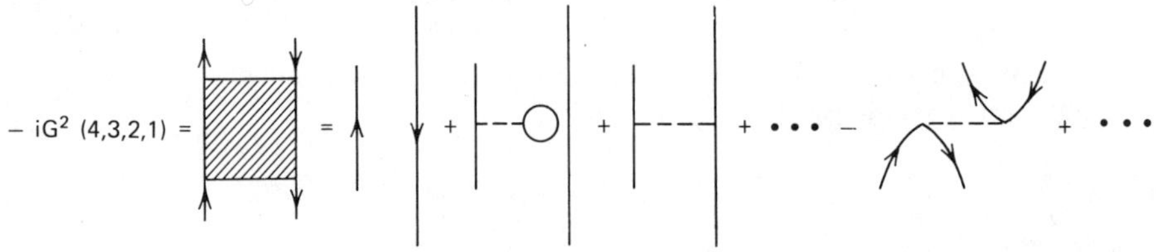

(27.16b)

or a hole-hole propagator. All such possible G^2 corresponding to all possible time orderings are summarized by

$$G^2(4,3,2,1) = -i\langle \Psi_0|T\{\psi(\mathbf{r}_4 t_4)\psi^+(\mathbf{r}_3 t_3)\psi(\mathbf{r}_2 t_2)\psi^+(\mathbf{r}_1 t_1)\}|\Psi_0\rangle \tag{27.16c}$$

which is to be compared to one-particle propagator definition, Eq. (27.1). Finally, we suppose our case to be for $T \neq 0$ with the appropriate analogy to Eq. (27.13).

28 Of Propagators, the Susceptibility and Its Diagrams

We will probably agree that there is a certain kinship between Eq. (26.20c) for the current commutator and Eq. (27.16c) for the two-particle propagator. Indeed the principle objective of a great deal of this maneuvering has been the expansion of the former in terms of the latter (plus one-particle complications). However, Ross wished to express the current commutator in terms of functions which he could generate, the straightforward two-particle propagator fulfilling such a requirement only indirectly. So what he did was appeal to a method due to Baym and Kadanoff (BK)(66) for generating this function through the agency of a related function. We shall obtain this function after a few additional remarks about the imaginary time, $T\neq 0$ propagator of Section 27.

Now if, in the $T\neq 0$ case, we transform our operators as

$$O(\tau)=e^{(H-\mu N)\tau}Oe^{-(H-\mu N)\tau},$$

where previously we have used

$$O(t)=e^{iHt}Oe^{-iHt},$$

we find that our one-particle, $T\neq 0$ propagator $G(\tau_2-\tau_1)$, $0<\tau_1,\tau_2<\beta$, is obtained by allowing

$$H\rightarrow H-\mu N \tag{28.1a}$$

$$it\rightarrow\tau \tag{28.1b}$$

In Eq. (28.1b) τ is real so that the time $t=-i\tau$, is imaginary. (It is common in the literature to speak of the "imaginary time" and display the symbol τ, which, of course, stands for a real quantity, albeit not the time. This appears to confuse no one, however, and any attempted redefinition would doubtless be disastrous.) The G which is thus obtained has an expansion identical to the real time, $T=0$ expansion [cf. Appendix B of Mattuck (29)]. As $0<\tau_1,\tau_2<\beta$, $\mathsf{G}(\tau)$ converges on $-\beta<\tau<\beta, \tau=\tau_2-\tau_1$. $\mathsf{G}(\tau)$ obeys the "quasi-periodic boundary condition," $\mathsf{G}(\tau)=\mathsf{G}(\tau+\beta)$. In order to avoid difficulties when going over to the $(\mathbf{p},\omega)$ form, a propagator, which is $\mathsf{G}(\mathbf{p},\tau)$ periodically repeated for τ between $-\infty$ and ∞, replaces $\mathsf{G}(\mathbf{p},\tau)$:

$$\mathsf{G}(k,\tau)=\frac{1}{\beta}\sum_{n=-\infty}^{\infty} e^{-i\omega_n\tau}\mathsf{G}(\mathbf{p},\omega_n) \tag{28.2a}$$

$$\mathsf{G}(p,\omega_n)=\tfrac{1}{2}\int_{-\beta}^{\beta} d\tau e^{i\omega_n\tau}\mathsf{G}(\mathbf{p},\tau) \tag{28.2b}$$

$$\omega_n=\pi n/\beta, \qquad n=0,\pm 1,\pm 2,\cdots \tag{28.2c}$$

where even-n terms disappear for Fermions, odd-n for Bosons. All of which should be sufficient justification for the BK boundary condition,

$$\mathsf{G}(\mathbf{r},-i\beta;\mathbf{r}_1',t_1')=\pm\mathsf{G}(\mathbf{r}_1,0;\mathbf{r}_1',t_1'), \qquad 0<it_1'<\beta \tag{28.3}$$

which is obeyed by the same time arguments of all the multiparticle Greens functions. BK then proceed to develop a method for determining the one-particle Greens function in the presence of an external disturbance $U(r_1t_1)n(r_1t_1)$ where U is an arbitrary scalar potential, n the number density operator. Before the disturbance is switched on, we suppose $\mathsf{G}(11';U)$ to be given by Eq. (27.1) ($T=0$). Then the effect of the disturbance is to induce a form of Heisenberg representation,

$$\psi(t,t)=\mathcal{U}^{-1}(t)\psi(\mathbf{r}_1,-\infty)\mathcal{U}(t) \tag{28.4a}$$

$$\mathcal{U}(t)=\mathsf{T}\left[\exp\left\{-\mathrm{i}\int_{-\infty}^{t}dt'\left[H(t')+\int d\mathbf{r}\,U(\mathbf{r},t')n(\mathbf{r},t')\right]\right\}\right] \tag{28.4b}$$

We expand $\mathcal{U}^{-1}$ and $\mathcal{U}$ in Eq. (28.3a), the resulting first term of course being that portion of ψ unaffected by the external disturbance, U. The next term will tell us the first-order change in ψ induced by U,

$$\delta\psi(r_1t_1)=\mathrm{i}\int_{-\infty}^{t}dt_2\,d\mathbf{r}_2[n(2),\psi(1)]U(2) \tag{28.5}$$

where the notation is the familiar one: $1=r_1t_1$.

By means of the definition of the propagator we may relate the change in the density of particles at a space-time point to the change in the propagator and hence obtain the former by computing the change in the latter. From Eq. (28.5) this may be evaluated as

$$\begin{aligned}\delta\langle n(1)\rangle &= \pm\mathrm{i}\delta\mathsf{G}(11^+)=-\mathrm{i}\langle\mathsf{T}[\delta\psi(1)\psi^+(1^+)+\psi(1)\delta\psi^+(1^+)]\rangle\\ &=-i\int_{-\infty}^{t_1}dt_2\,d\mathbf{r}_2\frac{\mathrm{Tr}\{e^{-\beta H}[n(1),n(2)]\}U(2)}{\mathrm{Tr}(e^{-\beta H})}\\ &=-\mathrm{i}\langle[n(1),n(2)]\rangle U(2)\end{aligned} \tag{28.6}$$

where $H=H-\mu N$, and the trace need not include the external disturbance, $U(2)$. From this BK obtain the linear response of the $\langle n\rangle$ to U as (infinitesimal initial U)

$$\frac{\delta\langle n(1)\rangle}{\delta U(2)}=\begin{matrix}-\mathrm{i}\langle[n(1),n(2)]\rangle_{U=0} & t_1>t_2\\ 0 & t_1\le t_2\end{matrix} \tag{28.7}$$

When the operators are written in the interaction representation vis-à-vis the external disturbance, the propagator becomes

$$\mathsf{G}(11';U)=-\mathrm{i}\,\mathrm{Tr}\{e^{-\beta H}\mathsf{T}[S\psi(1)\psi^+(1')]\}/\mathrm{Tr}(e^{-\beta H}\mathsf{T}[S]) \tag{28.8a}$$

$$S=\exp\left[-\mathrm{i}\int_0^{-\mathrm{i}\beta}d\mathbf{r}_2\,dt_2\,n(r_2t_2)U(r_2t_2)\right] \tag{28.8b}$$

S also in the interaction representation.

Consider the expansion of Eq. (28.8a)

$$G(11';U)=\frac{-i\langle\psi(1)\psi^+(1')\rangle+(-i)^2\int_0^{-i\beta}dt_2\,d\mathbf{r}_2\langle\psi(1)\psi^+(1')n(2)\rangle U(2)+\dots}{\langle 1\rangle+\int_0^{-i\beta}dt_2\,d\mathbf{r}_2\langle n(2)\rangle U(2)+\dots}\tag{28.9a}$$

where we have made use of the (a) cyclic nature of the trace,

$$\operatorname{Tr}(ABCD)=\sum_{i\cdots l}a_{ij}b_{jk}c_{kl}d_{li}=\sum_{i\cdots l}b_{jk}c_{kl}d_{li}a_{ij}\tag{28.10a}$$

and the (b) fact that the trace is not over the external coordinates of $U(2)$. Now expand the numerator in the familiar series and keep only terms linear in U,

$$G(11';U)=G(1-1')+(-i)^2\int_0^{-i\beta}dt_2\,d\mathbf{r}_2\{\langle T[\psi(1)\psi^+(1')n(2)\rangle$$

$$-\langle T[\psi(1)\psi^+(1')]\rangle\langle n(2)\rangle\}U(2)\tag{28.9b}$$

$$\frac{\delta G(11';U)}{\delta U(2)}=(-i)^2\{\langle T[\psi(1)\psi^+(1')n(2)]\rangle-\langle T[\psi(1)\psi^+(1')]\rangle\langle n(2)\rangle\}$$

$$=\pm\left[G^2(12,1^+2^+)-G(1-1^+)G(2-2^+)\right]_{U=0}\tag{28.10b}$$

Therefore, a function,

$$L(12,1'2')=\left[G^2(12,1'2')-G(1-1')G(2-2')\right]_{U=0}\tag{28.11a}$$

may be generated by

$$L(12,1'2')=\pm\left[\frac{\delta G(11')}{\delta U(2)}\right]_{U=0},\tag{28.11b}$$

where $G(11')$ is defined by Eq. (28.8a), S by

$$S=\exp\left[-i\int_0^{-i\beta}dt_2\,dt_2'\,d\mathbf{r}_2\,d\mathbf{r}_2'\psi^+(\mathbf{r}_2t_2)U(\mathbf{r}_2t_2\mathbf{r}_2't_2')\psi(\mathbf{r}_2't_2')\right],\tag{28.11c}$$

and such an L was used by Ross in order to express the current commutator. $U(22')$ represents a disturbance in which a particle is removed from the system at 2 and added at $2'$. When $U(22')=U(2)\delta(2-2')$, Eq. (28.11c) reduces to Eq. (28.8b). In $G(U)$ then we have located a generator for what has been called the "two-particle correlation function," L. As BK point out, the continuation of L to real time values yields an L which describes all the linear transport properties of the system.

If we use the familiar commutation relation, $[\psi_i(r),\psi_j^+(r)]=\delta_{ij}\delta(r-r')$, in order to rewrite Eq. (26.20c), we will obtain

$$\langle\psi_\beta(r_1)\psi_\delta(r_2')\psi_\gamma^+(r_2'')\psi_\alpha^+(r_1''')\rangle=\langle\psi_\beta(r_1)\psi_\alpha^+(r_1''')\psi_\delta(r_2')\psi_\gamma^+(r_2'')\rangle\tag{28.12}$$

where we have distinguished between particle 1 and particle 2 by the subscript. Equation (26.20c) will now have two first and two third terms, $H(i)$, one relating to each of the two particles, and a middle term relating to both. Because of the cyclic nature of the trace we may write

$$\begin{aligned}\mathrm{Tr}\{e^{-\beta H}\cdots\} &= \mathrm{Tr}\{\exp -[H(1,2)+H(1)]\beta\} \\ &\quad \times \psi_\beta(r_1)\psi_\alpha^+(r_1''')e^{-H(2)\beta}\psi_\delta(r_2')\psi_\gamma^+(r_2'')e^{-H(1,2)\beta}\} \\ &= \mathrm{Tr}\{e^{-\beta H}\psi_\beta(r_1)\psi_\alpha^+(r_1'')\}\,\mathrm{Tr}\{e^{-\beta H}\psi_\delta(r_2')\psi_\gamma^+(r_2'')\}\end{aligned}$$

Therefore, by Eq. (27.1) we may write Eq. (28.11a) as

$$\langle\psi_\beta(r_1)\psi_\delta(r_2')\psi_\gamma^+(r_2'')\psi_\alpha^+(r_1''')\rangle + \mathrm{G}_{\beta\alpha}(r_1-r_1''')\mathrm{G}_{\delta\gamma}(r_2'-r_2'') = L_{\alpha\beta\gamma\delta}(r_1r_2', r_1'''r_2'') \tag{28.13}$$

and Eq. (26.20c) may be rewritten accordingly. In order to obtain the most useful form of $\mathrm{G}_{\alpha\beta}$ and $L_{\alpha\beta\gamma\delta}$, however, we take the Fourier transforms

$$\mathrm{G}_{\alpha\beta}(\mathbf{k},\omega_k) = \int_0^{-\mathrm{i}\beta} dt \int_{-\infty}^{\infty} d\mathbf{r}\, e^{-\mathrm{i}(\mathbf{k}\cdot\mathbf{r}-\omega_k t)}\mathrm{G}_{\alpha\beta}(\mathbf{r},t) \tag{28.14a}$$

$$\begin{aligned}&L_{\alpha\beta\gamma\delta}(\mathbf{p}_1\omega_p, \mathbf{p}_2'\omega_{p'}; \mathbf{q}_1\omega_q, \mathbf{q}'\omega_{q'})\delta(\mathbf{p}_1+\mathbf{p}_2'-\mathbf{q}_1-\mathbf{q}_2')\delta(\omega_p+\omega_{p'}-\omega_q-\omega_{q'}) \\ &= \int_0^{-\mathrm{i}\beta} dt\,dt'\,dt''\,dt''' \int_{-\infty}^{\infty} d\mathbf{r}\,d\mathbf{r}'\,d\mathbf{r}''\,d\mathbf{r}'''L_{\alpha\beta\gamma\delta}(\mathbf{r}_1t, \mathbf{r}_2't'; \mathbf{r}_1''t'', \mathbf{r}_2''t''') \\ &\quad \cdot\exp\left[\mathrm{i}\left(\mathbf{q}_1\cdot\mathbf{r}_1''-\omega_q t''+\mathbf{q}_2'\cdot\mathbf{r}_2''-\omega_{q'}t'''-\mathbf{p}_1\cdot\mathbf{r}_1+\omega_p t-\mathbf{p}_2'\cdot\mathbf{r}_2'+\omega_{p'}t'\right)\right]\end{aligned} \tag{28.14b}$$

where the frequencies for Ross' boson case are given by Eq. (28.2c) with even n and β replaced by $\mathrm{i}\beta$. Remark the conservation of momentum $\mathbf{p}_1$ and $\mathbf{p}_2'$, respectively. At removal this delta function assures that the momenta $(\mathbf{q}_1+\mathbf{q}_2')$ will be the same.

Now substitute Eq. (28.14b) into the appropriate form of Eq. (28.2a) (n even for bosons),

$$\frac{1}{\mathrm{i}\beta}\sum_n L = \sum_n \frac{\omega_n}{2\pi}L \Rightarrow \int d\omega \frac{L}{2\pi}.$$

Thus L is substituted into

$$\frac{1}{(2\pi)^4}\int d\omega\, d\mathbf{k} \equiv \frac{1}{(2\pi)^4}\int dk$$

in Ross notation. That author also writes $(\mathbf{k},\omega_k)\equiv k$, so that the substitution of Eq. (28.13) into Eq. (26.20a), the substitution of this result into Eq. (26.7b) and the transformation of the result yields

$$\begin{aligned}\chi(\mathbf{k},\omega) &= \mathrm{i}\omega^{-2}\sum_{\alpha\beta\gamma\delta}\omega_{\alpha\beta}\omega_{\gamma\delta}\mathbf{d}_{\alpha\beta}\mathbf{d}_{\gamma\delta}\int\frac{dp}{(2\pi)^4} \\ &\quad\times\int\frac{dq}{(2\pi)^4}L_{\beta\delta\alpha\gamma}(p,q;p-k,q-k)\big|_{\omega_n\to\omega+\mathrm{i}\epsilon}\end{aligned} \tag{28.15a}$$

We may write Eq. (27.16b) as

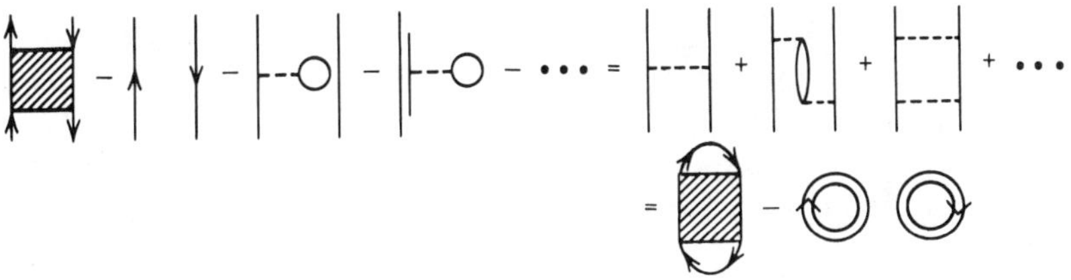

if $t_1=t_2$, $t_3=t_4$. This corresponds to Eq. (28.15a) (diagrams rotated)

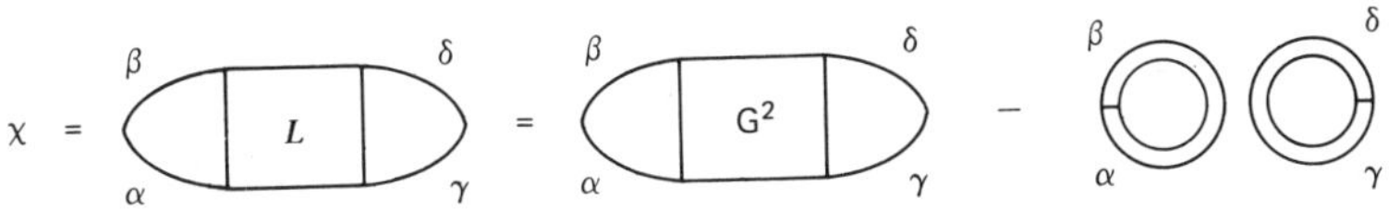

(28.15b)

It is clear from the diagrams that the second term removes the disconnected parts of the first.

We may recall the polarization part subsequent to Eqs. (27.15), an entity which, with considerable justification, may be dubbed the polarization propagator. This is essentially what Eq. (28.15b) amounts to, a fact which becomes obvious when we insert it into a free-photon propagator. The diagram describes the following physical process:

A photon enters the left vertex exciting the atom from state α to state β. Subsequently, the atom which has been excited interacts with the medium in which it exists as is exhibited by the boxed L. Later some atom, whether the originally excited one or some other, emits a photon as a consequence of the transition from state δ to state γ. Our particular interest is that situation wherein $\omega_{\alpha\beta}=\omega_{\delta\gamma}$, this describing a spectral line. In principle at least, a complete solution to the line broadening problem, subject of course to the restrictions which have been introduced, may be obtained by solving equations for the one-particle Greens function and the function L. (The problem as thus described includes photon scattering, although, as we shall see, it is, for all practical purposes, irrelevant whether we leave the left sides of these diagrams open, thus excluding the incoming photon.) The restrictions, we will recall, involve the assumptions of (1) point particles and (2) electric dipole transitions.

Vis-à-vis the restriction to point particles, it arises because Ross supposes the creation or annihilation of a radiator at a point. The fact that the particles are not actually created or annihilated at a point means that the commutation relations for these operators are not precise, a matter considered by Girardeau (134). The subject has not been given extensive consideration by those interested in line broadening.

29 The Equation of Motion for the Greens Functions

We now obtain the equation of motion for G using the Hamiltonian given by Eq. (26.20c). In doing so it is first necessary to obtain an equation of motion for $\psi_\alpha(r,t)$ from $i\dot{\psi}_\alpha = [H, \psi_\alpha]$. Let us illustrate this procedure with the first term in Eq. (26.20c), representing the square bracketed expression by $\nabla(r)$, supposing for the moment the ∇ to be diagonal in the ψ so that $\nabla\psi = E\psi$.

$$\begin{aligned}
[H,\psi_\alpha] &= -\int dr \sum_i \left[\psi_i^+(r)\nabla(r)\psi_i(r)\psi_\alpha(r') - \psi_\alpha(r')\psi_i^+(r)\nabla(r)\psi_i(r)\right] \\
&= -\int dr \sum_i E_i\left[\psi_i^+(r)\psi_i(r)\psi_\alpha(r') - \psi_i^+(r)\psi_i(r)\psi_\alpha(r')\right. \\
&\quad \left. - \delta_{i\alpha}\delta(r-r')\psi_i(r)\right]\delta(r-r_1) \\
&= \int dr \nabla(\mathbf{r})\psi_\alpha(r)\delta(r-r') = \nabla(\mathbf{r}')\psi_\alpha(r')\delta(r'-r_1)
\end{aligned} \tag{29.1}$$

In this fashion we obtain

$$\begin{aligned}
&\left(i\frac{\partial}{\partial t} + \frac{1}{2\mathrm{m}_\alpha}\nabla^2 - E_\alpha + \mu_\alpha\right)\psi_\alpha(\mathbf{r},t)\delta(r'-r_1) \\
&\qquad = \sum_{\beta\alpha\delta} \int dr' \psi_\alpha^+(\mathbf{r},t)\mathrm{V}_{\alpha\beta\gamma\delta}(\mathbf{r}_1-\mathbf{r}')\psi_\delta(\mathbf{r}',t)\psi_\gamma(\mathbf{r}',t)
\end{aligned} \tag{29.2}$$

Eq. (29.2) is now multiplied through on the right by $\psi_\beta^+(\mathbf{r}',t')$. When the expectation value of the factor on the left, for example, is then taken we can see that

$$-i\langle[\psi_\alpha(\mathbf{r},t)\psi_\beta^+(\mathbf{r}',t')]_+\rangle = G_{\alpha\beta}(r-r')\delta(r'-r_1)$$

Applying this procedure to both terms in Eq. (29.2), the commutation relations for ψ being applied to the second term is required, leads to the equation of motion for G,

$$\begin{aligned}
&\left(i\frac{\partial}{\partial t} + \frac{1}{2\mathrm{m}_\alpha}\nabla^2 - E_\alpha + \mu_\alpha\right)\mathrm{G}_{\alpha\beta}(r-r')\delta(r'-r_1) - i\sum_{\mu\nu\lambda}\int dr_1 \mathrm{V}_{\alpha\lambda\mu\nu}(\mathbf{r}-\mathbf{r}_1) \\
&\quad \cdot\left[L_{\mu\nu\beta\lambda}(\mathbf{r}t,\mathbf{r}_1t;\mathbf{r}'t',\mathbf{r}_1t^*) + \mathrm{G}_{\mu\beta}(r-r')\mathrm{G}_{\nu\lambda}(00^-)\right] = \delta(r'-r_1)
\end{aligned} \tag{29.3}$$

We will now obtain the vertex equation corresponding to Eq. (29.3), an exercise which was perhaps best described by Margenau, long before the equation existed, as a bookkeeping operation. A bookkeeping operation to which the adoption of a notation change is of considerable importance. First, the spatial, temporal, and state indices are combined as, say,

$$(\mathbf{r}_1 t_1 \alpha) = 1 \tag{29.4a}$$

Subsequently, the summation over state index together with whatever integration there may be are combined as

$$\sum_{\alpha}\int_{0}^{-i\beta}dt_1\int_{-\infty}^{\infty}d\mathbf{r}_1=\int d1 \tag{29.4b}$$

Let us illustrate the notation change as follows:

$$\sum_{\mu\nu\lambda}\int d\mathbf{r}_1 \mathsf{V}_{\alpha\lambda\mu\nu}(\mathbf{r}-\mathbf{r}_1)L_{\mu\nu\beta\lambda}(\mathbf{r}t,\mathbf{r}_1t;\mathbf{r}'t',\mathbf{r}_1t^+)$$

$$=\sum_{\mu\nu\lambda}\int d\mathbf{r}_1\,\mathsf{V}(\mathbf{r}\alpha\mu,\mathbf{r}_1\lambda\nu)L(\mathbf{r}t\mu,\mathbf{r}_1t\nu;\mathbf{r}'t'\beta,\mathbf{r}_1t^+\lambda)$$

$$=\int d\bar{1}\,\mathsf{V}(1\bar{1})L(1\bar{1};2\bar{1}^+).$$

Let us recall Eq. (27.10a) in the form

$$\Uparrow=\left[\uparrow^{-1}-\textcircled{\Sigma}\right]^{-1}\Rightarrow \mathsf{G}=\left[\mathsf{G}^{0^{-1}}-\Sigma\right]^{-1}\Rightarrow \mathsf{G}^{-1}+\Sigma=\mathsf{G}^{0^{-1}} \tag{29.5}$$

and such an equation may be obtained for the present situation, it merely being a question of finding a form for these operators. First of all, the augmentation of the operator on the left of Eq. (27.2a) by the constant $\mu_\alpha - E_\alpha$ will still yield a G^0. Multiplication of Eq. (27.2a) on the right by $(\mathsf{G}^0)^{-1}$ then tells us that

$$\left(i\frac{\partial}{\partial t}+\frac{1}{2\mathrm{m}_\alpha}\nabla^2-E_\alpha+\mu_\alpha\right)=(\mathsf{G}^0)^{-1}\delta(r-r') \tag{29.6}$$

If we designate the second term on the left of Eq. (29.3) by Ξ we may write that equation as

$$\delta_{\alpha\beta}\delta(1'-2)\mathsf{G}^{0^{-1}}(12)\mathsf{G}_{\alpha\beta}(12)-\Xi(11')=\delta(1'-2) \tag{29.7}$$

Multiplying through on the right by $G_{\alpha\beta}^{-1}$ yields

$$\mathsf{G}^{0^{-1}}(12)\delta_{\alpha\beta}\delta(1'-2)=\Xi(11')\mathsf{G}_{\alpha\beta}^{-1}(12)+\mathsf{G}_{\alpha\beta}^{-1}(12)\delta(1'-2)$$

which, when integrated over 2, yields

$$\mathsf{G}^{0^{-1}}(11')=\mathsf{G}^{-1}(11')+\Sigma(11') \tag{29.8a}$$

$$\Sigma(11')=\int d\bar{2}\,\Xi(\bar{2}1')\mathsf{G}^{-1}(\bar{2}1') \tag{29.8b}$$

where we have let $1\equiv\bar{2}$ under the integral. Making the necessary substitutions tells us that

$$\Sigma(11)=\mathrm{i}\int d\bar{1}\,d\bar{2}\,V(1\bar{1})\left[L(\bar{2}\bar{1};1'\bar{1}^+)+\mathsf{G}(1\bar{2})\mathsf{G}(\bar{1}\bar{1}^+)\right]\mathsf{G}^{-1}(\bar{2}1')$$

At this point Ross makes use of the BK conserving relations and their results. These relations assure the conservation of (1) the number of particles,

(2) the total momentum, (3) the total angular momentum, and (4) the total energy. When they are satisfied BK show that

$$\int d1\,d2\,\mathsf{V}(1\bar{1})L(\bar{2}\bar{1};1'\bar{1}^{+})\mathsf{G}^{-1}(\bar{2}1')=\int d1\,d2\,\mathsf{G}^{-1}(\bar{2}1')L(1\bar{1};\bar{2}\bar{1}^{+})\mathsf{V}(1\bar{1}).$$

In this fashion then it may be shown that

$$\Sigma(11')=\mathrm{i}\int d\bar{1}\,d2\,\mathsf{V}(1\bar{1})\mathsf{G}^{-1}(\bar{2}1')\Big[\,L(1\bar{1};\bar{2}\bar{1}^{+})+\mathsf{G}(1\bar{2})\mathsf{G}(\bar{1}\bar{1}^{+})\Big].\tag{29.8b$'$}$$

Perhaps we should remark that any apparent dissimilarity between Eq. (29.8b′) and the BK results arises from the use of L. The square-bracketed expression is simply G^2, but we are maintaining the L-form for direct comparison with Ross.

Finally, Ross makes use of the integral equation for L which BK obtain as their Eq. (43),

$$L(12;1'2')=\mathsf{G}(12')\mathsf{G}(21)+\int d3\,d4\,d5\,d6\,\mathsf{G}(1\bar{3})\mathsf{G}(\bar{4}1')\frac{\delta\Sigma(\bar{3}\bar{4})}{\delta\mathsf{G}(\bar{6}\bar{5})}L(\bar{6}2;\bar{5}2')\tag{29.8c}$$

In obtaining Eq. (29.8c) BK use a method reminiscent of that which yield Eq. (28.11), an external perturbing field U being introduced and utilized, the result eventuating when U is allowed to go to zero.

30 The *T*-Matrix Approximation and the General Like Shape Equations

As Ross remarks, Eqs. (29.8) contain all the many-body effects, and the next step is the determination of the mass operator, $\Sigma(11')$ appropriate to the spectral line broadening situation. The spectral line broadening situation is presumed to correspond to the interruption approximation.

We will recall that the interruption approximation is predicated on a time of collision short in comparison to the time between collisions. Such a situation will correspond to low densities, what is statistically known as a dilute system. It will of course also correspond to short-range forces, these forces defining the radius of the interaction sphere, residence time within this sphere corresponding to collision time. Such remarks on forces, however, should be subject to careful scrutiny. Van der Waals forces ($\sim r^{-6}$) are certainly short range. On the other hand, Stark-broadening forces ($\sim r^{-2}$) can hardly be so designated, although it is quite often appropriate to treat Stark broadening by electrons with an interruption-type formalism. Obviously such treatment is justified by the high velocity of these broadeners, high when compared to atomic broadeners, which results in short residence time in what are certainly large interaction spheres. The purpose of this aside is simply to emphasize the necessity for interpreting the remark relating to dilute systems and short-range forces according to the specific physical phenomenon under consideration. With this caveat, our system is dilute and our force short range.

Now suppose we designate the range of the force by a. What does this mean? In the Wigner hard-core force, $\mathsf{V}=\infty$ for $r<a$; $\mathsf{V}=0$ for $r>a$, and the definition of the range is straightforward. In the case of an exponential force, we may define it by $\exp[-r/a]$, certainly not quite as straightforward as the Wigner case. For forces of the form r^{-n} other aspects of the problem are often used to specify the range. For example, the Debye shielding radius tells us the separation at which, say, the positive ions shield out the electric field of a perturbing electron. This Debye radius is often used to define the range of this Coulomb force. In short, nature does not in general really define a range, but we presume that the physicist will be clever enough to justify one, and we take it as a. However, nature will define an average distance between particles, r_0. Therefore, our dilute and short-range criteria mean that $a/r_0 \ll 1$: the particles are considerably farther apart than the range of the forces acting between them. For such a criterion, let us now show that only those graphs having a minimum (one) of hole lines contribute substantially to the self-energy Σ.

In order to show this, let us concern ourselves with the $T=0$ case. The density of particles present—multiplying by two for spin—when we consider only their translational momentum will be

$$N=\frac{2}{V}\sum_{R}\langle\Psi_0|\psi_k^+\psi_k|\Psi_0\rangle \tag{30.1}$$

where $\psi_k^+\psi_k$ is the number operator, Ψ_0 the second-quantized eigenvector. If we are dealing with Fermions, there will be one particle for each k. Further, the system ground state will have all levels below the Fermi energy, which corresponds to k_F, filled, and all above empty. Therefore

$$N=\frac{2}{V}\sum_{R}1_k=\frac{2}{(2\pi)^3}\int_0^{k_F}dk=\frac{k_F^3}{3\pi^2} \tag{30.2}$$

by standard box normalization [cf. Sec. 1 of Källén (27)]. Now since $N=r_0^{-3}$, $r_0^{-1}\sim k_F$ so that our dilute, short-range criterion becomes $k_F a \ll 1$. This will have a marked effect on the importance of various contributions to a given diagram. Take the following self-energy diagram:

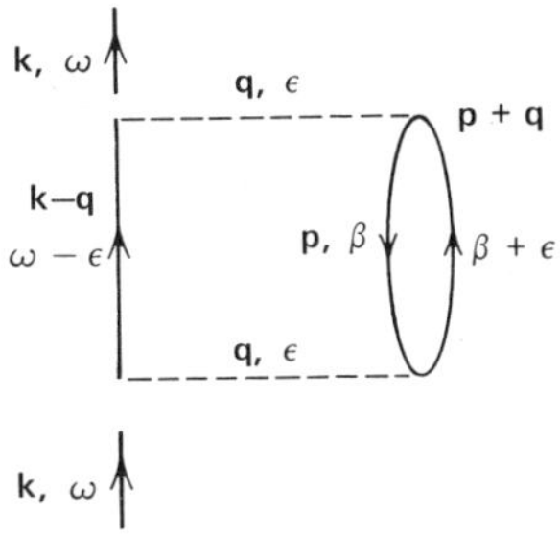

$$=(-)\sum_{pq}\int_{k_F}^{\infty}\frac{d\epsilon}{2\pi}\int_0^{k_F}\frac{d\beta}{2}\left[\mathrm{i}\mathrm{G}^0(\mathbf{k}-\mathbf{q},\omega-\epsilon)\right]\cdot\left[\mathrm{i}\mathrm{G}^0(\mathbf{p},\beta)\right]\left[\mathrm{i}\mathrm{G}^0(\mathbf{p}+\mathbf{q},\beta+\epsilon)\right]\left[-\mathrm{i}\mathrm{V}_q\right]^2 \tag{30.3}$$

First, we remark that the free-particle lines $\mathbf{k}, \omega$ do not form a part of the self-energy graph: they have simply been included so that our ground rules for momentum and frequency may be established. Otherwise Eq. (30.3) should be self explanatory, as should the effect of small k_F, for it is obvious that the integral over β, and hence the presence of a hole line, is of considerably less consequence than is the presence of a particle line. Therefore, diagrams of the form

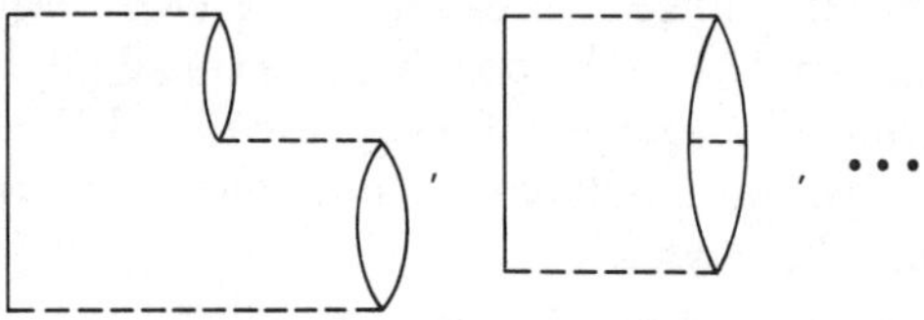

are dropped, and we are left with the ladder approximation,

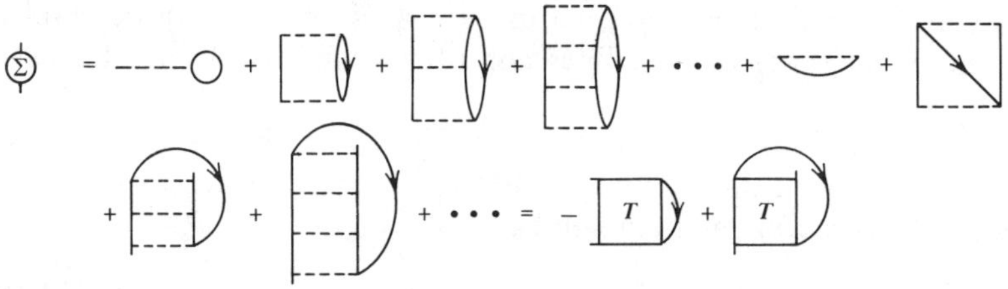

(30.4a)

where

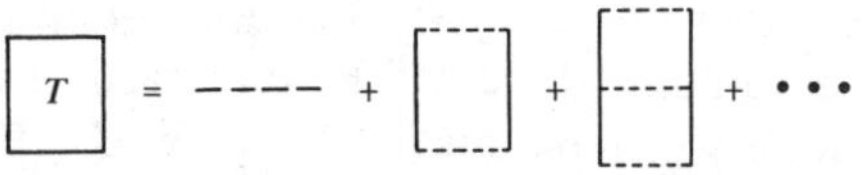

(30.4b)

With Eqs. (30.4), we have introduced the ladder approximation and the T matrix, sometimes called the K matrix.

Before we proceed with the application of the T matrix approximation to our broadening problem, we remark on the equivalent expression of this same approximation and certain ramifications of it with which it behooves us to be familiar.

With Ross we shall appeal to the T matrix because it provides a simplifying approximation which is admissible under interruption conditions and which, under such conditions, simplifies the self energy by eliminating graphs having more than one hole line. Such a statement applies to the self energy of a single particle interacting with the Fermi sea. When, however, we consider two real particles and ignore the Fermi sea, as did Bethe and Goldstone (78) in their introduction of the ladder approximation, we do essentially the same thing. Also it was to the Bethe-Goldstone equation that BK appealed in developing the T matrix form of the self energy which Ross used. There is, however, another avenue of approach to the T matrix as we have seen in Chapter 3.

If V is small—which of course it can never be for central collisions—the Greens function can be expanded in terms of it into a series which looks very much like that of Brillouin-Wigner,

$$\mathsf{G}=[E-H_0-\mathsf{V}]^{-1}=[E-H_0]^{-1}+[E-H_0]^{-1}\mathsf{V}[E-H_0]^{-1}$$
$$+[E-H_0]^{-1}\mathsf{V}[E-H_0]^{-1}\mathsf{V}[E-H_0]^{-1}+\cdots$$
$$=\mathsf{G}^0+\mathsf{G}^0\mathsf{V}\mathsf{G}^0+\mathsf{G}^0\mathsf{V}\mathsf{G}^0\mathsf{V}\mathsf{G}^0+\cdots$$

or

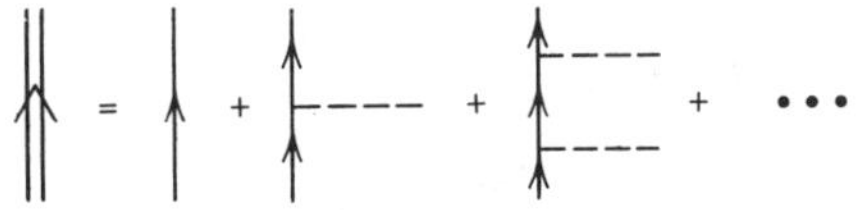

and which is substantially the ladder approximation for G, acceptable for distant, short-range collisions.

The T matrix or ladder approximation for the self energy is the mathematical expression of the interruption approximation to the broadening phenomenon.

BK use $\mathsf{G}^2(U)$, where U is still defined as before and G^2 is a sum of ladder diagrams, which is a solution to the Bethe-Goldstone equation as a starting point, and obtain an expression for the self energy and an expression for L as a function of the T matrix elements. We may obtain about the same self energy beginning with Eq. (29.3b′) in the form $\Sigma=\mathrm{i}\mathsf{V}\mathsf{G}^{-1}\mathsf{G}^2$, where we have simply substituted G^2 for the square-bracketed expression in the equation. From an equation such as Eq. (27.16a) we may take $\mathsf{G}^2=\mathsf{G}\mathsf{G}$, so that, when V is replaced by T, we obtain $\Sigma=\mathrm{i}T\mathsf{G}$. An analogous use of Eq. (29.8c) yields the BK L result.

Consider Eq. (29.8c). Since $\Sigma=\mathrm{i}T\mathsf{G}$, $\delta\Sigma/\delta\mathsf{G}\sim\mathrm{i}T$, and, from Eq. (29.8c), $L=\mathsf{GG}+\mathrm{i}\mathsf{GG}TL$. Substituting the entire right side for L on the right of this equation yields $L=\mathsf{GG}+\mathrm{i}\mathsf{GG}T\mathsf{GG}-\mathsf{GG}T\mathsf{GG}TL$. In the second term we suppose $L=\mathsf{GG}$ so that $L=\mathsf{GG}\,[1+\mathrm{i}TL-T\mathsf{GG}TL]$. BK take longer to obtain the same result in a more rigorous fashion using the external potential, U, of Section 25. Ross transforms the BK results to momentum space in order to obtain

$$\Sigma_{\alpha\beta}(p)=\mathrm{i}\sum_{\delta\gamma}\int\frac{dq}{(2\pi)^4}\langle pq|T_{\alpha\gamma\beta\delta}|pq\rangle\mathsf{G}_{\delta\gamma}(p) \tag{30.5a}$$

$$L_{\alpha\beta\alpha'\beta'}(p,p';p-k,p'+k)=\sum_{\lambda\mu}\mathsf{G}_{\alpha\lambda}(p)\mathsf{G}_{\mu\alpha'}(p-k)\big[\delta(p-p'-k)\delta_{\lambda\beta'}\delta_{\mu\beta}$$
$$+\mathrm{i}\sum_{\nu\eta}\int\frac{dq'}{(2\pi)^4}\langle pq'-k|T_{\lambda\eta\mu\nu}|p-kq'\rangle L_{\nu\beta\eta\beta'}(q',p;q'-k,p+k)$$
$$-\sum{}'\int\frac{dq'}{(2\pi)^4}\int\frac{dq}{(2\pi)^4}\langle pq|T_{\lambda\eta\lambda'\eta'}|q'q+p-q'\rangle\mathsf{G}_{\eta'\nu}(q+p-q')$$
$$\cdot\mathsf{G}_{\nu'\eta}(q)\langle q'-kq+q-q'|T_{\mu'\nu\mu\nu'}|p-kq\rangle L_{\lambda'\beta\mu'\beta'}(q',p';q'-k,p'+k)\big] \tag{30.5b}$$

$$\langle pp'|T_{\alpha\beta\alpha'\beta'}|p-kp'+k\rangle = \mathsf{V}_{\alpha\beta\alpha'\beta'}(\mathbf{k}) + \mathsf{V}_{\beta\alpha\alpha'\beta'}(\mathbf{p}-\mathbf{p}'-\mathbf{k})$$
$$+\sum_{\lambda\nu\mu\eta}\int\frac{dq}{(2\pi)^4}\langle pp'|T_{\alpha\beta\lambda\mu}|p+qp'-q\rangle \mathsf{G}_{\lambda\nu}(p+q)\mathsf{G}_{\mu\nu}(p'-q)\mathsf{V}_{\nu\eta\alpha'\beta'}(q+k). \tag{30.5c}$$

With Ross we can certainly agree that Eq. (29.8a) becomes

$$\mathsf{G}_{\alpha\beta}^{-1}(p)+\Sigma_{\alpha\beta}(p)=(\omega_k - p^2/2\mathsf{m}_\alpha - E_\alpha + \mu_\alpha)\delta_{\alpha\beta}, \tag{30.6}$$

the term on the right of Eq. (30.6) evolving as a direct result of a transform analogous to Eq. (37.14a).

Again letting $\mathsf{G}^0 = \longleftarrow$, $\mathsf{G} = \Longleftarrow$, $\mathsf{V} = ----$, Eq. (30.5a) becomes

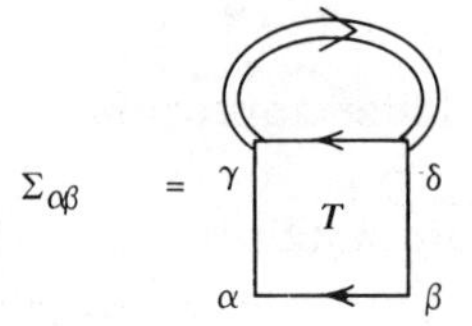

(30.5a′)

Multiply Eq. (30.6) through on the right by G^{-1} and on the left by $(\omega_k \cdots \mu_k)^{-1}$ in order to obtain

$$G_{\alpha\beta}(p)=\frac{\delta_{\alpha\beta}}{\omega - p^2/2\mathsf{m}_\alpha - E_\alpha + \mu_\alpha} + i\sum_{\lambda\delta}\int\frac{dq}{(2\pi)^4}\frac{\langle pq|T_{\alpha'\gamma\beta'\delta}|pq\rangle \mathsf{G}_{\delta\gamma}(q)\mathsf{G}_{\beta\beta'}(p)}{\omega_k - p^2/2m_\alpha - E_\alpha + \mu_\alpha}$$

or

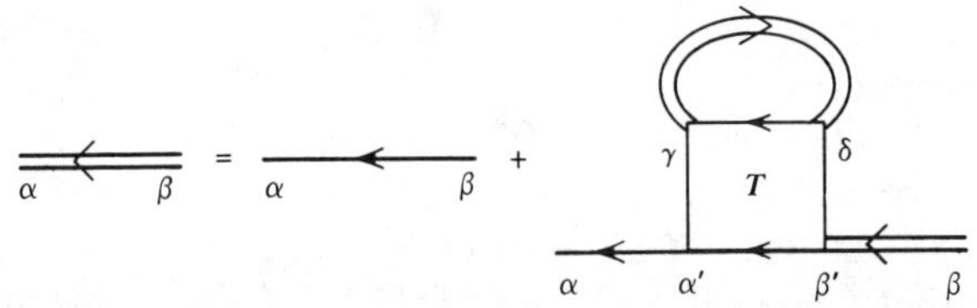

(30.6′)

The remaining Eqs. (30.5) may be written down as

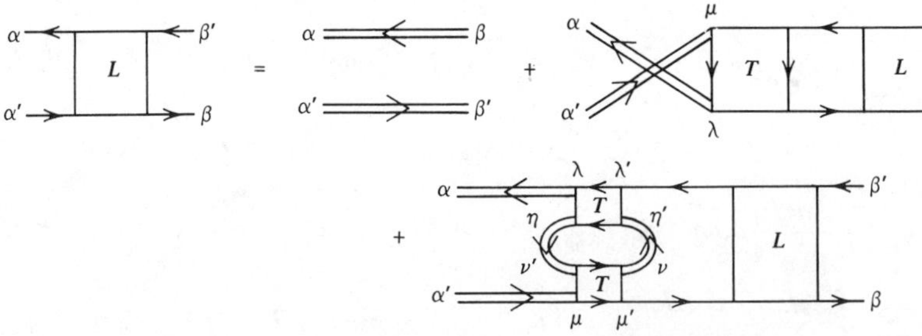

(30.5b′)

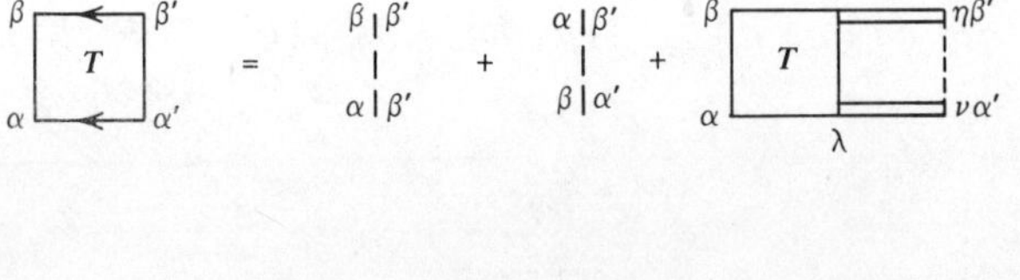

(30.5c′)

Let us ruminate for a moment on the idea of a vertex part as illustrated by Eq. (27.12) and Fig. 27.4. Fresh from such an exercise, let us rewrite the left side of Eq. (28.15) slightly

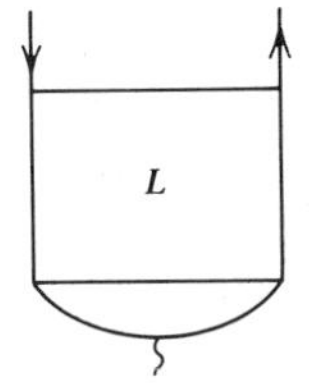

and L could certainly be construed as a vertex part. Now let us solve Eq. (30.5b) by iteration, that is, take the first term on the right hand side ($\mathsf{G}_{\alpha\lambda}\mathsf{G}_{\mu\alpha'}$) as a first approximation to L, substitute this for L in the second and third terms on the right as a second approximation, and so on. (The "so on" will never actually arise.) In doing so let us drop various integrals, arguments, and constants.

$$L_{\alpha\beta\alpha'\beta'} = \mathsf{G}_{\alpha\lambda}\mathsf{G}_{\mu\alpha'}\big[\,\delta_{\lambda\beta'}\delta_{\mu\beta} + T_{\lambda\eta\mu\nu}\mathsf{G}_{\nu\lambda}\mathsf{G}_{\mu\eta}\delta_{\lambda\beta}\delta_{\mu\beta}$$
$$+ T_{\lambda\eta\lambda'\eta'}\mathsf{G}_{\eta'\nu}\mathsf{G}_{\nu'\eta}T_{\mu'\nu\mu\nu'}\mathsf{G}_{\lambda'\lambda}\mathsf{G}_{\mu'\mu}\delta_{\lambda\beta'}\delta_{\mu\beta} + \cdots\,\big]$$

or

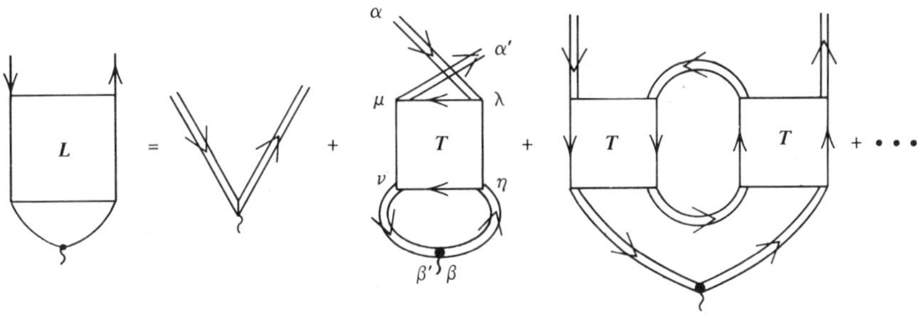

(30.7)

We might pause long enough to detail the construction of the second graph on the right of Eq. (30.7). In doing so we refer back to Eq. (30.5b) for the requisite momentum arguments. As we can see from the first diagram, $\mathsf{G}_{\alpha\lambda}$ describes momentum flow of magnitude p "inbound" from α to λ, $\mathsf{G}_{\mu\alpha'}$ momentum flow outbound, the difference being that associated with the photon. In like manner, $\mathsf{G}_{\nu\lambda}\delta_{\lambda\beta'}$ and $\mathsf{G}_{\mu\eta}\delta_{\mu\beta}$ describe momentum flow q' from ν to β' and $q'-k$ from β to η. The placement of the subscripts on T in conjunction with those on $\mathsf{G}_{\alpha\lambda}\mathsf{G}_{\mu\alpha'}$ then fix the upper Greens function lines.

We next consider the first term on the right of Eq. (30.7). This is the product of two one-particle Greens functions, the equation for each of which is Eq. (30.6′). Eq. (30.6′) may be solved by iteration. Take the first term on the right as the first approximation, the second term with $\leftarrow$ substituted for $\Leftarrow$ as the second term, and so on. We will surely agree that the result is

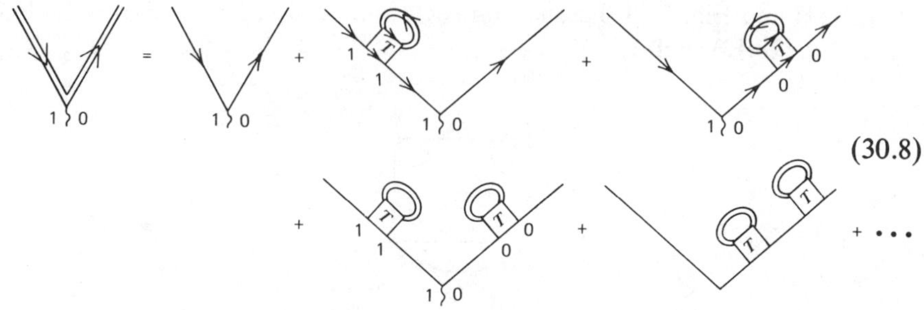

(30.8)

where we have, with Ross, taken the emitter to be in state 1 before photon emission, 0 after.

The results, which are displayed as Eqs. (30.7) and (30.8), are really rather impressively informative. First of all, Eq. (30.8) represents those situations where collisions occur either not at all, before emission, after emission, and all possible combinations of before and after. The second term in Eq. (30.7) corresponds to that situation wherein a pair of atoms exchange excitation before emission, a phenomenon which we recall as particularly important to self broadening. The presence of the diagram indicates that the phenomenon is present in foreign-gas broadening but inconsequentially so since there will be no possibility of resonance between the characteristic frequencies of the two atoms (or only the slight possibility associated with accidental resonance.) Finally, the third term on the right of Eq. (30.7) corresponds to that situation wherein a photon is emitted during a collision.

31 The Reduction of the General Diagrammatic Result to the Baranger Result

At this point then, Ross has set up a general, albeit complex expression for the spectral line shape, Eq. (28.15a). Note that this includes all broadening effects save those giving rise to the natural line shape, and these, as we see, could be included by including the particle-field interaction within the Hamiltonian. Broadening by interaction with other particles is obviously included, while Doppler broadening can be seen to be accounted for by the inclusion of the translational momentum of the emitter within the framework of the development. This general result has been specificized to the interruption result by Eqs. (30.4) wherein we replace the general expression for the mass operator by the ladder approximation. Even under such a simplification, however, we have see, through Eqs. (30.7) and (30.8), how many physical phenomena are described for us by this theory. Finally, it is of considerable importance to demonstrate that the theory reduces, under appropriate simplifications, to some previous theory which has, to some extent at least, proved itself. Ross chose the Baranger theory with which to make such a comparison, and we follow through his demonstration.

Before beginning, however, let us remark that there may be those of us who are not overly experienced in the machinations necessary for the extraction of something useful from equations such as (30.5). Therefore, we might ask what the generalized *modus operandi* may be. In attempting to answer such a query, we first note that what is really desired is some sort of relatively simple algebraic expression for Eq. (30.5b) or an equivalent expression in terms at least of a line-center frequency and a particular emitted or absorbed photon frequency. The techniques involved in obtaining this expression may be reduced to two:

1 Iterative procedures such as, for example, substituting G^0 for G in Eq. (30.5a) in order to find Σ, then utilizing this in order to determine G from Eq. (30.6). The same sort of procedure may also be applied to Eq. (30.5c).

2 The search for and use of poles in order to carry out (approximately) integrations over functions that are sought. Take the third term on the right of Eq. (30.5b). Here we have $\mathsf{GG}L$ within the integrand, L being the function sought. To a zeroth approximation, L is equal to GG, the first term on the right of this equation. Therefore, we can approximate the poles of L by the poles of GG and hence carry out certain of the integrations in the third term without knowledge of a specific functional form for L.

In a general way then, these two techniques are used: (1) to determine $\Sigma_{\alpha\beta}$ from Eq. (30.5a); (2) to utilize this result in order to determine $\mathsf{G}_{\alpha\beta}$ from Eq. (30.6); and (3) to utilize this result in order to determine, from Eq. (30.5b), a vertex equation so-called for a slightly disguised L. (4) When this equation is solved, a line shape expression results.

Let us begin with Eq. (28.15a) with the resonant assumption $\omega_{\alpha\beta}=\omega_{\gamma\delta}=\omega_0$ so that a spectral line is indeed what we are considering. Further, we suppose our emitter to be a two-state atom, as in the last section, having upper state 1, lower state 0. Thus only L_{1001} will appear. Equation (28.15a) may then be written

$$\chi(\mathbf{k},\omega)=i\mathbf{d}_{10}\mathbf{d}_{01}\frac{\omega_0^2}{\omega^2}\int\frac{dp}{(2\pi)^4}\int\frac{dp'}{(2\pi)^4}L_{1001}(p,p';p-k,p'+k) \quad (31.1)$$

where L is to be found from Eq. (30.5b). Because, as we have remarked, the second term on the right of Eq. (30.5b) describes an exchange of excitation between radiator and broadener before emission and is hence really only of interest to resonance braodening, we drop it. We specifically introduce the fact that we are describing the interruption broadening of a spectral line by perturbing electrons,

$$
\begin{aligned}
L_{1001}(p,p';p-k,p'+k) &= \mathrm{G}_{11}(p)\mathrm{G}_{00}(p-k)[\delta(p-p'-k) \\
&\quad -\sum_{\mu\nu\lambda}\int\frac{dq'}{(2\pi)^4}\int\frac{dq}{(2\pi)^4}\langle\mathbf{pq}|T_{1e\lambda\nu}(\omega_{p+q})|q'q+p-q'\rangle \mathrm{G}_{\nu\nu}(q+p-q') \\
&\quad \cdot \mathrm{G}_{ee}(p)\langle\mathbf{p}'-\mathbf{kq}+\mathbf{p}-\mathbf{q}'|T_{\mu\nu 0e}(\omega_{q+p-k})|\mathbf{p}-\mathbf{kq}\rangle \\
&\quad \cdot L_{\lambda 0\mu 1}(q',p';q'-k,p'+k)
\end{aligned}
\tag{31.2}
$$

The sums in Eq. (31.2) may be written out graphically as

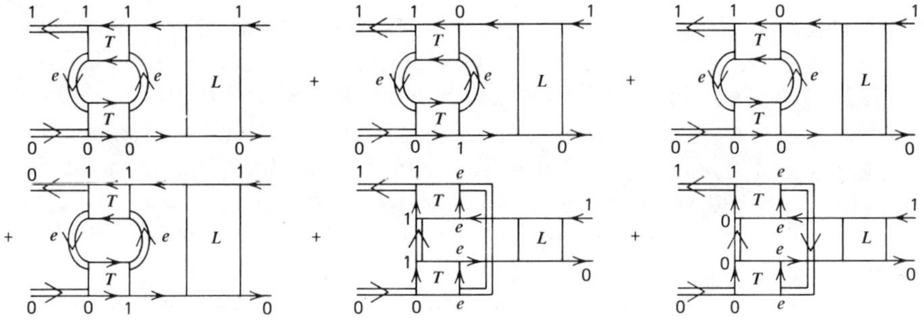

(31.2′)

The last two terms in Eq. (31.2′) correspond to the nonresonant energy transfer between atoms and electrons and, like that between different types of atoms, contribute negligibly and may be dropped. For density reasons the second, third, and fourth terms in Eq. (31.2′) may also be dropped, and it now becomes necessary for us to discuss why this is so before continuing the solution.

We will recall our intention of showing how the Ross theory reduces to the Baranger theory under the appropriate approximations. As the Ross theory will yield all powers of the perturber density in the expressions for the spectral line width and shift, while the Baranger theory provides these quantities linearly dependent on the density, an important simplification of the Ross theory will be the elimination of terms having a density dependence higher than the first. In order to carry out such an elimination we consider just how the density enters.

Consider Eq. (29.6) wherein $(\mathrm{G}^0)^{-1}$ is an operator. Now let us take the matrix element of the inverse of this operator, an operation which will of course involve the eigenvector of the atom. This eigenvector will generally be a product function, one portion of which involves the electronic states of the atom of energy E. However, there will also be a translational portion involving the atomic momentum, l, and the coordinates on which ∇ operates. Along with which there will be a factor involving ω_l thus: $\exp[\mathrm{i}\mathbf{k}\cdot\mathbf{r}-\mathrm{i}\omega t]$. We will agree then that our G^0 matrix is of the form

$$\mathsf{G}^0_{\alpha\beta}(1)=\delta_{\alpha\beta}\left[\omega_l - l^2/2\mathsf{m} - E_\alpha + \mu_\alpha\right]^{-1} \tag{31.3}$$

Eq. (30.5a), for example, implies a summation over ω_k which is replaced by a contour integral over ω. This integral will include a statistical factor which may be taken as the Bose factor, $[\exp(\beta\omega)-1]^{-1}$. Thus we will encounter

$$(\mathrm{i}\beta)^{-1}\sum_{\omega_k}\cdots=\int_C\frac{d\omega}{2\pi}(e^{\beta\omega}-1)^{-1}\cdots \tag{31.4}$$

By the "C" we indicate the contour about the imaginary axis and hence including the poles of the Bose factor. As it seems rather impractical to evaluate such a contour directly, we appeal to an alternate method.

In, say, Eq. (30.5c) let us replace G by G^0, and, from Eq. (31.3) for example, we can anticipate the integrand to have two simple poles other than those corresponding to the Bose factors in the complex ω-plane. Therefore, the contour integral of Eq. (31.4) may be redrawn as shown in Fig. 31.1. We suppose the summation in Eq. (31.4) converges. Therefore, an integration around C_1 which cuts in as C_2 to enclose the integrand poles, then as C to enclose the Bose poles will yield zero, that is $\int C_1+\int C+\int C_2=0$. If we now suppose the outer contour to be swept off to infinity, its integrand going to zero, the integration around the Bose poles, which we cannot evaluate, is equal to the integral around the integrand poles, which we can. To repeat ourselves then, we evaluate the integral of Eq. (31.4) by evaluating the

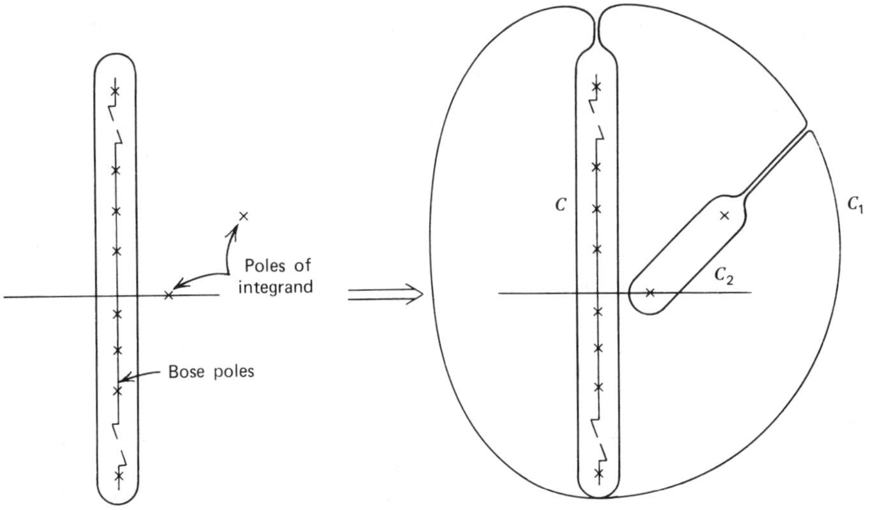

Figure 31.1

integrand at the poles of G^0. Next, we remark that Eq. (31.3) tells us that the contribution from the residue at a G^0 pole ($\omega_l = l^2/2m_\alpha + E_\alpha - \mu_\alpha$) will contain a statistical factor,

$$\left[\exp\beta(l^2/2m_\alpha + E_\alpha - \mu_\alpha) - 1\right]^{-1} \doteq \exp\left[-\beta(l^2/2m_\alpha + E_\alpha - \mu_\alpha)\right]$$
$$= N_\alpha(2\pi\beta/m_\alpha)^{3/2}\exp\left[-l^2/2m_\alpha\right] \tag{31.5}$$

where N_α is the number of atoms in state α when the atoms are unperturbed by interaction. Equation (31.5) is of particular interest as our first indication of density dependence.

We consider the lowest-order T matrix from Eq. (30.5c), that is, we consider the case where $G = G^0$. As a consequence, we obtain

$$\langle pp'|T_{\alpha\beta\alpha'\beta'}|p-kp'+k\rangle = V_{\alpha\beta\alpha'\beta'}(\mathbf{k}) + V_{\beta\alpha\alpha'\beta'}(\mathbf{p}-\mathbf{p}'-\mathbf{k}) + \int\frac{dq}{(2\pi)^4}$$
$$\cdot\frac{\langle pp'|T_{\alpha\beta\lambda\mu}|p+qp'-q\rangle V_{\lambda\mu\alpha'\beta'}(\mathbf{q}+\mathbf{k})}{\left[\omega_p+\omega_q-(p+q)^2/2m_\lambda - E_\lambda + \mu_\lambda\right]\left[\omega_{p'}-\omega_q-(p'-q)^2/2m_\mu - E_\mu + \mu_\mu\right]} \tag{31.6}$$

We have the previously discussed ω_q summation to be carried out. The potential, V, does not depend on the frequency, ω_l. Therefore, from Eq. (31.6), the T-matrix elements under the integral sign in this equation will have no ω_q-dependent poles so that the G^0 there are solely responsible for any poles in the ω_q-integration. These poles of course occur at ω_q values of $-\omega_p + (p+q)^2/2m_\lambda + E_\lambda - \mu_\lambda$ and $\omega_{p'} - (p'-q)^2/2m_\mu - E_\mu + \mu_\mu$ which, when used to evaluate the residues with the Bose factor included, yield for the integral,

$$-i\int\frac{d\mathbf{q}}{(2\pi)^3}\frac{\langle\mathbf{pp}'|T_{\alpha\beta\lambda\mu}(\omega_{p+p'})|\mathbf{p}+\mathbf{qp}'-\mathbf{q}\rangle V_{\lambda\mu\alpha'\beta'}(\mathbf{q}+\mathbf{k})}{\omega_{p+p'}-(p+q)^2/2m_\lambda-(p'-q)^2/2m_\mu - E_\lambda - E_\mu + \mu_\lambda + \mu_\mu}$$
$$\cdot\left\{\left[\exp\beta\left(\frac{-(p'-q)^2}{2m_\mu} - E_\mu + \mu_\mu\right) - 1\right]^{-1}\right.$$
$$\left. - \left[\exp\beta\left(\frac{(p+q)^2}{2m_\lambda} + E_\lambda - \mu_\lambda\right) - 1\right]^{-1}\right\}, \tag{31.7}$$

where the second Bose factor appears with minus sign because of the opposite sign of ω_q in the two Greens functions. There are two terms because there are two poles. As we have seen from Eq. (31.5), to lowest order in the density, the second square bracket in Eq. (31.7) is of order N. We could write the first square-bracketed term in Eq. (31.7) as $[N\exp(\cdots)-1]^{-1}$ which is 1 to lowest order in the density. Therefore, we replace the curly bracket by unity, and Eq. (31.6) becomes

$$\langle \mathbf{pp}'|T_{\alpha\beta\alpha'\beta'}(\omega_{p+p'})|\mathbf{p}-\mathbf{kp}'+\mathbf{k}\rangle = \mathrm{V}_{\alpha\beta\alpha'\beta'}(\mathbf{k})+\mathrm{V}_{\beta\alpha\alpha'\beta'}(\mathbf{p}-\mathbf{p}'-\mathbf{k})$$
$$+\mathrm{i}\int \frac{d\mathbf{q}}{(2\pi)^3}\frac{\langle \mathbf{pp}'|T_{\alpha\beta\lambda\mu}(\omega_{p+p'})|\mathbf{p}+\mathbf{qp}'-\mathbf{q}\rangle \mathrm{V}_{\lambda\mu\alpha'\beta'}(\mathbf{q}+\mathbf{k})}{\omega_{p+p'}-(p+q)^2/2\mathrm{m}_\lambda-(p'-q)^2/2\mathrm{m}_\mu-E_\lambda-E_\mu+\mu_\lambda+\mu_\mu} \tag{31.8}$$

As in Eq. (30.4d), when $\omega_{p+p'}$ is on the energy shell relative to the chemical potentials, that is,

$$\mu_\alpha+\mu_\beta=\mu_{\alpha'}+\mu_{\beta'}=\mu_\lambda+\mu_\mu, \tag{31.9}$$

we may write the energy-shell T matrix in terms of two-body scattering amplitudes f

$$\langle \mathbf{pp}'|T_{\alpha\beta\alpha'\beta'}\left(\frac{\mathrm{p}^2}{2\mathrm{m}_\alpha}+\frac{\mathrm{p}'^2}{2\mathrm{m}_\beta}+E_\alpha+E_\mu-\mu_\alpha-\mu_\beta+\mathrm{i}\epsilon\right)|\mathbf{p}+\mathbf{kp}'-\mathbf{k}\rangle$$
$$=2\pi\frac{\mathrm{m}_\alpha+\mathrm{m}_\beta}{\mathrm{m}_\alpha\mathrm{m}_\beta}\left[f_{\alpha\beta\alpha'\beta'}(\mathbf{pp}';\mathbf{p}+\mathbf{kp}'-\mathbf{k})+f_{\alpha\beta\beta'\alpha'}(\mathbf{pp}';\mathbf{p}'-\mathbf{kp}+\mathbf{k})\right]$$
$$\equiv\langle \mathbf{pp}'|T_{\alpha\beta\alpha'\beta'}|\mathbf{p}+\mathbf{kp}'-\mathbf{k}\rangle. \tag{31.10}$$

In Eq. (31.10) $f_{\alpha\beta\alpha'\beta'}$ represents the case where one atom enters in state α, leaves in state α', while the other atom involved in the collision enters in state β, leaves in state β'.

This development should give us a certain amount of insight into our attempt to follow the reduction of the general result to the Baranger interruption result. We have assumed a two-state atom and an electron so that three potentials will be of interest (1) $\mathrm{V}_{1e1e}(\mathbf{k})$, (2) $\mathrm{V}_{0e0e}(k)$, and (3) $\mathrm{V}_{1e0e}(k)$. That is, the atom is inbound to and outbound from the collision with the electron (1) in its excited state or (2) in its ground state or (3) the atom is inbound (outbound) to the collision in its excited state, outbound (inbound) from the collision in its ground state. Therefore, in this case, Eq. (31.8) reduces to three coupled equations for the scattering matrices T_{11}, T_{00}, and T_{10} where

$$\langle \mathbf{pp}'|T_{11}(\omega_{p+p'})|\mathbf{p}+\mathbf{kp}'-\mathbf{k}\rangle=\langle \mathbf{pp}'|T_{1e1e}(\omega_{p+p'})|\mathbf{p}+\mathbf{kp}'-\mathbf{k}\rangle$$
$$=\langle \mathbf{pp}'|T_{e1e1}(\omega_{p+p'})|\mathbf{p}'-\mathbf{kp}+\mathbf{k}\rangle \tag{31.11a}$$

$$\langle \mathbf{pp}'|T_{00}(\omega_{p+p'})|\mathbf{p}+\mathbf{kp}'-\mathbf{k}\rangle=\langle \mathbf{pp}'|T_{0e0e}(\omega_{p+p'})|\mathbf{p}+\mathbf{kp}'-\mathbf{k}\rangle$$
$$=\langle \mathbf{p}'\mathbf{p}|T_{e0e0}(\omega_{p+p'})|\mathbf{p}'-\mathbf{kp}+\mathbf{k}\rangle \tag{31.11b}$$

$$\langle \mathbf{pp}'|T_{10}(\omega_{p+p'})|\mathbf{p}+\mathbf{kp}'-\mathbf{k}\rangle=\langle \mathbf{pp}'|T_{1e0e}(\omega_{p+p'})|\mathbf{p}'-\mathbf{kp}+\mathbf{k}\rangle$$
$$=\langle \mathbf{p}'\mathbf{p}|T_{e1e0}(\omega_{p+p'})|\mathbf{p}'-\mathbf{kp}+\mathbf{k}\rangle$$
$$=\langle \mathbf{p}+\mathbf{kp}'-\mathbf{k}|T_{0e1e}(\omega_{p+p'})|\mathbf{pp}'\rangle^* \tag{31.11c}$$

Eqs. (31.11a) and (31.11b) describe scattering which is elastic by definition; Eq. (31.11c) describes inelastic scattering. Perhaps this is an appropriate point at which to recall just what we are about.

We wish to evaluate Eq. (31.2) for the interruption broadening by electrons of a spectral line emitted by a two-state atom. We have found the T-matrix elements which are to be retained, and it now devolves on us to determine the G's, which are likewise required by Eq. (31.2). This we shall do by (1) determining the mass operator from Eq. (30.5a) and (2) using this result for G determination.

We substitute G^0 in G in Eq. (30.5a) and insert a Bose factor. Then we follow precisely the procedure which led from Eq. (31.6) to Eq. (31.7) in order to obtain

$$\begin{aligned}\sum_{\alpha\beta}(\mathrm{p}) &= \sum_{\gamma}\int\frac{d\mathbf{q}}{(2\pi)^3}\langle\mathbf{pq}|T_{\alpha\gamma\beta\gamma}(\omega_p+q^2/2\mathrm{m}_\gamma+E_\gamma-\mu_\gamma)|\mathbf{pq}\rangle\\&\quad\cdot\left\{\left[\exp\beta(q^2/2\mathrm{m}_\gamma+E_\gamma-\mu_\gamma)-1\right]^{-1}\right\}\\&\doteq\sum_{\gamma}\int\frac{d\mathbf{q}}{(2\pi)^3}\exp\left[-\beta(q^2/2\mathbf{m}_\gamma+E_\gamma-\mu_\gamma)\right]\\&\quad\cdot\langle\mathbf{pq}|T_{\alpha\gamma\beta\gamma}(\omega_p+q^2/2\mathrm{m}_\gamma+E_\gamma-\mu_\gamma)|\mathbf{pq}\rangle\\&=\sum_{\gamma}N_\gamma\langle\langle\mathbf{pq}|T_{\alpha\gamma\beta\gamma}(\omega_p+q^2/2\mathrm{m}_\gamma+E_\gamma-\mu_\gamma)|\mathbf{pq}\rangle\rangle_{\mathrm{av}},\end{aligned}\tag{31.12}$$

wherein we have gone to the classical limit by replacing the Bose factor with the Boltzmann factor. The average indicated then is a classical one.

We may now substitute Eqs. (31.11) into Eq. (31.12), the result of this into Eq. (30.6) in order to obtain

$$\mathrm{G}^{-1}(\mathrm{p})=\begin{bmatrix}\begin{array}{c}\omega_p-\dfrac{p^2}{2\mathrm{m}}-\omega_0+\mu-N_e\langle\langle\mathbf{pq}|T_{11}(\omega_p\\+q^2/2\mathrm{m}_e-\mu_e)|\mathbf{pq}\rangle\rangle_{\mathrm{av}}\end{array} & -N_e\left\langle\left\langle\mathbf{pq}|T_{01}\left(\omega_p+\dfrac{q^2}{2\mathrm{m}_e}-\mu_e\right)|\mathbf{pq}\right\rangle\right\rangle_{\mathrm{av}}\\ -N_e\left\langle\left\langle\mathbf{pq}|T_{10}\left(\omega_p+\dfrac{q^2}{2\mathrm{m}_e}-\mu_e\right)|\mathbf{pq}\right\rangle\right\rangle_{\mathrm{av}} & \omega_p-\dfrac{p^2}{2\mathrm{m}}+\mu-N_e\langle\langle\mathbf{pq}|T_{00}\left(\omega_p+\dfrac{q^2}{2\mathrm{m}_e}-\mu_e\right)|\mathbf{pq}\rangle\rangle_{\mathrm{av}}\end{bmatrix}\tag{31.13}$$

wherein we have adjusted the energy origin slightly so that $E_0=0$ and $E_1=\omega_0$.

The off-diagonal elements of the inverse matrix G will be given by the off-diagonal cofactors divided by the determinant of G^{-1} so that these elements will be of order N_e. The diagonal elements of G will be given by the diagonal elements divided by the determinant so that, approximately,

$$G_{11}(p)=\left[\omega_p-p^2/2\mathrm{m}-\omega_0+\mu-N_e\langle\langle\mathbf{pq}|T_{11}(\omega_p+q^2/2\mathrm{m}_e-\mu_e)|\mathbf{pq}\rangle\rangle_{\mathrm{av}}\right]^{-1} \tag{31.14a}$$

$$\mathsf{G}_{00}(p)=\left[\omega_p-p^2/2\mathrm{m}+\mu-N_e\langle\langle\mathbf{pq}|T_{00}(\omega_p+q^2/2\mathrm{m}_e-\mu_e)|\mathbf{pq}\rangle\rangle_{\mathrm{av}}\right]^{-1} \tag{31.14b}$$

It will obviously be sufficient to use for G_{ee},

$$\mathsf{G}^0_{ee}(\mathrm{p})=\left[\omega_p-p^2/2\mathrm{m}_e+\mu_e\right]^{-1} \tag{31.14c}$$

We may now return to Eq. (31.2).

The first diagram in Eq. (31.2′) includes L_{1001} which, according to Eq. (30.5b), has a leading term containing $\mathsf{G}_{11}\mathsf{G}_{00}$. From Eqs. (31.15) such a term is proportional to $(1+N_e)^{-2}$ so that, to lowest order in the density, it is density independent. The second, third, and fourth terms in Eq. (31.2′), however, are dependent on $\mathsf{G}_{01}\mathsf{G}_{01}$, $\mathsf{G}_{01}\mathsf{G}_{11}$, and $\mathsf{G}_{11}\mathsf{G}_{01}$, respectively. The G_{01} are linearly proportional to the density so that these terms have a N_e or a N_e^2 dependence on that quantity. Therefore, to lowest order in the density, only the first term in Eq. (31.2) is retained. This means, of course, that we are left with an equation for L_{1001} which is uncoupled from the equations for $L_{\alpha\beta\alpha'\beta'}$. On the basis that p' is unessential to the equation, Ross is able to rewrite Eq. (31.2) in terms of the vertex $Q_{10}(p;k)$ as

$$Q_{10}(p;k)=\mathsf{G}_{11}(\mathrm{p})\mathsf{G}_{00}(p-k)\Bigg[1-\int\frac{dq'}{(2\pi)^4}\int\frac{dq}{(2\pi)^4}$$
$$\cdot\langle\mathbf{pq}|T_{11}(\omega_{p+q})|\mathbf{q'q+p-q'}\rangle\mathsf{G}_{ee}(q+p-q')\mathsf{G}_{ee}(q)$$
$$\cdot\langle\mathbf{q'-kq+p-q'}|T_{00}(\omega_{p+q-k})|\mathbf{p-kq}\rangle Q_{10}(q';k)\Bigg] \tag{31.15a}$$

where

$$Q_{10}(p;k)=\int\frac{dp'}{(2\pi)^4}L_{1001}(p,p';p-k,p'+k). \tag{31.15b}$$

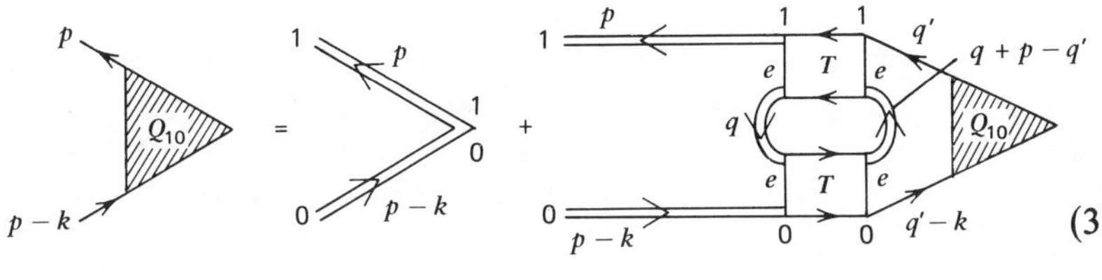

(31.15′)

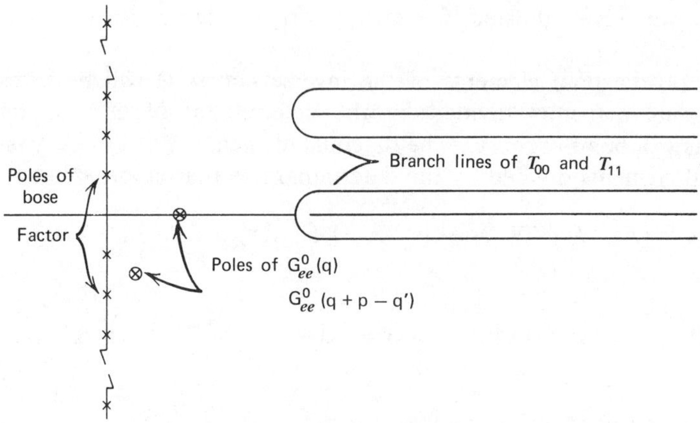

Figure 31.2 After Ross (236).

We first carry out the integration over ω_q as we did beginning with Eq. (31.4). In this case, however, the situation is somewhat complicated by the branch cuts of T_{00} and T_{11}. We reproduce Ross's Fig. 5 as Fig. 31.2.

Let us explain the origin of the branch lines which are encountered here by referring back to that encountered in Eq. (31.8). In considering the denominator on the right of the equation let us suppose that $\mathbf{p}\cdot\mathbf{q}=\mathbf{p}'\cdot\mathbf{q}$ for simplicity. If then, in $T(\omega_{p+p'})$ on the left of this equation, we substitute the value of $\omega_{p+p'}$ at the pole on the right as in Eq. (31.10), we find

$$\mathrm{q}=\left[2\mathrm{M}\left(\omega-p^2/2\mathrm{m}_\lambda-p'^2/2\mathrm{m}_\mu-E_\lambda-E_\mu+\mu_\lambda+\mu_\mu\right)\right]^{1/2}$$

We now have the familiar situation, $f(z)=z^{1/2}$ wherein the entire z plane maps into the upper $f(z)$ plane, as in Fig. 31.3, leading to a branch cut in the z plane with branch point at the origin. In our case the ω plane maps into the upper half of the q plane except that the branch point is displaced to the right along the real axis by the other factor. In order to maintain analyticity then, we must introduce a two-sheeted Riemann surface as in Fig. 31.4. In order to maintain analyticity in crossing the branch cut we must of course proceed from one sheet to the other which is to say that contours in the physical sheet must exclude the cut as indicated in Fig. 31.2. However, there is something more here than a branch cut; there is a singularity which may appear at any point along the cut. Let us write the integral in Eq. (31.8) as,

$$\int \doteq \text{const}\int\frac{dq}{c-q^2}=\frac{\text{const}}{c}\tan^{-1}\left(\frac{q}{c}\right),$$

where $c=\omega_{p+p'}-p^2-p'^2+\text{const}$, these results being approximate but instructive. In Eq. (31.15a), ω_{p+q}, q varies with the integration, p being fixed by the desired value of $Q_{10}(p;k)$. For the lower limit on the q-integration we will obviously have a singularity—through C above—in the ω-integration at

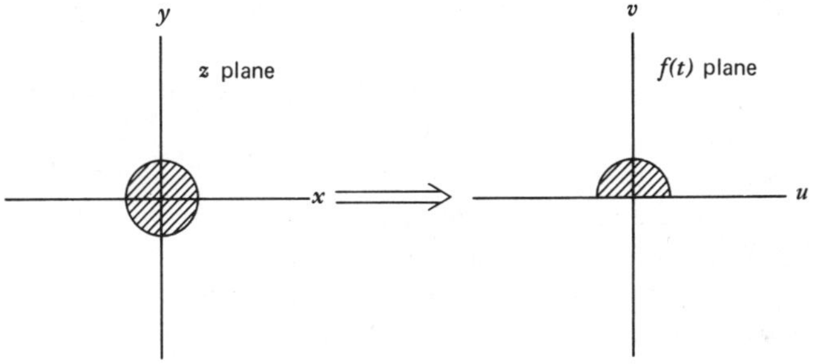

Figure 31.3

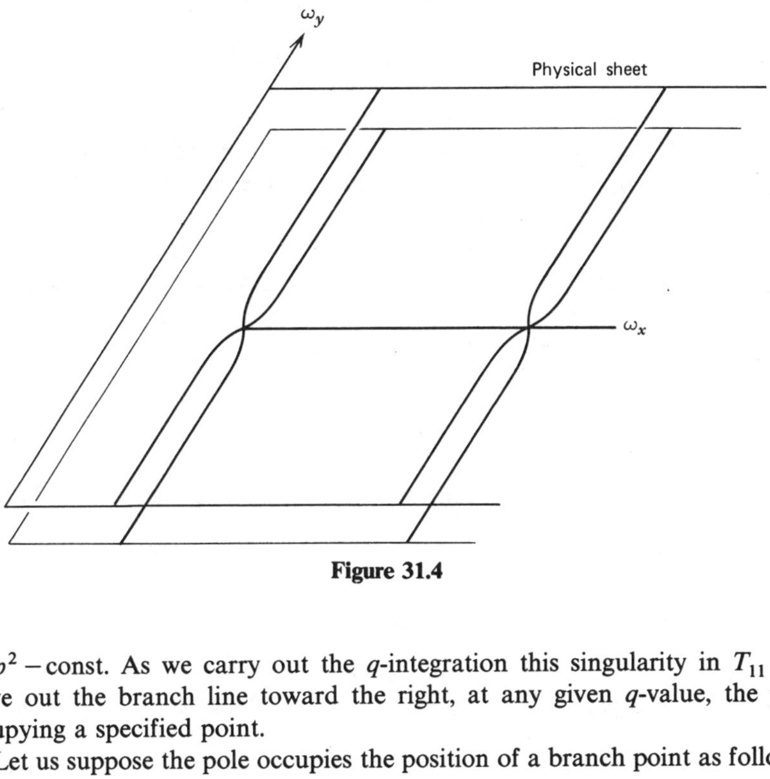

Figure 31.4

$\omega = p^2 - \text{const}$. As we carry out the q-integration this singularity in T_{11} will move out the branch line toward the right, at any given q-value, the pole occupying a specified point.

Let us suppose the pole occupies the position of a branch point as follows:

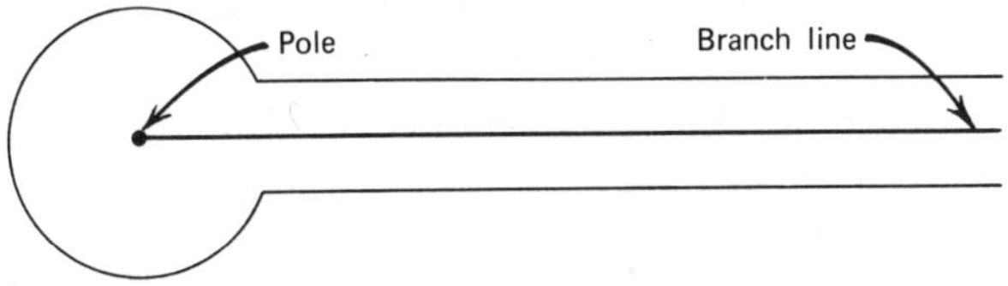

By analogy to Eq. (31.5) we would now expect a contribution from this singularity of NN_e. As the pole moves out, the branch line the contribution from it will not be changed. (We pick up one half of the residue as we move by the pole in one direction, the other half in the other direction.) T will always be multiplied by one or more Greens functions [cf. Eq. (31.15a)] which will yield zeroth- or first-order contributions to the density. Therefore, the contributions from the sort of contour considered here will be dropped as of higher order. With which we return to Fig. 31.2.

Once again the contour is initially around the poles of the Bose factor. Then, however, it is distorted to include not only the poles of the G^0 as was the case earlier but also the cuts of the T matrix, so that, when the overall contour is expanded without limit under the assumption that the ω_ν summation converges, we are left with the contours of Fig. 31.2.

Consider, for example, Eq. (31.8). The branch line begins at an ω value

$$\omega_{p+p'} = \left[(p+q)^2/2m_\lambda + (p'-q)^2/2m_\mu\right]_{\min} + E_\lambda + E_\mu - \mu_\lambda - \mu_\mu, \tag{31.16}$$

where "min" refers to the minimum value of q. Based on the arguments already presented, only the G_{ee} poles are considered. The result of the ω_q integration is then

$$\begin{aligned}
Q_{10}(p;k) = {} & G_{11}(p)G_{00}(p-k)\Bigg\{1 + i\int \frac{dq'}{(2\pi)^4}\int \frac{d\mathbf{q}}{(2\pi)^3} Q_{10}(q';k)\Big[\omega_p - \omega_{q'} \\
& -(q+p-q')^2/2m_e + q^2/2m_e\Big]^{-1} \exp\left[-\beta(q^2/2m_e - \mu_2)\right] \\
& \cdot\langle \mathbf{pq}|T_{11}\big(\omega_p + q^2/2m_e - \mu_e\big)|\mathbf{q'q+p-q'}\rangle\langle \mathbf{q'-kq+p-q'}|T_{00}\big(\omega_p \\
& -\omega_k + q^2/2m_e - \mu_e\big)|\mathbf{p-kq}\rangle - \exp\left[-\beta(q+p-q')^2/2m_e - \mu_e)\right] \\
& \cdot\langle \mathbf{pq}|T_{11}\big(\omega_{q'} + (q+p-q')^2/2m_e - \mu_e\big)|\mathbf{q'q+p-q'}\rangle\cdot\langle \mathbf{q'-kq} \\
& +\mathbf{p-q'}|T_{10}\left(\omega_{q'} - \omega_k + \frac{(q+p-q')^2}{2m_e} - \mu_e\right)|\mathbf{q-kq}\rangle\Bigg\}.
\end{aligned} \tag{31.17}$$

Ross next carries out the $\omega_{q'}$ integration which is rendered considerably more complex by virtue of the inclusion of $Q_{10}(q';k)$ within it. Ross utilizes the physical boundary conditions on the problem in adopting the following approximation: In solving Eq. (31.17) for the vertex he seeks small corrections to already existent modes, not new ones. Such an approach leads to the conclusion that Q_{10} will have about the same properties as $\mathsf{G}_{11}(p)\mathsf{G}_{00}(p-k)$: it will be peaked in the neighborhood of the poles of $\mathsf{G}_{11}\mathsf{G}_{00}$, its own singularities being branch lines. Once again the contributions associated with these branch cuts, as well as the T-matrix cuts, will be of order N. The $\omega_{q'}$ pole arising from the bracketed expression immediately following Q_{10} on the right of Eq. (31.17) introduces no new powers of the density. Only the residue at this pole is kept in obtaining

$$Q_{10}(p;k) \doteq \mathsf{G}_{11}(p)\mathsf{G}_{00}(p-k)\left\{1-\int\frac{d\mathbf{q}'}{(2\pi)^3}\int\frac{d\mathbf{q}}{(2\pi)^3}Q_{10}\big(q',\omega_p\right.$$

$$-(q+p-q')^2/2\mathrm{m}_e+q^2/2\mathrm{m}_e;\mathrm{k}\big)\exp\left[-\beta\left(q^2/2\mathrm{m}_e-\mu_e\right)\right]$$

$$\cdot\langle\mathbf{pq}|T_{11}\left(\omega_p+q^2/2\mathrm{m}_e-\mu_e\right)|\mathbf{q}'\mathbf{q}+\mathbf{p}-\mathbf{q}'\rangle$$

$$\left.\cdot\langle\mathbf{q}'-\mathbf{kq}+\mathbf{p}-\mathbf{q}|T_{00}\left(\omega_p-\omega_k+q^2/2\mathrm{m}_e-\mu_e\right)|\mathbf{p}-\mathbf{kq}\rangle\right\} \tag{31.18}$$

It is of interest to remark that, had Fermi statistics, $[\exp \beta\omega+1]^{-1}$, been used instead of Bose, the result would have been the same.

In Eq. (31.6) we have an equation for a vertex involving an emitter of momentum, $\mathbf{p}$, and a photon of wave number of momentum, $\mathbf{k}$. Our q integrations are essentially integrations over perturbing electron momenta. In carrying the solution further we are faced, as has been the case, with the necessity for some sort of approximate evaluation of these integrals. Contour integrals of some sort in the complex plane still offer a possibility if enabling approximations can be generated. Ross was able to generate such approximations and successfully reduce Eq. (31.18) to the Baranger result. His *modus operandi* was about as follows.

Q_{10} will assuredly peak near the poles of $\mathsf{G}^0_{11}\mathsf{G}^0_{00}$. So much is clear. Now the T matrices in Eq. (31.18)—this includes the T matrices in the mass operators of G_{11} and G_{00} Eq. (30.6)—are such that they have branch lines in the manner of Eq. (31.8). By a comparison of Eqs. (31.8) and (31.18) we see that T_{11} has a branch cut extending along the real axis $[-\infty,\infty]$, T_{00} a cut intersecting the imaginary axis at ω_k and paralleling the real axis over the same range. Ross then replaces the left side of Eq. (31.18) and the first term on the right side of that equation, respectively, by

$$Q_{10}(\mathbf{p};k) = -(\mathrm{i}\beta)^{-1}\sum_{\omega_p} Q_{10}(p;k) \tag{31.19a}$$

$$\bar{Q}^0_{10}(\mathbf{p};k) = -(\mathrm{i}\beta)^{-1}\sum_{\omega_p} \mathsf{G}_{11}(p)\mathsf{G}_{00}(p-k) \tag{31.19b}$$

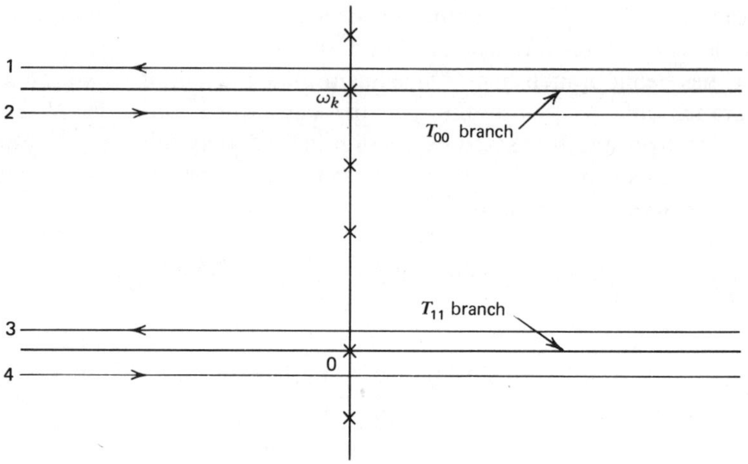

Figure 31.5 After Ross (236).

These summations are again replaced by integrations of the form Eq. (31.4), that is to say, contour integrations which include the Bose factors. For these are substituted contours surrounding the branch cuts as in Fig. 31.5.

Such a substitution may be made because the kernels of the equations for T now contain denominators,

$$T_{11}: \omega_p + q^2/2\mathrm{m}_e - p^2/2\mathrm{m} - (q' - p - p')^2/2\mathrm{m}_e + \mu,$$

$$= \omega_p + q^2/2\mathrm{m}_e - \mu_e \tag{31.20a}$$

$$T_{00}: \omega_p - \omega_k + q'^2/2\mathrm{m}_e - p^2/2\mathrm{m} - (q' + p - p')^2/2\mathrm{m}_e + \mu,$$

$$= \omega_p - \omega_k + q'^2/2\mathrm{m}_e - \mu_e \tag{31.20b}$$

On the branch lines indicated in Fig. 31.4 we may, using Eqs. (31.14), approximate the Greens functions by

$$\mathrm{G}_{00}(p-k) \doteq \Big[\omega_p - \omega_k - (p-k)^2/2\mathrm{m}$$

$$- N_2 \langle\langle \mathbf{p} - \mathbf{k}\mathbf{q}' | T_{00}\big((p-k)^2/2\mathrm{m} + q'^2/2\mathrm{m} - \mu - \mu_e \pm \mathrm{i}\epsilon\big) | \mathbf{p} - \mathbf{k}\mathbf{q}' \rangle\rangle_{\mathrm{av}} + \mu\Big]^{-1} \tag{31.21a}$$

$$\mathrm{G}_{11}(p) \doteq \Big[\omega_p - p^2/2m$$

$$- N_e \langle\langle \mathbf{p}\mathbf{q}' | T_{11}\big(p^2/2\mathrm{m} + q'^2/2\mathrm{m}_e - \mu - \mu_e \pm \mathrm{i}\epsilon\big) | \mathbf{p}\mathbf{q}' \rangle\rangle_{\mathrm{av}} + \mu\Big]^{-1} \tag{31.21b}$$

wherein we remark the factor $\pm \mathrm{i}\epsilon$ which establishes T either above or below the appropriate cut. From Eq. (9.17) we know that $-\mathcal{I}[T(+\mathrm{i}\epsilon)]$ is proportional to a cross section and hence greater than zero. It follows that $\mathrm{G}(\omega_p + \mathrm{i}\epsilon)$ has a pole in the lower half-plane, $\mathrm{G}(\omega_p - \mathrm{i}\epsilon)$ in the upper, and Q_{10} is evaluated by closing the paths of integration about two of these four poles.

The result is

$$Q_{10}(p;k) \doteq \bar{Q}^0_{10}(p;k)\left\{1 - i\exp\left[\beta(p^2/2m-\mu)\right] N_e\langle(2\pi)^{-2}\right.$$

$$\cdot \int dq_e \langle \mathbf{pq}|T_{11}|\mathbf{p}+\mathbf{q}-\mathbf{q}_e\mathbf{q}_e\rangle\langle \mathbf{p}+\mathbf{q}-\mathbf{q}_e-\mathbf{k}\mathbf{q}_e|T^*_{00}|\mathbf{p}-\mathbf{kq}\rangle \bar{Q}_{10}(\mathbf{p}+\mathbf{q}-\mathbf{q}_e;\mathbf{k})$$

$$\left.\cdot\delta\left(q_e^2/2m_e - q^2/2m_e + (p+q-q_e-k)^2/2m - (p-k)^2/2m\right)\rangle_{av}\right\} \quad (31.22a)$$

$$q_e = p+q-q' \quad (31.22b)$$

with the Boltzman average over $\mathbf{q}$. Since we are dealing with electrons, $\mathbf{q}$ and $\mathbf{q}_e$ are considerably smaller than $\mathbf{p}$, and $\mathbf{q}$ and $\mathbf{q}_e$ as well as $\mathbf{k}$ may be ignored in sums involving $\mathbf{p}$, that is, $\mathbf{p}+\mathbf{q}-\mathbf{q}_e \doteq \mathbf{p}$ and so on. By choosing $\mathbf{q}$ as a space-fixed axis we may refer $\mathbf{q}_e$ to it so that

$$\int d\mathbf{q}_e = m\int d\Omega\, d\left(\frac{q_e^2}{2m}\right)|\mathbf{q}_e| \quad (31.23)$$

With our assumptions vis-à-vis $\mathbf{p}$ and $\mathbf{q}$ the δ-function of Eq. (31.20) becomes $\delta(\mathbf{q}_e^2/2m_e - \mathbf{q}^2/2m)$ allowing the immediate evaluation of the integral over $(\mathbf{q}_e^2/2m)$. Therefore, Eq. (31.22) becomes

$$\bar{Q}_{10}(\mathbf{p};k) = \bar{Q}^0_{10}(\mathbf{p};k)\left\{1 - \bar{Q}_{10}(\mathbf{p};k)\exp\left[\beta\left(\frac{p^2}{2m}-\mu\right)\right]\right.$$

$$\left.\cdot iN_e\langle m_e|\mathbf{q}_e|\int \frac{d\Omega}{(2\pi)^2}\langle \mathbf{pq}|T_{11}|\mathbf{pq}_e\rangle\langle \mathbf{pq}_e|T^*_{00}|\mathbf{pq}\rangle_{|\mathbf{q}_e|=|\mathbf{q}|}\rangle_{av}\right\}$$

$$(31.24)$$

$\bar{Q}^0_{10}(\mathbf{p};k)$ is evaluated by evaluating residues at the poles of G_{00} and G_{11} as given by Eqs. (31.21), the method of this calculation being essentially that to which Ross appealed in obtaining Eq. (31.22).

$$\bar{Q}^0_{10}(\mathbf{p};k) = -i\exp\left[-\beta(p^2/2m-\mu)\right[\omega_k - \omega_0 - p^2/2m - (p-k)^2/2m$$

$$+N_e\langle\langle \mathbf{pq}|T^*_{00}|\mathbf{pq}\rangle\rangle_{av} - N_e\langle\langle \mathbf{pq}|T_{11}|\mathbf{pq}\rangle\rangle_{av}\Big]^{-1} \quad (31.25)$$

Eqs. (31.24) and (31.25) are combined in simple algebraic fashion to furnish Eq. (31.15a) which, when substituted into Eq. (31.1), yields

$$\chi(\mathbf{k},\omega) = \mathbf{d}_{10}\mathbf{d}_{01}\int \frac{d\mathbf{p}}{(2\pi)^3}\exp\left[-\beta\left(\frac{p^2}{2m}-\mu\right)\right]\left[\omega-\omega_0-\mathbf{p}\cdot\frac{\mathbf{k}}{m}\right.$$

$$-N_e\langle\langle \mathbf{pq}|T_{11}|\mathbf{pq}\rangle\rangle_{av} + N_e\langle\langle \mathbf{pq}|T^*_{00}|\mathbf{pq}\rangle\rangle_{av}$$

$$\left.-iN_e m_e(2\pi)^{-2}\langle|\mathbf{q}|\int d\Omega\langle \mathbf{pq}|T_{11}|\mathbf{pq}_e\rangle\langle \mathbf{pq}_e|T^*_{00}|\mathbf{pq}\rangle_{|\mathbf{q}_e|=|\mathbf{q}|}\rangle_{av}\right]^{-1}$$

$$(31.26)$$

where terms of greater than linear dependence on density have been dropped. That contact with the Baranger result has indeed been established may be seen by reference to Eqs. (14.20).

The spectral line shape is of course given by the imaginary part of the susceptibility so that

$$I(\omega)=\left\langle \frac{2\delta/\pi I_0}{[\omega-\omega_0-\mathbf{v}\cdot\mathbf{k}-\Delta]^2+\delta^2} \right\rangle_{\mathbf{p}} \tag{31.27a}$$

$$\delta=N_e \mathrm{m}_e(2\pi)^{-2}\left\langle|\mathbf{q}|\int d\Omega\langle\mathbf{pq}|T_{11}|\mathbf{pq}_e\rangle\langle\mathbf{pq}_e|T^*_{00}|\mathbf{pq}\rangle_{|\mathbf{q}_e|=|\mathbf{q}|}\right\rangle_{\mathrm{av}} \tag{31.27b}$$

$$\Delta=N_e\left[\langle\langle\mathbf{pq}|T_{11}|\mathbf{pq}\rangle\rangle_{\mathrm{av}}+\langle\langle\mathbf{pq}|T^*_{00}|\mathbf{pq}\rangle\rangle_{\mathrm{av}}\right] \tag{31.27c}$$

wherein we have supposed that $\mathbf{v}=\mathbf{p}/\mathrm{m}$, and I_0 simply normalizes the result to the proper integrated intensity.

32 Certain Neglected Terms to Include the Natural Width

These results essentially duplicate those of Baranger except that, through $\mathbf{v}\cdot\mathbf{k}$, Doppler broadening effects are included and the dependence of the scattering amplitudes on the emitter momenta are retained. Rather obviously, the presence of the $\mathbf{v}\cdot\mathbf{k}$ term together with the averaging over the emitter momenta amounts to folding the interruption contour into a Doppler contour as, for example, occurs in the Voigt profile. The following is a brief recapitulation.

In Eqs. (29.8) we have the general results which comprehend only the approximations (1) point particles, (2) dipole transitions, and (3) two-body potentials. These include all broadening phenomena save natural broadening. Ross next introduces the interruption approximation which leads to the result displayed as Eqs. (30.5). In proceeding and establishing contact with work such as that of Baranger, Ross ignores terms involving powers of the density higher than the first. We recall the discard of diagrams such as the second, third, and fourth in Eq. (31.2′) as a consequence of such density considerations. As an example of the many terms that are thus discarded, let us draw a diagram for a three-body collision and, in so doing, illustrate the manner in which the diagrams can contribute to our understanding of the physical phenomena. If we continue the iterative procedure which led to Eq. (30.7) by taking the second term on the right of Eq. (30.7) and substituting it for L in the third term on the right of Eq. (30.5b) we will obtain

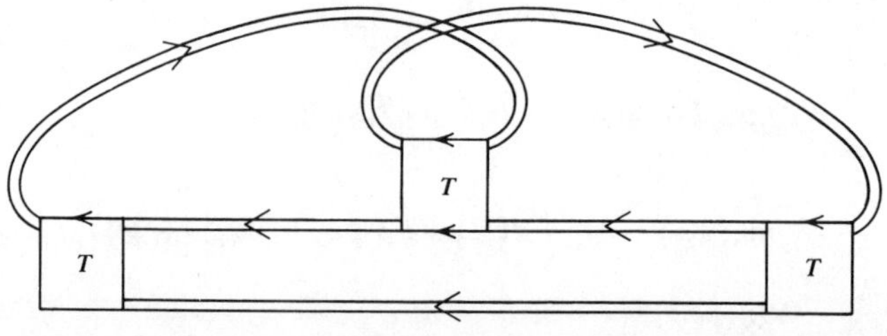

Quite obviously from the figure, particle 1 will interact with particle 2 which, after interacting or colliding with particle 3, will again interact with particle 1.

In addition to ignoring the higher powers of the density, it is of course necessary for Ross to make the approximations appropriate to electrons ($\mathsf{m}_e \ll \mathsf{m}$, $\mathbf{p}_e \ll \mathbf{p}$) so as to arrive at the Baranger expression, albeit modified for the better, Eqs. (31.17).

As we have seen, Ross' general formulation is, in theory at least, capable of dealing with any broadening situation, although he largely restricts himself to contact establishment with proven theory for the case of interruption broadening. Nevertheless, he indicates the direction to be followed in applying the theory to (1) foreign-gas broadening and (2) resonance broadening. (That he likewise makes an important point vis-à-vis natural line shape we shall touch on later.) First of all, we consider (1).

The interaction potential will of course change in proceeding from electrons to foreign-gas molecules. However, as we have not considered the specifics of V in any case, this will have no practical effect. The important change will be that the broadener mass will no longer be much smaller than the emitter mass so that the broadener momentum may no longer be ignored. This can of course have rather important consequences. Let us recall that $\mathbf{p}$ and $\mathbf{p}'$ referred to the emitter, $\mathbf{q}$ and $\mathbf{q}'$ and, later, $\mathbf{q}_e$ to the broadener, specifically, to electrons. Let us also recall that, if $\mathbf{p}'$ is the molecular momentum after a collision, $\mathbf{p}$ the momentum before, forward scattering will mean that direction $(\mathbf{q})$ = direction $(\mathbf{q}')$. If the scattering is "peaked in the forward direction," an integration over $\mathbf{q}'$ will yield a maximum for $\mathbf{q}'$ roughly in the direction of $\mathbf{q}$. It is of some consequence to remark this since, for emitter mass about equal to perturber mass, and when scattering peaks in the forward direction, Ross is able to show that a relatively simple result is obtained for Eq. (31.24):

$$\bar{Q}_{10}(\mathbf{p};\mathbf{k}) = e^{-\beta(p^2/2\mathsf{m}-\mu)}\left[\omega-\omega_0-\mathbf{p}\cdot\mathbf{k}/\mathsf{m}-\Delta_0(\mathbf{p})\right]^{-1} \tag{32.1a}$$

$$\begin{aligned}\Delta_0(\mathbf{p}) = {} & N_p\langle\langle\mathbf{pq}|T_{11}|\mathbf{pq}\rangle\rangle_{\text{av}} - N_p\langle\langle\mathbf{pq}|T^*_{00}|\mathbf{pq}\rangle\rangle_{\text{av}} \\ & + i\frac{\mathsf{m}\mathsf{m}_p}{\mathsf{m}+\mathsf{m}_p}\int\frac{d\mathbf{q}}{(2\pi)^3}\int\frac{d\Omega}{(2\pi)^3}|\mathbf{p}_e|\exp\left[-\beta\left(\frac{q^2}{2\mathsf{m}_p}-\mu_p\right)\right] \\ & \cdot\left\langle\mathbf{pq}|T_{11}|\frac{\mathsf{m}}{\mathsf{m}+\mathsf{m}_p}(\mathbf{q}+\mathbf{p})+\mathbf{p}_p\frac{\mathsf{m}_p}{\mathsf{m}+\mathsf{m}_p}(\mathbf{q}+\mathbf{p})-\mathbf{p}_p\right\rangle \\ & \cdot\left\langle\frac{\mathsf{m}}{\mathsf{m}+\mathsf{m}_p}(\mathbf{q}+\mathbf{p})+\mathbf{p}_p\frac{\mathsf{m}_p}{\mathsf{m}+\mathsf{m}_p}(\mathbf{q}+\mathbf{p})-\mathbf{p}_p|T^*_{00}|\mathbf{pq}\right\rangle\end{aligned} \tag{32.1b}$$

Eq. (32.1b) to be evaluated for $|\mathbf{p}_p| = |\mathbf{q}\mathsf{m}/(\mathsf{m}+\mathsf{m}_p)+\mathbf{p}\mathsf{m}_p/(\mathsf{m}+\mathsf{m}_p)|$.

$$\mathbf{p}_p = \mathbf{q}_p + (\mathbf{p}+\mathbf{q}-\mathbf{k})\mathsf{m}_p/(\mathsf{m}+\mathsf{m}_p). \tag{32.1c}$$

which reduces to the Baranger result for $\mathsf{m}_p \ll \mathsf{m}$. For the more general case we obtain Eq. (32.1a) with Δ_0 replaced by

$$\Delta(\mathbf{p},\omega)=\Delta_0(\mathbf{p})-i\int\frac{d\mathbf{q}}{(2\pi)^3}\frac{\mathsf{m}\mathsf{m}_p}{\mathsf{m}+\mathsf{m}_p}\int\frac{d\Omega}{(2\pi)^2}|\mathbf{p}_p|\exp\left[-\beta\left(\frac{q^2}{2\mathsf{m}_p}-\mu_p\right)\right]$$
$$\cdot\left\langle pq|T_{11}|\frac{\mathsf{m}}{\mathsf{m}+\mathsf{m}_p}(q+p)-p_p\frac{\mathsf{m}_p}{\mathsf{m}+\mathsf{m}_p}(q+p)-p_p\right\rangle$$
$$\cdot\left\langle\frac{\mathsf{m}}{\mathsf{m}+\mathsf{m}_p}(q+p)-p_p\frac{\mathsf{m}_p}{\mathsf{m}+\mathsf{m}_p}(q+p)-p_p|T^*_{00}|pq\right\rangle\left(\mathbf{p}_p+\frac{\mathbf{q}\mathsf{m}+\mathbf{p}\mathsf{m}_p}{\mathsf{m}+\mathsf{m}_p}\right)\cdot\frac{\mathbf{k}}{\mathsf{m}}$$
$$\cdot\frac{\Delta(\mathbf{p}_p+(\mathbf{q}+\mathbf{p})\mathsf{m}/(\mathsf{m}+\mathsf{m}_p),\omega)-\Delta(\mathbf{p},\omega)}{\omega-\omega_0-\left[\mathbf{p}_p/\mathsf{m}+(\mathbf{q}+\mathbf{p})/(\mathsf{m}+\mathsf{m}_p)\right]\cdot\mathbf{k}-\Delta(\mathbf{p}_p+(\mathbf{q}+\mathbf{p})\mathsf{m}/(\mathsf{m}+\mathsf{m}_p),\omega)}\qquad(32.1\text{d})$$

$$|\mathbf{p}_p|=|-\frac{\mathbf{m}}{\mathbf{m}+\mathbf{m}_p}\mathbf{q}+\frac{\mathbf{m}_p}{\mathbf{m}+\mathbf{m}_p}\mathbf{p}|.\qquad(32.1\text{e})$$

Equation (32.1d) rather obviously reduces to Eq. (32.1b) in the wings of the spectral line where $|\omega-\omega_0|\gg\Delta$, and the embarrassing necessity for solving the integral equation evaporates.

As we have remarked in connection with Eq. (30.7), those terms involving the exchange of excitation between emitter and perturber prior to emission, that process so crucial to resonance broadening, are present in our *L*-expansion. This process is present but not important to the varieties of line broadening considered hitherto. Ross remarks on the subject, but we shall postpone any detailed discussion until our introduction of the Zaidi efforts in this direction. This, in considering the Ross labors, leaves us with natural line broadening.

The point that is of principal interest and that Ross makes vis-à-vis natural line shape is that, to first order, the linewidth and line shift are the sum of the natural and perturber-induced widths and shifts. In pointing this out, let us begin by adding the particle-field perturbation to the Hamiltonian Eq. (26.20c). This will eventually affect the spectral line through the *T* in Eq. (31.26) by adding the particle-field potential to Eq. (30.5c). If, in Eq. (30.5a), we replace G by G^0, *T* by the first term in Eq. (30.5c), the particle-field V, we will change the mass operator from Eq. (30.5a′) to

Σ =

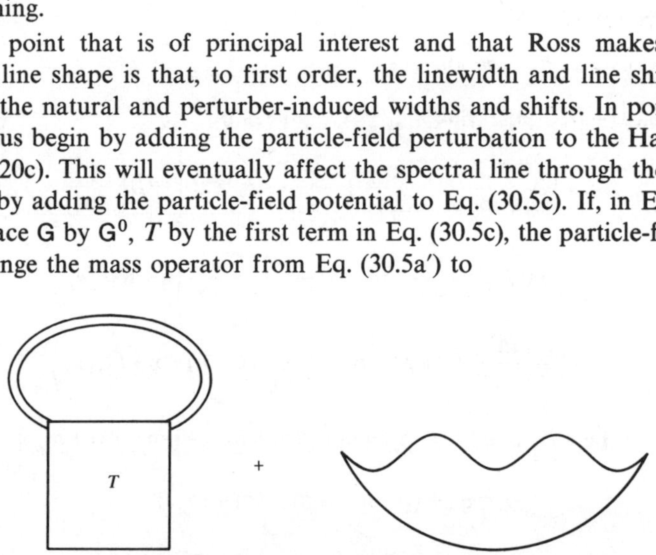

(32.2)

Such a change in the mass operator will of course lead to a change in the vertex of Eq. (31.15′), specifically, to the addition of

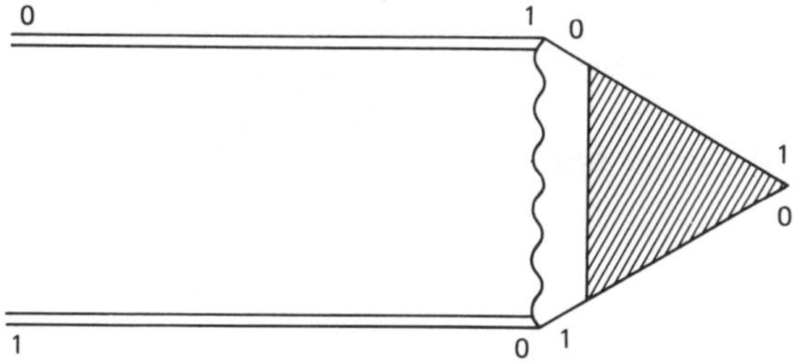

(32.3)

It is reasonably obvious that the first few terms in the expansion of Eq. (32.3) —G replaced by G^0 and so on—will be

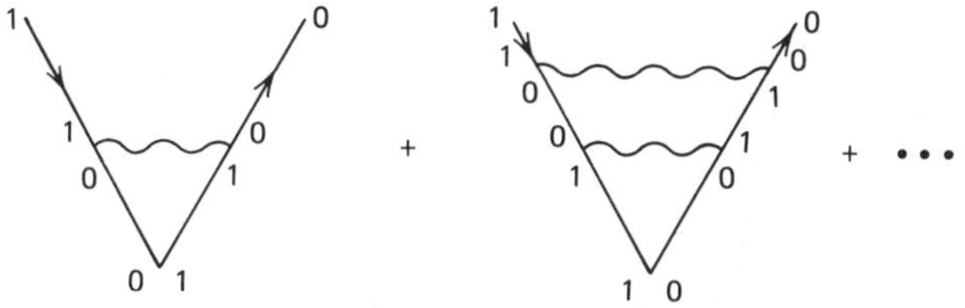

(32.4)

That Ross is well justified in neglecting terms higher than the first in radiation damping—all save the first term in Eq. (32.4)—we have seen in our study of natural line shape. The first term in Eq. (31.4) is antiresonant and therefore small enough so that Eq. (31.3) may simply be dropped. The vertex Eq. (31.15′) then remains ostensibly identical to Eq. (31.15′), although now the Greens functions are still calculated using Eqs. (31.2). Thus we see that, to the denominator or within the square-bracketed expression of Eq. (31.25), we simply add $\gamma_1 - \gamma_0^*$, where

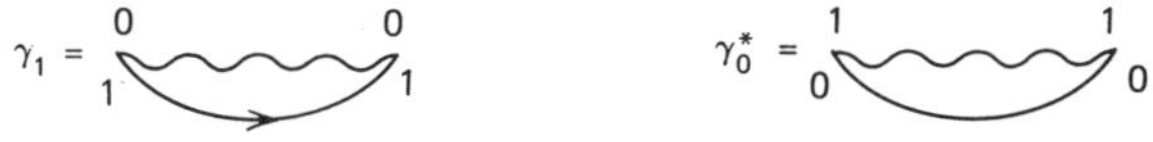

(32.5)

This has a very important consequence: the linewidth and line shift are the sum of the natural and pressure-induced line widths and shifts. That we are thus well justified in treating the two phenomena independently of each other has the virtue of allowing a considerable simplification of the theoretical treatment.

We also see that, considering the natural width alone, the total width is the sum of the widths or, from one point of view, of the uncertainties of the upper and lower radiator levels. In the pressure-broadening case this is not true which accounts, for example, for the difference between the results of Lewis (180), who assumed upper and lower effects to be simply additive, and Balling, Hanson, and Pipkin (62), who realized that they were not. Once again the diagrams are of assistance to us in determining why this is so.

Equation (30.8), referring to the first term of Eq. (30.7), details those processes wherein collisions occur either with the emitter in its upper or in its lower state. Were these the only sort of collisions to be anticipated, then the additivity assumption for upper and lower state broadenings would lead to the proper result. However, as we have already remarked, the third term in Eq. (30.7) is typical of those terms relating to collisions during emission, and this and similar terms negate, in a obvious fashion, the additivity assumption.

7

On the Doppler-Pressure Synthesis and Binary Collisions

Christian Doppler of Prague enunciated a principle in 1842 with which we presume ourselves familiar. We also presume ourselves familiar with the work of Lord Rayleigh (259), who enlightened us on the relationship, through the Boltzmann distribution, that this principle evoked between the thermal motion of a (low density) collection of radiators and the distribution of intensities in the spectral line. The intensity distribution in a purely Doppler-broadened spectral line is an exponential dependence on the square of the frequency separation from line center. The distribution in, say, the interruption approximation to foreign-gas broadening is an inverse square dependence on the frequency separation from line center. Now suppose neither thermal nor foreign-gas effects may be ignored. It would, at first blush, seem logical to treat the situation by taking the intensity at each frequency in the interruption-broadened line and smearing it over the Gaussian distribution of the Doppler-broadened line. This, of course, is precisely what was done long ago when the so-called Voigt profile was introduced. This classic folding of an interruption into the Doppler distribution was effected by Voigt in 1912 (275). That the Voigt profile is anything but the ultimate answer we shall see. That it remains of interest and importance is indicated by its continued tabulation by, for example, Posener (219), Davies and Vaughan (108), and Hummer (149).

In Section 33 we shall consider a classical treatment of combined Doppler and pressure broadening that leads, in the proper limit (but only in the proper limit), to the Voigt profile. In another range of the Doppler and pressure broadening parameters, the theory leads to the narrowing effect which Dicke predicted in the early fifties. The quantum treatment of Section 34 is not limited to the Dicke effect as the section title might indicate, but that phenomenon is of sufficient importance to warrant this sort of emphasis. Section 35 is another quantum Doppler-pressure treatment, this one having, as Mr. Holmes was wont to remark, certain points of interest, enough indeed to warrant its inclusion.

Section 36–38 and portions of Section 39 are concerned with the careful study by Roney of Doppler and pressure broadening, generally under the purview of the Liouvillian, specifically with the introduction of, to us at least, a new operator, the reduced density matrix. In Section 39 we consider the time of collision and its connection with the off-shell elements of the T matrix, which was also one of Roney's concerns.

33 Classical Treatment of Doppler-Pressure Broadening

We will recall that, as Lord Rayleigh pointed out to us in 1889, the purely Doppler-broadened spectral line is one having an exponential dependence on the square of the frequency separation from line center, $\exp[-A(\nu-\nu_0)^2]$. (Although we have no specific interest in line profiles associated with relativistic velocities, it is worthwhile to remark that they have been investigated by Gerbal and Prud'Homme (130). These velocities, which may be encountered in quasar winds and collapsed objects, yield both blue shifts and asymmetries.) It was not long after 1889 before the question was raised as to what sort of spectral line is to be expected when the effects of radiator velocities and foreign-gas perturbations are both included. Indeed, Godfrey (135) was studying the folding of an interruption distribution into a Doppler distribution five years before Lorentz generated his treatment of the former. Apparently however, the only individual to read Godfrey's article is Breene (9), whose motivation is somewhat obscure. In any event, Voigt's profile followed in 1912 and has been the object of a great deal of attention over the years. However, that the Voigt profile cannot be expected to describe the Doppler-pressure broadened line for a broad spectrum of the parameters descriptive of the broadening phenomena is now well known. In order to begin our investigation of why this is so, let us return first to the classical idea of Doppler broadening.

A molecule moving with respect to the observer will emit a spectral line shifted by the Doppler effect by an amount proportional to its translational velocity. Now let us observe the emission process by all the (excited) molecules of a gas in equilibrium with the velocities of the various molecules in the assemblage ordered by a Gaussian distribution. It follows that the emitted spectral line will have a Gaussian distribution as we are, of course, aware. Forty odd years after Lord Rayleigh enunciated this classical description of Doppler broadening, Fermi (117) told us how the effect may be viewed quantum mechanically.

The molecule will emit a photon which will have a momentum related to its frequency. In order that momentum may be conserved, the radiating molecule will acquire a recoil momentum which will obviously change its preemission translational momentum. This change in the momentum corresponds directly to a change in the molecule kinetic energy, a change that is directly reflected in a change in the photon energy from that corresponding to the separation of the two internal states of the molecule between which the photon-producing transition took place. Thus, we arrive at the same result as that yielded by the classical treatment, and so much for pure Doppler broadening.

If the Voigt profile is properly to represent the combined effects of Doppler and collision broadening, the latter must not alter the basic physical phenomenon which yields the former. However, such, as Dicke (109) and Mizushima (204, 205) pointed out, is anything but the case. We shall return to Dicke; we now consider the non sequitur which Mizushima found in the Voigt profile.

In the Voigt profile we smear $[(\omega-\omega_0)^2+\delta^2]^{-1}$, δ a constant, over a Boltzmann distribution of velocities. As we have seen in, say, Section 15, the width, δ, is proportional to $N\sigma v$. Mizushima shows quite straightforwardly that σ is proportional to v^2 so that, instead of a constant, δ, we have a v^{-1} to be smeared over a Boltzmann distribution of velocities. The consequence of all this is that, while the Voigt profile may be satisfactory for negligible pressure broadening, the true result is some 30% lower than the Voigt in the line wings or tail region. Because it is calculationally helpful to use an interruption distribution, Mizushima (205) shows that we may obtain the proper result by the superposition of two or three interruption profiles. [Berman (72) contends that the Mizushima result is incorrect, a contention predicated on a temporally noncommutative, no-cutoff numerical treatment of the Schrödinger equation by Berman and Lamb (73). These authors arrive at what amounts to a velocity dependence additively inserted in the width, and, since Mizushima hardly started with this, he could scarcely have obtained the correct result. From our point of view, however, the Mizushima ideas remain valid.] We now turn to Dicke. That author considered the situation wherein the mean free path is small compared to the wavelength of the emitted radiation. Now he supposed the broadening of the line to be purely Doppler, but Doppler under conditions such that the mean free path of the radiator was less than the wavelength of the emitted radiation. This is an apparent non sequitur which he circumvented by having the molecule collide with the walls of a potential well, thus simulating the effects on the Doppler distribution without including the collision broadening per se. He obtained what is quite clearly described as a "collision narrowed" spectral line. This narrowing has been called both the Dicke effect and the Mossbauer effect [cf. Cattani (93)]. The Doppler-broadened line under these circumstances consisted of an interruption contour centrally surmounted by a narrow and high spike. Sobelman (252) and Rautian and Sobelman (222) later treated this same subject. Galatry (128) reexamined the work of Dicke and Wittke and Dicke (286), in particular the probability function used by these authors, and found it somewhat wanting. The probability involved is that for a displacement Δx of a particle after time t. In short, Galatry obtained roughly the Dicke results, albeit considerably improved upon and expanded. Finally, we remark that Herbert (140) has recently published a generalized Voigt function which includes collision narrowing and provides a dimensionless standardization of the Galatry result. With which we turn to a rather more specific consideration of the work of Gersten and Foley (GF) (131).

Classically, we will recall, a radiating atom consists of a bound electron whose (perhaps harmonic) oscillation leads to the emission or absorption of radiation of the frequency of its oscillation. Now obviously our negative electron in the presence of the positive atomic core constitutes a dipole at least reasonably analogous to the dipole of the quantum treatment, the dipole which enters the line-broadening correlation function. Therefore, our correlation function may be written in terms of the separation, $x(t)$, of the electron from the atom core, thus

$$\Phi(t)=\langle x(t)x(0)\rangle \tag{33.1}$$

and so GF wrote it. We take $\eta(t)$ as the phase shift induced in the electronic oscillation by collisions between time 0 and time, t. Next, we take the frequency of electronic oscillation as ω_0, the observed frequency as $\omega'(t)$, the former differing from the latter by the fact of the Doppler effect, GF accounting for this by

$$\omega'(t)=\omega_0+\omega_0 b(t), \tag{33.2a}$$

b being the ratio of the z-component of the velocity to the speed of light. The reader will surely agree that the displacement of such an oscillator will be given by

$$x(t)=A\cos\left[\int_0^t dt'\,\omega'(t')+\eta(t)\right] \tag{33.2b}$$

where A is the amplitude. Eq. (33.1) becomes

$$\Phi(t)=\tfrac{1}{2}A^2\mathfrak{R}\left[e^{i\omega_0 t}f(t)\right] \tag{33.3a}$$

where

$$f(t)=\left\langle \exp\left[i\left(\omega_0\int_0^t dt'\,b(t')+\eta(t)\right)\right]\right\rangle. \tag{33.3b}$$

The averaging envisaged is over (1) a distribution of collision times, (2) the phase shift distribution, and (3) the distribution of velocities. These averages GF carry out after the introduction of certain details of the collision process are specified.

During the time interval 0–t, n collisions occur, a given collision being responsible both for trajectory deflection and radiation interruption. The time interval from $t=0$ to $t=t_1$, where t_1 is the time of the first (obviously instantaneous) collision, we call Δ_0, that from the first to the second, Δ_1, and so on, that from the $(n-1)$st to the nth, Δ_{n-1}. Finally, we call the time interval between the n-th collision and the end of the time period, t, Δ_n.

We may recall that the classical Doppler width varies as $T^{1/2}/\lambda$. Therefore, since the width here varies as $(l/\lambda)^2$, $\lambda \gg l$, we can see that this result yields a line considerably narrowed from the classic result.

As one would expect, the other extreme, $l \gg \lambda$, leads to the classic Doppler result, and GF then study the poles of $W(z)$ via Abramowitz and Stegun in order to show the sort of result that is to be expected in the intermediate case. They find that the simple superposition of a number of contours of the type Eq. (33.12) obtains.

Finally, GF replace Eq. (33.8) by ϵ in order to study the Doppler-interruption synthesis. We do not repeat the straightforward treatment which produces the following result. For $\epsilon \neq 0$, the correlation function is

$$f(t) = \exp\{(t/\tau_0)[(1-\epsilon) + (3\epsilon - 2\epsilon^2)/2\zeta^2]\} \tag{33.14}$$

with $\delta = [1 - \Re(\epsilon)]/\tau_0$. Thus (1) for the value of ϵ not close to one,

$$f(t) = \exp[-(t/\tau_0)(1-\epsilon)]$$

and we see that the pressure-broadening parameters dominate the line shape. (2) For $\epsilon^2 \doteq \epsilon$, it is apparent that Eq. (33.14) is roughly

$$f(t) = \exp\{-(t/\tau_0)[(1-\epsilon) + 1/2\zeta^2]\},$$

a simple folding of the interruption shape with that obtained as Eq. (33.11), the linewidth being the addition of the interruption and Dicke-Doppler widths. (3) For $\epsilon = 0$, we obtain what is in essence a Voigt profile.

34 The Quantum Treatment of the Dicke Effect. Dicke-Vertex Equivalence

The correlation function, whose Fourier transform is the spectral line intensity, is a familiar starting point for line shape ruminations and the one used by Cattani (92, 93) in his treatment of the Dicke effect:

$$\Phi(t) = \sum_{ij} \Upsilon_1 \langle \psi_i^{(1)} | H_F^+ | \psi_j^{(1)} \rangle [\langle \psi_j^{(1)} | \mathfrak{T}^{-1}(t) H_f \mathfrak{T}(t) | \psi_i^{(1)} \rangle]_{\text{av}} \tag{34.1a}$$

$$[\cdots]_{\text{av}} = \sum_i \Upsilon_2 \langle \psi_i^{(2)} | [\cdots] | \psi_i^{(2)} \rangle \tag{34.1b}$$

$$i\dot{\mathfrak{T}} = (H_1 + H_2 + \mathsf{V}_{12})\mathfrak{T} \tag{34.1c}$$

$$H_F = \mathsf{d}_f \exp[i\mathbf{k} \cdot \mathbf{r}] \tag{34.1d}$$

In Eqs. (34.1) $\psi_i^{(1)}$ is the wave function of the radiator, $\psi_i^{(2)}$ that of the perturber. H_F is of course the particle-field interaction, d_f being the component of the radiator dipole moment in the field direction, r the location of its center of mass. Finally, Υ_1 is, of course, the density matrix. One could have begun with the radiator internal and translational states indicated, but it is perhaps somewhat less confusing to introduce them at this point: $\psi_i^{(1)} = |ip_i\rangle$. When this substitution is made in Eq. (34.1a), and a slight rearrangement made in the result, we obtain

$$\Phi(t)=\sum_{ijkl}\Upsilon_{1i}\Upsilon_{1p_i}\langle ip_i|H_F^+|jp_j\rangle\langle kp_k|H_F|lp_l\rangle$$
$$\cdot\left[\langle jp_j|\mathfrak{T}^{-1}(t)|kp_k\rangle\langle lp_l|\mathfrak{T}(t)|ip_i\rangle\right]_{\text{av}} \tag{34.2}$$

wherein the kl matrix element over H_F has been moved out of the average, as it may be since it depends only on radiator coordinates. Note also that the summation over i includes an integration over $\mathbf{p}_i$, the momentum of the radiator in internal state i; the summation over k includes an integration over $\mathbf{p}_k$. At this point it is convenient to replace the TDO in use by an interaction TDO, $U(t)$, where

$$\mathfrak{T}(t)=\exp[-\mathrm{i}t(H_1+H_2)]\mathfrak{U}(t) \tag{34.3a}$$

$$\mathrm{i}\dot{\mathfrak{U}}(t)=\exp[\mathrm{i}t(H_1+H_2)]\mathrm{V}_{12}\exp[-\mathrm{i}t(H_1+H_2)] \tag{34.3b}$$

Next, we remark that

$$\langle ip_i|H_F^+|jp_j\rangle=\langle ie^{\mathrm{i}\mathbf{p}_i\cdot\mathbf{r}}|e^{-\mathrm{i}\mathbf{k}\cdot\mathbf{r}}|e^{-\mathrm{i}\mathbf{p}_j\cdot\mathbf{r}}j\rangle$$

so that $\mathbf{p}_i-\mathbf{k}=\mathbf{p}_j$. Similarly, $\mathbf{p}_k+\mathbf{k}=\mathbf{p}_l$ from the second matrix element in Eq. (34.2), and Eq. (34.2) becomes

$$\Phi(t)=\sum_{ijkl}\Upsilon_{1i}\Upsilon_{1p_i}\langle ip_i|H_F^+|jp_j\rangle\langle kp_j|\exp\left(\frac{\mathrm{i}tp_j^2}{2\mathrm{m}_1}\right)H_F\exp\left(\frac{-\mathrm{i}tp_i^2}{2\mathrm{m}_1}\right)|lp_i\rangle$$
$$\cdot\exp\left(\frac{-\mathrm{i}tk^2}{\mathrm{m}_1}+\mathrm{i}t\omega_{k1}\right)\exp\left[\mathrm{i}t(\mathbf{p}_i-\mathbf{p}_k)\cdot\frac{\mathbf{k}}{\mathrm{m}_1}\right]\left[\langle jp_i-k|\mathfrak{U}^{-1}(t)|kp_k\rangle\right.$$
$$\left.\cdot\langle lp_1+k|\mathfrak{U}(t)|ip_i\rangle\right]_{\text{av}} \tag{34.4}$$

where $\omega_{kl}=E_k-E_l$, the E the internal energies of the radiator.

We now consider an isolated line so that $k\to F$, $l\to I$, F and I referring to final and initial states for the line. Eq. (34.4) now reduces to

$$\Phi(t)=\exp[\mathrm{i}\omega_{IF}t]\Upsilon_{1I}|\langle I|\mathbf{d}_f|F\rangle^2\Phi'(t) \tag{34.5a}$$

$$\Phi'(t)=e^{-\mathrm{i}tk^2/\mathrm{m}_1}\int d\mathbf{p}_k\int d\mathbf{p}_i\,\Upsilon_{1p_i}\exp\left(\mathrm{i}t(\mathbf{p}_i-\mathbf{p}_k)\cdot\frac{\mathbf{k}}{\mathrm{m}_1}\right)$$
$$\cdot\langle p_i|e^{-\mathrm{i}\mathbf{k}\cdot\mathbf{r}}e^{\mathrm{i}\mathbf{k}\cdot\mathbf{r}(t)}|p_i\rangle\left[\langle Fp_i-k|\mathfrak{U}^{-1}(t)|Fp_k\rangle\right.$$
$$\left.\cdot\langle Ip_k+k|\mathfrak{U}(t)|Ip_i\rangle\right]_{\text{av}} \tag{34.5b}$$

since

$$\sum_j\langle p_i|e^{-\mathrm{i}\mathbf{k}\cdot\mathbf{r}}|p_j\rangle\langle p_j|e^{\mathrm{i}tp_j^2/\mathrm{m}_1}e^{\mathrm{i}\mathbf{k}\cdot\mathbf{r}}e^{-\mathrm{i}tp_i^2/\mathrm{m}}|p_i\rangle$$
$$=\sum_j\langle p_i|e^{-\mathrm{i}\mathbf{k}\cdot\mathbf{r}}|p_j\rangle\langle p_j|e^{\mathrm{i}H't}e^{\mathrm{i}\mathbf{k}\cdot\mathbf{r}}e^{-\mathrm{i}H't}|p_i\rangle$$
$$=\langle p_i|e^{-\mathrm{i}\mathbf{k}\cdot\mathbf{r}}e^{-\mathrm{i}\mathbf{k}\cdot\mathbf{r}(t)}|p_i\rangle,\qquad H'=K^{(1)}+K^{(2)}+\mathrm{V}$$

Here K is the translational Hamiltonian for (1) the emitter and (2) the broadeners; V is the pair potential averaged over internal states.

With Cattani we now consider the situation wherein there is negligible perturbation of the internal levels of the radiator. Such a condition should not be construed as equivalent to pure Doppler broadening, although of course it may be. It does, however, correspond to $V_{12}=0$ which, in turn, means $\mathfrak{U}(t)=1$. Thus $[\cdots]_{av}$ becomes δ_{p_i-k,p_k}, and

$$\Phi'(t)=\int dp_i \Upsilon_{1p_i}\langle p_i|e^{-i\mathbf{k}\cdot\mathbf{r}}e^{i\mathbf{k}\cdot\mathbf{r}(t)}|p_i\rangle \tag{34.6}$$

We will recall that, if the commutator $[\mathbf{r},\mathbf{r}(t)]=0$, the average that is being carried out in Eq. (34.6) to include the matrix element over the exponential is equal to the average in the exponential. If the classical path treatment is acceptable, the commutator is zero. Thus the matrix element and Boltzmann average might be moved into the exponential and some means of evaluation subsequently sought. However, it is not in this fashion that Cattani goes about it. Instead, (1) taking $k\equiv e_x k$ and (2) assuming that the resulting x and $x(t)$ are random Gaussian variables, Cattani obtains

$$\begin{aligned}\Phi'(t)&=\exp\left\{-\tfrac{1}{2}k^2\langle[x-x(t)]^2\rangle+\tfrac{1}{2}k^2[x,x(t)]\right\}\\ &\Rightarrow\exp\left\{-\tfrac{1}{2}k^2\langle[x-x(t)]^2\rangle\right\},\end{aligned} \tag{34.7}$$

where now the brac operator infers both Boltzmann averaging and matrix element. Using a more or less standard, normalized-in-a-box wave function for matrix element evaluation, Cattani obtains

$$\langle[x-x(t)]^2\rangle=\tfrac{2}{3}vlt \tag{34.8}$$

for the case where the radiator collides many times during the lifetime of its upper level. Here $\bar{v}$ is the mean relative velocity, l the mean free path. Therefore, from Eqs. (34.5) and (34.8) we see that

$$\begin{aligned}&\Phi(t)\propto\exp\left[it(\omega_{IF}-k^2vl/3)\right]\Rightarrow\\ &I(\omega)\propto\frac{\omega^2vl/3}{(\omega_{IF}-\omega)^2+(\omega^2\bar{v}l/3)^2}\end{aligned} \tag{34.9}$$

Cattani inserts the natural damping factor, but this is of no interest to us.

From Eq. (34.9) we see that, since the pressure of the buffer gas is proportional to the mean free path, the half width decreases with increasing pressure, (the Dicke effect). The half width decreases within reason, of course, for in this Case I we have supposed $V_{12}=0$ so that true "pressure broadening" effects are absent. We see also that the linewidth is dependent on the frequency, a situation common to Doppler broadening in the most elementary of treatments. For many collisions during the lifetime of the upper level this situation leads to Eq. (34.8).

Now when there are no collisions during the lifetime of the upper level, that is, when the intercollision time becomes infinite, Cattani finds that

$$\lim_{T\to\infty} \langle [x - x(t)]^2 \rangle = (\bar{v}t)^2/3$$

and his treatment reduces to the classic Doppler effect as it should. From immediately subsequent to Eqs. (34.5) we have been considering the case where the collisional perturbation of the internal levels is negligible, $\mathrm{V}_{12}=0$.

When the Doppler effect is taken as negligible, $\mathbf{k}=0$, $\mathrm{V}_{12}\neq 0$, Eqs. (34.5) reduce to the interruption result as we would anticipate; so much for the second of the cases considered by Cattani.

In the general case Cattani finds that the factor which controls the Dicke effect—a factor effectively identical to Eq. (34.6)—is strongly correlated with the perturbation of the internal states of the radiator. The correlation is so strong, in fact, that when we take the emitter mass so large in comparison to the perturber mass that the emitter is at rest—and hence the Doppler effect disappears—we find the correlation function is

$$\Phi'(t) = \exp\left[-t\omega^2\bar{v}l/3\right]\exp\left[-\mathrm{i}t(\Delta_{IF} - \mathrm{i}\delta_{IF})\right] \tag{34.10}$$

wherein we recognize that the first factor describes the Dicke effect. Therefore, the Dicke effect contribution is present even in the absence of the Doppler effect per se.

Zaidi (290) has shown that the Dicke narrowing effect is described by what he calls the vertex function of the hole-particle propagator. Consider, for example, Eqs. (31.15). Clearly, Q_{10} here was a vertex. What Zaidi calls the "vertex function" is everything under the integral sign—$T_{11}\mathrm{G}_{ee}\mathrm{G}_{ee}T_{00}$—save the vertex.

Zaidi's approach is, of necessity, the same as that of Ross, although he carries it through using a notation wherein the self-energy operator, Σ, remains clearly in evidence. In Eq. (31.15a), for example, we have G_{11} and G_{00}. From Eq. (20.6) G_{11}^{-1} may rather obviously be written as

$$\mathrm{G}_{11}^{-1}(p) = \omega_p - p^2/2\mathrm{m} - E_1 + \mu_1 - \Sigma_1 \tag{34.11}$$

where $\Sigma_1 \equiv \Sigma_{11}$. Now we may state Zaidi's entire thesis: the collision broadening arises from the SEP's of the propagators whereas the Dicke narrowing arises from the vertex function. This may be illustrated by considering only the first term in Eq. (31.15a) which, by Eq. (31.1), must be integrated over $\mathbf{p}$ in order to obtain the line shape, thus:

$$\chi \sim \int dp\, f(p)\left(\omega_p - \frac{p^2}{2\mathrm{m}} - E_1 + \mu - \Sigma_1\right)^{-1}\left(\omega_p - \omega_k - \frac{(p-k)^2}{2\mathrm{m}} - E_0 + \mu - \Sigma_0\right)^{-1}$$

$$= \text{const}\int d\mathbf{p}\, f(p)\left(\omega_k - \omega_0 + \frac{k^2}{2\mathrm{m}} - \frac{\mathbf{p}\cdot\mathbf{k}}{\mathrm{m}} - \Sigma_1 + \Sigma_0\right)^{-1} \tag{34.12}$$

where we have evaluated the ω-integral at the ω_p pole [cf. Eq. (31.18)]. We see that the line shift and width must arise from the self-energy operator.

We next consider Eq. (31.15a) in its entirety. (The G_{ee} we suppose replaceable by some other propagator when the perturbers are not electrons.) We are quite safe in presuming that some argument may be generated which allows us to bring vertex, Q_{10}, and vertex function, Γ_{10}, out from under the integrals on the right side of Eq. (31.15a). Therefore, assuming that the ω_p-integration that led to Eq. (34.12) has been carried out again and ignoring the **p**-integration, we see that

$$\chi = A^{-1}\{1-\chi\Gamma_{10}\}$$

where A is the expression in parentheses on the extreme right of Eq. (34.12), the p-integral over Q_{10} having wrought the change $Q \rightarrow \chi$. Thus we obtain a line shape

$$\chi \sim \left(\omega_k - \omega_0 + \frac{k^2}{2\mathrm{m}} - \frac{\mathbf{p}\cdot\mathbf{k}}{\mathrm{m}} - \Sigma_1 + \Sigma_0 - \Gamma_{10}\right)^{-1} \tag{34.13}$$

The collision width will be given by $\mathcal{I}[\Sigma_1 - \Sigma_0 - \Gamma_{10}]$. It is apparent then that, if $\Sigma_1 = \Sigma_0$ so that the collisional perturbation of the two states involved in the radiation is the same, the pressure broadening effects will disappear (to this approximation), leaving only the Dicke narrowing given by Γ_{10}. (It would admittedly be something of an accomplishment to narrow a line of zero width which we seem to have here.) Even if the effects of the collisional perturbations do not disappear, we see that the vertex function will induce a collisional narrowing of the spectral line critically dependent of course on the T-matrix elements and perturber propagators through, say, Eq. (30.5b).

35 An Interruption Theory Treatment of the Doppler-Pressure Effect

Smith, Cooper, Chappell, and Dillon (SCCD) (251) developed an essentially Liouvillian treatment of an interruption theory of combined pressure and Doppler broadening which, since it includes the Fano approximation relating to density matrix separability, is of perhaps more than average interest. That these authors seek a correlation function of the following form is not particularly surprising

$$\Phi(t) = \mathrm{Tr}\{\mathbf{d}\cdot\mathbf{D}(t)\} \tag{35.1a}$$

where

$$\begin{aligned}\mathbf{D}(t) &= \mathrm{Tr}_{BS_t}\{e^{i\boldsymbol{\kappa}\cdot\mathbf{r}}e^{-itH}(\Upsilon e^{i\boldsymbol{\kappa}\cdot\mathbf{r}}\mathbf{d})e^{itH}\} \\ &= \mathrm{Tr}_{BS_t}\{e^{-i\boldsymbol{\kappa}\cdot\mathbf{r}}e^{-it\mathfrak{L}_0}\mathfrak{U}_{BS}(t)[\Upsilon e^{i\boldsymbol{\kappa}\cdot\mathbf{r}}\mathbf{d}]\}\end{aligned} \tag{35.1b}$$

In Eqs. (35.1b) $\mathfrak{L}_0$ is the Liouvillian for one unperturbed radiator plus the n unperturbed perturbers sharing the "cell" assigned to that radiator. $\mathfrak{U}_{BS}$ is a TDO in Liouville space which we should perhaps clarify somewhat. Given an interaction Liouville operator, $\mathfrak{V}$, such that, say

$$e^{-it\mathfrak{V}}(\Upsilon e^{i\boldsymbol{\kappa}\cdot\mathbf{r}}\mathbf{d})e^{it\mathfrak{V}} = e^{-it\mathfrak{V}}(\Upsilon e^{i\boldsymbol{\kappa}\cdot\mathbf{r}}\mathbf{d})$$

in familiar form, $\mathfrak{U}=\exp[-\mathrm{i}t\mathfrak{V}]$. [The reader should keep in mind, in obtaining Eq. (35.1b), the cyclical commutability of the operators in a trace.] It is important to remark here that the interparticle interaction, $\mathfrak{V}$, only contemplates interactions between radiator and perturbers; there is no interperturber interaction.

These authors, or some of them (250), concluded earlier that the separable density matrix is quite acceptable as long as $\Delta\omega<kt$, a conclusion that appears intuitively acceptable. In this development then, they took Υ as $\Upsilon_S\Upsilon_{Bt}\Upsilon_B$ where the bath density matrix has been further split into a translational Υ_{Bt}, and an internal portion Υ_B. This of course reduces $D(t)$ somewhat:

$$\mathbf{D}(t)=\mathrm{Tr}_{St}\{e^{-\mathrm{i}\kappa\cdot\mathbf{r}}e^{-\mathrm{i}t\mathfrak{L}_s}\langle\mathfrak{U}_{BS}(t)\rangle[\Upsilon_S\Upsilon_{Bt}e^{\mathrm{i}\kappa\cdot\mathbf{r}}\mathbf{d}]\} \tag{35.2a}$$

where

$$\langle\mathfrak{U}_{BS}\rangle=\mathrm{Tr}_B\{\mathfrak{U}_{BS}(t)\Upsilon_B\} \tag{35.2b}$$

Here the authors have used the fact that H_B is diagonal in the states in which the trace is being evaluated so that $\exp[-\mathrm{i}t\mathfrak{L}_B]=1$ in these states. Thus for an arbitrary operator

$$\mathrm{Tr}_B\{\exp[-\mathrm{i}t\mathfrak{L}_0]A\}=\exp[-\mathrm{i}t\mathfrak{L}_S]\,\mathrm{Tr}_B\{A\} \tag{35.2c}$$

At this point it becomes necessary for us to define more precisely our Liouvillian TDO.

Because the radiator-perturbers (system-bath) interaction is a sum over pair potentials,

$$\begin{gathered}\mathfrak{U}_{SB}(t)=\mathsf{T}\prod_j\mathfrak{U}(j,t),\ \mathfrak{U}(j,t)=\mathsf{T}\exp\left\{-\mathrm{i}\int_0^t\hat{\mathfrak{V}}(j,t')\,dt'\right\}\\ \hat{\mathfrak{V}}(j,t')=\exp[\mathrm{i}t\mathfrak{L}_0]\mathfrak{V}(j)\exp[-\mathrm{i}t\mathfrak{L}_0]\end{gathered} \tag{35.3}$$

and so on, where j runs over the perturbers in the cell associated with the radiator. Advantage is taken of the form by SCCD in order to introduce the operator

$$\varphi(j,t)=\mathfrak{U}(j,t)-1\Rightarrow \tag{35.4a}$$

$$\mathfrak{U}_{SB}(t)=\mathsf{T}\prod_j[1+\varphi(j,t)] \tag{35.4b}$$

The perturbers have been assumed statistically independent so that each interaction $\hat{\mathfrak{V}}$ will be statistically independent of all the others. (We will recall the assumption that there was no interperturber interaction.) Therefore, the $\mathfrak{U}(j,t)$ operators and hence the φ operators may be averaged separately. All perturbers are the same so that, by the arguments leading to Eq. (14.6)

$$\begin{aligned}\langle\mathfrak{U}_{BS}(t)\rangle&=\mathsf{T}[1+\langle\varphi(t)\rangle]^n=\mathsf{T}\exp\{n\langle\varphi(t)\rangle\}\\&=\mathsf{T}\exp\{n\langle\mathfrak{U}(t)-1\rangle\}\end{aligned} \tag{35.5a}$$

If we suppose the translational average to be over a Boltzmann momentum distribution, a factor $Nf(p)/n$, N being perturber density, is going to enter. Therefore,

$$\langle \mathfrak{U}_{BS}(t)\rangle = \mathsf{T}\exp\left\{N\sum\langle pi|[\mathfrak{U}(t)-1]\Upsilon_\alpha f|pi\rangle\right\} \tag{35.5b}$$

Here $|p\rangle$ is the translational, $|i\rangle$ the internal state of a perturber. Υ_α is diagonal in the $|p_i\rangle$, its ith element telling us the probability for finding the perturber in state i.

Now we introduce a function Ξ that will serve as a vehicle for the introduction of the interruption approximation. In order to introduce it, SCCD first rewrite Eq. (35.3) for $\mathfrak{U}(t)$.

$$\begin{aligned}\mathfrak{U}(t)-1 &= -\mathrm{i}\int_0^t \hat{\mathfrak{V}}(t')\,dt' - \int_0^t \hat{\mathfrak{V}}(t')\,dt' \int_0^{t'} \hat{\mathfrak{V}}(t'')\,dt'' + \ldots \\ &= -\mathrm{i}\int_0^t \hat{\mathfrak{V}}(t')\mathfrak{U}(t')\,dt' \end{aligned} \tag{35.6a}$$

Since $\mathfrak{U}$ is a Liouville-space TDO, it follows that

$$\mathfrak{U}(t,0) = e^{\mathrm{i}t\mathfrak{L}_0}\mathfrak{U}(0,-t)e^{-\mathrm{i}t\mathfrak{L}_0} \tag{35.6b}$$

Eqs. (35.6) are used to rewrite the curly-bracketed factor on the right of Eq. (35.5b) as

$$[\mathfrak{U}(t)-1]_{\mathrm{av}} = -\mathrm{i}\int_0^t e^{\mathrm{i}t'\mathfrak{L}_s}[\mathfrak{V}(0)\mathfrak{U}(0,t')]_{\mathrm{av}}e^{-\mathrm{i}t'\mathfrak{L}_s}\,dt' \tag{35.7}$$

wherein Eq. (35.2c) and $[\mathfrak{L}_B, \Upsilon_\alpha f]=0$ have been used. We are in a position to assume that there is an average collision time τ_c such that, for $t>\tau_c$,

$$[\mathfrak{V}(0)\mathfrak{U}(0,-t)]_{\mathrm{av}} = [\mathfrak{V}(0)\mathfrak{U}(0,-\tau_c)]_{\mathrm{av}} = [\mathfrak{V}(0)\mathfrak{U}(0,-\infty)]_{\mathrm{av}} \tag{35.8}$$

a familiar assumption. At this point we define the operator

$$\Xi = -\mathrm{i}[\mathfrak{V}(0)\mathfrak{U}(0,-\infty)]_{\mathrm{av}} \tag{35.9}$$

which yields the following approximation for Eq. (35.7):

$$[\mathfrak{U}(t)-1]_{\mathrm{av}} = \int_0^t e^{\mathrm{i}t'\mathfrak{L}_s}\Xi e^{-\mathrm{i}t'\mathfrak{L}_s}\,dt' \tag{35.10a}$$

The interruption approximation amounts to letting the correction term

$$\int_0^{\tau_c}\{\mathrm{i}[\mathfrak{V}(0)\mathfrak{U}(0,-t')]_{\mathrm{av}} - \Xi\}\,dt' \tag{35.10b}$$

(which would of course render Eq. (35.10a) precise) be zero. For reasons that will shortly become apparent, the object now is to write Eq. (35.10a) as Ξt. The reader may wish to use the interruption approximation in order to justify the attainment of this objective. We now write Eq. (35.10a) as

$$\begin{aligned}\langle \mathfrak{U}_{SB}(t)\rangle &= \mathsf{T}\exp\left\{\int_0^t e^{\mathrm{i}t'\mathfrak{L}_s}\Xi e^{-\mathrm{i}t'\mathfrak{L}_s}\,dt'\right\} \\ &= e^{\mathrm{i}t\mathfrak{L}_s}\exp\{t\Xi - \mathrm{i}t\mathfrak{L}_s\}. \end{aligned} \tag{35.11}$$

Where it is supposed that $\exp(\tau_c\Xi)\sim 1$.

When we substitute Eq. (35.11) into Eq. (35.2a), the reason for desiring the former in the form in which we obtained it is that the factor $\exp[it\mathfrak{L}_s]$ happily cancels a corresponding factor in Eq. (35.2a) and we are going to obtain a line shape expression in the familiar resolvant form. In any case,

$$\mathbf{D}(t)=\mathrm{Tr}_{St}\{e^{-i\kappa\cdot\mathbf{r}}\exp[t\Xi-it\mathfrak{L}_s][\Upsilon_S\Upsilon_{Bt}e^{i\kappa\cdot\mathbf{r}}\mathbf{d}]\} \tag{35.12}$$

SCCD could follow the semi-standard technique and take the Fourier transform of this. If they did, however, they would have a Ξ in the denominator of the line shape expression that is not diagonal in the translational states of the radiator. Thus they would have an infinite matrix to invert in order to obtain the matrix elements of Ξ in the denominator as we have noted. As a result, a different *modus operandi* is chosen.

As $K_s=p^2/2\mathrm{m}=-\nabla^2/2\mathrm{m}, |p\rangle\sim\exp[i\mathbf{p}\cdot\mathbf{r}], |p+\kappa\rangle\sim\exp[i(\mathbf{p}+\boldsymbol{\kappa})\cdot\mathbf{r}]$ we will agree that

$$e^{-i\kappa\cdot\mathbf{r}}K_s e^{i\kappa\cdot\mathbf{r}}|p+\kappa\rangle=(p+\kappa)^2|p+\kappa\rangle/2\mathrm{m}=(K_s+\boldsymbol{\kappa}\cdot\mathbf{p}/m+\kappa^2/2\mathrm{m})|p+\kappa\rangle$$

and that therefore,

$$e^{-i\kappa\cdot\mathbf{r}}\mathfrak{L}_s e^{i\kappa\cdot\mathbf{r}}=\mathfrak{L}_r+\boldsymbol{\kappa}\cdot\mathbf{p}/\mathrm{m}+\kappa^2/2\mathrm{m} \tag{35.13}$$

Next we let

$$\langle i|D(t)|j\rangle=\sum_p F_{ij}(p,t) \tag{35.14a}$$

where

$$F_{ij}(p,t)=\langle ip|F(t)|jp\rangle \tag{35.15b}$$

and

$$F(t)=e^{-i\kappa\cdot\mathbf{r}}\exp[t\Xi-it\mathfrak{L}_s][\Upsilon_s\Upsilon_{Bt}e^{i\kappa\cdot\mathbf{r}}\mathbf{d}] \tag{35.15c}$$

We take the temporal derivative of $F(t)$, apply Eq. (35.13) to the result, and rearrange slightly in order to obtain

$$\left[\frac{\partial}{\partial t}+i\left(\mathfrak{L}_s+\frac{\boldsymbol{\kappa}\cdot\mathbf{p}}{\mathrm{m}}+\frac{\kappa^2}{2\mathrm{m}}\right)\right]F(t)=\Lambda(F(t)) \tag{35.16a}$$

where

$$\Lambda(F(t))=(e^{-i\kappa\cdot\mathbf{r}}\Xi e^{i\kappa\cdot\mathbf{r}})F(t) \tag{35.16b}$$

SCCD now take matrix elements of Eq. (35.16a) in order to obtain their general expression,

$$\left[\frac{\partial}{\partial t}+i\left(\omega_{ij}+\frac{\boldsymbol{\kappa}\cdot\mathbf{p}}{\mathrm{m}}\right)\right]F_{ij}(p,t)=\langle ip|\Lambda|jp\rangle \tag{35.17}$$

Eq. (35.17) may be considered as the equation for an F_{ij} having a natural frequency ω_{ij} plus the Doppler shift $\boldsymbol{\kappa}\cdot\mathbf{p}/\mathrm{m}$ and perturbed by a collision term, the matrix element of Λ on the right side, brought about obviously enough by radiator-perturber collisions. SCCD call Δ the "collision integral."

In order to obtain the spectral line shape then, we determine F_{ij} from Eq. (35.17) and take the Fourier transform of the result. As is universally true, however, this can only be done for certain limiting cases. SCCD, after reexpressing the collision integral in terms of Moller operators, consider the limiting cases of (1) pure Doppler broadening, (2) small perturbers, and (3) no lower state interaction. (1) leads to a Fredholm equation for F which agrees with the earlier results of Rautian and Sobelman (222), that is, with those for collisional narrowing. (2) specifically refers to electron perturbers, $\mathrm{m}_e \ll \mathrm{m}$, m being the mass of the radiator, and it means that the radiator trajectory will be effectively unaffected by the broadening collisions. The Voigt profile is the result obtained in this case. (3) has the practical value of being often encountered and the theoretical advantage of rendering the equation for F considerably more tractable.

The approach to Doppler-foreign-gas-interruption broadening in this section is perhaps of greatest interest for the form in which what is essentially the correlation function is approached, that is, through Eq. (35.17). We now turn to still another approach to the Doppler-pressure problem; this time the reduced density matrix is the principal object of our attentions.

36 The Reduced Density Matrix and the Liouvillian Operator

That the frequency-dependent absorption coefficient and hence the spectral line shape may be obtained from the imaginary part of the susceptibility we know. We also know that this susceptibility is in turn a function of the statistical average of the dipole moment and hence of whatever the appropriate density matrix may be. In a series of papers Roney (232–234) has studied the low-density Doppler-foreign-gas broadening by analyzing the one-particle reduced density matrix which appropriately enters this dipole moment average. Since we have not hitherto encountered a "reduced" density matrix per se, let us begin by inquiring as to just what it may be. [We remark in passing the interesting kinetic theory treatment of Doppler-pressure by Hess (142). Many of the ideas are to be encountered in what follows.]

We will agree that the n-particle density matrix obeys the equation,

$$\mathrm{i}\dot{\Upsilon}^{(n)} = \left[H^{(n)}, \Upsilon^{(n)} \right] \tag{36.1a}$$

where $H^{(n)}$ is the n-particle Hamiltonian,

$$H^{(n)} = \sum_{i=1}^{n} H_0^{(1)}(1) + \frac{1}{2} \sum_{i \neq j} \mathrm{V}(i,j) + \sum_{i=1}^{n} H_F^{(1)} e^{-\mathrm{i}\omega t} \tag{36.1b}$$

where $H_0^{(1)}$ is the one-particle Hamiltonian involving translational and internal motions, $\mathrm{V}(i,j)$ the interaction between particles, and $H_F^{(1)}$ the interaction of the electromagnetic field with one particle. We may always extract an $\exp[-\mathrm{i}\omega t]$ from the vector potential in the particle-field interaction, and Roney finds it convenient to do just this.

Now we are going to restrict ourselves to binary collisions, which is just another way of saying that the radiator will not interact with more than one perturber at a time. Thus the following definition of a one-particle reduced density matrix will be quite appropriate:

$$\Upsilon^{(1)}(1)=(n-1)^{-1}\mathrm{Tr}_2\Upsilon^{(2)}(1,2) \tag{36.2a}$$

The trace is over both the translational and internal coordinates of particle 2. The two-particle reduced density matrix, on the other hand, will be

$$\Upsilon^{(2)}(1,2)=\left[(n-2)!\right]^{-1}\mathrm{Tr}_{3,\ldots,n}\Upsilon^{(n)}(1,2,\ldots,n) \tag{36.2b}$$

$\Upsilon^{(n)}$ is taken as normalized to $n!$.

The superscript n in Eq. (36.1a) can take on any value, beginning with $n=2$,

$$\begin{aligned}\Upsilon^{(2)}&=\left[H^{(2)},\Upsilon^{(2)}\right]\\&=\left[\left(H_0^{(1)}(1)+H_0^{(1)}(2)\right),\Upsilon^{(2)}\right]+\left[\mathsf{V}(1,2),\Upsilon^{(2)}\right]\\&\quad+\left[\left(H_F^{(1)}(1)+H_F^{(1)}(2)\right),\Upsilon^{(2)}\right]e^{-i\omega t}\end{aligned}$$

We take the trace of this equation over the unperturbed states of particle 2. Because the $H_0^{(1)}$ and $H_F^{(1)}$ are all diagonal in the unperturbed states.

$$\mathrm{Tr}_2\left[H_0^{(1)}(i),\Upsilon^{(2)}\right]=\left[H_0^{(1)}(i),\mathrm{Tr}_2\Upsilon^{(2)}\right]=\left[H_0^{(1)}(i),\Upsilon^{(1)}\right]$$

and so on. By definition $H_0^{(1)}(2)$ commutes with $\Upsilon^{(1)}(1)$, so that we obtain

$$\begin{aligned}\mathrm{i}\dot{\Upsilon}^{(1)}(1)&=\left[H_0^{(1)}(1),\Upsilon^{(1)}(1)\right]-\left[H_F^{(1)}(1),\Upsilon^{(1)}(1)\right]e^{-i\omega t}\\&=\mathrm{Tr}_2\left[\mathsf{V}(1,2),\Upsilon^{(2)}(1,2)\right]\end{aligned} \tag{36.3a}$$

Similarly,

$$\begin{aligned}\mathrm{i}\dot{\Upsilon}^{(2)}(1,2)&=\left[H_0^{(2)}(1,2),\Upsilon^{(2)}(1,2)\right]+\left[\left(H_F^{(1)}(1)\right.\right.\\&\quad\left.\left.+H_F^{(2)}(2)\right),\Upsilon^{(2)}(1,2)\right]e^{-i\omega t}+\left[\mathsf{V}(1,2),\Upsilon^{(2)}(1,2)\right]\\&\quad+\mathrm{Tr}_3\left[\left(\mathsf{V}(1,3)+\mathsf{V}(2,3)\right),\Upsilon^{(3)}(1,2,3)\right]\end{aligned} \tag{36.3b}$$

where $H_0^{(2)}(1,2)=H_0^{(1)}(1)+H_0^{(1)}(2)$.

Because only low-pressure binary collisions are to be considered, the last term in Eq. (36.3b), which involves particle 3, is dropped. Further, in consonance with the restriction to foreign-gas broadening, particle 1 is taken as the radiator, particle 2 as the perturber, the density of the former, N_1, being much less than that of the latter, N_2. Particle 2 is supposed quite incapable of absorbing any frequencies near that of the field, ω, so that terms involving $H_F^{(1)}(2)$ may be dropped. As has been our custom, we consider the thermal bath of perturbers to remain in equilibrium, the effect of exchanges of energy between a perturber and the radiator being immediately dissipated throughout the bath. We now proceed to the Liouvillian formulation of Eq.s (36.3):

$$\mathrm{i}\dot{\Upsilon}^{(1)}(1,t)-\mathfrak{L}_0^{(1)}\Upsilon^{(1)}(1,t)-\mathfrak{L}_F\Upsilon^{(1)}(1,t)e^{-\mathrm{i}\omega t}=\mathrm{Tr}\left[\mathfrak{B}\Upsilon^{(2)}(1,2,t)\right] \quad (36.4\mathrm{a})$$

$$\mathrm{i}\dot{\Upsilon}^{(2)}(1,2,t)-\mathfrak{L}^{(2)}\Upsilon^{(2)}(1,2)=\mathfrak{L}_F^{(1)}\Upsilon^{(2)}(1,2,t)e^{-\mathrm{i}\omega t} \quad (36.4\mathrm{b})$$

where $\mathfrak{L}^{(2)}=\mathfrak{L}_0^{(2)}+\mathfrak{B}$.

Roney follows the method of Snider (251a) in solving Eq. (36.4b) for $\Upsilon^{(2)}(1,2,t)$ and substituting this result into Eq. (36.4a), the ultimately desired quantity being $\Upsilon^{(1)}$. Now we note that the form of H_F utilized by Roney is

$$H_F^{(1)}(1)e^{-\mathrm{i}\omega t}=\mathbf{d}\cdot\mathbf{e}e^{-\mathrm{i}\omega t+\mathrm{i}\boldsymbol{\kappa}\cdot\mathbf{r}} \quad (36.5)$$

where $\mathbf{e}$ is of course the polarization of the electric vector and $\mathbf{d}$ is the electric moment. Roney wishes no restrictions on ω but only terms linear in $\mathbf{e}$.

$\Upsilon^{(2)}(1,2,t)$ is to be expressed in terms of $\Upsilon^{(2)}(1,2,t_0)$, t_0 being the time earlier than t at which the electric field was switched on. Prior to switching and, indeed, at $t_0, \Upsilon^{(2)}(1,2,t_0)$ is an equilibrium two-particle density matrix, an important point to keep in mind. Low densities have been presumed; hence the separability of the density matrix, $\Upsilon^{(2)}(1,2)$, into the product form, $\Upsilon^{(1)}(1)\Upsilon^{(1)}(2)$, a far from unfamiliar procedure, should hardly be difficult to justify.

As Roney points out, H_F may not be strictly time independent during a collision since induction effects could lead to a variation in $\mathbf{d}$ and hence H_F. Nevertheless, it may quite accurately be assumed so in obtaining the following solution to Eq. (36.4b):

$$\Upsilon^{(2)}(t)=\exp\left\{-\mathrm{i}\left[(t-t_0)\mathfrak{L}+\mathrm{i}\mathfrak{L}_F(e^{-\mathrm{i}\omega t}-e^{-\mathrm{i}\omega t})/\omega\right]\right\}\Upsilon^{(2)}(t_0) \quad (36.6)$$

The exponential in Eq. (36.6) is now to be expressed as a time development about $\mathfrak{L}_F$ as a perturbation. In order to do this we put $t-t_0=\alpha$, $\mathrm{i}\mathfrak{L}=-\mathfrak{A}$, $\mathrm{i}L_F=-\mathfrak{B}$ and $\Upsilon^{(2)}=P$, and Eq. (36.6) becomes

$$P(\alpha,t_0)=\exp\left[\mathfrak{A}\alpha+i\mathfrak{B}(e^{-\mathrm{i}\omega(\alpha+t_0)}-e^{-\mathrm{i}\omega t_0})/\omega\right],$$

so that $P(0,t_0)=1$. Differentiating with respect to α yields

$$\frac{d}{d\alpha}P(\alpha,t_0)=\left[\mathfrak{A}+\mathfrak{B}e^{-\mathrm{i}\omega(\alpha+t_0)}\right]P(\alpha,t_0)$$

If we let $Q(\alpha,t_0)=e^{-\mathfrak{A}\alpha}P(\alpha,t)$, $Q(0,t_0)=1$, we may obtain

$$\frac{d}{d\alpha}Q(\alpha,t_0)=e^{-\mathrm{i}\omega(\alpha+t_0)}e^{-\mathfrak{A}\alpha}\mathfrak{B}e^{\mathfrak{A}\alpha}Q(\alpha,t_0) \quad (36.7)$$

If we integrate Eq. (36.7) and then carry out an iterated solution of the result through first order, we obtain

$$Q(\alpha,t_0)=1+\int_0^\alpha e^{-\mathrm{i}\omega(\alpha'+t_0)}e^{-\mathfrak{A}\alpha'}\mathfrak{B}e^{\mathfrak{A}\alpha'}Q(\alpha',t_0)\,d\alpha'$$

$$\doteq 1+\int_0^\alpha e^{-\mathrm{i}\omega(\alpha'+t_0)}e^{-\mathfrak{A}\alpha'}\mathfrak{B}e^{\mathfrak{A}\alpha'}\,d\alpha' \quad (36.8)$$

Equation (36.8) is multiplied through on the left by $\exp(\mathfrak{A}\alpha)$ in order to obtain $P(\alpha, t_0)$, the resulting expression then transformed back to the original notation. We have obtained the density matrix to first order in e and hence in $\mathfrak{L}_F$ as

$$\Upsilon^{(2)}(t) = \left(e^{-\mathrm{i}\mathfrak{L}t} - \mathrm{i}\int_0^t e^{-\mathrm{i}\mathfrak{L}t} e^{-\mathrm{i}\mathfrak{L}(t-t')} \mathfrak{L}_F e^{-\mathrm{i}\mathfrak{L}t'}\, dt' \right) \Upsilon^{(2)}(0) \tag{36.9}$$

where $t_0 = 0$.

If, in Eq. (36.9) we let $\mathfrak{L} \to \mathfrak{L}_0$, there is obviously no particle interaction in the bracketed expression. We have already supposed that $\Upsilon^{(2)}(0) = \Upsilon^{(1)}(1,0)\Upsilon^{(1)}(2,0)$, so that at some later time, t, $[\ ]\Upsilon^{(2)}(0) = \Upsilon^{(1)}(1,t)\Upsilon^{(1)}(2,t)$ with $\mathfrak{L} = \mathfrak{L}_0$ in $[\ ]$. Now Roney is in a position to expand Eq. (36.9) using $\mathfrak{V}$ as the perturbation. Straightforward integration should convince the reader that the following is an identity:

$$e^{-\mathrm{i}\mathfrak{L}t} = e^{-\mathrm{i}\mathfrak{L}_0 t} - \mathrm{i}\int_0^t dt'\, e^{-\mathrm{i}\mathfrak{L}t'} \mathfrak{V} e^{-\mathrm{i}\mathfrak{L}_0(t-t')} \tag{36.10}$$

Equation (36.10) is substituted for the first term within the square bracket of Eq. (36.9) and for the exponential following $\mathfrak{L}_F$ within that equation. A four-term expression for Eq. (36.9) results. Eq. (36.10) is applied to the third of these four terms, and, after a certain amount of regrouping and integration variable transformation, we obtain

$$\Upsilon^{(2)}(t) = \Upsilon^{(1)}(1,t)\Upsilon^{(1)}(2,t) - \mathrm{i}\int_0^t dt'\, e^{-\mathrm{i}\mathfrak{L}t} \mathfrak{V} \Upsilon^{(1)}(1, t-t')\Upsilon^{(1)}(2, t-t'')$$

$$- e^{-\mathrm{i}\omega t}\int_0^t dt'\, e^{\mathrm{i}\omega t'} e^{-\mathrm{i}\mathfrak{L}t'} \mathfrak{L}_F \int_0^{t-t'} dt''\, e^{-\mathrm{i}\mathfrak{L}t''} \mathfrak{V} \exp[-\mathrm{i}\mathfrak{L}_0(t-t'-t'')]^z \Upsilon^{(2)}(0) \tag{36.11}$$

At which point Royer goes over to the resolvant formalism by expressing the exponential as

$$e^{-\mathrm{i}\mathfrak{L}t} = -(2\pi\mathrm{i})^{-1}\int dz\, e^{-\mathrm{i}zt}(z-\mathfrak{L})^{-1} \tag{36.12a}$$

where the z-plane contour is

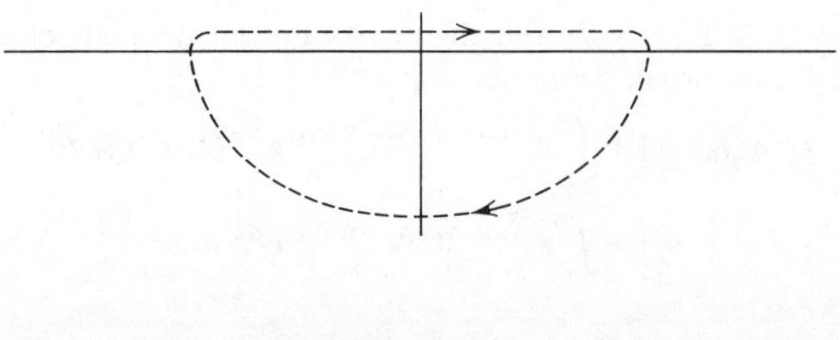

It is convenient to rewrite the first term on the right of Eq. (36.11) as an integral, an endeavor in which the Dirac delta function may be quite helpful

$$\Upsilon^{(1)}(1,t)\Upsilon^{(2)}(2,t)=\int_0^t \delta(t-t')\Upsilon^{(1)}(1,t')\Upsilon^{(1)}(2,t')\,dt'$$

$$=\int_0^t dt'\,\Upsilon^{(1)}(1,t')\Upsilon^{(1)}(2,t')(2\pi \mathrm{i})^{-1}\int_{\eta-\mathrm{i}\infty}^{\eta+\mathrm{i}\infty} e^{z(t-t')}\,dz$$

$$=(2\pi)^{-1}\int_0^t dt'\int_{\mathrm{i}\eta-\infty}^{\mathrm{i}\eta+\infty} dz\, e^{-\mathrm{i}z(t-t')}\Upsilon^{(1)}(1,t')\Upsilon^{(1)}(2,t')$$

$$=(2\pi)^{-1}\int_0^t dt'\int dz\, e^{-\mathrm{i}zt'}\Upsilon^{(1)}(1,t-t')\Upsilon^{(1)}(2,t-t') \tag{36.12b}$$

the contour being that of Eq. (36.12a). Equations (36.12) now allow us to write Eq. (36.11), after a certain amount of rearrangement, as

$$\Upsilon^{(2)}(1,2,t)=(2\pi)^{-1}\int_0^t dt'\int dz\, e^{-\mathrm{i}zt'}\Upsilon^{(1)}(i,t-t')\Upsilon^{(1)}(2,t-t')$$

$$+(2\pi)^{-1}\int dt'\int dz\, e^{-\mathrm{i}zt'}(z-\mathfrak{L})^{-1}\mathfrak{B}\Upsilon^{(1)}(i,t-t')\Upsilon^{(1)}(2,t-t')$$

$$+(2\pi \mathrm{i})^{-1}e^{-\mathrm{i}\omega t}\int dt'\int dz(z+\omega-\mathfrak{L})^{-1} \tag{36.13}$$

where $\Upsilon^{(2)}(0)$ is taken as $\Upsilon^{(1)}_{\mathrm{eq}}(1)\Upsilon^{(2)}_{\mathrm{eq}}(0)$.

By which time the reader may well have forgotten that his interest was originally in the spectral line shape which is related to the averaged dipole moment—and hence the radiator density matrix $\Upsilon^{(1)}(1,t)$—through the susceptibility. Having recalled this interest, our reason for substituting Eq. (36.13) into Eq. (36.3a) will be obvious. The result of this substitution is

$$\mathrm{i}\dot{\Upsilon}^{(1)}(1,t)-\mathfrak{L}_0^{(1)}\Upsilon^{(1)}(1,t)-\mathfrak{L}_F\Upsilon^{(1)}_{\mathrm{eq}}(1)e^{-\mathrm{i}\omega t}$$

$$=(2\pi)^{-1}\mathrm{Tr}\left\{\int_0^t dt'\int dz\, e^{-zt}\mathfrak{B}\left[(z-\mathfrak{L})^{-1}(z-\mathfrak{L})\Upsilon^{(1)}(i,t-t')\Upsilon^{(1)}(2,t-t')\right.\right.$$

$$\left.\left.-\mathrm{i}(z+\omega-L)^{-1}\mathfrak{L}_F\int_0^{t'} dt''e^{-\mathrm{i}\mathfrak{L}t''}\mathfrak{B}\Upsilon^{(1)}_{\mathrm{eq}}(1)\Upsilon^{(1)}_{\mathrm{eq}}(2)\right]\right\}, \tag{36.14}$$

wherein $\Upsilon^{(1)}_{\mathrm{eq}}$ simply refers to the equilibrium and hence density independent density matrix. If Eq. (36.14) does not have the following stationary state solutions in the limit, $t\to\infty$, we will certainly ascertain as much,

$$\Upsilon^{(1)}(1,t)=\Upsilon^{(1)}_{\mathrm{eq}}(1)+\overline{\Delta\Upsilon}(1)e^{-\mathrm{i}\omega t} \tag{36.15a}$$

$$\Upsilon^{(1)}(2,t)=\Upsilon^{(1)}_{\mathrm{eq}}(2) \tag{36.15b}$$

although we are already aware of Eq. (36.15b). We now substitute Eqs. (36.15) into Eq. (36.14) and multiply through by $\exp[i\omega t]$ in order to obtain a straightforward result which we do not write down. The integration over t' and z, taken as $z=\omega+i\epsilon$, are carried out in the limit $t\to\infty$ with the result

$$(\omega-L_0+i\epsilon)\overline{\Delta\Upsilon}(1)-\mathfrak{L}_F\Upsilon_{eq}^{(1)}(1)=$$

$$-\mathrm{Tr}_2\left\{\mathfrak{V}(\omega-\mathfrak{L}+i\epsilon)^{-1}\left[(\omega-L+i\epsilon)\overline{\Delta\Upsilon}^{(1)}(1)\Upsilon_{eq}^{(1)}(2)\right.\right.$$

$$\left.\left.-i\mathfrak{L}_F\int_0^\infty dt'\,e^{-i\mathfrak{L}t'}\mathfrak{V}\Upsilon_{eq}^{(1)}(1)\Upsilon_{eq}^{(1)}(2)\right]\right\} \quad (36.16)$$

The pair correlation operator, $g_{eq}^{(2)}(1,2)$, is now introduced, and Eq. (36.16) takes the form

$$(\omega-\mathfrak{L}_0+i\epsilon)\overline{\Delta\Upsilon}-\mathfrak{L}_F\Upsilon_{eq}^{(1)}(1)=-\mathrm{Tr}_2\left\{\mathfrak{V}(\omega-\mathfrak{L}+i\epsilon)^{-1}\left[(\omega-\mathfrak{L}_0+i\epsilon)\overline{\Delta\Upsilon}\,\Upsilon_{eq}^{(1)}(2)\right.\right.$$

$$\left.\left.+\mathfrak{L}_F g_{eq}^{(2)}(1,2)\right]\right\} \quad (36.17a)$$

Tip (268), whom we now follow in obtaining an expression for $g_{eq}^{(2)}$, does not drop the three-particle term in proceeding from Eq. (36.3) to Eqs. (36.4) but begins with

$$\dot{\Upsilon}^{(1)}(1,t)=-i\mathfrak{L}_0^{(1)}\Upsilon^{(1)}(1,t)-i\mathrm{Tr}_2\mathfrak{V}\Upsilon^{(2)}(1,2,t) \quad (36.18a)$$

$$\dot{\Upsilon}^{(2)}(1,2,t)=-i\mathfrak{L}^{(2)}\Upsilon^{(2)}(1,2,t)-i\mathrm{Tr}_3\mathfrak{V}(1,2/3)\Upsilon^{(3)}(1,2,3,t) \quad (36.18b)$$

where we have partially adopted Tip's notation $\mathfrak{V}(1,2/3)=\mathfrak{V}(1,3)+\mathfrak{V}(2,3)$. The interaction with the radiation field does not enter this discussion.

Four assumptions are made: (1) Reduced density operators remain well behaved in the limit $V\to\infty$, $n\to\infty$, and n/V remains constant. (2) Bound collision pair states are excluded by supposing the potential short range and repulsive, and thus, the gas being dilute, only first-order terms in the density are of interest. (3) Boltzmann statistics are obeyed. (4) The correlations represented by $g^{(2)}$ and $g^{(3)}$ (not yet introduced) are of finite range of order the potential range, a.

The pair of three-particle correlation operators are now introduced

$$g^{(2)}(1,2,t)=\Upsilon^{(2)}(1,2,t)-\Upsilon^{(1)}(1,t)\Upsilon^{(1)}(2,t) \quad (36.19a)$$

$$g^{(3)}(1,2,3,t)=\Upsilon^{(3)}(1,2,3,t)-\Upsilon^{(1)}(1,t)\Upsilon^{(1)}(2,t)\Upsilon^{(1)}(3,t)$$

$$-\Upsilon^{(1)}(1,t)g^{(2)}(2,3,t)$$

$$-\Upsilon^{(1)}(2,t)g^{(2)}(1,3,t)-\Upsilon^{(1)}(3,t)g^{(2)}(1,2,t) \quad (36.19b)$$

From Eqs. (36.18) and (36.19) we see that

$$\begin{aligned}\dot{g}^{(2)}(1,2,t) = & -i\mathfrak{L}^{(2)}g^{(2)}(1,2,t) - i\mathfrak{B}\Upsilon^{(1)}(1,t)\Upsilon^{(1)}(2,t)\\ & -i\mathrm{Tr}_3\big[\mathfrak{B}\{\Upsilon^{(1)}(1,t)g^{(2)}(2,3,t)+\Upsilon^{(1)}(3,t)g^{(2)}(1,2,t)\\ & +g^{(3)}(1,2,3,t)\}+\mathfrak{B}\{\Upsilon^{(1)}(2,t)g^{(2)}(1,3,t)\\ & +\Upsilon^{(1)}(3,t)g^{(2)}(1,2,t)+g^{(3)}(1,2,3,t)\}\big] \end{aligned} \tag{36.20}$$

From assumptions (2) and (4), the contributions to the trace over the coordinates of particle 3 will vanish unless all three particles are simultaneously within a distance a of each other. In the low-density gas then this term may be discarded with the attendant simplification of Eq. (36.20). The resulting equation may be integrated, and we obtain

$$\begin{aligned}g^{(2)}(1,2,t) = & \, e^{-i\mathfrak{L}^{(2)}t}g^{(2)}(1,2,0)\\ & -i\int_0^t dt'\, e^{-i\mathfrak{L}^{(2)}t'}\mathfrak{B}\Upsilon^{(1)}(1,t-t')\Upsilon^{(1)}(2,t-t')\end{aligned} \tag{36.21}$$

In order to shorten matters somewhat, we will simply suppose that assumption (4) will yield zero for $g^{(2)}(1,2,0)$. If at some finite time, t, $g^{(2)}$ is nonzero, we can expect that, at $t=0$, the pair will have sufficient separation so that this follows. Now we see that $g^{(2)}$ is precisely the integral within the square bracket of Eq. (36.16), Υ replaced by Υ_{eq}. We therefore replace $g^{(2)}$ by $g_{\text{eq}}^{(2)}$,

$$\begin{aligned}g_{\text{eq}}^{(2)} &= -i\int_0^\infty dt'\, e^{-i\mathfrak{L}^{(2)}t'}\mathfrak{B}(N/Z)^2 e^{-\beta H_0^{(2)}}\\ &= (N/Z)^2\int_0^\infty dt'\, e^{-i\mathfrak{L}^{(2)}t'}(-i\mathfrak{B})e^{i\mathfrak{L}^{(2)}t'}e^{-\beta H_0^{(2)}}\end{aligned}$$

on the basis that

$$\exp\big[i\mathfrak{L}_0^{(2)}t'\big]\exp\big[-\beta H_0^{(2)}\big] = \exp\big[-\beta H_0^{(2)}\big]$$

Since, as the reader may wish to demonstrate, for an operator $A^{(2)}$ which commutes with $H_0^{(2)}$, $\mathfrak{L}_0^{(2)}A^{(2)}=0$, this becomes

$$\begin{aligned}g_{\text{eq}}^{(2)} &= \left(\frac{N}{Z}\right)^2\int_0^\infty dt'\,\frac{d}{dt'}\left\{e^{-i\mathfrak{L}^{(2)}t'}e^{i\mathfrak{L}_0^{(2)}t'}\right\}e^{-\beta H_0^{(2)}}\\ &= \frac{N}{Z}\left[\lim_{t\to\infty} e^{-i\mathfrak{L}^{(2)}t}e^{i\mathfrak{L}_0^{(2)}t}-1\right]e^{-\beta H_0^{(2)}}\end{aligned}$$

then

$$\lim_{t\to\infty} e^{-i\mathfrak{L}^{(2)}t}e^{i\mathfrak{L}_0^{(2)}t}e^{-\beta H_0^{(2)}} = \lim_{t\to\infty} e^{-i\mathfrak{L}^{(2)}t}e^{-\beta H_0^{(2)}} = e^{-\beta H_0^{(2)}}$$

when we view $\exp[-i\mathfrak{L}^{(2)}t]$ as a TDO. Thus we find

$$g_{\text{eq}}^{(2)}(1,2)=(N/Z)^2\left[e^{-\beta H^{(2)}}-e^{-\beta H_0^{(2)}}\right] \tag{36.17b}$$

with which we return to Eq. (36.17a).

At which point Roney introduces the notion of degenerate internal states and the off-diagonal elements of $\Upsilon_{\text{eq}}^{(1)}(1)$, a notion that was apparently first enunciated by Snider (251a). Let us suppose that the isolated—from both fields and particles—atom possesses a set of states in which its density matrix is diagonal. Then the introduction of a colliding particle, for example, may mix these states and hence introduce off-diagonal elements through, say, the rotational diabaticity described by Spitzer long ago. The result is of course that only structureless particles having no internal states could be taken as having no off-diagonal elements of $\Upsilon_{\text{eq}}^{(1)}(1)$. Hence

$$\Upsilon_{\text{eq}}^{(1)}(1)={}^{d}\Upsilon_{\text{eq}}^{(1)}(1)+{}^{\text{od}}\Upsilon_{\text{eq}}^{(1)}(1) \tag{36.22a}$$

where

$${}^{d}\Upsilon_{\text{eq}}^{(1)}(1)=C_1 e^{-\beta_0^{(1)}(1)} \tag{36.22b}$$

$${}^{\text{od}}\Upsilon^{(1)}(1)=C_1C_2\,\text{Tr}_2\left\{e^{-\beta H^{(2)}(1,2)}-e^{-\beta H_0^{(2)}(1,2)}\right\} \tag{36.22c}$$

with $\Upsilon_{\text{eq}}^{(1)}(2)$ taken as diagonal.

Let us suppose we wish to rewrite $[z-\mathfrak{L}]^{-1}A$ in terms of H operators. Recall that

$$[z-\mathfrak{L}]^{-1}A=[z-(HI^*-IH^*)]^{-1}A$$

The reader will therefore agree that

$$(z-\mathfrak{L})^{-1}A=(2\pi i)^{-1}\int dz'(z'-HI^*)^{-1}A(z'-z-IH^*)^{-1} \tag{36.23a}$$

with the contour about the simple pole at $z'=z+IH^*$.

Eq. (10.4) may be written as

$$(z-H)^{-1}=(z-H_0)^{-1}+(z-H_0)^{-1}t(z)(z-H_0)^{-1} \tag{36.23b}$$

$t(z)$ the two-particle T matrix, and we recall that

$$V(z-H)^{-1}=t(z)(z-H_0)^{-1} \tag{36.23c}$$

Eqs. (36.23) are inserted into Eq. (36.17a) with the result

$$\begin{aligned}(\omega-\mathfrak{L}_0+i\epsilon)\overline{\Delta\Upsilon}-\mathfrak{L}_F\Upsilon_{\text{eq}}(1)=&-(2\pi i)^{-1}\text{Tr}_2\Big\{\int dz\Big[t(z)(z-H_0)^{-1}\big[(\omega-\mathfrak{L}_0\\
&+i\epsilon)\overline{\Delta\Upsilon}\,\Upsilon_{\text{eq}}(2)+\mathfrak{L}_F g_{\text{eq}}^{(2)}\big](z-\omega-H_0-i\epsilon)^{-1}\big[1+t^+(z-\omega-i\epsilon)(z-\omega-H_0-i\epsilon)^{-1}\big]\\
&-\big[1+(z-H_0)^{-1}t(z)\big](z-H_0)^{-1}\big[(\omega-\mathfrak{L}_0+i\epsilon)\overline{\Delta\Upsilon}\,\Upsilon_{\text{eq}}(2)+\mathfrak{L}_F g_{\text{eq}}^{(2)}\big]\\
&\cdot(z-\omega-H_0-i\epsilon)^{-1}t^+(z-\omega-i\epsilon)\Big]\Big\}\end{aligned} \tag{36.24}$$

where the integration contour is now over the upper half-plane to include the real axis.

Roney now takes matrix elements of Eq. (36.24) over the states which diagonalize $H_0^{(1)}(1)+H_0^{(2)}(2)$ and carries out a partial integration of the result over a selected contour in the complex plane in order to obtain an equation which compares quite closely with Eq. (24.10). The remainder of his formal development goes as follows.

The transformation is made from whatever the distribution function (density matrix) may happen to be—it has not really been specified—to the Wigner distribution function (283). This we have not previously encountered. $\overline{\Delta \Upsilon}$ is expressed as the Wigner distribution function, Δf, whose matrix element in turn is written as a product of functions known and to be determined. A function of the photon recoil momentum and the momenta of the collision partners is the one to be determined. For this function a quite complex equation, whose solution would complete the problem, is obtained. (Remember that this is so because an intimate knowledge of the distribution function would, in principle at least, allow us to find the average of **d** and hence the absorption coefficient.) Roney's development of this equation is reasonably straightforward, but the equations themselves are quite lengthy, and it is not clear that we would benefit overmuch by their reproduction in detail.

The bases for conversion to the Wigner distribution function for the external coordinates, position and momentum, while maintaining the density matrix for the internal coordinate distribution are the following: (1) the dependence of $\Delta\Upsilon$ on **r** (position) is clarified; (2) the quantum mechanical Doppler broadening mechanism (photon recoil momentum) is brought out explicitly (3) as is its relation, through the external field term, to Wigner's quantum mechanical corrections to classical thermodynamic equilibrium. Irving and Zwanzig (151) have discussed the properties of the Wigner distribution function, matrix elements of which over the internal states may be written as

$$f_{ij}(\mathbf{r},\mathbf{p})=\int \exp[i\mathbf{q}\cdot\mathbf{r}]\,\Upsilon_{ij}(\mathbf{p}+\mathbf{q}/2,\mathbf{p}-\mathbf{q}/2)\,d\mathbf{q} \tag{36.25a}$$

$$\Upsilon_{ij}(\mathbf{p},\mathbf{p}')=\int \exp[-i(\mathbf{p}-\mathbf{p}')\cdot\mathbf{r}]\,f_{ij}[\mathbf{r}',(\mathbf{p}+\mathbf{p}')/2]\,d\mathbf{r}' \tag{36.25b}$$

In the equation, which we have not obtained from Eq. (36.24), Δf_{ij} now replaces $\overline{\Delta\Upsilon}_{ij}$. The form of the resulting equation tells us that $\Delta f \sim \exp[i2\mathbf{x}\cdot\mathbf{r}]$, where $2\mathbf{x}$ is the photon recoil momentum, if indeed we are not already aware that this should be so. It is now but the work of a moment to conclude [cf. Eq. (36.5)] that Δf_{ij} will be of the form

$$\Delta f_{ij}(\mathbf{r}',\mathbf{k})=e^{i2\mathbf{x}\cdot\mathbf{r}}\varphi_{ij}(\mathbf{k},\mathbf{x}) \tag{36.26}$$

It is then tedious but reasonably straightforward to obtain a lengthy equation that we do not reproduce in its entirety and that yields a set of coupled integral equations for the φ_{ij}. Roney lists the additional complications rendering a solution more formidable than it might otherwise be, and the list is impressive. Thus as we have seen repeatedly, it becomes necessary to try to determine just what the equation can tell us through its reasonably tractable approximate solutions.

37 Isolated Line Profiles in the Reduced Density Matrix Treatment

First we consider a purely Doppler-broadened spectral line. The fact of pure Doppler broadening infers the absence of perturbers with the concomitant disappearance of the t matrices and pair-correlation operators, $g_{\mathrm{eq}}^{(1)}$, in the equation for φ that we have not reproduced. The result is a much simplified equation,

$$\left[(\omega-\omega_{ij})-2p_z x_z/\mathrm{m}_1+i\epsilon\right]\varphi_{ij}^z(p,x_z)-\mathrm{d}_{ij}^z\left[\Upsilon_i^0 f_1^0(p+x_z)-\Upsilon_j^0 f_1^0(p-x_z)\right]=0 \tag{37.1a}$$

where $f_1^0(k)=e^{-\beta k^2/2\mathrm{m}_1}$. In Eqs. (37.1) and in future the electric field polarization is taken as in the z direction, which in turn will eliminate all dipole moment components save that in the z direction. This restriction to the z direction is indicated by the superscripts, φ^z and d^z. The solution to Eq. (37.1a) is immediate.

Now in order to obtain the susceptibility we require the average value of the dipole moment. Originally this would have been the trace over the operator product of the dipole moment. This, however, has been altered by Eqs. (36.25) so that this average consists of the trace over the internal coordinates of the radiator and an integration over the external or translational coordinates of, in the case of Doppler broadening, the radiator. With these facts in mind, the absorption coefficient may be written down immediately.

$$\alpha_D=4\pi\omega\mathcal{G}\left[\sum_{ij}\mathrm{d}_{ij}^z\int\varphi_{ji}^z(p,x_z)\,d\mathbf{p}\right]$$

$$=4\pi\omega N\left(\frac{\beta}{2\pi\mathrm{m}}\right)^{3/2}\left[\sum_n e^{-\beta E_n}\right]^{-1}\mathcal{G}\left\{\sum_{ij}|\mathrm{d}_{ij}^z|^2\int\left[e^{-\beta E_i}\exp\left(\frac{-\beta(p^2+2p_zx_z+x_z^2)}{2\mathrm{m}}\right)\right.\right.$$

$$\left.\left.-e^{-\beta E_j}\exp\left(\frac{-\beta(p^2-2p_zx_z+x_z^2)}{2\mathrm{m}}\right)\right]\left[\omega-\omega_{ij}-2p_zx_z/\mathrm{m}+i\epsilon\right]^{-1}dp_x\,dp_y\,dp_z\right\}$$

$$=4\pi\omega N\left(\frac{\beta}{2\pi\mathrm{m}}\right)^{1/2}Z^{-1}\mathcal{G}\left\{\sum_{ij}|\mathrm{d}_{ij}^z|^2\int dp_z\left[\exp\left(-\beta\left(\frac{E_i+x_zp_z}{\mathrm{m}}\right)\right)\right.\right. \tag{37.2}$$

$$\left.\left.-\exp\left(-\beta\left(\frac{E_j-x_zp_z}{\mathrm{m}}\right)\right)\exp\left[-\beta(p_z^2+x_z^2)/2\mathrm{m}\right]/\left[(\omega-\omega_{ij})-2p_zx_z/\mathrm{m}+i\epsilon\right]\right]\right\}$$

after integration over p_x and p_y.

Letting $A = \mathrm{m}(\omega-\omega_{ij})/2x_z$, we agree that

$$[(A-p_z)+\mathrm{i}\epsilon]^{-1} = \mathcal{P}[A-p_z]^{-1} - \mathrm{i}\pi\delta(A-p_z) = -\mathrm{i}\pi\delta(A-p_z) \quad (37.3)$$

if $(\omega-\omega_{ij})$ remains small enough so that we never actually encounter $A=p_z$. Once again we are restricting ourselves from the line wing. However, we provide ourselves with a method of evaluating the integral in Eq. (37.2). We easily obtain

$$\alpha_D = 4\pi\left(\frac{\beta\mathrm{m}}{2\pi}\right)^{1/2} z^{-1} e^{-\beta\omega^2/8\mathrm{m}} \sum_{ij} |\mathrm{d}_{ij}^z|^2 \exp\left[-\left(\frac{\omega_{ij}(\omega-\omega_{ij})}{\omega\delta_D}\right)^2\right]$$

$$\cdot\left[N\exp\{-\beta[E_j-\tfrac{1}{2}(\omega-\omega_{ij})]\} - N\exp\{-\beta[E+\tfrac{1}{2}(\omega-\omega_{ij})]\}\right] \quad (37.4a)$$

where $\delta_D = (2/\mathrm{m}\beta)^{1/2}\omega_{ij}$ is what we might call the classic Doppler width, and the fact that $\kappa_z = \frac{1}{2}\omega$ has been used. (The factor of two has entered simply by way of definition.) We remark that the square-bracketed factor is the population factor. This result is to be compared to the familiar expression for the Doppler shape, an expression traceable back to Lord Rayleigh:

$$\alpha_D' = 4\left(\frac{\beta\mathrm{m}}{2\pi}\right)^{1/2} Z^{-1} \sum_{ij} |\mathrm{d}_{ij}^z|^2 \exp\left[-\left(\frac{(\omega-\omega_{ij})}{\delta_D}\right)^2\right]\left[Ne^{-\beta E_b} - Ne^{-\beta E_a}\right] \quad (37.4b)$$

A comparison of these two expressions may be greatly simplified by reexpressing them in terms of an S, which the reader may determine, for an isolated line such that $\omega \doteq \omega_{ij} = \omega_0$ in α_D and α_D' are the same, and the factor $\exp[\frac{1}{2}\beta(\omega-\omega_{ij})]$ induces the asymmetry.

$$\alpha_D = SAe^{-\beta\omega^2/8\mathrm{m}}\left[\exp\{-\beta[\omega_0+\tfrac{1}{2}(\omega-\omega_{ij})]\} - e^{\beta(\omega-\omega_{ij})/2}\right]$$

$$= SAe^{-\beta\omega^2/8\mathrm{m}} e^{\beta(\omega-\omega_{ij})/2}\left[\exp\{-\beta[\omega_0+(\omega-\omega_{ij})]\} - 1\right]$$

$$\alpha_D' = SA\left[e^{-\beta\omega_0}-1\right], \qquad A = \exp\left[-\omega_0(\omega-\omega_{ij})/\omega\delta_D\right]^2$$

Thus Roney has obtained an expression for the Doppler-broadened line which differs from the classic shape in two regards: (1) the factor, $\exp[-\beta\omega^2/8\mathrm{m}]$, will render the originally symmetric line asymmetric; (2) the square bracket will yield an additional asymmetry in the wing, that is, for $\omega_0 \gg \frac{1}{2}(\omega-\omega_{ij})$, which will disappear near line center. The factor (1) becomes of importance for the Lyman α line of hydrogen ($\lambda = 1215$ Å) at temperatures of 0.1 °K and below as may readily be calculated. At temperatures of the order of 300 °K wavelengths as short as the x ray are required, and, since the Compton effect has not been included in our considerations, the theory does not apply at these wavelengths anyhow, Factor (2) may be evaluated by taking $(\omega-\omega_{ij})$ as δ_D. The results for this factor differ from those for factor

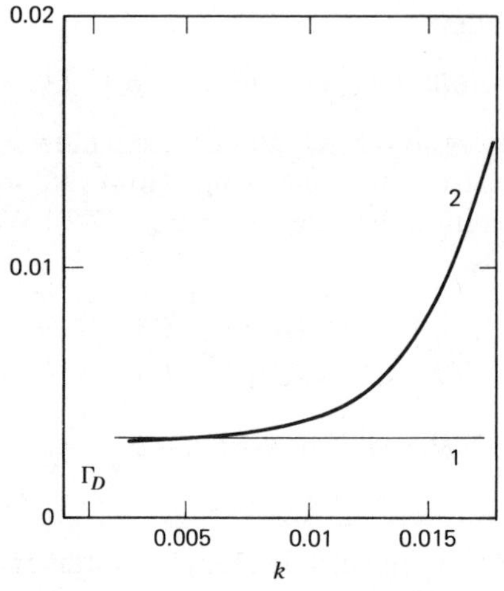

Figure 37.1 After Czuchaj (107).

(1) by a factor of four. This leads us to conclude that the classic Doppler broadened intensity distribution remains acceptable, and the photon recoil momentum, x, may be neglected in the external field term. Such a result for pure Doppler broadening with no collisions is not in serious conflict with earlier results, nor is it apparently at variance with subsequent work such as that of Czuchaj (107).

Czuchaj carried out a Liouvillian pressure-Doppler treatment which he specifically applied to the 3889 Å line of He. The results are of some little interest and are displayed as Fig. 37.1. The line width obtained is about 60% greater than the Doppler width and is a constant in the entire pressure range considered, $0.005 \leq k \leq 0.02$ corresponding to a few Torr. For comparison the width yielded by the conventional Voigt profile is given as Curve 2 of Fig. 37.1.

We have seen that Roney's Doppler limit has a factor $\exp[-\beta\omega^2/8\mathsf{m}]$ that is not to be found in the classic Doppler result. Thus we anticipate a width increase with decreasing mass from the Roney theory, perhaps approaching that obtained by Czuchaj. Having encountered these quantum effects in this limit of the Roney theory, we turn to the other limiting cases for the theory.

38 Reductions of the Reduced Density Matrix Result

Since Δf in Eq. (36.22) is a scalar and ϵ a vector, φ must surely behave as a vector, in particular, it must transform under a rotation as does the dipole moment **d**. We have restricted the electric field polarization to the z direction, so it should follow that

$$\langle i, m_i | \varphi_z(k, x) | j, m_j \rangle = \langle i, m_i | \mathsf{d}_z | j, m_j \rangle F_{ij}(k, x) \tag{38.1}$$

where F_{ij} is independent of the magnetic quantum number m. We shall now write down a very abbreviated form of the equation for F from the equation for φ which we did not write down in Section 37. Under these conditions the equation for F_{ij} should be basically unintelligible but hopefully useful.

$$\begin{aligned}((\omega-\omega_{ij})+2p_z x_z/\mathsf{m}_1+i\epsilon)F_{ij}(p,x)-\big[\Upsilon_i^0 f_1^0(p+x)\\ -\Upsilon_j^0 f_1^0(p-x)\big]-N_{ij}(p,x)-C_{ij}(p,x,\omega)\\ =\sum_k\big[F_{kj}(p,x)T_{ik}(p,x,\omega)-F_{ik}(p,x)T_{kj}^+(p,x,\omega)\big]\\ +\sum_{kl}\int\int d\bar{\mathbf{p}}\,d\mathbf{k}\,F_{kl}(p-\bar{p}+k)K_{ikjl}(p,\bar{p},k,\omega)\end{aligned} \tag{38.2a}$$

where $\Upsilon_i^0=e^{\beta\epsilon_i}/Z_i$, $f_l^0(\mathrm{k})=C_l e^{-\beta k^2/2\mathsf{m}_i}$. The momentum arguments can cause confusion and are surely worth a certain amount of discussion. We begin with $\mathbf{p}$ as the radiator momentum, obviously an arbitrary quantity. Then $\mathbf{k}_2$ is taken as the perturber momentum when the radiator is in internal states i and j, $\mathbf{k}_2'$ the perturber momentum for k and $\mathbf{k}_2''$ the perturber momentum for radiator state l. Given this set of designations, Roney subsequently changes variables so that $\mathbf{k}_2=(\mathsf{m}_2\mathbf{p}-\mathsf{m}\bar{\mathbf{p}})/\mathsf{m}_1$, $\mathbf{k}_2'=\mathsf{m}_2(\mathbf{p}+\mathbf{k}_2)/\mathsf{m}-\mathbf{k}$, and $\mathbf{k}_2''=\mathsf{m}_2(\mathbf{p}+\mathbf{k}_2)/\mathsf{m}-\mathbf{k}'$. We shall not devote too much effort to the various momentum integrations, but we note the following. An integration over $\bar{\mathbf{p}}$ is an integration over the (weighted) difference between emitter and perturber momenta while an integration over $\mathbf{k}$ is over the difference between the (momentum of the emitter plus the difference between the) perturber momenta corresponding to two different internal emitter states.

Returning to Eq. (38.2a), we first remark that the T_{ij} are linear functions of the matrix elements of the two-particle T matrices. N_{ij} is a function of the difference between two matrix elements of the equilibrium pair-correlation operator, $g_{\mathrm{eq}}^{(2)}$, of Eq. (36.17b). C_{ij} is a rather lengthy expression involving matrix elements of the t matrices and of $g_{\mathrm{eq}}^{(2)}$, as well as the functions

$$\begin{aligned}P_{ijkl}(z,\mathbf{p},\mathbf{k},\mathbf{k}')=(2\pi i)^{-1}\Big\{\big[z-E_l^{(-)}-\omega-(p^2-k'^2)/2\mathsf{M}\\ +\mathbf{x}\cdot(\mathbf{p}-\mathbf{k}')/\mathsf{m}_1-i\epsilon\big]^{-1}-\big[z-E_k^{(+)}-(p^2-k^2)/2\mathsf{M}-\mathbf{x}\cdot(\mathbf{p}-\mathbf{k})/\mathsf{m}_1\big]^{-1}\Big\}\\ \times\Big\{\big[z-\omega-E_j^{(-)}-i\epsilon\big]^{-1}-\big[z-E_i^{(+)}\big]^{-1}\Big\}\end{aligned} \tag{38.2c}$$

where M is the reduced mass of the collision pair. Finally, the kernel, K_{ijkl}, is a function of a t-matrix element product and a P_{ikjl}.

$$E^{(\pm)}=\epsilon_n+|\mathbf{p}+\mathbf{x}|^2/2\mathsf{m}_1+|(\mathsf{m}_2\mathbf{p}-\mathsf{m}\bar{\mathbf{p}})/\mathsf{m}_1|^2/\mathsf{m}_1 \tag{38.2d}$$

If we neglect the Doppler recoil momentum as our investigation of the pure Doppler broadening has indicated that we may, $E^{(\pm)}\to E$. We consider the second square bracket on the right of Eq. (38.2c). For frequencies near line center $\omega \doteq \omega_{ij}$ so that this bracket may be written

$$[z-\omega-E_j-\mathrm{i}\epsilon]^{-1}-[z-\omega-E_j+\mathrm{i}\epsilon]^{-1}=2\mathrm{i}\pi\delta(z-\omega-E_j) \tag{38.3}$$

by Eq. (37.3), and the reader will note that we have inserted the $\mathrm{i}\eta$ for rather transparent reasons. Which provides us with a somewhat simplified P_{ijkl} but calls for a diversion before further simplification.

$T_{ik}(\mathbf{p},\mathbf{x},\omega)$ may be written out as follows:

$$T_{ik}(\mathbf{p},\mathbf{x},\omega)=(\mathrm{m}/\mathrm{m}_1)^3 \sum_{m_i m_j m_k} \langle km_k|\mathrm{d}^z|jm_j\rangle\langle jm_j|\mathrm{d}^z|im_i\rangle$$

$$\cdot\left\{\sum_{m_i m_j} |\langle im_i|\mathrm{d}^z|jm_j\rangle|^2\right\}^{-1}$$

$$\cdot\int d\bar{\mathbf{p}}\, f_2^0((\mathrm{m}_2\mathbf{p}-\mathrm{m}\bar{\mathbf{p}})/\mathrm{m})\left\langle {im_i \atop \mathbf{p}+\mathbf{x}'}\right|t\left(E_j^{(-)}+\omega\right)\left|{km_k \atop \mathbf{p}+\mathbf{x}}\right\rangle \tag{38.2e}$$

In Eq. (38.2e) $\mathbf{p}$ is the radiator momentum. Note that, in the matrix elements over t, the momentum before collision (and hence the direction of the motion), $\bar{\mathbf{p}}+\mathbf{x}'$ is the same as that after collision. Thus these matrix elements refer to forward scattering wherein the radiator does not change direction as a consequence of collision, not a new consideration but one worthy of emphasis. On the other hand, in the expressions for C_{ij} and K_{ijkl} we encounter matrix elements of the form

$$\left\langle {im_i \atop \mathbf{p}+\mathbf{x}'}\right|t(z)\left|{km_k \atop \mathbf{k}+\mathbf{x}'}\right\rangle \tag{38.4}$$

that is, matrix elements in which the scattering is not necessarily forward scattering. We now return to Eq. (38.2c), as modified by Eq. (38.3) and preceding equations, which may be written out as

$$P_{ijkl}(z,\bar{\mathbf{p}},\mathbf{k},\mathbf{k})\doteq\left\{\left[z-E_j+(\omega_{jl}-\omega)-\frac{(k^2-\bar{p}^2)}{2\mathrm{M}}-\mathrm{i}\epsilon\right]^{-1}\right.$$

$$\left.-\left[z-E_j-\omega_{kj}-\frac{(k^2-\bar{p}^2)}{2\mathrm{M}}\right]^{-1}\right\}\begin{Bmatrix}\delta(z-E_j-\omega)\\ \delta(z-E_i)\end{Bmatrix}$$

$$=\left\{\left[\omega_{jl}-(\mathrm{k}^2-p^2)/2\mathrm{M}-\mathrm{i}\epsilon\right]^{-1}-\left[\omega_{ik}-(k^2-p^2)/2\mathrm{M}+\mathrm{i}\eta\right]^{-1}\right\}$$

$$\cdot\begin{cases}\delta(z-E_j-\omega)\text{ acting on } t(z)\\ \delta(z-E_i)\text{ acting on } t^+(z-\omega-\mathrm{i}\epsilon)\end{cases} \tag{38.5}$$

Roney has selected the delta functions so that, say, the argument of the t_{ik} in Eq. (38.5) corresponds to the argument of that t-matrix element in Eq. (38.2e), that is, in the forward scattering T_{ik}.

If we now proceed to an isolated spectral line, $k \to i$, $l \to j$, $\omega_{jl} \to \omega_{jj} = 0$, and so on, and we obtain

$$P_{ijij}(z,\bar{\mathbf{p}},\mathbf{k},\mathbf{k}) \doteq 2\pi i \delta\left(\frac{(k^2-\bar{p}^2)}{2M}\right)\begin{cases}\delta(z-E_j-\omega)\\ \text{or}\\ \delta(z-E_i)\end{cases} \tag{38.6}$$

It is interesting to remark, with Roney, that, to a reasonable level of approximation, the same result may be utilized for P_{ijkl}. As we would suspect, this leads, on the one hand, to considerable simplification in the summation over certain of the indices of P_{ijkl} and, on the other hand, to simplification in the integration over its arguments.

With Eq. (38.6) we have arrived at a comparatively simplified expression for the intensity distribution in a set of foreign-gas, binary-collision broadened spectral lines. Our final step will be to inquire as to the result to which this reduces for an isolated spectral line. This means that only terms F_{ij} will be retained in Eq. (38.2a) and that equation will reduce in an obvious fashion to an equation that we call Eq. (38.2a′) but do not write down. The term on the right of Eq. (38.2a′) will be

$$\int\int d\bar{\mathbf{p}}\, d\mathbf{k}\, F_{ij}(\mathbf{p}-\bar{\mathbf{p}}+\mathbf{k},\mathbf{x})K_{iijj}(\mathbf{p},\bar{\mathbf{p}},\mathbf{k},\omega) \tag{38.7}$$

Since an iterated solution for F_{ij} is contemplated, it is useful to generate a scattering direction parameter by letting $F_{ij}=1$ in Eq. (38.7).

$$\Gamma_{iijj}(p,x,\omega) = \int\int d\bar{\mathbf{p}}\, d\mathbf{k}\, K_{iijj}(\mathbf{p},\bar{\mathbf{p}},\mathbf{k})$$

$$\doteq 2\pi i\left(\frac{m}{m_1}\right)^3\left\{|\langle im_i|d_z|jm_j\rangle|^2 / \sum_{m_i m_j}|\langle im_i|d_z|jm_j\rangle|^2\right\}$$

$$\cdot\int\int d\mathbf{p}\, d\mathbf{k}\, f_2^0\left(\frac{(m_2\mathbf{p}-m\bar{\mathbf{p}})}{m}\right)\exp\left(\frac{-\beta(\mathbf{p}-\bar{\mathbf{p}})\cdot(\bar{\mathbf{p}}-\mathbf{k})}{m_1}\right)$$

$$\cdot\delta\left(\frac{(k^2-\bar{p}^2)}{2M}\right)\cdot\left\langle \begin{matrix} im_i \\ \bar{\mathbf{p}}+\mathbf{x}'\end{matrix}\right| t(E_j+\omega)\left|\begin{matrix} im_i \\ \mathbf{k}+\mathbf{x}'\end{matrix}\right\rangle\left\langle \begin{matrix} jm_j \\ \mathbf{k}-\mathbf{x}'\end{matrix}\right| t^+(E_i-\omega)\left|\begin{matrix} jm_j \\ \mathbf{p}-\mathbf{x}\end{matrix}\right\rangle \tag{38.8}$$

Roney now carries out an iterated solution to Eq. (38.2a′) the details of which we do not reproduce. However, it is of interest to remark the form which he chooses for the zero-order solution. One could, with a certain amount of justification, drop the last term in Eq. (38.2a) when setting up a zeroth-order equation for F_{ij}. The resulting equation for F_{ij} and its solution would then contain no off-energy shell t-matrix elements, that is to say, only forward scattering effects would enter. If instead we choose $F_{ij}=1$ in the last term on the right of Eq. (38.2a), our zeroth-order solution is obviously

$$F_{ij}^{(0)}(\mathbf{p},\mathbf{x}) = \{[\Upsilon_i^0 f_1^0(\mathbf{p}+\mathbf{x}) - \Upsilon^0 f_1^0(\mathbf{p}-\mathbf{x})] + N_{ij}(\mathbf{p},\mathbf{x}) + C_{ij}(\mathbf{p},\mathbf{x},\omega)\}\{(\omega-\omega_{ij}) - 2p_z x/\mathrm{m}_1 - [T_{ii}(\mathbf{p},\mathbf{x}) - T_{jj}^{+}(\mathbf{p},\mathbf{x}) + \Gamma_{iijj}(\mathbf{p},\mathbf{x})] + \mathrm{i}\epsilon\}^{-1} \tag{38.9}$$

Now the zeroth-order F_{ij} still contains only forward scattering since $F(\mathbf{p}-\bar{\mathbf{p}}+\mathbf{k})$ has already reduced to $F(p)$. However, when we carry out an iterated solution of Eq. (38.2a′) by substituting $F^{(0)}$ under the integral sign, Γ_{iijj} may certainly be adjusted to include non-forward scattering. The reader is doubtless familiar with the sort of iterated solution which could now be carried out. Roney carried out such a solution, it being of interest to remark that, since he was dealing with binary collisions, he dropped terms involving powers of the density greater than N_1N_2, the product of radiator and perturber densities. The absorption coefficient that evolved is

$$\alpha(\omega,x) = 4\pi\omega \sum_{m_i m_j} |\langle im_i|\mathrm{d}_z|jm_j\rangle|^2 \int d\mathbf{p}\{[\Upsilon_i^0 f_1^0(\mathbf{p}+\mathbf{x}) - \Upsilon_j^0 f_1^0(\mathbf{p}-\mathbf{x})] - [1+\vartheta_{ij}(\mathbf{p},\mathbf{x},\omega)] + N_{ij}(\mathbf{p},\mathbf{x}) + C_{ij}(\mathbf{p},\mathbf{x},\omega)\}/\{(\omega-\omega_{ij}) - 2(\mathbf{p}-\bar{\mathbf{p}}+\mathbf{k})_z x/\mathrm{m}_1 - [T_{ii}(\mathbf{p}-\bar{\mathbf{p}}+\mathbf{k},x,\omega) - T_{jj}^{+}(\mathbf{p}-\bar{\mathbf{p}}+\mathbf{k},x,\omega) + \Gamma_{iijj}(\mathbf{p}-\bar{\mathbf{p}}+\mathbf{k},x,\omega)] + \mathrm{i}\epsilon\} \tag{38.10a}$$

$$\vartheta_{ij}(p,x,\omega) = \int\int d\bar{\mathbf{p}}\,d\mathbf{k}\,K_{iijj}(\mathbf{p},\bar{\mathbf{p}},\mathbf{k})\exp[-\mathrm{i}\beta(\bar{\mathbf{p}}-\mathbf{k})\cdot(\bar{\mathbf{p}}-\mathbf{p})/\mathrm{m}_1]/\{(\omega-\omega_{ij}) - 2(\mathbf{p}-\bar{\mathbf{p}}+\mathbf{k})_z x/\mathrm{m}_1 - [T_{ii}(\mathbf{p}-\bar{\mathbf{p}}+\mathbf{k},x,\omega) - T_{jj}^{+}(\mathbf{p}-\bar{\mathbf{p}}+\mathbf{k},x,\omega) \cdot + \Gamma_{iijj}(\mathbf{p}-\bar{\mathbf{p}}+\mathbf{k},x,\omega)] + \mathrm{i}\epsilon\} - \Gamma_{iijj}(\mathbf{p},\mathbf{x},\omega)/\{(\omega-\omega_{ij}) - 2p_z x/\mathrm{m}_1 \cdot - [T_{ii}(\mathbf{p},\mathbf{x},\omega) - T_{jj}^{+}(\mathbf{p},\mathbf{x},\omega) + \Gamma_{iijj}(\mathbf{p},\mathbf{x},\omega)] + \mathrm{i}\epsilon\} \tag{38.10b}$$

Roney refers to $T_{ii} - T_{jj}^{+} + \Gamma_{iijj}$ as "line-center parameters" and quite aptly so, the real part of this sum furnishing the line shift, the imaginary part the linewidth. Eq. (38.10a) proceeds through an interesting and appropriate series of stages as we proceed from very low densities. (1) At very low densities the distribution is Gaussian and classically descriptive of the Doppler broadening. (2) At somewhat higher densities the interruption shape evolves provided, of course, that $(T_{ii} - T_{jj}^{+} + \Gamma_{iijj})$ is at most weakly dependent on **p**. (3) By the time the density has increased to what Roney calls "intermediate" the Voigt profile appears. (4) For "high" densities the entire Eq. (38.10a) must be used, "high" being not so high as to vitiate the binary collision approximation. A comparison of Roney's results for stage (2) to those of Baranger as given by Eq. (14.19) is immediate.

In stage (2) shift and width are given by $T_{ii} - T_{jj} + \Gamma_{iijj}$. We see that $T_{ii} - T_{jj}$ is the difference in scattering amplitudes constituting the square-bracketed term on the right of Eq. (14.19). Because Baranger considers a stationary radiator, we may eliminate the integration over p in Eq. (38.8) for purposes of our comparison. Then we are left with the k integration which, since $\mathbf{k}'_2 = \mathrm{m}_2(\mathbf{p}+\mathbf{k}_2)-\mathbf{k}$ with $\mathbf{p}=0$, is simply the integration over possible changes in perturber momentum. The delta function in Eq. (38.8) converts this into an angular integration indicated in the Baranger result, Γ_{iijj} of course introducing the effects of non-forward scattering. In sum then, we see that Eq. (14.19) differs from $T_{ii} - T_{jj}^{+} + \Gamma_{iijj}$ only in the factor $\exp[-(\beta/\mathrm{m}_1)(\mathbf{p}-\bar{\mathbf{p}})\cdot(\mathbf{p}-\mathbf{k})]$, a result which we would certainly anticipate since Roney has considered Doppler and related translational effects while Baranger has not. The reduction then is favorable to the Roney theory.

Finally, one may rather readily investigate the coupling of two lines arising from three levels, i, j, and k, the line corresponding to transitions from i to j and from j to k. Two coupled equations of the form

$$\left[(\omega-\omega_{ij})-(T_{ii}-T_{jj}^{+}-\Gamma_{iijj})\right]F_{ij}$$
$$=\left[\Upsilon_i^0 f_1^0(p+x_2)-\Upsilon_j^0 f_1^0(p-x_z)\right]+\left[T_{ik}-\Gamma_{ikjj}\right]F_{kj}$$

now evolve for F_{ij} and F_{kj}. These equations may be solved, and Roney does so in order to obtain a rather lengthy expression, albeit a reasonbly straightforward one, for the absorption coefficient. Whether the considerable additional complexity of this expression as compared to the Baranger result yields commensurate additional accuracy has apparently not been studied in detail.

39 The Time of Collision

We have deferred any extensive discussion of the off-energy shell t-matrix elements to which we now turn our attention. Fano originally pointed out that such elements referred to the "transient stages of collision," that is to say, to phenomena taking place during the actual time of collision. Such off-shell transients are to be contrasted with the on-shell elements which refer to completed events in the collision process (e.g., a collision has taken place, and the effect on the radiation is so and so). This is perhaps qualitatively informative but hardly anything more. For something more quantitative, we turn to the work of Branson (82).

Branson's avowed purpose in perpetrating this paper was to "introduce time into S-matrix theory." That author uses the Stapp (257) definition of the S matrix which, though obvious, we have not specifically stated: Given a particle system having energy E and total momentum $a, b, \cdots$, we take the outcome of a measurement on the initial system configuration to have a probability $|A_i(E', a)|^2$ and that in the final system configuration to have probability $|A_f(E, b)|^2$. Then the S matrix is defined by the relationship which we presume to exist between the quantities thus defined as

$$A_f(E, b) = \sum_{E'a} S(E, b; E', a) A_i(E', a) \tag{39.1}$$

Given the existence of S, we may use our ruminations of Section 18 in order to infer the existence of a T, thus

$$S(E,a;E',a)=\delta(E-E')\,\delta(b-a)+\mathrm{i}\delta(E-E')\,\delta^3(P-P')T_{ab} \quad (39.2)$$

P and P' being total three-momenta in the final and initial states, respectively. We will certainly agree that Eq. (39.2) may be substituted into Eq. (39.1) with the result

$$A_f(E,b)=A_i(E,b)+\mathrm{i}\sum_a T_{ab}(E)A_i(E,a) \quad (39.3)$$

the sum being over those values of a such that $\mathbf{P}=\mathbf{P}'$, a result induced by $\delta^3(\mathbf{P}-\mathbf{P}')$.

Branson phrases his introduction of time in a careful fashion by remarking that "Stapp believes that in S-matrix theory also, time and energy are related by Fourier transformation." Apparently Branson believes so too, for he introduces t by means of

$$A_i(t,a)=\int_{-\infty}^{\infty} dE\, e^{-\mathrm{i}Et}A_i(E,a) \quad (39.4)$$

As we have seen, the Fourier transform relates Heisenberg-conjugate variables such as time and energy, momentum and position, and so on. From one point of view then, we may relate the Fourier transform to the uncertainty principle, although there is nothing particularly enlightening about doing so. Similar remarks may be made about the Dirac delta function. In any event, the sign in the exponential is arbitrary. If the opposite sign is chosen, the results are reversed, which corresponds to the reversal accompanying the change of sign in time in field theory. The Fourier transform of Eq. (39.3) is

$$\tilde{A}_f(t,b)=\tilde{A}_i(t,b)+\mathrm{i}\int_{-\infty}^{\infty} dt'\sum_a \tilde{T}_{ab}(t-t')\tilde{A}_i(t',a) \quad (39.5a)$$

$$\tilde{T}_{ab}(\tau)=(2\pi)^{-1}\int_{-\infty}^{\infty} dE\, e^{-\mathrm{i}E\tau}T_{ab}(E) \quad (39.5b)$$

$$T_{ab}(E)=\int_{-\infty}^{\infty} d\tau\, e^{\mathrm{i}E\tau}\tilde{T}_{ab}(\tau) \quad (39.5c)$$

Thus we have established a connection between the final amplitude at time t, $\tilde{A}_f(t,b)$, and the initial amplitude at time t', $\tilde{A}_i(t',a)$. If, that is, t is indeed time, which, at this point, Branson assumes that it is. Of greatest importance to us, Branson assumes that,

"The probability of a reaction taking place with the appropriate variables having values a, b and such that the particles are interacting for a time between τ and $\tau+d\tau$ is proportional to"

$$|\tilde{T}_{ab}(\tau)|^2\,d\tau \quad (39.6)$$

Therefore, we have taken a time of collision, $d\tau$, which is finite because the forces of interaction are taken as having a finite range. As we have had any number of occasions to remark, the question of range is an iffy one on which specific assumptions—usually all that is possible—must be made for specific cases. In any event, what occurs during this time of collision is described by the off-shell T-matrix elements, $\tilde{T}_{ab}$. However, matters must be specificized more precisely, for $\tilde{T}_{ab}(\tau)$ is completely independent of the energy due to the infinite limits on the integral. Thus, in accordance with the uncertainty principle, all knowledge of the energy has been sacrificed to a precise determination of the collision time. By accepting an uncertainty in the collision time the energy dependence may be recaptured, this being accomplished by replacing Eq. (39.5b) with

$$T_{ab}(\tau, E_0, \sigma) = (2\pi)^{-1}\int_{-\infty}^{\infty} dE\, e^{-iE\tau} e^{-(E-E_0)^2/\sigma^2} T_{ab}(E) \qquad (39.7a)$$

Eq. (17.6) is replaced by

$$|\tilde{T}_{ab}(\tau, E_0, \sigma)|^2\, d\tau \qquad (39.7b)$$

the probability for the reaction where the energy is approximately E_0 with an uncertainty σ, and τ is the collision time with a corresponding uncertainty. The most probable value for the collision time τ will be

$$\langle\tau\rangle = \int_{-\infty}^{\infty} d\tau\,\tau\,|\tilde{T}_{ab}(\tau, E_0, \sigma)|^2 \Big/ \int_{-\infty}^{\infty} d\tau\,|\tilde{T}_{ab}(\tau, E_0, \sigma)|^2$$

$$= -\tfrac{1}{2}i\int_{-\infty}^{\infty} dE\left[T_{ab}^{*}(E)T'_{ab}(E) - T'^{*}_{ab}(E)T_{ab}(E)\right]e^{-2(E-E_0)^2/\sigma^2}$$

$$\cdot\left[\int_{-\infty}^{\infty} dE\, T_{ab}^{*}(E)T_{ab}(E)e^{-2(E-E_0)^2/\sigma^2}\right]^{-1} \qquad (39.8)$$

by Eq. (39.7a) and where $T' = dT/dE$. For σ, the uncertainty small, the Gaussian, sharply peaked Eq. (39.8) is approximately

$$\mathcal{R}\{-iT'(E_0)/T(E_0)\}$$

and hence similar to the Wigner time-delay formula (284). In his work Roney used the following variation of this:

$$\tau(E) = -it'(E+i0)/t(E+i0) \qquad (39.9)$$

We will also use this variation after manipulating the Fadeev equations of Section 17 in such a way that use may be made of them. However, even before doing this let us remark on the approach to collision time introduced by Futrelle (127), an approach using entities termed sum rules and spectral moments.

We will certainly be familiar with the following form for the correlation function:

$$\Phi(t) = \langle \mathsf{d}(0)\, e^{i\mathfrak{L}t}\mathsf{d}(0)\rangle \Rightarrow$$

$$I(\omega) \sim (2\pi)^{-1}\int_{-\infty}^{\infty} e^{-i\omega t}\Phi(t)\, dt = \langle(2\pi)^{-1}\mathsf{d}(0)\int_{-\infty}^{\infty} e^{i(\mathfrak{L}-\omega)t}\, dt\, \mathsf{d}(0)\rangle$$

$$= \langle \mathsf{d}\,\delta(\mathfrak{L}-\omega)\,\mathsf{d}\rangle$$

which allows us to define the nth moment as

$$M_n = \int_{-\infty}^{\infty} d\omega\, \omega^n I(\omega) = \langle \mathsf{d}\Omega^n \mathsf{d} \rangle$$

In passing, we remark that, the existence of

$$\frac{d^n}{dt^n} \Phi(t)|_{t=0} = (\mathrm{i})^n M_n$$

allows the expression of $\Phi(t)$ as

$$\Phi(t) = \sum_{n=1}^{\infty} \frac{\mathrm{i}^n}{n!} M_n t^n$$

an expression which Futrelle classifies as of critical importance to his theory. (The Futrelle theory is an isolated-line, foreign-gas theory of pressure broadening which reduces to the interruption theory in one limit, the statistical theory in the other.)

Now in order to satisfy requirements relating to detailed balancing [cf. Berne, Jortner, and Gordon (75)], Futrelle adds a term relating to $-\omega$ to his correlation function for ω, $I(\omega)$ thus being related to $I(-\omega)$. This leads to a two-term correlation function only the positive-frequency term of which he uses in computing moments from a precise analogy to the above equation for M_n. In the third moment a characteristic length associated with the potential is encountered, this length leading, as is intuitively obvious, to the collision time. With which we return to the Roney treatment and manipulate the Fadeev equation of Section 17 in such a way that use may be made of it.

Equation (17.11) is the expression with which Roney begins although he uses the lower case t for the double-index subscript, a convention in accordance with using t for the two-particle T matrix. We do the same from this point. We take 1 and 2 as referring to the collision pair, 3 to the photon. With this convention only matrix elements of the t_{12} are retained, the t_{13} and t_{23} being obviously particle-photon terms and dropped as such. This means we may let $t_{12} \equiv t$. Further $\mathsf{V}_{ij} \to \mathsf{V}$, there being no interaction of the sort in mind between particle and photon. Roney uses two equations in seeking an expression for the t matrix, the first of which is obtained, for example, from Eq. (12.5),

$$t(z_1) = \mathsf{V} + \mathsf{V}[z_1 - H_0]^{-1} t(z_1) \tag{39.10a}$$

wherein the reversal in G^0 changes the sign of the second term. From Eq. (12.5) we see that this may also be written as

$$t(z_2) = \mathsf{V} + t(z_2)\mathsf{V}[z_2 - H_0]^{-1} \tag{39.10b}$$

where we have changed the yet to be detailed argument. By eliminating V from Eqs. (39.10) and rewriting in a quite straightforward fashion we may solve for

$$\begin{aligned} t(z_2) &= t(z_1) + t(z_2)\{[z_2 - H_0]^{-1} - [z_1 - H_0]^{-1}\} t(z_1) \\ &= t(z_1) + t(z_2)[z_2 - H_0]^{-1}(z_1 - z_2)[z_1 - H_0]^{-1} t(z_1) \end{aligned} \tag{39.11}$$

Now the arguments are chosen as $z_1 = E+\Delta E+i\epsilon$, $z_2 = E+i\epsilon$, where $E+\Delta E$ is the total three-particle energy, E the total two-particle transition energy, and Eq. (39.11) becomes

$$t(E+\Delta E+i\epsilon)=t(E+i\epsilon)-\Delta E t(E+i\epsilon)[E-H_0+i\epsilon]^{-1} \cdot [E+\Delta E-H_0+i\epsilon]^{-1} t(E+E+i\epsilon) \qquad (39.12)$$

Taking matrix elements of Eq. (39.12) over states in which H_0 is diagonal, we obtain

$$t_{ij}(E+\Delta E+i\epsilon)=t_{ij}(E+i\epsilon)-\Delta E\sum_k t_{ik}(E+i\epsilon) \cdot \frac{t_{kj}(E+\Delta E+i\epsilon)}{(E-E_k+i\epsilon)(E-E_k+\Delta E+i\epsilon)} \qquad (39.13)$$

Therefore, for transition $i \to j$ where $E_i = E_j = E$, the t-matrix element may be reduced to the on-energy shell $t_{ij}(E+i\epsilon)$ plus a term which contains off-shell t's. For $E_i \neq E_j$, the off-shell t is irreducible. This is of interest, but perhaps the form of Eq. (39.12) is more so. Let us rearrange it somewhat and take a limit:

$$\lim_{\Delta E\to 0}\{[t(E+\Delta E+i\epsilon)-t(E+i\epsilon)]/\Delta E\}=-t(E+i\epsilon)[E-H_0+i\epsilon]^{-1}[E+\Delta E -H_0+i\epsilon]^{-1}t(E+\Delta E+i\epsilon)\Rightarrow$$

$$\frac{\partial}{\partial E}t(E+i\epsilon)=-t(E+i\epsilon)[E-H_0+i\epsilon]^{-2}t(E+i\epsilon) \qquad (39.14)$$

which will allow the evaluation of Eq. (39.9). That the off-shell t matrix is a function of the time of collision may be seen from the formal solution of Eq. (39.12):

$$\begin{aligned} t(E+\Delta E+i\epsilon)&=t(E+i\epsilon)\left[1+\Delta E(E+\Delta E-H_0+i\epsilon)^{-1}(E-H_0+i\epsilon)^{-1}t(E+i\epsilon)\right]^{-1}\\ &=t(E+i\epsilon)\left[1-i\Delta E(E+\Delta E-H_0+i\epsilon)^{-1}(E-H_0+i\epsilon)\tau(E+i\epsilon)\right]^{-1} \end{aligned} \qquad (39.15)$$

by Eqs. (39.14) and (39.9).

As we have seen in, say, Eqs. (38.9) and (38.10), the line shift and linewidth for an isolated line in the forward scattering approximation will involve the following off-shell t elements:

$$T_{ii}\sim\left\langle {i \atop \mathbf{p}+\mathbf{x}'}\right|t[E+(\omega-\omega_{ij})]\left|{i \atop \mathbf{p}+\mathbf{x}'}\right\rangle \qquad (39.16a)$$

$$T_{jj}^{+}\sim\left\langle {j \atop \mathbf{p}-\mathbf{x}'}\right|t^{+}[E_j-(\omega-\omega_{ij})]\left|{j \atop \mathbf{p}-\mathbf{x}'}\right\rangle \qquad (39.16b)$$

$$\Gamma_{iijj}\sim\left\langle {i \atop \mathbf{p}+\mathbf{x}'}\right|t[E_i+(\omega-\omega_{ij})]\left|{i \atop \mathbf{k}+\mathbf{x}'}\right\rangle\delta\left(\frac{\mathrm{k}^2-\mathrm{p}^2}{2\mathrm{M}}\right) \cdot\left\langle {j \atop \mathbf{k}-\mathbf{x}'}\right|t^{+}(E_j-(\omega-\omega_{ij}))\left|{j \atop \mathbf{p}-\mathbf{x}'}\right\rangle \qquad (39.16c)$$

Eqs. (39.16) may be used to investigate the limits of validity of the interruption approximation. This approximation demands that shift and width be independent of frequency, among other things. The familiar requirement for validity, $\Delta E\tau = (\omega - \omega_{ij})\tau \ll 1$, will be satisfied in Eqs. (39.16) for the t-matrix elements on-shell, and shift and width will be independent of frequency since these equations will be. Thus this development yields the familiar criterion for interruption validity, although these results may be applied to a more quantitative determination of validity limits than this simple statement regarding collision times.

8

Resonance Broadening

Resonance broadening is a precise analogy to the shell game which, incidentally, is not a game of chance. In fact there is no chance involved. For those who are not aficionados of this particular exercise we should perhaps explain what transpires.

The manipulator uses three empty walnut half-shells. (Some of the truly great used as many as five, but the days of such greatness are, alas, far behind us.) The manipulator also uses what is called a "pea" but is actually a dried bean. The reader who recalls that a blackeyed pea is actually a bean will understand this apparent anomaly. The mark is placed across a flat surface from the manipulator in front of whom rest the three shells, one covering the pea. The mark places a bet—without a bet, there is of course no game as is only proper—and the manipulator raises a shell in order to show the mark the apparent position of the pea. We say "apparent" intentionally. If the reader will grasp two shells with the thumbs and first fingers of his two hands and move them about, raising the rear of each shell occasionally, he will remark the possibility of moving the pea from one to another or of simply picking it up with his little finger. He will hardly be able to accomplish this, but he will see how a maestro will find it child's play to do so. In any event, after the mark has placed his bet and the supposed location of the pea has been indicated, the manipulator performs apparently intricate convolutions with the three empty half-shells. The mark diligently follows the motion of the shell under which he supposes the pea to reside until these covering operations have concluded. He then indicates to the legerdemainian the shell under which the elusive pea is to be found. Of course nothing is there; indeed, it has been some little time since anything has been there. Another bet is placed or, more precisely, another admission is charged for the opportunity of witnessing this slight of hand display.

The resonance broadening game is played with Mother Nature as the manipulator, a collection of molecules as the empty half-shells, and an excitation as the pea. In this game, unfortunately, it is impossible to lay a wager save perhaps on some physicist who is busying himself with the problem. However, our inability to locate the excitation is, in a word, the basis for the uniqueness of the resonance-broadening phenomenon. Such has of course been known for the almost 60 years since Holtsmark used the notion in describing the phenomenon classically with the molecules represented by coupled oscillators. Any number of authors have apparently "rediscovered" the phenomenon in the interim. However, it is not our intention to rediscuss what we have already covered in detail. We may, however, remark that the fifties drew toward their close with our descriptions relatively well established: A simple interruption shape usually described a resonance-broadened line rather well, a line with no apparent shift and whose width is linearly dependent on the density. However, this happy state of affairs was not to obtain for long.

For in 1963 Kuhn and Vaughan (172) encountered a puzzling—at least from the point of view of the last paragraph—phenomenon when observing the resonance broadening of certain helium lines. Let us define something called the "residual width."

Suppose we observe the resonance broadening of a particular spectral line as a function of the gas density. We begin at an upper density of 10^{19} atoms per cubic centimeter or some other number which can be considered without excessive complications. We then measure at lower and lower densities until it has become infeasible to measure at any lower ones (say 10^{15}, although the exact number is an irrelevancy). Suppose further we find that the half width varies linearly with the density over the entire range of densities observed. We now extrapolate the curve to zero density using the same density-width dependence. The width at zero density we call the "residual width."

When there are no other atoms present we would expect the width of a line to be given by its natural width, and this may surely be determined from the measured oscillator strength of the line. It would also seem logical to suppose that we could determine this natural width by determining the residual width. What Kuhn and Vaughan found for the certain helium lines and what indeed seemed curious was that the residual width from resonance broadening was not the same as the experimentally determined natural width. This was considered sufficiently interesting that Kuhn and Lewis (170) looked for and found a similar effect in neon; Vaughan (274) looked for and found one in krypton.

One intriguing explanation was offered: the density dependence changed from linear to something else at densities below those measurable, so that the curve really should have ended up at the natural width. This is an ideal argument as it is irrefutable, although it may have no validity at all. Indeed, the argument proved so appealing that certain authors convinced themselves that they had observed density dependences differing from the linear. Recently, however, Exton (111) has investigated these claims and seems to have demonstrated they arise from improper interpretation of the observations. This is not to say that the question is settled, although it may well be. As we shall see, linear density dependence theories which explain the residual width anomalies do exist. These will be considered in this chapter, although our purpose is really not the investigation of specific aspects of resonance broadening. Rather our purpose is to see what, if anything, the various theoretical approaches have to say about the problem.

First, we shall consider the work of Mead and his collaborators on resonance broadening, work which covered a period of years and might perhaps be best described as resolvant oriented. Following this, we consider the problem from the diagrammatic Greens function point of view and then from the Liouville viewpoint. We conclude with the work of Byron and Foley, work which they described as based on the "sudden approximation."

40 The Resonance Resolvant

In a series of papers covering a temporal period of more than ten years Mead (197, 199), sometimes in collaboration (226), concerned himself with the resonance broadening of spectral lines. If one feels constrained to apply descriptive phrases to his formalism for the purpose of classification, one might decide to include a reference to the resolvant operator therein. In any event, in accordance with our policy of beginning at the beginning, we begin by remarking that Mead approaches the spectral line shape through the susceptibility, a *modus operandi* by now familiar to us. An expression for this quantity was obtained by Reck, Takebe, and Mead (RTM) (226) in terms of the matrix elements of the resolvant operator and in a reasonably straightforward fashion. We shall consider this expression beginning, as did these authors, with a careful description of the notation employed, a notation essentially introduced by Mead (196).

A number n of identical atoms of mass m are placed in a volume V. The initial state is the ground state, $|0\rangle$, in which all the atoms of the assemblage are taken as in ground S states, each atom A having translational momentum $\mathbf{q}_A$, assigned on the basis of a Boltzmann distribution at the assumed temperature T. States other than $|0\rangle$ are now designated by the fashion in which such states differ from the system ground state. Obviously they may so differ (1) by the presence of a photon, (2) by having one or more of the atoms in an excited state, and (3) by having one or more of the atoms with translational momentum differing from the norm provided by the Boltzmann distribution. These situations are indicated as follows:

1 A Greek letter within the vector symbol, say $|\lambda\rangle$, represents the presence of what RTM call a "bare" photon of wave number κ_λ and polarization $\mathbf{e}_\lambda$. The use of the adjective "bare" implies the existence of the participles "clothed" or "dressed." To this we shall return in a moment.

2 $|A_i\rangle$ means that atom A is present in an excited p state with m_l value such that the d_i component of the electric dipole moment exists. Thus for example, for $l=1$, $m_l=0$ in the upper state, $l=0$, $m_l=0$ in the lower, only the d_z component exists.

3 If atom A has a translational momentum $\mathbf{q}_A + \boldsymbol{\kappa}_A$, where $\mathbf{q}_A$ remains its Boltzmann distribution momentum, this is indicated by $|\boldsymbol{\kappa}(A)\rangle$. It should be obvious that symbols arising from the three situations described may appear in various combinations within the eigenvector symbol. We now turn to the dressed and undressed photons.

Let us suppose a photon to be present amongst our n absorbing atoms in their ground state, a photon whose frequency is close to the "resonance" frequency of the assemblage, that is, close to the frequency difference between atomic ground and excited states. Mead (196) then views this situation as follows: We suppose there to exist a stationary state to which this photon $+\,n$ atoms is a zeroth-order approximation. This state will consist of the admixture of such zeroth-order functions as follows:

1 What we may call "real" states, which may be:
   (a) states with no photon present and one atom excited to the level with which the photon is in resonance;
   (b) states with a photon of slightly different frequency present, the result of an excitation and emission.

2 What we may call "virtual" states, which have energies considerably different from our originally postulated state and which Mead typifies by:
   (a) atoms excited to a level not in resonance with the original photon frequency;
   (b) atoms excited to the resonance level with the emission of another photon rather than the absorption of the one originally present.

It may be anticipated that the "virtual" states make a considerably smaller contribution to the stationary state than do the "real," or at least so Mead presumed. On this basis, he used perturbation theory through second order to eliminate these "virtual" states much as the virtual states of the electron were eliminated by Arnous and Heitler (Section 7) in clothing the electron states. This of course leads to the dressed photon state idea, and only "real" transitions among those dressed states need be considered. In the earlier paper, Mead (196) used the Arnous-Bleuler transformation of Section 7, in the later, RTM used a somewhat simpler approach in order to accomplish the same purpose.

RTM supposed that these dressed photon states, $|\bar{\lambda}\rangle$, could be built up of (1) states with one photon present, no excited atoms and (2) states having a single excited atom and no photons present. Specifically, we are saying

$$|\bar{\lambda}\rangle = |\lambda\rangle + \sum_b |b\rangle\langle b|\bar{\lambda}\rangle \tag{40.1a}$$

where $|\lambda\rangle$ is (1) the state with one photon present and (2) $|b\rangle$ are simply the general run of zeroth-order states. Remark now that our diagonality requirement leads to

$$H_{\text{tot}}|\bar{\lambda}\rangle = \nu_\lambda|\bar{\lambda}\rangle \tag{40.1b}$$

We see that this essentially eliminates the virtual states by the rather simple expedient of stating that they do not exist, an approximation that RTM account for by restricting their considerations to frequencies near line center. The reader will see the logic of this in that strongly nonresonant states are going to assume importance in the line wings.

We will recall from Eq. (12.20) that

$$\mathbf{P}/\mathbf{E} = (\epsilon - 1)/4\pi = \chi_e \tag{40.2}$$

Further application of our recollective facilities tells us that the refractive index, n, is related to the dielectric constant and the magnetic permeability by $n^2 = \epsilon\mu \doteq \epsilon$ for gases of high refractivity. Therefore, Eq. (40.2) may be written as $4\pi\mathbf{P} = (n^2 - 1)\mathbf{E}$. It should be apparent that, when we replace the classical quantities by quantum operators, we obtain Mead's definition of the refractive index,

$$4\pi\langle 0|P_i(\mathbf{r})|\bar{\lambda}\rangle = \left[n^2(\nu_\lambda) - 1\right]\langle 0|E_i(\mathbf{r})|\bar{\lambda}\rangle \tag{40.3a}$$

In Eq. (40.3a), $P_i(\mathbf{r})$ is the operator for the ith component of the dipole moment density at the point $\mathbf{r}$, $n(\nu_\lambda)$ now of course being the frequency-dependent refractive index.

RTM find it useful to rewrite Eq. (40.3a) somewhat by first recalling that the electric field, $\mathbf{E}$, may always be expressed in terms of a longitudinal Coulomb field and a transverse vector potential: Now the vector potential, $\mathbf{A}$, will have a time dependence of the form $\exp[-i\nu_\lambda t]$, and the photon annihilation and creation operators which appear in it will zero out everything except our single allowed photon $|\lambda\rangle$ so that

$$\langle 0| - \dot{A}|\bar{\lambda}\rangle = i\nu_\lambda\langle 0|A|\lambda\rangle$$

which leads RTM to use a form of Eq. (40.3a)

$$\langle 0|4\pi P_\alpha|\bar{\lambda}\rangle = F(\nu_\lambda, \boldsymbol{\kappa}_\lambda, \boldsymbol{\kappa}_\lambda\cdot\boldsymbol{\epsilon}_\lambda)\langle 0|E_\alpha|\bar{\lambda}\rangle \tag{40.3b}$$

where

$$E = E_{\text{Coul}} + i\nu_\lambda A \tag{40.3c}$$

In Eq. (40.3b) F is the susceptibility, transverse or longitudinal depending on the component α under consideration.

We have seen that the spectral line contour may be taken as given by the imaginary portion of the susceptibility. In a word then, the manipulations of RTM are aimed at using the dressed photon states in some fashion in order to obtain χ from Eqs. (40.2) and (40.3) as a function of the resolvant matrix elements. The development is, as is to be expected, largely concerned with the algebraic manipulations of various straightforward matrix elements, the thread of development running through two rather well defined if somewhat detail-obscured steps:

1 First, the transverse susceptibility is determined. In order to evaluate Eq. (40.3a) for this quantity, we are going to require matrix elements of the polarization. If atom A has electric dipole moment $\mathbf{d}_A$, we can agree that the polarization may be written

$$\mathbf{P}(\mathbf{r}) = \sum_A \mathbf{d}_A \delta(\mathbf{r} - \mathbf{r}_A) \Rightarrow P_\eta = V^{-1} \int \epsilon_\eta \cdot \mathbf{P}(\mathbf{r}) e^{-i\kappa_\eta \cdot \mathbf{r}} d\mathbf{r}$$

$$= V^{-1} \sum_A \epsilon_\eta \cdot \mathbf{d}_A e^{-i\kappa_\eta \cdot \mathbf{r}} \tag{40.4}$$

wherein we suppose η to be a transverse mode. Recall now that we have defined these dressed photon eigenvectors as eigenvectors of the total Hamiltonian having eigenvalues ν_λ, that is, $H_{\text{tot}}|\bar{\lambda}\rangle = \nu_\lambda|\bar{\lambda}\rangle$, $|\bar{\lambda}\rangle$ given by Eq. (40.1a). If we agree that

$$\langle 0|4\pi P_\eta|\bar{\lambda}\rangle = \sum_A \langle 0|4\pi P_\eta|A_j, \kappa_\eta(A)\rangle\langle A_j, \kappa_\eta(A)|\bar{\lambda}\rangle$$

$$= 4\pi V^{-1} \mathsf{d} \sum_A \epsilon_{kj} \langle A_j, \kappa_\eta(A)|\bar{\lambda}\rangle \tag{40.5a}$$

and that, from Eqs. (40.1),

$$\langle a|\bar{\lambda}\rangle = (\nu_\lambda - \nu_a + i\xi)^{-1} \sum_{b \neq a} \langle a|H_{\text{tot}}|b\rangle\langle b|\bar{\lambda}\rangle \tag{40.5b}$$

which may be obtained by a straightforward substitution of Eq. (40.1a) into Eq. (40.1b), we obtain an expression for the right side of Eq. (40.5a). It is at this point that it is no longer obvious why we should reproduce the lengthy expressions that follow. Reasonably straightforward matrix element manipulation leads to the required

$$\langle A_j, \kappa_\eta(A)|\bar{\lambda}\rangle = -\mathsf{d}\langle 0|E_\eta|\bar{\lambda}\rangle\left[\omega - K(\mathbf{q}_A, \kappa_\eta) - \Delta\right]^{-1}_{jk} \epsilon_{\eta k} \tag{40.6a}$$

where

$$K(q_A, \kappa_\eta) = \left[2\mathbf{q} \cdot \kappa_\eta + \kappa_\eta^2\right]/2\mathsf{m}, \quad \omega = \nu_\lambda - \nu_0 \tag{40.6b}$$

Recall now that $|A_j, \boldsymbol{\kappa}_\eta(A)\rangle$ is the eigenvector for a system state wherein atom A is excited with m_j value appropriate to dipole moment polarization i and wherein atom A has translational momentum in excess of Boltzmann by an amount $\boldsymbol{\kappa}_\eta$. As we might suspect, Δ is an operator of considerable consequence and considerable complication whose definition arises naturally during the evaluation leading to Eq. (40.6a). We make no attempt to detail it.

Equation (40.6a) is substituted into Eq. (40.5a), the result summed over A using a Boltzmann distribution. Finally, the latter result is utilized in the solution of Eq. (40.3b) for the transverse susceptibility:

$$F_t(\omega,\kappa_\eta)=4\pi d^2N\left(\frac{\beta}{\pi}\right)^{3/2}\int e^{-\beta q^2}\epsilon_{\eta j}\left[\omega-K(q,\kappa_\eta)-\Delta(\omega,\boldsymbol{\kappa}_\eta,q)\right]^{-1}_{jk}dq \tag{40.7}$$

with an analogous equation for F_l (longitudinal). This might be called the conclusion of step (1).

2 Second, the transverse susceptibility is reexpressed in terms of the matrix elements of the resolvant operator. This is accomplished by:

(a) finding a series expansion for the matrix elements, Δ_{jk},

(b) finding a series expansion for the matrix elements of the resolvant operator, $R(\nu)=(\nu-H_{\text{tot}})^{-1}$,

(c) comparing these two series in order to conclude that

$$\langle A_j,\boldsymbol{\kappa}(A)|R(\nu)|A_k,\boldsymbol{\kappa}(A)\rangle=\left[\nu-E_A-\Delta(\omega,\boldsymbol{\kappa},q_A)\right]^{-1}_{jk} \tag{40.8}$$

so that Eq. (40.7) becomes

$$F_t(\omega,\kappa_\eta)=-4\pi d^2N\left(\frac{\beta}{\pi}\right)^{3/2}\int e^{-\beta q_A^2}\epsilon_{\eta j}$$
$$\times\langle A_j,\boldsymbol{\kappa}_\eta(A)|R(\nu)|A_k,\boldsymbol{\kappa}_\eta(A)\rangle\epsilon_{\eta k}\,d\mathbf{q}_A \tag{40.9}$$

Equation (40.9) is the general RTM result for the spectral line shape. We do not follow these authors further for the moment but turn to the later work of Mead (199). That author proceeded from Eq. (40.9) by setting up a differential equation for the resolvant, then considering the limiting cases involving the classical path, the two-body statistical and interruption approximations and the corrections to these considerations arising from the quantum and many-body effects.

As his first step Mead writes out the integration over the translational degrees of freedom which of course form a part of a determination of the matrix elements of the resolvant. The translational eigenvectors will surely be

$$|\boldsymbol{\kappa}(A)\rangle=(V^n)^{-1/2}\exp[i\Gamma],\qquad \Gamma=\boldsymbol{\kappa}\cdot\mathbf{r}_A+\hbar^{-1}\sum_D\mathbf{p}_D\cdot\mathbf{r}_D \tag{40.10}$$

since atom A will have momentum $\boldsymbol{\kappa}$ in addition to the momentum $\mathbf{p}_A$ which it possesses by virtue of the Boltzmann distribution. (At this point we begin the insertion of $\hbar$, although $\hbar=1$ still holds in our system of units, in order to follow the expansion which uses this as a parameter of smallness. $\hbar$ will be dropped after this expression has been obtained.) Therefore,

$$\langle B_j,\boldsymbol{\kappa}(A)|R(\nu)|C_k,\boldsymbol{\kappa}(A)\rangle=V^{-n}\int d\mathbf{r}_1\cdots\int d\mathbf{r}_n\langle B_j|\overline{R}(\nu)|C_k\rangle \tag{40.11a}$$

$$\overline{R}(\nu)=\exp[-i\Gamma]R(\nu)\exp[i\Gamma] \tag{40.11b}$$

Mead takes the Hamiltonian entering the resolvant as that for the matter alone,

$$H=\sum_A \frac{p_A^2}{2\mathrm{m}}+H_{\mathrm{int}}+\hbar\mathrm{V}-\hbar^2\sum_A \frac{q_A^2}{2\mathrm{m}} \tag{40.12a}$$

where the last term has been added so that $H|0\rangle=0$. H_{int} is the Hamiltonian for the internal (electronic) energies of the atom, the ground rules for our problem yielding eigenvalues for this of ν_0 (one atom excited). Therefore, with $\omega=\nu-\nu_0$, we see that Eq. (40.11b) becomes

$$\bar{R}(\nu)=\exp[-\mathrm{i}\Gamma]\left[\omega+\hbar\sum_D \frac{\nabla_D^2}{2\mathrm{m}}+\sum_D \frac{p_D^2}{2\mathrm{m}}-\mathrm{V}\right]^{-1}\exp[\mathrm{i}\Gamma] \tag{40.11c}$$

Multiplying Eq. (40.11c) through on the left, first by exp[iΓ], then by the bracketed expression, we obtain

$$\left(\omega+\hbar\sum_D \frac{\nabla_D^2}{2\mathrm{m}}+\sum_D \frac{p_D^2}{2\mathrm{m}\hbar}-\mathrm{V}\right)\{\exp[\mathrm{i}\Gamma]\bar{R}(\nu)\}=\exp[\mathrm{i}\Gamma] \tag{40.12b}$$

We will surely agree that

$$\sum_D \nabla_D^2\frac{\{\exp[\mathrm{i}\Gamma]\bar{R}(\nu)]\}}{2\mathrm{m}}=\sum_D \frac{\{\nabla^2\exp[\mathrm{i}\Gamma]\}\bar{R}(\nu)}{2\mathrm{m}}$$

$$+\sum_D \frac{\{\nabla\exp[\mathrm{i}\Gamma]\}\cdot\nabla\bar{R}(\nu)}{2\mathrm{m}}+\sum_D \frac{\exp[\mathrm{i}\Gamma]\nabla^2\bar{R}(\nu)}{2\mathrm{m}}$$

where

$$\sum_D \frac{\nabla_D^2\exp[\mathrm{i}\Gamma]}{2\mathrm{m}}=\frac{-\kappa^2}{2\mathrm{m}}-\boldsymbol{\kappa}\cdot\mathbf{v}_A-\sum_D \frac{p_D^2}{2\mathrm{m}},\qquad \mathbf{v}_A=\frac{\mathbf{p}_A}{\mathrm{m}}$$

so that Eq. (40.12b) becomes

$$\left(\omega-\boldsymbol{\kappa}\cdot\mathbf{v}_A+\frac{\hbar\kappa^2}{2\mathrm{m}}+\mathrm{i}\sum_D \mathbf{v}_D\cdot\nabla_D+\frac{\mathrm{i}\hbar\boldsymbol{\kappa}\cdot\nabla_A}{\mathrm{m}}+\hbar\sum_D \frac{\nabla_D^2}{2\mathrm{m}}-\mathrm{V}\right)\bar{R}=1 \tag{40.13}$$

If we suppose R to be expanded using powers of $\hbar$,

$$\bar{R}=\bar{R}_0+\hbar\bar{R}_1+\cdots \tag{40.14}$$

we may obtain the following from Eq. (40.13):

$$\left(\lambda+\mathrm{i}\sum_D \mathbf{v}_D\cdot\nabla_D-\mathrm{V}\right)\bar{R}_0=1, \tag{40.15a}$$

$$\left(\lambda+\mathrm{i}\sum_D \mathbf{v}_D\cdot\nabla_D-V\right)\bar{R}_1=-\left\{\frac{\kappa^2}{\mathrm{m}}+\frac{\mathrm{i}\boldsymbol{\kappa}\cdot\nabla_A}{\mathrm{m}}+\frac{\sum_D \nabla_D^2}{2\mathrm{m}}\right\}\bar{R}_0 \tag{40.15b}$$

and so on where

$$\lambda = \omega - \boldsymbol{\kappa}\cdot\mathbf{v}_A \tag{40.15c}$$

Equation (40.15a) corresponds to the situation $\hbar \to 0$, that is to say, to, among other things, the classical path, Eq. (40.15b) the first-order correction to it.

A change of coordinates from r_D to

$$t = \left(\sum_D v_D^2\right)^{-1} \sum_D \mathbf{v}_D \cdot \mathbf{r}_D \tag{40.16}$$

and $(3n-1)$ additional orthogonal coordinates, η, for which we shall not require specific expressions, leads to the replacement of Eq. (40.15a) by

$$\left[\lambda + \mathrm{i}\frac{\partial}{\partial t} - \mathsf{V}(t,\eta)\right]\bar{R}_0(t,\eta) = 1 \tag{40.17}$$

$$\bar{R}_0(t,\eta) = \mathfrak{U}_\eta(t,t_0)\Xi(t,\eta)e^{\mathrm{i}\lambda t} \tag{40.18}$$

where $\mathfrak{U}_\eta$ is a TDO satisfying

$$i\frac{\partial}{\partial t}\mathfrak{U}_\eta(t,t_0) = V(t,\eta)\mathfrak{U}_\eta(t,t_0) \Rightarrow \mathfrak{U}_\eta(t,t_0) = \mathsf{T}\exp\left(-\mathrm{i}\int_{-t_0}^{0}\mathsf{V}(t',\eta)\,dt'\right) \tag{40.19}$$

The unitarity of $\mathfrak{U}_\eta$ allows us to solve Eqs. (30.17) and (30.18) for

$$\Xi(t,\eta) = -\mathrm{i}\int_{-\infty}^{t}\mathfrak{U}_\eta(t_0,t')e^{-\mathrm{i}\lambda t'}\,dt'$$

which of course leads to

$$\bar{R}_0(t,\eta) = -\mathrm{i}\int_{-\infty}^{t}\mathfrak{U}_\eta(t,t')e^{\mathrm{i}\lambda(t-t')}\,dt' = -\mathrm{i}\int_0^{\infty}\mathfrak{U}_\eta(t,t-t'')e^{\mathrm{i}\lambda t''}\,dt'' \tag{40.20a}$$

where $t'' = t - t'$. Transformation back to the original coordinates yields

$$\bar{R}_0(r) = -\mathrm{i}\int_0^{\infty}\mathfrak{U}(r, r - vt)e^{\mathrm{i}\lambda t}\,dt \tag{40.20b}$$

wherein $\mathfrak{U}$ is the TDO for internal states of a system being translated with velocity, $\mathbf{v}$, $\mathbf{v}_D$ for each atom D, from an initial configuration $\mathbf{r} - \mathbf{v}t$ to a final configuration $\mathbf{r}$.

When Eq. (40.20b) is substituted into Eq. (40.11) we obtain the Mead result in the classical path approximation. The next approximation considered by that author is the binary or two-body approximation. In such an approximation, if the single excitation (among the n atoms) is associated with atom A at the beginning of a broadening collision, it is supposed to be found on A at the end of the process, and only matrix elements $\langle A_i|\bar{R}|A_j\rangle$, not $\langle B_j|\bar{R}|C_k\rangle$ as in the general Eq. (40.16), need be considered. One binary collision at a time.

Thus the TDO, $\mathfrak{U}$, becomes a time-ordered—corresponding to collision ordering—product of $\mathfrak{U}$'s for each successive collision. At which point one has reasoned oneself into a rather familiar viewpoint: The effect of n atoms may be replaced (or almost) by the nth power of the effect of one atom.

$$\langle A_i|\mathfrak{U}(r,r-vt)|A_j\rangle=\delta_{ij}\prod_{D\neq A}\mathfrak{U}_D(\mathbf{r}_{AD},\mathbf{v}_{AD},t) \tag{40.21a}$$

$$\mathfrak{U}_D(\mathbf{r}_{AD},\mathbf{v}_{AD},t)=\langle A_k|\mathfrak{U}_{AD}(\mathbf{r}_{AD},\mathbf{v}_{AD},t)|A_k\rangle/3 \tag{40.21b}$$

$$\mathfrak{U}_{AD}=\mathrm{T}\exp\left\{-\mathrm{i}\int_0^t \mathsf{V}_{AD}(\mathbf{r}_{AD}-\mathbf{v}_{AD}(t-t'))\,dt'\right\}$$

$$=\mathrm{T}\exp\left(-\mathrm{i}\int_0^{vt}\frac{\mathsf{V}_{DD}(x,y,z-vt+\zeta)\,d\zeta}{v}\right),\qquad \zeta=vt' \tag{40.21c}$$

Obviously, one may now go to the statistical approximation by supposing v very small, to the interruption by supposing it very large. For v and hence vt small, the integral in Eq. (40.21c) may be replaced by $\mathsf{V}t$ so that Eq. (40.21a) becomes

$$\mathfrak{U}(r,r-vt)=\exp\left[-\mathrm{i}\alpha t/(\rho^2+v^2t^2)^{3/2}\right]\doteq\exp\left[-\mathrm{i}\alpha t/\rho^3\right] \tag{40.22}$$

since $V=\alpha/r^3$ for the dipole-dipole interaction. (We do not follow what some may consider the superior treatment of Mead in this step.) For the statistical case we take the optical collision diameter, ρ, very large so that $\rho^3\sim V$, the volume. Therefore, Eq. (40.21c) becomes

$$\langle A_i|\mathfrak{U}(\mathbf{r},\mathbf{r}-\mathbf{v}t)|A_j\rangle=\left\{\exp\left[-\mathrm{i}\alpha t/\rho^3\right]\right\}^n=\exp\left[-\mathrm{i}\alpha Nt\right]$$

which leads to the following result for Eq. (40.20b):

$$\overline{R}_0(r)=-\mathrm{i}\int_0^\infty \exp\left[\lambda-\mathrm{i}\alpha N\right]t\,dt=\left[\lambda-\mathrm{i}\alpha N\right]^{-1} \tag{40.23}$$

The line shape is of course given by the imaginary part of the susceptibility so that, from Eq. (40.9), we obtain

$$I(\nu)=\text{const}\int dv\,N^2e^{-\beta v^2}\left[(\nu-\nu_0-\boldsymbol{\kappa}\cdot\mathbf{v})^2+(\alpha N)^2\right]^{-1} \tag{40.24}$$

for the classical path, binary collision, statistical result.

As we know [cf. Chapter 5 of Breene (9)], it so happens that, in the classical treatments, the interruption treatment of resonance broadening yields the same line shape as does the statistical treatment, this due to the form (r^{-3}) of the interaction potential. Therefore, we would expect and Mead obtains a result essentially similar to Eq. (40.24) for the interruption result. For v very large, Eq. (40.21c) and hence Eq. (40.19) take on the aspect of the S matrix. We will recall that the T matrix is related to the cross section, and this fact is utilized in a straightforward treatment leading to the reduction which is sought.

RTM had developed a density expansion for the shift and width of the line based largely on dimensional arguments. As Royer has pointed out (cf. Section 23), the convergence of these expansions offer certain rather knotty problems, problems to which RTM did not address themselves. Nevertheless, the qualitative conclusions at which they arrive, although perhaps obvious, are worthy of repetition. The term linear in the density is found to be of greatest importance in the line wing, a reflection of the well-known ascendancy of binary collisions in that region. RTM further conclude that the higher and higher order density dependences become successively more important as we approach closer to line center.

An investigation by Mead of the applicability of the classical path leads him to conclude that an N-dependent term arises as the first quantum correction to the classical approximation. Mead found that this leads to a quantum deviation of "several percent" from the classical result for the $3^1S \to 2^1P$ (7281 Å) He line measured by Kuhn and Vaughan (172) and Vaughan (274).

Finally, we remark on the square root dependence of the line width on the density obtained by RTM. Such a square-root dependence was originally obtained by Mead (198), apparently as a consequence of the introduction of a cutoff, originally introduced earlier [Eq. (31) of Mead (197)]. The parameter, a, characterizing the cutoff is

$$a = \pi r_0^3 N/6$$

the width $\delta \sim a^{1/2}\Gamma\sigma$, $\Gamma = 3\pi^2/4$. Mead (196) states "··· two absorbers are not permitted to approach one another more closely than a distance of the order of an atomic radius··· This can be accounted for at least approximately without introducing any explicit correlation (which would spoil our formalism) by simply doctoring the interaction in such a way··· This means that r_0 should be of the order of an atomic radius and independent of N, and that a be proportional to $1/N$."

Hynne (150) has published an application of a statistical and essentially macroscopic many-body theory of the complex refractive index of molecular fluids to thin vapors. He obtains a square-root dependence of width on density below a density of 10^{14} cm^{-3}, a linear dependence above. The author himself feels that the results are semiquantitative.

41 Diagrammatic Treatment of Resonance Broadening

As we have seen in Chapter 6, the generalized diagrammatic or Greens function treatment of spectral line broadening developed by Ross carries within it the details of resonance broadening, although Ross has sufficient with which to occupy himself without investigating this area in detail. Somewhat later, Bezzerides (79) and Zaidi (288) did consider this particular broadening situation using these methods. We shall consider a portion of the latter author's work somewhat later. First, however, we shall consider the extrapolation of the Ross methods of Chapter 6 to resonance broadening.

As we shall assume a two-state atom, Eq. (31.1) is again applicable, L_{1001} is again the object of our search. This means the evaluation of Eq. (30.5b). Now the third term on the right of this equation should be important, if not as important as it was in the earlier situation, but we shall not consider it. It could certainly be introduced later. We now concentrate on the second term which is unique to resonance broadening except in the rare case of accidental resonance with a foreign perturber. [It is of course true that the last two terms in Eq. (31.2′) relate to similar phenomena.] For Eq. (30.5b) then we obtain

$$L_{1001}(p,p';p-k,p'+k)=\mathsf{G}_{11}(p)\mathsf{G}_{00}(p-k)\delta(p-p'-k)+i\sum_{\substack{\nu\eta\\ \lambda\mu}}\mathsf{G}_{1\lambda}(p)$$
$$\cdot\mathsf{G}_{\mu 0}(p-k)\int\frac{dq'}{(2\pi)^4}\langle p,q'-k|T_{\lambda\eta\mu\nu}(\omega_{p+q'-k})|p-k,q'\rangle$$
$$\cdot L_{\nu 0\eta 1}(q',p';q'-k,p'+k) \qquad (41.1\text{a})$$

All summation indices run over 0 and 1, and the summation in Eq. (41.1a) contains 16 terms. Now we shall suppose, correctly as it turns out, that the G_{ij} will have a form as given by Eqs. (31.13) and (31.14), that is, the off-diagonal elements will be linearly dependent, the diagonal elements dependent on a density independent factor plus a factor linearly dependent on the density. Thus, terms involving G_{10} will be of higher order in the density and will be dropped as of higher order. Therefore, Eq. (41.1a) becomes

$$L_{1001}(p,p';p-k,p'+k)=\mathsf{G}_{11}(p)\mathsf{G}_{00}(p-k)\delta(p-p'-k)+i\mathsf{G}_{11}(p)$$
$$\cdot\mathsf{G}_{00}(p-k)\int\frac{dq'}{(2\pi)^4}\{\langle p,q'-k|T_{1001}(\omega_{p+q'-k})|p-k,q'\rangle$$
$$\cdot L_{1001}(q',p';q'-k,p'+k)$$
$$+\langle p,q'-k|T_{1000}(\omega_{p+q'-k})|p-k,q'\rangle$$
$$\times L_{0001}(q',p';q'-k,p'+k)\} \qquad (41.1\text{b})$$

Let us consider Eq. (30.5b) in conjuction with Eq. (41.1b). The first-order solution for L_{1001} will be $\mathsf{G}_{11}\mathsf{G}_{00}$, that for L_{0001} will be $\mathsf{G}_{01}\mathsf{G}_{00}$. Thus the arguments of the last paragraph eliminate the second term on the right of Eq. (41.1b). It is now worthwhile to display the thus modified Eq. (41.1b):

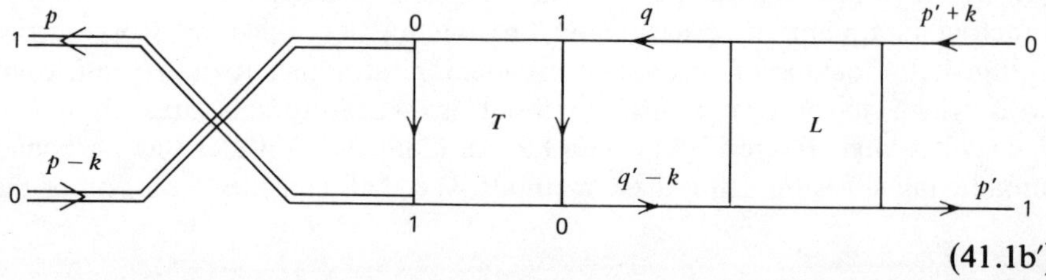

(41.1b′)

It is of interest to remark that by integrating over p we close the left end, by integrating over p' the right, and we are left with a polarization part. Such would doubtless have a certain aesthetic appeal, but, in the footsteps of Ross, we merely close the right end in order to obtain a vertex equation as in Eqs. (31.15).

$$Q_{10}(p;k)=\int\frac{dp'}{(2\pi)^4}L_{1001}(p,p';p-k,p'+k) \tag{41.2a}$$

$$Q_{10}(p;k)=\mathsf{G}_{11}(p)\mathsf{G}_{00}(p-k)\left[1+\mathrm{i}\int\frac{dq'}{(2\pi)^4}\cdot\langle p,q'-k|T_{1001}(\omega_{p+q'-k})|p-k,q'\rangle Q_{10}(q';k)\right] \tag{41.2b}$$

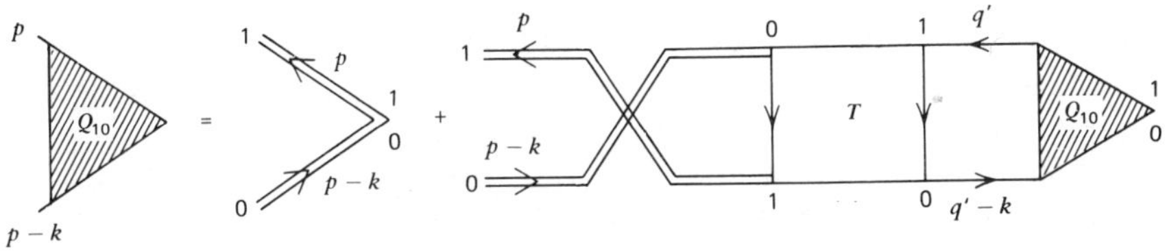

(41.2b′)

For our future reference, we remark that this compares directly with Fig. 1 of Zaidi (288) with the right side closed—no integration over q' there—and the vertex of L, Q, replaced by the susceptibility. Now if we close the left side of Eq. (41.2b′) by integrating over p we obtain the polarization part or polarization operator, Π, as in Fig. 41.1. This will be of interest to us in considering Zaidi's work in Section 42.

There is no necessity for repeating the steps which led to Eq. (31.12) for the mass operator; they will be the same. If we suppose there to be N_0 atoms in the lower state, N_1 in the upper, we obtain

$$\begin{aligned}\Sigma_{\alpha\beta}&=N_1\langle\langle\mathbf{pq}|T_{\alpha1\beta1}(\omega_p+q^2/2\mathrm{m}+E_1-\mu)|\mathbf{pq}\rangle\rangle_{\mathrm{av}}\\&+N_0\langle\langle\mathbf{pq}|T_{\alpha0\beta0}(\omega_p+q^2/2\mathrm{m}+E_0-\mu)|\mathbf{pq}\rangle\rangle_{\mathrm{av}}\end{aligned} \tag{41.3}$$

From Eqs. (31.14) and (41.3) the G_{11} and G_{00} follow. Now let us introduce the

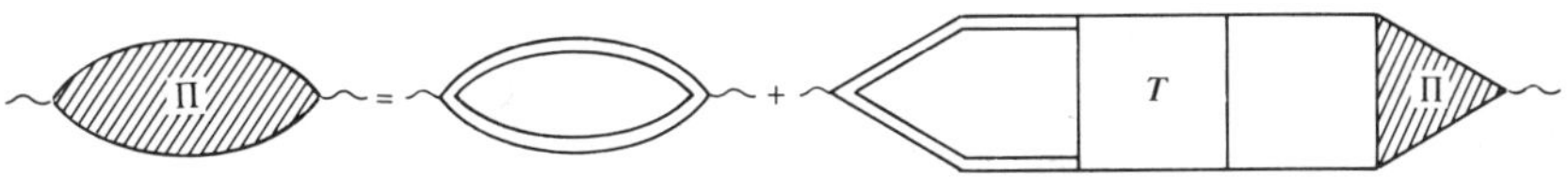

Figure 41.1

following shorthand:

$$\Sigma_{\alpha\alpha}(\mathrm{p}) = N_1 T_{\alpha 1 \alpha 1} + N_0 T_{\alpha 0 \alpha 0} \tag{41.4}$$

Thus

$$\mathrm{G}_{11}(p) = \left[\omega_p - p^2/2\mathrm{m} - \omega_0 + \mu - N_1 T_{1111} - N_0 T_{1010}\right]^{-1} \tag{41.5a}$$

$$G_{00}(p) = \left[\omega_p - p^2/2\mathrm{m} + \mu - N_1 T_{0101} - N_0 T_{0000}\right]^{-1} \tag{41.5b}$$

The β summation of Eq. (31.15) may now be carried out. Note that the integration to which this corresponds is over ω_p. Therefore, as in Eq. (31.25), we simply evaluate the residue, which is $\mathrm{G}_{00}(p-k)$ with a Boltzmann factor that we ignore, at the pole which is defined by $\omega_p = p^2/2\mathrm{m} + \omega_0 - \mu + N_1 T_{1111} + N_0 T_{1010}$ by $\mathrm{G}_{11}(p)$. The result

$$Q_{10}^0(\mathbf{p}; k) = \mathrm{i}\ \mathrm{const}\, f(p)\big[\omega_0 - \omega_k + p^2/2\mathrm{m} - (p-k)^2/2\mathrm{m} + N_1(T_{1111} - T_{0101}) \\ - N_0(T_{0000} - T_{1010})\big]^{-1} \tag{41.6}$$

where $f(p)$ is a Boltzmann factor.

If we follow the lead of Ross, the idea here is to get $Q_{10}(q'; \mathrm{k})$ outside the integral sign in Eq. (41.2b) so that this quantity may then be obtained from a simple algebraic equation. This of course means a careful consideration of the second term on the right of Eq. (41.2b). First of all, it obviously behooves us to evaluate the integral over $\omega_{q'}$ in that equation. We do so by supposing that Q_{10} will possess the same poles as Q_{10}^0, that is, the poles of $\mathrm{G}_{11}(q)\mathrm{G}_{00}(q'-k)$. From Eqs. (41.5) these poles will appear at $\omega_{q'} = q'^2/2\mathrm{m} + \omega_0 - \mu + N_1 T_{1111} + N_0 T_{1010}$ and $\omega_{q'} = \omega_k + (q'-k)^2/2\mathrm{m} - \mu + N_1 T_{0101} + N_0 T_{0000}$. We would thus obtain a pair of terms, one from each pole. The Bose factors are replaced by Boltzmann factors, and these Boltzmann factors are the only portions of the integrand which will specifically concern us. We shall approximate G_{11} and G_{00} now by G_{11}^0 and G_{00}^0, that is, we will evaluate these Boltzmann factors at the poles of the free-particle Greens function. Now by Eq. (31.5) the evaluation of the Boltzmann factor leads to

$$\left(\frac{2\pi\beta}{\mathrm{m}}\right)^{3/2} \exp\left[\frac{-\beta k^2}{2\mathrm{m}}\right] \int \frac{d\mathbf{q}'}{(2\pi)^2} \{N_0 + N_1\} \langle \mathbf{p}, \mathbf{q}' - \mathbf{k} | T_{1001} | \mathbf{p} - \mathbf{k}, \mathbf{q}' \rangle Q_{10}(q'; k)$$

Now we shall restrict our result to the energy shell. In order to see what we mean by this, let us consider Eq. (41.2b′). The energy incoming to the T matrix (at the top) is $(\mathbf{p}-\mathbf{k})^2/2\mathrm{m} + \mathbf{q}'^2/2\mathrm{m}$; that outgoing is $\mathbf{p}^2/2\mathrm{m} + (\mathbf{q}' - \mathbf{k})^2/2\mathrm{m}$. We may therefore restrict ourselves to the energy shell by inserting a Dirac delta function that will assure the equality of incoming and outgoing.

$$\mathrm{const}(N_1 + N_0)\exp\left(\frac{-\beta k^2}{2\mathrm{m}}\right) \int \frac{d\mathbf{q}'}{(2\pi)^3} \langle \mathbf{p}, \mathbf{q}' - \mathbf{k} | T_{1001} | \mathbf{p} - \mathbf{k}, \mathbf{q}' \rangle \\ Q_{10}(\mathbf{q}'; k)\delta(\mathbf{p}\cdot\mathbf{k} - \mathbf{q}'\cdot\mathbf{k}) \doteq \mathrm{const}(N_1 + N_0) Q_{10}(\mathbf{p}; k)$$

We substitute this into Eq. (41.11b) as modified in the manner of Eqs. (30.19) in order to obtain

$$Q_{10} = Q_{10}^{0}\left[1 + \text{const}(N_1 + N_0)Q_{10}\right] \Rightarrow$$

$$\begin{aligned} Q_{10} &= \left[\left(Q_{10}^{0}\right)^{-1} + \text{const}(N_1 + N_0)\right]^{-1} \\ &= \left\{ \mathrm{i}C_1\left[\omega_0 - \omega + \mathbf{v}\cdot\mathbf{k} + N_0(T_{1111} - T_{0101}) - N_1(T_{0000} - T_{1010})\right] \right. \\ &\quad \left. + C_2(N_0 + N_1)T_{1001}\right\}^{-1} \end{aligned} \tag{41.7}$$

We may obviously replace $N_0 + N_1$ by N, the density multiplied by the sum of a pair of Boltzmann factors, say $[1 + \exp(-\beta E_1)]$. Thus we have arrived at the not very unusual result that there is a linear dependence of the width on the density. The factor within the square bracket, $N_0(T_{1111} - T_{0101}) - N_1(T_{0000} - T_{1010})$, tells us the line shift which is also linearly dependent on the density. It is obvious, however, that this shift, involving nothing but differences, should be quite small and possibly unobservable.

The familiar and useful if unrealistic assumption to the effect that $N_1 = 1$, $N_0 = N$, a single excited atom, would lead to a width dependent on $N + 1 = N$ for any densities. (Remark that, if, instead of 1 we have some sensible constant so that $N + C$ is the density dependence over a substantial range of densities, the residual width mystery ceases to be any mystery at all. More on this later, but this is probably the direction in which the explanation lies.) Such an assumption would thus have no effect for the width, but it could lead to a deceptively large line shift. If the lower level involved in the transition is an excited state, the width should include a temperature dependence. We shall turn to certain of Zaidi's diagrammatic studies after a remark on the brief but interesting study by Nowotny (213).

As we are by now aware, the third term in Eq. (30.5b′) is of greater consequence than the second in resonance broadening. We may therefore write out a vertex equation of the form Eq. (31.15′) with the second term on the right replaced by the second term on the right of Eq. (30.5b′) appropriately modified. In doing so, however, let us close the left sides of all diagrams in Eq. (31.15′) by integration. This means, of course, that we go from Q to the susceptibility or line shape. We obtain Fig. 41.1 which may then be expanded as

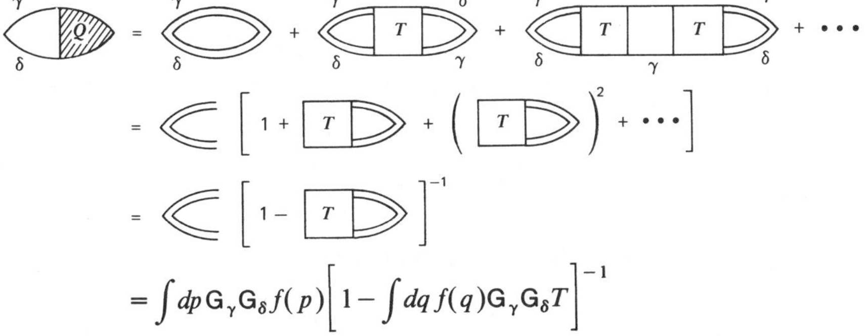

$$= \int dp\, \mathrm{G}_\gamma \mathrm{G}_\delta f(p)\left[1 - \int dq\, f(q)\mathrm{G}_\gamma \mathrm{G}_\delta T\right]^{-1}$$

This is not precisely what Nowotny did, but it will serve. Now we are going to suppose that $\mathsf{G}^{-1}=\omega_k-\omega_p-E_\gamma-\Sigma_\gamma$, neglecting Doppler effects and supposing γ to be the ground state. We shall allow the reader to use the methods at the end of Section 34 in order to obtain for this

$$N\left[\omega-\omega_0+\Sigma_{\gamma\delta}+NT\right]^{-1}$$

Nowotny's conclusion is that the density dependence of the linewidth is linear.

42 A Constant in the Half Width

Zaidi began by taking the first term on the right of Eq. (30.5b) as what he termed the zeroth-order line shape. Note that none of the terms unique to the resonance broadening phenomenon are included save those that we can anticipate as arising from the details of the interaction potential. Eq. (41.1b′), for example, is specifically excluded. Therefore, the line shape is obtained by inserting the first term on the right of Eq. (30.5b) into Eq. (28.15a). We may simply ignore the integration over q in Eq. (28.15a) on the basis that (1) the variable will not appear in the Greens functions of Eq. (30.5b) and (2) the delta function will deal with it. Therefore, we obtain what Zaidi calls $\bar{\chi}$:

$$\bar{\chi}^0_{\mu\nu}=\int\frac{dp}{(2\pi)^4}L_{\mu\nu;\,\nu\mu}(p,p-k;p-k,p)=\int\frac{dp}{(2\pi)^4}\langle \mathsf{G}_{\mu\mu}(p)\mathsf{G}_{\nu\nu}(p-k)\rangle_{\text{av}} \tag{42.1}$$

At this point Zaidi's state selection must be described. The radiant transition is from state ν to state μ. The ground state is designated $\bar{\nu}$.

From Eq. (30.6) we may write $\mathsf{G}_{\mu\mu}(p)$ as

$$\mathsf{G}_{\mu\mu}(p)=\left[\omega_p-p^2/2\mathsf{m}-E_\mu+\mu-\Sigma_{\mu\mu}\right]^{-1} \tag{42.2a}$$

Zaidi supposes the broadening of the spectral line as primarily due to the "broadening" of state μ which leads him to suppose the mass operator relating to the other state, $\Sigma_{\nu\nu}$, to be small. Such is logically the case as we may expect the mass operator here to describe the broadening. Therefore, in the $\mathsf{G}_{\nu\nu}$ analogy to Eq. (42.2a) we may drop the $\Sigma_{\nu\nu}$ and obtain

$$\mathsf{G}_{\nu\nu}(p-k)=\mathsf{G}^0_{\nu\nu}(p-k)=\left[\omega_p-\omega_k-(p-k)^2/2\mathsf{m}-E_\nu+\mu\right]^{-1} \tag{42.2b}$$

We may now evaluate the ω_p integration in Eq. (42.1) over the pole of $G^0_{\nu\nu}$ with the result

$$\bar{\chi}^0_{\mu\nu}=\int\frac{d\mathbf{p}}{(2\pi)^3}\frac{f_\nu(\mathbf{p}-\frac{1}{2}\mathbf{k})}{\left[\omega-\omega_{\mu\nu}-\mathbf{v}-\mathbf{k}-\Sigma_{\mu\mu}\right]} \tag{42.3a}$$

which is Zaidi's Eq. (12b). He chooses Fermi-Dirac statistics so that

$$f(\mathbf{p}-\tfrac{1}{2}\mathbf{k})=\left[1+\exp\{\beta(E_\nu+p^2/2\mathsf{m}-\mu)\}\right]^{-1} \tag{42.3b}$$

The problem is now almost exclusively a question of the mass operator, Σ, and the particular choice which we make for it. The choice which Zaidi makes is as follows [compare Eq. (32.2)]:

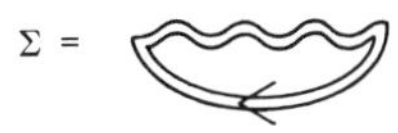

(42.4a)

where

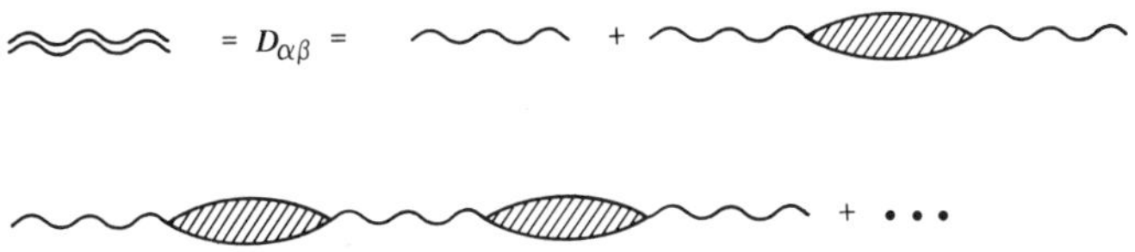

(42.4b)

where $D_{\alpha\beta}$ is the dressed photon propagator of Greens function, 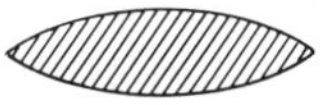 is the polarization operator, Π, and  is, as always, the particle propagator or Greens function.

From Section 21 we should be reasonably familiar with the free photon propagator and the dressed photon propagator or, simply, photon propagator. In particular, Eqs. (12.16) and (12.18) relate the photon propagators, the dielectric constant, and the polarization operator. Zaidi utilizes essentially these equations in order to obtain the following expression for the mass operator in the dipole approximation:

$$\Sigma_{\mu\mu} = \mathrm{i}\sum_{ij\nu'}\left(\frac{\mathrm{i}}{\beta}\right)^{-1}\sum_{\Delta}\int\frac{d\mathbf{q}}{(2\pi)^3}\,\mathbf{d}^{i}_{\mu\nu'}\mathbf{d}^{*j}_{\mu\nu'}E_{ij}(\mathbf{q},\Delta)\mathrm{G}_{\nu\nu'} \qquad (42.5\mathrm{a})$$

$$E_{ij} = \Delta^2 D_{ij} + q_i q_j D_{00} \qquad (42.5\mathrm{b})$$

$$D_{ij} = 4\pi\left[\Delta^2 - q^2 - 4\pi\Pi\right]^{-1}\left(\delta_{ij} - q_i q_j/q^2\right), \qquad i, j = 1,2,3 \qquad (42.5\mathrm{c})$$

$$D_{00} = (4\pi/q^2)\left[1 - 4\pi\Pi/\Delta^2\right]^{-1} \qquad (42.5\mathrm{d})$$

$$\Delta = \tfrac{1}{3}\mathrm{i}2\pi n$$

In Eqs. (42.5) $D_{\alpha\beta}$ becomes the free photon propagator, $D^0_{\alpha\beta}$, the first term in Eq. (42.4b), as Π approaches zero. Here we recall that α and β run from 0 to 3, i and j from 1 to 3. The Δ of course is a somewhat different notation from the ω_k of, say, Eq. (31.4). In order to see that Eq. (42.5c) indeed corresponds with Eq. (42.4b), let us write the latter as follows

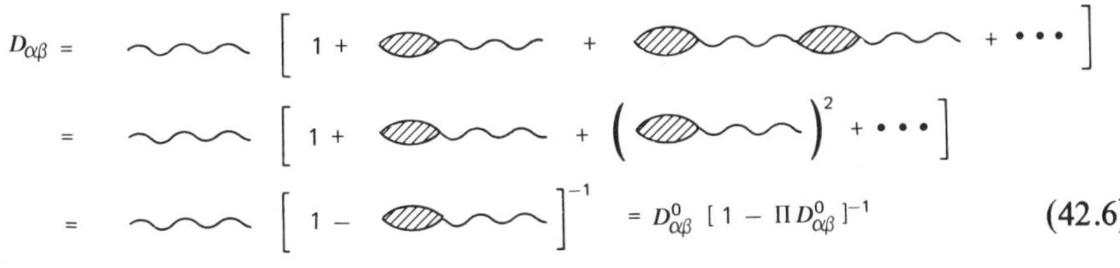

(42.6)

which may be algebraically verified using Eqs. (42.5).

We should by now be familiar with the replacement of the summation over Δ by an integration over ω in the complex plane and its subsequent evaluation at whatever poles present themselves for our edification. Zaidi follows such a procedure in order to obtain

$$\Sigma_{\mu\nu} \doteq \mathrm{i}\int \frac{d\mathbf{q}}{(2\pi)^3} \sum_{ij\nu'} (e^{-\beta\omega}-1)^{-1} \mathsf{d}^{i}_{\mu\nu'} \mathsf{d}^{*j}_{\mu\nu'} \mathcal{I}\left[E_{ij}(\mathbf{q}, -\omega' + \mathrm{i}\delta)\right] \tag{42.7a}$$

where

$$\omega' = \omega + E_{\nu} - E_{\nu'} + \left(|\mathbf{p} - \tfrac{1}{2}\mathbf{k}|^2 - |\mathbf{p} + \tfrac{1}{2}\mathbf{k} + \mathbf{q}|^2\right)/2\mathrm{m} \tag{42.7b}$$

At this point the mass operator is split into a transverse part, Σ^t, and a longitudinal or Coulomb part, Σ^c,

$$\Sigma_{\mu\mu} = \Sigma^{c}_{\mu\mu} + \Sigma^{t}_{\mu\mu}, \tag{42.7c}$$

and the longitudinal portion is used to obtain a width dependence on the root of the density, the transverse a width independent of the density over a certain range of densities. First, take the Coulomb mass operator.

The Coulomb part of E_{ij} will of course be the second term on the right of Eq. (42.5b). Very straightforward substitution indeed yields

$$\Sigma^{c}_{\mu\mu} = -\mathrm{i}\sum_{\nu'} v|d_{\mu\nu'}|^2 \int \frac{dq}{2\pi^2} \mathcal{I}\{[1 - S(\mathbf{q}, \omega, \beta)]^{-1}\} \tag{42.8a}$$

where

$$S(\mathbf{q}, \omega, \beta) = 4\pi \Pi(\mathbf{q} - \omega' + \mathrm{i}\delta, \beta)/\omega'^2 \tag{42.8b}$$

As we have remarked so often, the polarization operator or polarization part is what one inserts into a free photon propagator ()$\Rightarrow$() in order to determine what effect the medium may be expected to have on such a photon. We may, for example, close the left side of Eq. (42.1b′) by integration over p as we more or less did in obtaining Eq. (42.13a). [Perhaps the reader would prefer to go back and close both sides of Eq. (30.5b′) by integrations.] As one might hope, the dipole approximation introduces a frequency factor which cancels the denominator on the right of Eq. (42.8b). With Zaidi then, we obtain for Eq. (42.8b)

$$S(q, \omega, \beta) = 4\pi v|\mathsf{d}_{\mu\nu'}|^2 \int \frac{dp'}{(2\pi)^3} \frac{f_{\nu'}(\mathbf{p} + \frac{1}{2}\mathbf{q})}{[\omega - \omega_{\mu\nu'} - \mathrm{i}\gamma]} \tag{42.8c}$$

$$\omega_{\mu\nu'} = \omega_{\mu\nu} + (\mathbf{p} + \tfrac{1}{2}\mathbf{q} - \mathbf{p}')\cdot\mathbf{q}/\mathrm{m} + (\mathbf{p} + \tfrac{1}{2}\mathbf{q})\cdot\mathbf{k}/\mathrm{m} \tag{42.8d}$$

where we have substituted $\mathrm{i}\gamma$ for $\Sigma_{\mu\mu}$. More on this anon.

One particularly important point has now arisen. By comparing Eq. (42.8c) to Eq. (42.1a), we see that we have a susceptibility associated with the virtual transition $\mu \to \nu'$. Therefore, $\Sigma^{c}_{\mu\mu}$ in this equation represents the broadening—or damping—of the intermediate state during the collision. We may recall from Chapter 2 that this mass operator now corresponds to the self energy of the state μ.

As we should be aware by now, the real part of Σ^c is associated with the spectral line shift, the imaginary part with the width. We know that shift in self broadening is generally not observed. Zaidi also knew this which is the basis on which he replaced $\Sigma^c_{\mu\mu}$ by $i\gamma$. Next, that author considered this Coulomb component in the interruption limit and the opposite or statistical limit.

Not surprisingly, the interruption limit rather unambiguously leads to an N-dependent Σ and hence linewidth. Such is, however, not necessarily the case for the statistical limit. In this latter limit the kinetic factors in Eq. (42.8d) may surely be neglected. The contribution of v will depend on specific assumptions as to the states involved—Zaidi supposes state μ to have angular momentum unity, state $\bar{\nu}$ zero and gets $\frac{1}{9}$—and $f_{\nu'}$ will, as in Eq. (31.5), yield $N_{\bar{\nu}}$, the density of atoms in state $\bar{\nu}$. Thus Eq. (42.8c) becomes

$$S(\omega)\doteq \text{const}\, N_{\bar{\nu}}|\mathsf{d}_{\mu\bar{\nu}}|^2\left[\omega-\omega_{\mu\nu}+\Sigma^c_{\mu\mu}(\omega)\right]^{-1} \tag{42.9}$$

Letting $S=S_r+iS_i$, S_r and S_i being real and imaginary parts, respectively, we may simply determine the imaginary part indicated in Eq. (42.8a) and carry out the integration in order to obtain

$$\Sigma^c_{\mu\mu}(\omega)=-i\,\text{const}\,q_0^3|\mathsf{d}_{\mu\bar{\nu}}|^2 S_i(\omega)\left\{\left[1-S_r(\omega)\right]^2+S_i^2(\omega)\right\}^{-1} \tag{42.10}$$

For $|S|^2\ll 1$, S_i^2 and S_r^2 in the denominator of Eq. (42.10) may be set equal to zero, and the $S_i(\omega)$ that results substituted into Eq. (42.24) to obtain

$$\Sigma^c_{\mu\mu}=-i\,\text{const}\left(N_{\bar{\nu}}q_0^3\right)^{1/2}|\mathsf{d}_{\mu\bar{\nu}}|^2 \tag{42.11}$$

We have thus obtained the square-root density dependence for the particular condition,

$$|S|^2\ll 1 \Rightarrow 18\pi^2 N_{\bar{\nu}}/q_0^3\ll 1 \tag{42.12}$$

In Eq. (42.12) q_0 is an upper cutoff for q, the momentum transfer during the existence of the intermediate states of the collision. This cutoff Zaidi takes as the inverse of the gas-kinetic collision diameter for the colliding atoms. When the criterion given by Eqs. (42.12) is not met, we see that the linewidth is rather complexly dependent on the density and, further, that a convoluted interruption-Doppler or Voigt profile is no longer to be anticipated. We now turn very briefly to the transverse part of the mass operator.

In this case we are dealing with Eq. (42.8c) in Eq. (42.10a) with the result

$$\Sigma^t_{\mu\mu}(\omega)\doteq -i\,\text{const}\sum_{\nu'}\Upsilon|\mathsf{d}_{\mu\nu'}|^2\int_0^\infty dq\, q^2\omega'^4 S(\omega)\left[(\omega'^2-q^2)+\omega'^4S^2(\omega)\right] \tag{42.13a}$$

where

$$\omega'=\omega+E_\nu-E_{\nu'} \tag{42.13b}$$

$$S(\omega)\doteq 4\pi\mathcal{I}\left\{(\omega-\omega_{\mu\nu}-i\gamma)^{-1}\sum_{\nu'}\Upsilon N_{\nu'}|\mathsf{d}_{\mu\nu'}|^2\right\} \tag{42.13c}$$

where $\Re[S]$ has been dropped as small. Contour integration yields

$$\Sigma'_{\mu\mu} = -\mathrm{i}\sum \Upsilon\omega'^3 |\mathsf{d}_{\mu\nu'}|^2 2^{1/2} S\left[(1+S^2)^{1/2}-1\right]^{-1/2}$$
$$\doteq -\mathrm{i}\sum_{\nu'} 2\Upsilon\omega'^3 |\mathsf{d}_{\mu\nu'}|^2 (1+S^2/8) \tag{42.14}$$

for $S \ll 1$. The first term on the right of Eq. (42.14) gives the natural width, the second the resonance. Zaidi's further arguments go about as follows:

At very low densities, with perhaps an emphasis on the "very low," we suppose γ to be a constant of the order of the natural width. Then, from Eq. (42.13c), S is linearly dependent on $N_{\nu'}$ so that the resonance correction to $\Sigma'_{\mu\mu,}$ through Eq. (42.14), is quadratically dependent on the density. Actually, it is difficult to see that, at densities sufficiently low so that these arguments follow, anything but the natural width will be observed. Since no observations at densities this low would appear conceivable in the foreseeable future, the discussion is rather moot.

At densities where γ is no longer the constant natural width and at frequencies near line center for which $\omega - \omega_{\mu\nu} \doteq 0$, S will be density independent so long as we can suppose γ linearly dependent on density. If we suppose $\gamma \propto N$ in the interruption regime, then $\Sigma'_{\mu\mu}$ would be a constant there. Therefore, in this interruption regime the half width of the resonance line may be given by

$$\delta = C_1 + C_2 N \tag{42.15}$$

which is of course what Breene obtained by a Liouvillian treatment.

Is all this "true"? The next question is of course Pilate's: "What is truth?" This may be answered: truth is that which is revealed by experiment. Experiment, since it extrapolates a linear dependence of the width on the density at measurable densities to the natural width plus a constant, seems to arrive at the same conclusion. Yet the matter does not really appear to be settled.

In Section 34 we discussed Zaidi's demonstration of the source of Dicke narrowing in the Greens function treatment. In several subsequent papers (291, 239, 256) that author considered the application of this theory to the resonance broadening case, with particular emphasis on the effects of considering a collision of finite duration. For example (256) the collision duration for the $2\,^1p$ level of He at $T = 80°$K was taken as 10^{-12} sec. Further, methods of excitation exchange T-matrix element computation which had been developed earlier (289) were applied. These calculations of Srivastava and Zaidi yielded a splitting of the theoretical spectral line profile at its intensity maximum, a splitting that became pronounced at a density of 5×10^{17} cm^{-3} and was present at higher densities. This rather interesting result arose from their inclusion of the duration of collision.

43 The Liouvillian Approach to Resonance Broadening

During the same period that the Ben-Reuven formulation, which we have discussed in Chapter 5, was developing, Smith and Hooper (SH) (248) were considering the broadening of atomic spectral lines in plasmas. Breene (85) applied their techniques to the broadening of molecular spectral lines in plasmas. This brings us to the work we wish to consider in this section, that of Breene (87), who applied the Liouville operator formalism to the resonance broadening phenomenon. [The prolific Ben-Rueven (70) dealt with the same phenomenon.]

Breene begins by claiming that the use of this approach eschews any necessity for introducing the interruption approximation, a remark that is true in a mechanical sense. However, in the next paragraph he introduces the dilute approximation in order to untangle the internal and translational coordinates of the atoms, and this really amounts to the same thing. Be this as it may, we begin with a Hamiltonian

$$H=\sum K_i+\sum H_i+\sum \mathsf{V}_{ij} \tag{43.1}$$

wherein K_i is the translational kinetic energy of the ith atom, H_i the Hamiltonian for the isolated ith atom and V_{ij} the pair interaction potential.

When the dilute approximation is made, that is, when we take $\mathsf{V}_{ij}=0$, the density matrix will split into product from as follows:

$$\Upsilon=\exp[-\beta H]=\exp\left[-\beta\sum K_i\right]\exp\left[-\beta\sum H_i\right]=\Upsilon^a\Upsilon^t \tag{43.2}$$

where Υ^a refers to the internal coordinates of the atoms, Υ^t to their translational coordinates. The simplification thus brought about is obviously of immense value.

We begin with Eqs. (21.1) and (21.4), although we do not combine them as yet, and we carry out the integration over time in order to get Eq. (21.5). Before carrying out this integration we shall use a projection operator in order to recast the trace in a more tractable form. As a first step it is helpful to rewrite Eq. (21.4) as

$$\Phi(t)=\operatorname{Tr}\left\{\mathbf{d}\cdot e^{-\mathrm{i}\Omega t}\mathbf{D}(0)\right\} \tag{43.3a}$$

where

$$\mathbf{D}(t)=e^{-\mathrm{i}Ht}\Upsilon^a\Upsilon^t\mathbf{d}e^{\mathrm{i}Ht}=e^{-\mathrm{i}Ht}\mathbf{D}(0)e^{\mathrm{i}Ht} \tag{43.3b}$$

where H is given by Eq. (43.1).

Now we define a projection operator P as having the following effect on an arbitrary operator M:

$$\mathsf{P}M=\Upsilon^t\operatorname{Tr}_t\{M\} \tag{43.4}$$

where Tr_t means the trace over the translational coordinates.

Equation (21.1) we now write down as

$$\int e^{i\omega t}\mathrm{Tr}_{\mathrm{at}}\{\mathbf{d}\cdot\mathbf{D}(t)\}\,dt \tag{43.5}$$

Since **d** is a function only of the atomic coordinates, it will be true that

$$\mathrm{Tr}_{\mathrm{at}}\{\mathbf{d}\cdot\mathbf{D}(t)\}=\mathrm{Tr}_a\{\mathbf{d}\cdot\mathrm{Tr}_t(\mathbf{D}(t))\}$$

Since $\mathrm{Tr}_t\{\Upsilon'\}=1$, this may be written as

$$\begin{aligned}\mathrm{Tr}_a\{\mathbf{d}\cdot\mathrm{Tr}_t\{\Upsilon\}\mathrm{Tr}_t(\mathbf{D}(t))\}&=\mathrm{Tr}_{\mathrm{at}}\{\mathbf{d}\cdot\Upsilon'\mathrm{Tr}_t(\mathbf{D}(t))\}\\&=\mathrm{Tr}_{\mathrm{at}}\{\mathbf{d}\cdot\mathsf{P}\mathbf{D}(t)\},\end{aligned} \tag{43.6}$$

by Eq. (43.4). We now determine an expression for $\mathsf{P}\mathbf{D}(t)$. To begin with, it is trivially true that

$$\mathbf{D}(t)=\mathsf{P}\mathbf{D}(t)+(1-\mathsf{P})\mathbf{D}(t)\equiv\mathbf{D}_1(t)+\mathbf{D}_2(t) \tag{43.7a}$$

and true, if not trivial, that $\mathbf{D}(t)$ will satisfy an equation of motion [cf. Eq. (1.4)]

$$i\dot{\mathbf{D}}(t)=\mathfrak{L}\mathbf{D}(t) \tag{43.7b}$$

Applying first P and then $(1-\mathsf{P})$ to both sides of Eq. (43.6b) yield

$$i\dot{\mathbf{D}}_1=\mathsf{P}\mathfrak{L}[\mathbf{D}_1+\mathbf{D}_2] \tag{43.8a}$$

and

$$i\dot{\mathbf{D}}_2=(1-\mathsf{P})\mathfrak{L}[\mathbf{D}_1+\mathbf{D}_2] \tag{43.8b}$$

The method of procedure here is originally due to Zwanzig who essentially projected out the diagonal elements of the density matrix. What we are seeking is of course $\mathbf{D}_1(t)\equiv\mathsf{P}\mathbf{D}(t)$ for use in equations of the form Eq. (43.6), that is, a specific expression for the effects of projection operator application. What SH refer to as the irrelevant part of **D**, $\mathbf{D}_2$, may be taken as the following solution of Eq. (43.8b):

$$\mathbf{D}_2=-i\int_0^t e^{-it'(1-\mathsf{P})\mathfrak{L}}(1-\mathsf{P})\mathfrak{L}\mathbf{D}_1(t-t')\,dt' \tag{43.9}$$

Let us recall that one form of the Laplace transform of a function, $F(t)$, is

$$\mathfrak{L}(F(t))=\int_0^\infty e^{-i\omega t}F(t)\,dt,\qquad \Re[\omega]=0 \tag{43.10a}$$

and that the convolution or faltung theorem for Laplace integrals is

$$\int_0^\infty F(y)h_+(x-y)\,dy=\int_{-it_0}^{it_0}A_+(k)e^{-ikx}\,dx \tag{43.11a}$$

where

$$A_+(k)=(2\pi)^{-1/2}\int_0^\infty F(x)e^{ikx}\,dx \tag{43.11b}$$

Equation (43.9) is substituted into the right side of Eq. (43.8a). We now carry out a most straightforward series of operations on this equation: (1) both sides of it are Laplace transformed; (2) the faltung theorem is applied to the second term on the right of the result; (3) the left side of the equation is integrable by parts, the first integral in the second term on the right directly integrable. We carry out these steps and collect terms in order to obtain

$$\mathrm{i}\mathbf{D}_1(0)=\{\omega-\mathsf{P}\mathfrak{L}-\mathsf{P}[\omega-(1-\mathsf{P})L]^{-1}(1-\mathsf{P})\mathfrak{L}\}\mathbf{D}_1'(\omega) \tag{43.12a}$$

where

$$\mathbf{D}_1'(\omega)=\int_0^\infty e^{\mathrm{i}\omega t}\mathbf{D}_1(t)\,dt \tag{43.12b}$$

We now define the operator

$$\mathfrak{T}(\omega)=\mathfrak{L}+\mathfrak{L}[\omega-(1-\mathsf{P})\mathfrak{L}]^{-1}(1-\mathsf{P})\mathfrak{L}=\mathfrak{L}\sum_n\left[\frac{(1-\mathsf{P})}{\omega}\mathfrak{L}\right]^n \tag{43.13}$$

which is to be compared, for example, to Eqs. (21.13). Eq. (43.12a) now becomes

$$\mathrm{i}\mathbf{D}_1(0)=[\omega-\mathsf{P}\mathfrak{T}(\omega)]\mathbf{D}_1'(\omega) \tag{43.14}$$

It is time to recall that the necessity for an expression for $\mathsf{P}\mathbf{D}(t)$ in equations such as Eq. (43.6) furnished the motivation for our search which has brought us to Eq. (43.14). We may substitute Eq. (43.6) for the trace in Eq. (43.5),

$$\int_0^\infty e^{\mathrm{i}\omega t}\mathrm{Tr}_{\mathrm{at}}(\mathbf{d}\cdot\mathsf{P}\mathbf{D}(t))=\mathrm{Tr}_{\mathrm{at}}\left\{\mathbf{d}\cdot\int_0^\infty e^{\mathrm{i}\omega t}\mathbf{D}_1(t)\,dt\right\}=\mathrm{Tr}_a\{\mathbf{d}\cdot\mathrm{Tr}_t(\mathbf{D}_1'(\omega))\} \tag{43.15}$$

We now determine an expression for $\mathrm{Tr}_t(\mathbf{D}_1'(\omega))$.

In order to do so we substitute in Eq. (43.14) for $\mathbf{D}_1$ and $\mathbf{D}(0)$ from Eqs. (43.7a) and (43.3b), respectively. Taking the trace over the translational coordinates of both sides of the result yields

$$\begin{aligned}\mathrm{i}\Upsilon^a\mathbf{d}&=\omega\,\mathrm{Tr}_t\{\mathbf{D}_1'(\omega)\}-\mathrm{Tr}_t(\mathsf{P}\mathfrak{T}(\omega)\mathbf{D}_1'(\omega))\\&=\omega\,\mathrm{Tr}_t(\mathbf{D}_1'(\omega))-\mathrm{Tr}_t\{\Upsilon^t\}\mathrm{Tr}_t(\mathfrak{T}(\omega)\mathbf{D}_1'(\omega))\\&=\omega\,\mathrm{Tr}_t(\mathbf{D}_1'(\omega))-\mathrm{Tr}_t(\mathfrak{T}(\omega)\mathbf{D}_1'(\omega))\end{aligned} \tag{43.16}$$

by the definition of P. Using the idempotent character of a projection operator, $\mathsf{P}^2=\mathsf{P}$, we may show that $\mathbf{D}_1'(\omega)=\mathsf{P}\mathbf{D}_1'(\omega)$. When this quality of $\mathbf{D}_1'$ is used in the second term on the right of Eq. (43.16), $\mathrm{Tr}_t(\mathfrak{T}(\omega)\mathbf{D}_1'(\omega))$ becomes $\mathrm{Tr}_t(\mathfrak{T}(\omega)\Upsilon^t)\mathrm{Tr}_t(\mathbf{D}_1'(\omega))$. This result is substituted into Eq. (43.16) and an expression for $\mathrm{Tr}_t(\mathbf{D}_1'(\omega))$ obtained. When this in turn is substituted into Eq. (43.15), we have but to multiply by π^{-1} [compare Eq. (21.5)] in order to obtain the expression for the intensity with which Breene began

$$I(\omega)=\pi^{-1}\mathrm{Tr}_a\{\mathbf{d}\cdot[\omega-\langle\mathfrak{T}(\omega)\rangle]^{-1}(\Upsilon^a\mathbf{d})\} \tag{43.17a}$$

where

$$\langle\mathfrak{T}(\omega)\rangle = \mathrm{Tr}_t(\mathfrak{T}(\omega)\Upsilon') \tag{43.17b}$$

an expression which, although its roots reach back to the Zwanzig efforts, is only a trivial modification of the SH line shape expression for the Stark broadening problem.

$\mathfrak{T}(\omega)$, as defined by Eq. (43.13), obviously has a denumerably infinite number of terms, very few of which, with equal obviousness, did Breene evaluate. However, the specific evaluation has not been obtained yet.

Following Eq. (21.5) we pointed out that $(\omega-\mathfrak{L})^{-1}$ is the resolvant operator, the Laplace transform of $\exp[-i\mathfrak{L}t]$, the Greens function. SH make use, as a consequence of Eq. (30.17a), of what they term an "effective resolvant,"

$$\mathfrak{R}(\omega) = [\omega - \langle\mathfrak{T}(\omega)\rangle]^{-1} \tag{43.18}$$

wherein, by analogy, $\langle\mathfrak{T}(\omega)\rangle$ is dubbed the effective Liouville operator. With complete unanimity, we will agree on the wisdom of dividing the Liouville, as in Eq. (21.6b), into an unperturbed $\mathfrak{L}_0$, and a perturbed part, $\mathfrak{L}_1$. Having done this, we can express the effective Liouville as

$$\langle\mathfrak{T}(\omega)\rangle = \langle\mathfrak{L}_0\rangle + \langle\mathfrak{M}_c(\omega)\rangle, \tag{43.19}$$

SH call $\langle\mathfrak{M}_c(\omega)\rangle$ an "effective interaction tetradic" (EIT). We have followed through the obtention of the EIT in detail in Section 30; an emphasis on a few of the principal features of it should suffice here.

In Eq. (43.13) there will be a portion of a term that may be written as follows:

$$\begin{aligned}
[\omega-(1-\mathsf{P})\mathfrak{L}]^{-1} &= [\omega-(1-\mathsf{P})(\mathfrak{L}_0+\mathfrak{L}_1)]^{-1} \\
&= [\omega-(1-\mathsf{P})\mathfrak{L}_0]^{-1}\{1-[\omega-(1-\mathsf{P})\mathfrak{L}_0]^{-1}(1-\mathsf{P})\mathfrak{L}_1\}^{-1} \\
&= [\omega-(1-\mathsf{P})\mathfrak{L}_0]^{-1}\sum_{n=0}^{\infty}\{1-[\omega-(1-\mathsf{P})\mathfrak{L}_0]^{-1}(1-\mathsf{P})\mathfrak{L}_1\}^{n}
\end{aligned} \tag{43.20}$$

In their Appendix B, SH use the matrix elements of $\mathsf{P}M$, M an arbitrary matrix, and those of $\mathsf{P}(\mathfrak{L}_0 M)$ in order to show that P and $\mathfrak{L}_0$ commute. This commutation, together with Eq. (43.20), may be used in a rather tedious derivation in order to obtain

$$\langle\mathfrak{M}_c(\omega)\rangle = \sum_{n=0}^{\infty}\langle\mathfrak{L}_1[\mathfrak{R}^0(\omega)(1-\mathsf{P})\mathfrak{L}_1]^n\rangle \tag{43.21a}$$

where

$$\mathfrak{R}^0(\omega) = -i\int_0^{\infty} e^{i\omega t}e^{-i\mathfrak{L}_0 t}\,dt = [\omega-\mathfrak{L}_0]^{-1} \tag{43.21b}$$

$\mathfrak{R}^0(\omega)$ is thus the unperturbed resolvant.

Specifically, Breene uses the first two terms of Eq. (43.21a) so that, for Eq. (43.17a) we are to evaluate

$$[\omega-\langle\mathfrak{T}(\omega)\rangle]\doteq\omega-\langle\mathfrak{L}_0\rangle-\langle\mathfrak{L}_1\rangle-\langle\mathfrak{L}_1\mathfrak{R}^0(\omega)(1-\mathsf{P})\mathfrak{L}_1\rangle \quad (43.21c)$$

in order to develop a Liouvillian description of resonance broadening.

We now hypothesize a collection of n identical two-state atoms, one of which is in its upper state, the other $n-1$ of which are in the ground state. The wave functions for the lower and upper states of the n-atomic system are then

$$\Psi_1=\psi_0(1)\psi_0(2)\cdots\psi_0(n) \quad (43.22a)$$

$$\Psi_u=n^{-1/2}P\psi_+(1)\psi_0(2)\cdots\psi_0(n) \quad (43.22b)$$

where P is the permutation operator without change of sign, the subscripts 0 and + indicating lower and upper states, respectively. The interaction Hamiltonian leading to the resonance broadening is

$$\mathsf{U}=\sum_{j>i=1}^{n}\mathsf{V}_{ij} \quad (43.23a)$$

where

$$\mathsf{V}_{ij}=e^2r_{ij}^{-3}(y_iy_j+z_iz_j-2x_ix_j) \quad (43.23b)$$

this V_{ij} being the dipole-dipole result developed by Margenau (188). Here r_{ij} is the interatomic distance, x_i and so on, the internal atomic coordinates. In the appendix Breene begins with the potential,

$$\mathsf{V}_{ij}=-e^2r_{iJ}^{-1}-e^2r_{jI}^{-1}+e^2r_{ij}^{-1}+e^2R^{-1} \quad (43.23b')$$

for a pair of hydrogen atoms, the lower case subscripts referring to electronic positions, the upper to nuclear. He then applies Eq. (1.7) to the evaluation of the matrix elements of the Liouville operator as given by Eqs. (43.23a) and (43.23b′),

$$\langle u\alpha l\alpha|\mathfrak{L}|u\alpha' l\beta'\rangle=\langle u\alpha|\mathsf{U}|u\alpha'\rangle\delta_{\alpha\beta}-\langle l\alpha|\mathsf{U}|l\beta'\rangle\delta_{\alpha\alpha'} \quad (43.24)$$

wherein u and l refer to the internal atomic states, α, β,$\cdots$ to translational states. That author then shows that this reduces to the Margenau result for large atomic separations, a byway down which we need not accompany him.

With Eq. (43.24) the obviously important translational states have entered. They are defined as

$$|\alpha\rangle=|p_1\rangle|p_2\rangle\cdots|p_n\rangle \quad (43.25a)$$

where

$$|p_j\rangle=V^{-1/2}\exp[i\mathbf{r}_j\cdot\mathbf{p}_j] \quad (43.25b)$$

and Eqs. (43.18) must be multiplied by the appropriate translational state to obtain the complete wave function or eigenvector.

We apply Eqs. (43.22) and (43.13) to the evaluation of the following hydrogenic matrix elements:

$$\langle l\alpha|\sum \mathsf{V}_{ij}|l\alpha'\rangle = 0 \tag{43.26a}$$

$$\langle u\alpha|\sum \mathsf{V}_{ij}|u\alpha'\rangle = C_u\langle\alpha|\sum r_{ij}^{-3}|\alpha'\rangle n^{-1} \tag{43.26b}$$

where

$$C_u = e^2 f_{ul}/8\pi^2 m\nu_0 \tag{43.26c}$$

where f_{ul} is the oscillator strength for a dipole transition between the levels u and l, ν_0 the frequency of such a transition. We are going to require the following matrix element combinations:

$$\begin{aligned}\langle\alpha|\sum r_{ij}^{-3}|\alpha'\rangle\langle\alpha'|\sum r_{ij}^{-3}|\alpha\rangle n^{-2} &= \tfrac{1}{2}n(n-1)|\langle\alpha|r_{ij}^{-3}|\alpha'\rangle|^2 n^{-2} + n(n-1)(n-2)\\ &\cdot\langle\alpha|r_{ij}^{-3}|\alpha'\rangle\langle\alpha'|r_{jk}^{-3}|\alpha\rangle n^{-2} + \tfrac{1}{4}n(n-1)\\ &\cdot(n-2)(n-3)\langle\alpha|r_{ij}^{-3}|\alpha'\rangle\langle\alpha'|r_{kl}^{-3}|\alpha\rangle n^{-2}\end{aligned} \tag{43.27}$$

A little rumination on Eq. (43.27) tells us that the first term refers to a two-particle interaction, the second to a three-particle interaction, and the third term to a four-particle interaction. The simpler diagonal element of Eq. (43.26b) is

$$\langle\alpha|\sum r_{ij}^{-3}|\alpha\rangle n^{-1} = \tfrac{1}{2}n(n-1)\langle\alpha|r_{ij}^{-3}|\alpha\rangle n^{-1} \tag{43.28}$$

For all practical purposes, n will be of the order of 10^{19}, perhaps 10^{15}, but surely such that $n \doteq n-1 \doteq n-2 \doteq n-3$. Then Eq. (43.28) becomes

$$\begin{aligned}\tfrac{1}{2}n\langle\alpha|r_{ij}^{-3}|\alpha\rangle &= \tfrac{1}{2}nV^{-2}\int\int d\mathbf{r}_i\, d\mathbf{r}_j r_{ij}^{-3} = \tfrac{1}{2}nV^{-1}\int d\mathbf{r}_j r_j^{-3}\\ &= 2\pi nV^{-1}\int d\mathbf{r}_j r_j^{-3} = 2\pi N\Gamma_2\end{aligned} \tag{43.29}$$

where we have transformed the coordinate origin to the ith particle before integrating over r_i. N is the particle density. Γ_2 is a constant which would of course diverge at the lower limit if the integral over r_j is simply carried out. This constitutes no particular objection, however, as our potential will certainly not hold for small interparticle separations. Such difficulties are overcome by introducing a cutoff whose value is established by some sort of physical arguement or by introducing a short-range potential of some variety. In any case, the apparent divergence of Γ_2 we take as irrelevant to the development.

The off-diagonal elements may be written down with the aid of Eq. (43.21b) as

$$\langle\alpha|r_{ij}^{-3}|\alpha'\rangle = V^{-2}\int\int \exp[i\mathbf{r}_i\cdot(\mathbf{p}_i-\mathbf{q}_i)]\exp[i\mathbf{r}_j\cdot(\mathbf{p}_j-\mathbf{q}_j)]|r_{ij}|^{-3} d\mathbf{r}_i\, d\mathbf{r}_j$$

wherein the $\mathbf{p}_k$ and the $\mathbf{q}_k$ are the momenta of the kth particle before and after collision. Let us again move the origin to, say, particle i. Let us specify momentum difference $\Delta\mathbf{p}_i = \mathbf{p}_i - \mathbf{q}_i$. Then the momentum of particle j with respect to this new origin will be the difference between the momenta of the two particles, that is, $\Delta\mathbf{p}_j - \Delta\mathbf{p}_i$. Therefore, we obtain

$$\begin{aligned}\langle\alpha|r_{ij}^{-3}|\alpha'\rangle &= V^{-2}\int\int \exp\left[\mathrm{i}\mathbf{r}_j\cdot(\Delta\mathbf{p}_j - \Delta\mathbf{p}_i)\right] r_j^{-3}\, d\mathbf{r}_i\, d\mathbf{r}_j \\ &= V^{-1}\int \exp\left[\mathrm{i}\mathbf{r}\cdot(\Delta\mathbf{p}_j - \Delta\mathbf{p}_i)\right] r^{-3}\, d\mathbf{r} \\ &= 4\pi V^{-1}\int \sin z\, z^{-2}\, dz = 4\pi V^{-1}\Gamma_2(\Delta p_i, \Delta p_j), \end{aligned} \qquad (43.30)$$

wherein we have transformed to the coordinate $z = |\Delta\mathbf{p}_i - \Delta\mathbf{p}_j| r$. The result of integration, Γ_2, is now a function of the momenta difference for the two particles. We may now turn to the evaluation of the matrix elements of Eq. (43.17c), basic to the line shape as given by Eq. (43.17a).

Using Eq. (43.4) we may specify one portion of the last term on the right of Eq. (43.17c) as

$$\langle\mathfrak{L}_1\mathfrak{R}^0(\omega)\mathsf{P}\mathfrak{L}_1\rangle = \mathrm{Tr}_t\{\mathfrak{L}_1\mathfrak{R}^0(\omega)\Upsilon'\mathrm{Tr}_t\{\mathfrak{L}_1\Upsilon'\}\} = \langle\mathfrak{L}_1\mathfrak{R}^0(\omega)\rangle\langle\mathfrak{L}_1\rangle \qquad (43.31)$$

Equation (43.31) is canceled by the four-particle portion of $\langle\mathfrak{L}_1\mathfrak{R}^0(\omega)\mathfrak{L}_1\rangle$, the other portion of the last term on the right of Eq. (43.17c), after a fashion that will become clear somewhat later. Thus the last term in that equation reduces to the two- and three-particle portions of $\langle\mathfrak{L}_1\mathfrak{R}^0(\omega)\mathfrak{L}_1\rangle$. In evaluating this term we begin with

$$\begin{aligned}\langle\mathfrak{L}_1\mathfrak{R}^0(\omega)\mathfrak{L}_1\rangle_{ul,ul} = \sum_{\alpha\gamma\alpha'\beta'} &(\mathfrak{L}_1)_{u\alpha l\alpha,u\alpha' l\beta'}(\mathfrak{R}^0)_{u\alpha' l\beta',u\alpha' l\beta'} \\ &\cdot(\mathfrak{L}_1)_{u\alpha' l\beta',u\gamma l\gamma} f_\gamma, \end{aligned} \qquad (43.32a)$$

where

$$\Upsilon'_{\alpha\beta} = f_\gamma\delta_{\alpha\beta} = \delta_{\alpha\beta}\prod_{j=1}^{n} f(p_j) \qquad (43.32b)$$

with

$$f(p_j) = (2\pi)^{3/2}V^{-1}(\mathrm{m}kT)^{-3/2}\exp\left[-p_j^2/2\mathrm{m}kT\right] \qquad (43.32c)$$

From Eq. (1.7b)

$$(\mathfrak{L}_1)_{u\alpha l\alpha,u\alpha' l\beta'} = \langle u\alpha|\mathsf{U}|u\alpha'\rangle\delta_{\alpha\beta'} - \langle l\alpha|\mathsf{U}|l\beta'\rangle\delta_{\alpha\alpha'} \qquad (43.33)$$

Using Eqs. (43.21b), (43.32), and (43.33), we obtain

$$\begin{aligned}\langle\mathfrak{L}_1\mathfrak{R}^0(\omega)\mathfrak{L}_1\rangle_{ul,ul} = -\mathrm{i}\int \exp\left[\mathrm{i}(\omega-\omega_{ul})t\right] dt \sum_{\alpha\alpha'}\langle u\alpha|\mathsf{U}|u\alpha'\rangle \exp\left[-\mathrm{i}\omega_{\alpha\alpha'}t\right] \\ \cdot\langle u\alpha'|\mathsf{U}|u\alpha\rangle f_\alpha \end{aligned} \qquad (43.34)$$

For the two-particle case the summation in eq. (43.34) may be written

$$C_u^2 n^{-2} \sum_{\{p,q\}} \exp[-i\omega_{\alpha\alpha'} t] \langle \alpha | \sum r_{ij}^{-3} | \alpha' \rangle |^2 f_\alpha \tag{43.35}$$

where

$$\{p,q\} = \{(p,q) : p \in \alpha, q \in \alpha'\}$$

Now $\omega_{\alpha\alpha'} = \sum_j (p_j^2 - q_j^2)/2\text{m}$, so that, by Eqs. (43.32b) and (43.32c), Eq. (43.35) becomes

$$C_u^2 n^{-2} \sum_{\{p,q\}} \left[-i \sum_j \frac{(p_j^2 - q_j^2)t}{2\text{m}} \right] \left| \langle \alpha | \sum r_{ij}^{-3} | \alpha' \rangle \right|^2 \prod_{k=1}^{n} f(p_k)$$

$$= \tfrac{1}{2} C_u^2 n(n-1) n^{-2} \sum_{\substack{p_i p_j \\ q_i q_j}} 2^2 \pi^3 V^{-2} \Gamma_2^2(\Delta p_i, \Delta p_j) \prod_{m \neq i,j} \delta_{p_m q_m} \exp\left(\frac{-i(p_i^2 - q_i^2)t}{2\text{m}} \right)$$

$$\cdot \exp\left(\frac{-i(p_j^2 - q_j^2)t}{2\text{m}} \right) f(p_i) f(p_j) \prod_{k \neq ij} \sum_{k \neq ij} f(p_k) \tag{43.36}$$

where the Kronecker delta product arises from the matrix element. Normalization assures that

$$\prod \sum f(p_k) = 1$$

so Eq. (43.36) becomes

$$2\pi^3 C_u^2 V^{-2} \sum_{p_i q_i p_j q_j} \Gamma_2^2 f(p_i) f(p_j) \exp\left(\frac{-i(p_i^2 - q_i^2)t}{2\text{m}} \right) \exp\left(\frac{-i(p_j^2 - q_j^2)t}{2\text{m}} \right) \tag{43.37}$$

As we allow the volume to go to infinity, $\sum_{p_i}$ goes to $V(2\pi)^{-3} \int dp_i$ in the familiar fashion, and the substitution of Eq. (43.32c) into Eq. (43.37) yields

$$C_u^2 \pi^{-1} (2\pi \text{m} kT)^{-3} \left\{ \int \int \Gamma_2 \exp[-(ibt + a)p^2] \exp[-ibtq^2] \, d\mathbf{p} \, d\mathbf{q} \right\}^2 \tag{43.38a}$$

$$a = (2\text{m}kT)^{-1}, \qquad b = (2\text{m})^{-1} \tag{43.38b}$$

The evaluation of the three-particle portion of Eq. (43.34) is entirely analogous and yields

$$2C_u^2 N \pi^{-1} (2\pi \text{m} kT)^{-3/2} \int \int \Gamma'_2 \exp[-(ibt + a)p^2] \exp[-ibtq^2] \, d\mathbf{p} \, d\mathbf{q} \tag{43.39}$$

We have now obtained

$$\langle \mathfrak{L}_1 K^0(\omega)\mathfrak{L}_1\rangle_{ul,ul} = \mathrm{i}C_1 T^{-3}F_1(T) + \mathrm{i}C_2 T^{-3/2}F_2(T)N \tag{43.40a}$$

$$C_1 = C_u^2\pi^{-1}(2\pi \mathrm{m}kT)^{-3}, \qquad C_2 = 2C_u^2\pi^{-1}(2\pi \mathrm{m}kT)^{-3/2} \tag{43.40b}$$

$$F_1(T) = \int \exp[\mathrm{i}(\omega-\omega_{ul})t]\,dt \int\int \Gamma_2 \exp[-(\mathrm{i}bt+a)p^2]\exp[-\mathrm{i}btq^2]\,d\mathbf{p}\,d\mathbf{q} \tag{43.40c}$$

$$F_2(T) = \int \exp[\mathrm{i}(\omega-\omega_{ul})t]\,dt \int\int \Gamma_2' \exp[-(\mathrm{i}bt+a)p^2]\exp[-\mathrm{i}btq^2]\,d\mathbf{p}\,d\mathbf{q} \tag{43.40d}$$

Immediately following Eq. (43.31) we remarked that this equation was cancelled by the four-particle portion of $\langle \mathfrak{L}_1 K^0(\omega)\mathfrak{L}_1\rangle$. That such is the case may be readily demonstrated in precisely the fashion by which we have arrived at Eqs. (43.38) and (43.39). Our line shape expression as given by Eqs. (43.17) now only requires

$$\langle \mathfrak{L}_0\rangle_{ul,ul} = \omega_{ul} \tag{43.41a}$$

$$\langle \mathfrak{L}_1\rangle_{ul,ul} = 2\pi C_u \Gamma^2 \tag{43.41b}$$

which are obtained using the same procedures.

The width of the spectral line is now immediately obtainable from Eq. (43.40a) as

$$\delta = C_1 T^{-3}F_1''(T) + C_2 T^{-3/2}F_2''(T)N \tag{43.42}$$

where F_1'' and F_2'' are the real parts of F_1 and F_2, respectively. Several comments are in order.

Although the momentum of the particles in the assemblage appears in the line shape expression, we do not really account for the Doppler effect in this result, this particular phenomenon not after all being the object of this exercise. The Doppler effect enters treatments such as that of Zaidi by virtue of the fact that the argument of $\mathsf{G}_{\mu\alpha'}(p-k)$ is the difference between particle and photon momenta. That there is such a difference present may be traced back to the presence of the photon vector potential A in the radiation-inducing interaction which, in turn, brings in the photon momentum k. In order to include the Doppler effect herein, the photon vector potential should be introduced in Eq. (43.3).

We will recall that the frequency separation from line center actually enetered the Zaidi expression for the halfwidth through Eqs. (42.13) and (42.14). This is also true here, for we see that the frequency spearation from line center $(\omega-\omega_{ul})$, appears both in Eq. (43.40c) and (43.40d). One may be to some extent justified in ignoring this near line center, but it will certainly mean a deviation from the interruption shape in the line wings.

The principal result of the Breene paper was of course the constant that appears in the expression for the linewidth, Eq. (43.42). This apparently explains the fact that extrapolation of the observed half width versus density curve to zero density does not yield the natural width while at the same time the width varies linearly with density in the observed nonzero region. Zaidi of course obtained what can be interpreted as the same result. Finally, we consider the approximation involved in Eq. (43.23b) and which, for example, through Eq. (43.28), has such a profound effect on the result.

In order to obtain Eq. (43.28b) we supposed that one out of n atoms is excited, not a very realistic supposition. However, we could still arrive at the constant in the linewidth with what is perhaps a more realistic assumption, to wit, that there are n_1 atoms in the upper state, n_1 being a constant over the observed range of densities, n. This means, of course, that the temperature-invoked Boltzmann distribution does not control the number of atoms in the upper state, for, if it did, the constant vanishes in a simple linear density dependence as was the case in Section 41. Therefore, if the excitation conditions are taken as constant and, say, radiant in origin over the observed range of densities, we may suppose a constant number of upper-state atoms, a constant in the linewidth. [It is true that the reader will have to rewrite Eq. (43.28b) in order to account for n_1 instead of 1 excited atoms and rework that which follows, but he will find it no particular problem.]

Although by Eq. (43.18) it is too late for the emergence of the constant, two-body t-matrix terms unique to resonance broadening do appear in that equation. Pasmanter and Ben-Reuven (218) have considered such terms in detail and conclude that a blue shift in the spectral line, rather than observable effect on the width constitutes the principal consequence of such terms. Their study is after the fashion of Anderson and makes extensive use of the often pleasing Wigner and $9J$ coefficients, although there is no obvious reason for detailing it here.

44 Resonance Broadening in the Sudden Approximation

The work of Byron and Foley (90) (BF) is quite often remarked on in the literature, a fact that would render us remiss in failing to discuss it were its intrinsic interest not as great as it is. Basically, these authors discuss the broadening of the microwave spectral line, which arises by virtue of a transition between two of the magnetic sublevels of an atomic energy level. As the reader will see, however, the results will apply to certain, much more generalized transitions for which the two-level approximation is applicable. The term "sudden approximation" is their way of saying that the time of collision is much smaller than the time between collisions, a figure of 10^5 being quoted as an exemplar. This means in essence that we are dealing with the interruption approximation. The method of approach is rooted in that of Anderson and Baranger. Thus although the basic methods are by now familiar to us, these authors introduce certain specific calculations that are both interesting and informative.

BF begin with the level-flipping formula of Majorana (195), although there is no reason for us to do so as they wend their way to what is essentially Eq. (13.12) a page or two thereafter. We begin with their Hamiltonian,

$$H = H_0 + H_c(t) - \tfrac{1}{2}B\left[\mu_+ e^{-i\omega t} + \mu_- e^{i\omega t}\right] \tag{44.1a}$$

where

$$\mu_\pm = \mu_x \pm i\mu_y. \tag{44.1b}$$

In Eq. (44.1a) $H_c(t)$ is the perturbing portion of the Hamiltonian arising from collisions and giving rise to pressure broadening. Here the absorbing or radiating transition is induced by $\boldsymbol{\mu}\cdot\mathbf{B}$, μ being the magnetic dipole moment of the atom, $\mathbf{B}$ the magnetic field. If we write $\boldsymbol{\mu} = \gamma_j \mathbf{J}$, where $\mathbf{J}$ is the total angular momentum of the excited atom, and $\mathbf{B} = \mathbf{B}(i\cos\omega t + j\sin\omega t)$, we will obtain the square bracketed expression in Eq. (44.1a), μ_+ being a raising, μ_- a lowering operator resulting from this selection of the *xy* plane for that of $\mathbf{B}$. At this point then the reader should see a reasonable analogy between Eq. (13.12) and that of BH

$$I(\omega) = \Re\left\{\tfrac{1}{2}B^2 \sum_{cd} \int_0^\infty e^{-i(\omega-\omega_{cd})\tau}\Phi_{cd}(\tau)\,d\tau\right\} \tag{44.2a}$$

$$\Phi_{cd} = \sum_{ab} \int_{-\infty}^{\infty} \langle a|\Upsilon(t)|a\rangle\langle a|\mu_-|b\rangle\langle b|\mathfrak{T}^{-1}(t+\tau,t)|c\rangle$$
$$\cdot\langle c|\mu_+|d\rangle\langle d|\mathfrak{T}(t+\tau,t)|a\rangle\,dt \tag{44.2b}$$

$$\mathfrak{T} = \mathfrak{U}_0^{-1}\mathfrak{U}_p \tag{44.2c}$$

$\mathfrak{U}_0$ being the TDO relating to H_0, $\mathfrak{U}_p$ being that relating to $H_0 + H_c$. Various reasons are introduced by various authors for reducing the density matrix to the diagonal form we have just encountered in Eq. (44.2b). BH assume that "the effect of collisions will destroy any coherence (off-diagonal terms)." This amounts to Ben-Reuven's SRPA (statistical random phase approximation).

The radiative lifetime, $\tau_{\text{rad}} = \gamma^{-1}$, is now simply inserted, and a steady-state situation is assumed to prevail so that the time dependence of the diagonal elements of the density matrix may be ignored. Eqs. (44.2) now become

$$I(\omega) = \tfrac{1}{2}B^2 \sum_{cd} \Re\left\{\int_0^\infty \exp\left[-i(\omega-\omega_{cd}-i\gamma)\tau\right]\Phi_{cd}(\tau)\,d\tau\right\} \tag{44.3a}$$

$$\Phi_{cd} = \sum_{ab} \langle a|\bar{\Upsilon}(t)|a\rangle\langle a|\mu_-|b\rangle\langle c|\mu_+|d\rangle \int_{-\infty}^{\infty} \langle b|\mathfrak{T}^{-1}(t+\tau,t)|c\rangle$$
$$\cdot\langle d|\mathfrak{T}(t+\tau,t)|a\rangle\,dt \tag{44.3b}$$

The next step is the reduction of the correlation function to diagonality in the TDO's, a task which is begun by multiplying

$$\mathfrak{T}(t+\tau,t) = \mathrm{T}\exp\left\{-i\int_t^{t+\tau} \mathfrak{U}_0^{-1}(t',t_0)H_c(t')\mathfrak{U}_0(t',\tau)\,dt'\right\}$$

T on the right the TOO, by $I=\mathfrak{U}_0^{-1}(t,t_0)\mathfrak{U}_0(t,t_0)$ on the left and right, thus:

$$\mathfrak{T}(t+\tau,t)=\mathsf{T}\exp\left\{-\mathrm{i}\int_t^{t+\tau}\mathfrak{U}_0^{-1}(t,t_0)\mathfrak{U}_0(t,t_0)\mathfrak{U}_0^{-1}(t',t_0)H_c(t')\mathfrak{U}_0(t',t_0)\right.$$

$$\left.\cdot\mathfrak{U}_0^{-1}(t,t_0)\mathfrak{U}_0(t,t_0)\,dt'\right\}$$

$$=\mathsf{T}\exp\left\{\mathfrak{U}_0^{-1}(t,t_0)(-\mathrm{i})\int_t^{t+\tau}\left[\mathfrak{U}_0(t',t_0)\mathfrak{U}_0^{-1}(t,t_0)\right]^{-1}H_c(t')\right.$$

$$\left.\cdot\left[\mathfrak{U}_0(t',t_0)\mathfrak{U}_0^{-1}(t,t_0)\right]dt'\,\mathfrak{U}_0(t,t_0)\right\}$$

$$=\mathfrak{U}_0^{-1}(t,t_0)\mathsf{T}\exp\left\{-\mathrm{i}\int_t^{t+\tau}\mathfrak{U}_0^{-1}(t',t)H_c(t')\mathfrak{U}_0(t',t)\,dt'\right\}\mathfrak{U}_0(t,t_0)$$

$$=\mathfrak{U}_0^{-1}(t,t_0)\mathsf{T}\exp\left\{-\mathrm{i}\int_0^{\tau}\mathfrak{U}_0^{-1}(x,0)H_c(x+\tau)\mathfrak{U}_0(x,0)\,dx\right\}\mathfrak{U}_0(t,t_0)$$

$$=\mathfrak{U}_0^{-1}(t,t_0)\mathfrak{T}_t(\tau)\mathfrak{U}_0(t,t_0), \tag{44.4}$$

where $x=t-t'$. When Eq. (44.4) is substituted into the matrix element of Eq. (44.3b) and the fact that $\mathfrak{U}_0(t,0)=\exp[-\mathrm{i}H_0t]$ is kept in mind, we obtain for the integral in the latter equation

$$\int_{-\infty}^{\infty}e^{\mathrm{i}(\omega_{bc}+\omega_{da})t}\langle b|\mathfrak{T}_t^{-1}(\tau)|c\rangle\langle d|\mathfrak{T}_t(\tau)|a\rangle\,dt \tag{44.5}$$

The argument goes something like this: the collisions are random so that $H_c(x)$ can be taken as translation invariant, $\mathfrak{T}_t(\tau)$ hence essentially independent of t. Therefore, the integration in Eq. (44.5) is really only over the exponentials, that is to say, it produces a delta function, $\delta(\omega_{bc}+\omega_{da})$, so that, in an integration over t, $\omega_{bc}=-\omega_{da}$. From the matrix elements over μ_+ and μ_- in Eq. (44.3b) we see that $a\neq b$, $c\neq d$, if we are to have any radiation. Therefore, this delta function has the effect of requiring that $b=c$ and $a=d$. This is half the argument. The remaining half consists in saying that this diversion now demonstrates the effective diagonality of the TDO matrix elements in Eq. (44.3b). A trivial notation change then leads to

$$I(\omega)=\tfrac{1}{2}B^2\sum_{if}\mathfrak{R}\left\{\int_0^{\infty}\exp\left[\mathrm{i}(\omega-\omega_{if}-i\gamma)\tau\right]\Phi_{if}(\tau)\,d\tau\right\} \tag{44.6a}$$

$$\Phi_{if}(\tau)=\langle i|\bar{v}(t)|i\rangle|\langle f|\mu_+|i\rangle|^2\int_{-\infty}^{\infty}\langle f|\mathfrak{T}^{-1}(t+\tau,t)|f\rangle$$

$$\cdot\langle i|\mathfrak{T}(t+\tau,t)|i\rangle\,dt \tag{44.6b}$$

We call the integral of Eq. (44.6b) $F(\tau)$. Then, using the interruption approximation and the arguments of Eq. (13.17) and subsequent, a differential equation for $F(\tau)$ is obtained as

$$\frac{dF(\tau)}{F(\tau)}=-N\left\{\int v\mathfrak{R}\left[1-\langle f|\mathfrak{T}^{-1}(\rho)|f\rangle\langle i|\mathfrak{T}(\rho)|i\rangle\right]d\rho\right\}d\tau \tag{44.7}$$

wherein ρ represents the spatial collision parameters over which the thermal motion average is taken. Therefore,

$$I_{if}(\omega)=\tfrac{1}{2}\langle i|\overline{\Upsilon}(t)|i\rangle|\langle f|\mu_+|i\rangle|^2(\gamma_j B)^2$$

$$\cdot\mathfrak{R}\left[\int_0^{\infty}\exp\left[-\mathrm{i}\{\omega-\omega_{if}-i(\gamma+N\bar{v}\sigma)\}\tau\right]d\tau\right]$$

$$=\frac{f(\Upsilon,\mu_+,B)}{\left(\omega-\omega_{if}+N\bar{v}\sigma_I\right)^2+\left(\tau_{\text{rad}}^{-1}+N\bar{v}\sigma_R\right)^2} \qquad (44.8a)$$

$$\sigma=\mathfrak{R}\left\{\int\left[1-\langle f|\mathfrak{T}^{-1}(\rho)|f\rangle\langle i|\mathfrak{T}(\rho)|i\rangle\right]d\rho\right\} \qquad (44.8b)$$

Eqs. (44.8) apply to both self and foreign-gas broadening; for the latter the real part is not taken in Eq. (44.8b). For self broadening only the real part is taken, and, since σ_R and σ_I are the real and imaginary parts of σ, respectively, σ_I does not exist and there is no line shift. That the real part is taken arises during the derivation of Eq. (44.7). There is an average over the product of the TDO matrix elements that is over the symmetric and antisymmetric states of the two colliding atoms taken as a system. With like atoms and the dipole-dipole interaction, the matrix elements of the TDO averaged over the symmetric states will be the complex conjugate of those over the antisymmetric states. Thus taking the average as half the sum of the symmetric and antisymmetric elements leads to the real part. Since the symmetric and antisymmetric elements are the same for foreign-gas broadening, the real part simply does not enter.

The ad hoc insertion of the radiative damping into Eq. (44.3a) of course led to the additivity of the natural and resonance line widths. However, we are already aware that they are additive to a quite high level of approximation so that the result is acceptable.

Finally, we remark that BF emphasize the following restrictions on the theory: (1) $\gamma_j B \ll \tau_{\text{rad}}^{-1}$ and (2) the splitting of the magnetic sublevels must be large with respect to the width of the lines.

BF demonstrated the applicability of their theory to the self broadening of certain spectral lines of mercury, cadmium, and zinc, and Byron et al. (91) demonstrated this applicability in considerably more detail for the optical double resonance lines of cadmium. All this is of interest to us only insofar as it emphasizes the valuable role that the theory may play. However, these comparisons with experiment call forth evaluations of the TDO matrix elements, and there are certain points of interest in connection with these evaluations.

45 Evaluation of Certain TDO Matrix Elements

The reader will be quite capable of developing the following equation of motion:

$$\mathrm{i}\dot{\mathfrak{T}}=\mathfrak{U}_0^{-1}(t,t_0)H_c^{\rho}(t)\mathfrak{U}_0(t,t_0)\mathfrak{T} \qquad (45.1)$$

when he is informed that H_c^ρ is the Hamiltonian for a single collision parametrized by ρ. (By parametrized we mean that the optical collsion diameter or distance of closest approach will be so and so, the relative velocity of the collsion partners will be so and so, etc.) Taking matrix elements of both sides of Eq. (45.1) yields

$$\begin{aligned} i\langle n|\dot{\mathfrak{T}}|n\rangle &= \sum_{n'n''n'''} \langle n|\mathfrak{U}_0^{-1}(t,t_0)|n'\rangle\langle n'|H_c^\rho(t)|n''\rangle \\ &\quad\cdot\langle n''|\mathfrak{U}_0(t,t_0)|n'''\rangle\langle n'''|\mathfrak{T}|n\rangle \\ &= \sum_{n'} e^{i\omega_{nn'}t}\langle n|H_c^\rho(t)|n'\rangle\langle n'|\mathfrak{T}|n\rangle \end{aligned} \tag{45.2}$$

BF consider the two limiting cases: (1) the time collision is long with respect to all the $\omega_{nn'}$; (2) the time of collision is short compared to the periods of interest to the problem, that is, compared to 10^{-7} to 10^{-8} sec. For case (1) the rapid oscillation of $\exp[i\omega_{nn'}t]$ will yield a zero result unless $\omega_{nn'}=0$, that is, unless $n=n'$ so that Eq. (45.2) becomes

$$\begin{aligned} i\langle n|\dot{\mathfrak{T}}|n\rangle &= \langle n|H_c^\rho(t)|n\rangle\langle n|\mathfrak{T}|n\rangle \Rightarrow \\ \langle n|\mathfrak{T}|n\rangle &= \exp\left[-i\int_{t_0}^{t}\langle n|H_c^\rho(t')|n\rangle\,dt'\right] \Rightarrow \\ \langle n|\mathfrak{T}(\rho)|n\rangle &= \exp\left[-i\int_{-\infty}^{\infty}\langle n|H_c^\rho(t)|n\rangle\,dt\right] \end{aligned} \tag{45.3a}$$

For case (2) we recall that the time of collision will be about the time requisite to perturber transit of the interaction sphere. This sphere can be taken as of the order of the distance of closest approach, ρ, which means the time of collision $\delta t=\rho/\bar{v}$, $\bar{v}$ being the mean velocity. Taking ρ as 10^{-6}, $\bar{v}$ as 10^5, δt is a factor 10^{-3} to 10^{-4} smaller than the period of interest to the problem, 10^{-7} to 10^{-8} sec. Under these conditions the exponential in Eq. (45.2) may be taken equal to unity. The equation thus becomes

$$\begin{aligned} i\langle n|\dot{\mathfrak{T}}|n\rangle &= \langle n|H_c^\rho(t)\mathfrak{T}|n\rangle \Rightarrow \\ \mathfrak{T} &= \exp\left[-i\int_{-\infty}^{\infty}H_c(t)\,dt\right] \end{aligned} \tag{45.3b}$$

Vis-à-vis Eq. (45.3b), BF want $H_c(t)$ to be "thought of" as a matrix involving the magnetic sublevels. At which point we remark that their figure is reproduced as Fig. 45.1. In this figure, $\bar{v}t$ is the straight-line trajectory of the colliding atom, ρ the distance of closest approach of the collision pair. The collision Hamiltonian is taken as the dipole-dipole interaction.

$$H_c(t)=R^{-3}(t)\{3[\mathbf{p},\mathbf{R}(t)][\boldsymbol{\pi},\mathbf{R}(t)]R^{-2}-(\mathbf{p},\boldsymbol{\pi})\}, \tag{45.4}$$

where we have adopted the BF convention of designating one atom by Latin, the other by Greek letters, and where $\mathbf{p}=e\sum\mathbf{r}_i$, i running over the elctrons in

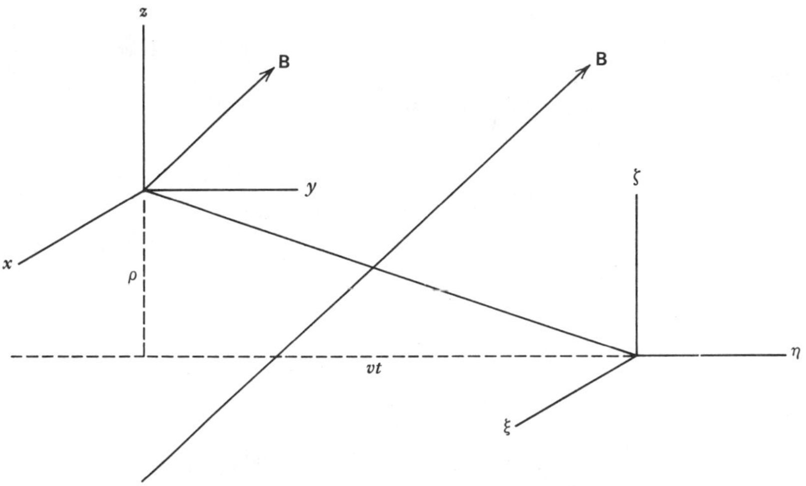

Figure 45.1 After Byron and Foley (90).

the atom. (The level splitting is a consequence of the atom's magnetic dipole moment, the interatomic interaction as a consequence of its electric dipole moment.) By making use of Fig. 45.1, from which we see that $R^2(t)=\rho^2+v^2t^2$, we may rewrite Eq. (45.1) as

$$H_c(t)=A\rho^2R^{-5}(t)+B\rho\bar{v}tR^{-5}(t)+CR^{-3}(t) \tag{45.5a}$$

where

$$A=3(p_z\pi_\zeta-p_y\pi_\eta) \tag{45.5b}$$

$$B=-3(p_z\pi_\zeta+p_y\pi_\eta) \tag{45.5c}$$

$$C=(2p_y\pi_\eta-p_x\pi_\xi-p_z\pi_\zeta) \tag{45.5d}$$

Equations (45.5) are now substituted into Eq. (45.3b), the integration carried out with the result

$$T=\exp\left(-\frac{2i}{\bar{v}\rho^2}(p_z\pi_\zeta-p_x\pi_\xi)\right) \tag{45.6}$$

whose matrix element must now be evaluated over wave functions symmetric or antisymmetric in the two atoms. We now return our attention to the figure and the $2J+1$ by $2J+1$ matrix $H_c(t)$.

We will probably agree that, were the magnetic field, **B**, along the z axis of the atom suffering the Zeeman splitting, the wave functions would be considerably simplified. Rather in the spirit of moving the mountain to Mahomet, BF discuss rotating the magnetic field to coincide with the z axis. Although it is perhaps largely a matter of viewpoint, we shall say that we are going to rotate the wave functions that are to be introduced later. This is precisely what the three-dimensional pure rotation group does as we have seen in Section 17.

Now let us call the square braket on the right of Eq. (45.3b) $M(\rho)$. Since H_c is—or will be when wave functions are applied—a $2J+1$ by $2J+1$ matrix, we here require the Jth irreducible representation of $O^+(3)$ [Eq. (15.27) of Wigner (48)]. Therefore, $M(\rho)$ will be multiplied on the right by $D^{(J)}$, on the left by $D^{(J)^{-1}}$ and

$$\mathfrak{T}(\rho)=\exp\left[-\mathrm{i}(D^{(J)})^{-1}M(\rho)\,D^{(J)}\right] \tag{45.7}$$

Suppose the matrix C diagonalized $M(\rho)$. Then $(D^{(J)})^{-1}C$ diagonalizes the bracketed expression on the right of Eq. (45.7). Letting $M_D=C^{-1}M(\rho)C$, we then see that

$$\begin{aligned}\mathfrak{T}(\rho)&=(D^{(J)})^{-1}C\exp[-\mathrm{i}M_D](D^{(J)})^{-1}C^{-1}\\&=(D^{(J)})^{-1}C(\exp[-\mathrm{i}\mu_{MM}])\left[(D^{(J)})^{-1}C\right]^{-1}\end{aligned} \tag{45.8}$$

In Eq. (45.8), $(\exp[-\mathrm{i}\mu_{mm}])$ is a diagonal matrix, each of whose elements is an exponential.

In order to complete our computation of the matrix elements of $\mathfrak{T}$ we must (1) form a wave function for the two-atom system, (2) evaluate the matrix $M(\rho)$ for the various magnetic quantum numbers, (3) diagonalize $M(\rho)$ in order to determine M_D and hence $(\exp[-\mathrm{i}\mu_{MM}])$, and (4) evaluate Eq. (32.8) using (3).

The wave function (symmetric) will surely be $2^{-1/2}[\varphi_g^0(\mathbf{r})\varphi_e^M(\rho)+\varphi_e^M(\mathbf{r})\varphi_g^0(\rho)]$ so that a matrix element of the $M(\rho)$ matrix will be

$$\begin{aligned}M^{MM'}&=\tfrac{1}{2}\langle\varphi_g^0(\mathbf{r})\varphi_e^M(\rho)+\varphi_e^M(\mathbf{r})\varphi_g^0(\rho)|\,p_z\pi_\zeta-p_x\pi_\xi\,|\varphi_g^0(\mathbf{r})\varphi_e^{M'}(\rho)+\varphi_e^{M'}(\mathbf{r})\varphi_g^0(\rho)\rangle\\&=\langle\varphi_g^0(\mathbf{r})|\,p_z\,|\varphi_e^{M'}(\mathbf{r})\rangle\langle\varphi_e^M(\rho)|\,\pi_\zeta\,|\varphi_g^0(\rho)\rangle\\&\quad-\langle\varphi_g^0(\mathbf{r})|\,p_x\,|\varphi_e^{M'}(\mathbf{r})\rangle\langle\varphi_e^M(\rho)|\,\pi_\xi\,|\varphi_g^0(\rho)\rangle\end{aligned} \tag{45.9}$$

when we take φ_g^0 as a $^1S_0$ state and φ_e^M as the lowest $J=1$ state as we might just as well take them for purposes of this evaluation. (Such states were also of specific interest to BF.) Equation (45.9) may be evaluated in terms of the reduced matrix elements by Eqs. (17.7) with the result

$$M^{MM'}=|\langle g\|P\|e\rangle|^2\begin{bmatrix}a&0&c\\0&b&0\\c&0&a\end{bmatrix},\qquad a=-\tfrac{1}{6},\quad b=\tfrac{1}{3},\quad c=\tfrac{1}{6} \tag{45.10}$$

Values of M and M' decrease from the upper left ($M=1$, $M'=1$) element.

A matrix of the form Eq. (45.10) may be diagonalized by

$$C=\begin{bmatrix}0&2^{-1/2}&2^{-1/2}\\1&0&0\\0&2^{-1/2}&-2^{-1/2}\end{bmatrix} \tag{45.11}$$

Thus we obtain M_D from the $M^{MM'}$ of Eq. (45.10) by the application of Eq. (45.11). In order then to obtain Eq. (45.8) we require only $D^{(J)}=D^{(1)}$.

Thus our TDO is given by

$$\mathfrak{T}=\left[(D^{(1)})^{-1}C\right]\begin{bmatrix} e^{ib} & 0 & 0 \\ 0 & e^{i(a+c)} & 0 \\ 0 & 0 & e^{i(a-c)} \end{bmatrix}\left[(D^{(1)})^{-1}C\right]^{-1} \tag{45.12}$$

There is no obvious reason for us to carry out the indicated multiplications, but, when we do so, we obtain the diagonal elements of $\mathfrak{T}$. These begin with $\langle M=1|\mathfrak{T}|M'=1\rangle$ in the upper left corner of the resulting matirx and then decreases diagonally to $\langle -1|\mathfrak{T}|-1\rangle$.

Of course the double-barred element $\langle \mathrm{g}\|\mathbf{P}\|\mathrm{e}\rangle$ has not been evaluated. This might be accomplished, for example, by means of the experimentally detemined oscillator strength. Some ab initio calculation might be carried out were sufficiently reliable wave functions available. Our interest has been solely one of illustration, however, and we consider this no further.

9

Satellites of Spectral Lines

"The problem of satellite-band formation on spectral-line profiles has intrigued experimental and theoretical physicists for 45 years." This is the opening gambit of Atakan and Jacobson (60) in one of their papers on this subject, and a most appropriate entrée it is. During this now over 50 year period a number of authors have "shot once and run away," as Henry Margenau once phrased it with regard to another facet of line broadening theory, many convinced that their shaft had penetrated a bull's-eye until then pristine before the assaults of earlier marksmen. In spite of this, it is not generally agreed that the entire phenomenon is even yet understood, although it has occupied a number of authors, some of whose work it is our intention to peruse in the present chapter. However, let us first presume ourselves to be as pristine as the mark and inquire as to what these satellites may be.

Let us suppose that we are observing an isolated spectral line, that is to say, an atomic spectral line whose upper and lower levels are sufficiently well separated from all other levels that the sort of coupling which we have encountered may be neglected. We might anticipate that such a spectral line, although broadened by atoms of the same or other species, would display a single intensity maximum. Some isolated atomic lines do just this. However, when certain other isolated atomic lines are broadened by certain foreign gases the elementarily anticipated intensity maximum is joined by one or more subsidary maxima which do not correspond to transitions among the levels of the isolated radiator. (By "subsidiary maximum" we mean a local maximum in the radiant intensity that does not correspond to the original (anticipated) intensity maximum in the spectral line and that is generally of lesser intensity than the original. The language is thus not very precise but is hopefully interpretable.) These subsidiary maxima are called satellites.

Oldenberg (216,217) first observed the phenomenon in the 2536.7 Å emission line of Hg broadened by He and certain heavier gases, he and Moore (213) in the 2528 Å absorption line of Hg. To Oldenberg's original observation of a Hg satellite—and, incidentally, his observation of violet satellites associated with the resonance lines of certain alkali metals broadened by noble gases—Kuhn and Oldenberg (171) soon added that of sharper but still violet or short-wavelength satellites. If not immediately, Kuhn (169) and Preston (220) soon added quantitative measurements to these qualitative observations. The cat, as it were, was squarely among the pigeons, large numbers of studies of both absorption and emission satellites on the noble-gas-broadened principal-series lines of the alkali metals (101, 133, 137, 100) following. Nor was the satellite phenomenon limited to this series in the alkali metals, the sharp, diffuse, and fundamental series (139) as well as certain forbidden lines (77) in these metals likewise displaying subsidary maxima. Examples of other radiators displaying satellites are Hg (231), Tl (102), In (102), various alkaline earths (97) and inert gases (97, 138). According to Atakan and Jacobson (60), for example, "satellites are induced not only by the inert gases but also by perturbers such as H_2, D_2, N_2, and many hydrocarbons (158)." It is not our intention to attempt an even approximate survey of the rather extensive experimental work that has been carried out in this area, but it does seem only proper to remark that, beginning with his 1936 observations (214), the field has been dominated by the experimental work of Ch'en and his many collaborators.

The earliest attempts at an explanation of the satellite phenomenon were made by Preston (220) and Kuhn (169), both based on the statistical broadening theory. Now obviously there will be a potential interaction curve corresponding to the interaction between perturber and radiator when the radiator is in, say, its upper *p* state, another corresponding to that interaction when the radiator is in, say, its lower *s* state. We suppose the Franck-Condon principle in operation so that transitions take place with sufficient rapidity that the radiator-perturber separation does not change during the transition. Then the statistical theory tells us (1) the frequency in the spectral line will correspond to the separation of the upper and lower state potential curves at transition and (2) the intensity will correspond to the probability for the radiator-perturber separation at transition. Let us recall that (1) the statistical theory is most often applicable in the line wings (satellites are a wing phenomenon) and (2) to slow collisions (the case here). In any event, this is the theory used by Preston and Kuhn.

Preston assumed a single interaction curve for the upper *p* state. If the upper and lower curves are parellel or nearly so for a substantial range of radiator-perturber separations he would anticipate a satellite at a frequency corresponding to the curve separations over this range. This is so, Preston felt, because of the increase in probability and hence intensity which, say, a random distribution of separations will yield for that frequency.

Kuhn, on the other hand, ascribed the main spectral line and a single satellite to the existence of two different *p* state potential curves. Two different curves might well give rise to what amounts to two overlapping lines.

Preston's essentially statistical approach, which looks to certain potential curve vagaries for its explanation of satellites, has often been appealed to during the years since he first applied it, and it has yielded varying levels of success. The reader will find examples provided by Jefimenko (156, 157), Kielkopf and Gwinn (161), Kielkopf et al. (162), and Hindmarsh and Farr (143) among others. Of course, with the exception of interference phenomena, the potential curve plays an essential role in all the satellite studies, although we shall be more concerned with theory formulation than potential detail.

Even though such is the case, the heart of the violet satellite explanation studied in the first section is a potential vagary, this being the splitting of the repulsive portion of the potential curve. For the red satellite explanation, a bound diatomic upper state comes into play.

In many line broadening situations the radiator and the perturber may well be considered as maintaining their atomic identities during the collision that leads to the line broadening, the collision being described by the interatomic potential. In such situations the appearance of a satellite should probably be attributed to certain unique details of the potential curve. In Sections 46, 47, and 51, we consider such satellite origins, although, in justice to precision, the work of Section 51 is considered more with the specific details of the potential as a satellite source than with anything notably unique. But, even in these first two sections, the authors only appeal to such a physical environment for an explanation of the violet satellites. For their red brethren the appeal is made to the only other obvious possibility.

If the two collision partners do not maintain their atomic identities they must perforce merge in the collective identity of a diatomic molecule. Given a potential with an attractive well, vibrational levels are to be anticipated if the capture resulting in well population occurs. This will of course yield a few intensity maxima. Such is the red satellite view of Section 46. In Section 48 we encounter a diatomic—albeit a dissociated diatomic—phenomenon that is related if somewhat distantly.

If we take the collision partners as a diatomic molecule having an attractive potential well but having energy such as to be in a dissociated state immediately above this well, we encounter a phenomenon that is quite appropriately called "quantum oscillations," these appearing under appropriate circumstances in the line wing. These are considered in Section 48. In Section 49, we are again considering our collision molecule, this time, however, from a particularly interesting point of view that includes not only the phenomena already mentioned in the intriguing context of path interference but also the so-called rainbow spike. In Section 51 we consider what have been called "unified" theories of pressure broadening, the reason for the verbiage to be discussed with the treatment itself. That the approach has been fruitful can hardly be gainsaid. But that it comprehends the Born-Oppenheimer approximation is fundamental to it.

It is all very well to ruminate on diatomic identity, but we have to this point done so under the aegis of the Born-Oppenheimer approximation. This approximation, we will recall, allows us to write the molecular electronic-vibrational wave function as the product of an electronic and a vibrational wave function. When we cannot justify this approximation, when the approximation breaks, down various additional satellite effects are going to emerge. These effects we discuss in Section 51.

46 The Statistical Theory of Satellites

Henry Margenau has been a towering figure in spectral line broadening theory for very nearly half a century; with all deference to Holtsmark, he must be considred the father of the statistical theory. It is to Margenau's statistical theory that he and Klein (KM) (163) appeal in the treatment of satellites which we consider now. These authors find a cruical role in this phenomenon to be played by Spitzer's rotational adiabaticity (255 and § 2.11 of 9), the details of which we might now recall.

The van der Waals forces of the broadening interaction establish a direction in the configuration space of the radiator and thus render the magnetic quantum number m meaningful. For the p state that will be considered here $m=-1,0,1$. We suppose the radiator in the p state when, at some finite separation from the perturber, it becomes aware of the perturber's existence. The orientations of the radiators may be taken as random, and, as a consequence, m, which depends on orientation, may suddenly find itself with, say, a value of 1. Thus when contact between radiator and perturber is established, the angular momentum vector of the former is directed toward the latter. As the perturber proceeds from this initial location to and through its point of closest approach, one of two things will occur: (1) the radiator will swing around with perturber passage such that its angular momentum vector will always point toward the perturber. In this case $m=1$ throughout the adiabatic collision. (2) The radiator will remain more or less stationary during perturber passage. The interparticle field will thus rotate during this perturber passage so that m changes during this diabatic collision.

In order to establish a criterion for adiabaticity KM appeal to the uncertainty principle in the form $\Delta E\Delta t>1$, taking Δt as the time of passage, ΔE as the difference in energy between, say a state with $m=0$ and one with $m=1$. For van der Waals forces, $E=Cr^{-6}$ so that $\Delta t \gg r^6/\Delta C$. Taking r as the distance of closest approach, ρ and v as the perturber velocity yields a typical time $\Delta t \sim \rho/v$. Thus, the adiabatic criterion is

$$\rho^5 \ll \Delta C/v \tag{46.1}$$

ΔC may be obtained from an earlier calculation of Margenau (186) as

$$\Delta C = C_{\pm}(P) - C_0(P) = \frac{3}{4}\frac{e^2}{\mathrm{m}}\alpha^{\mathrm{II}}E_{01}^{\mathrm{II}} \times\left[\frac{3}{10}\frac{f_{12}^{\mathrm{I}}}{\Delta E_{12}^{\mathrm{I}}(\Delta E_{12}^{\mathrm{I}}+\Delta E_{01}^{\mathrm{II}})}+3\frac{f_{10}^{\mathrm{I}}}{\Delta E_{10}^{\mathrm{I}}(\Delta E_{10}^{\mathrm{I}}+\Delta E_{01}^{\mathrm{II}})}\right] \tag{46.2}$$

wherein α is the polarizability, f the oscillator strength, ΔE_{ij} the energy of the $i-j$ transition, I and II referring to radiator and perturber, respectively. KM find that, for $v \sim 10^5$, for Hg perturbed by Ar the ρ must be no greater than 4.5 Å for adiabaticity, for K perturbed by Ar no greater than 8 Å. Thus they conclude that, if the density is such that the interatomic spacing is less that these limits, two upper state potential curves must be used, if less, some sort of average of the two curves.

These two authors were interested in pressure effects of alkali doublet lines. Now as we know, the true potential curve for the alkali-noble gas interaction has an inner, repulsive part at relatively small value of the atomic separation, a minimum at larger values, and an attractive part at large values. In a manner of speaking, the cr^{-6} form for the van der Waals is an asymptotic form for the potential, and it is this form that KM used in their careful treatment of the statistical broadening of these doublets which we discuss in Section 47. As we shall see, their results are in satisfactory agreement with experiment insofar as shifts and widths are concerned so that their treatment is probably quite good as far as it goes. However, the treatment does not go so far as to include the (unknown) inner repulsive part of the interaction potential, nor does the treatment yield any satellites of the spectral line. KM conclude that the theory's failure to produce satellites is to be attributed to its failure to include the repulsive part of the potential curve. What they specifically conclude is: (1) satellites arise in some cases from transitions to the two curves formed by the adiabatic splitting of the repulsive part of the excited curve; (2) the addition of the repulsive portion of the potential will lead to a potential well and the temporary, collisional formation of a diatomic molecule. This transient molecule will have a set of vibrational energy levels appropriate to its potential well. Satellites in certain cases will originate from transitions to or from these levels. (Here and hereafter a violet satellite originates from a transition to the high-frequency side of the parent line; a red satellite from a transition to the low-frequency side.)

A careful study of the available experimental data convinced KM that the violet satellites of the alkalis broadened by rare gases arise from transitions to an adiabatically split excited state. The red satellites arise from transitions involving the vibrational levels of the collisional quasimolecule. We consider their conclusions in somewhat more detail.

Violet Satellites The specific arguments whereby KM support their assignment of these satellites to the split excited state are:

1 The upper states associated with the doublet are $^2P_{3/2}$ and $^2P_{1/2}$. As we shall see in the next section, the van der Waals interaction will split the $^2P_{3/2}$ but not the $^2P_{1/2}$ state. Therefore, we should expect a satellite associated with the $^2P_{3/2}$ member of the doublet but not with the $^2P_{1/2}$. This is exactly what was observed for the alkali doublets broadened by noble gases. (The Hg 2537 Å line has three violet satellites and may not be so explained.)

2 This is a high pressure phenomenon. High pressures mean small collision diameters which in turn favor the inner repulsive portion of the potential curves, these portions being where the splitting occurs. On the other hand, high pressures tend to rule out the formation of diatomic collision molecules, favoring instead the formation of polyatomic molecules which exhibit features of considerably more complexity than those involved in a single violet satellite.

3 It has been observed that the satellite moved toward the parent line with increasing perturber or alkali mass or size. (Here size refers indirectly to atomic number and can be confused with mass as we move from, say, Na to K to Rb, etc. However, Ch'en's observation that the $Cs-H_2$ satellite has the same position as the $Cs-D_2$ satellite militates toward size rather than mass as the controlling consideration.) That this is in agreement with the split-state idea may be demonstrated after we recall a notion whose discussion is facilitated by Fig. 46.1.

The classical turning points in a potential well correspond to the walls of the well itself. The vibrational motion of our diatomic molecule is (classically) most rapid over the deepest portion of the potential well, slowing down as it approaches the well wall and the potential rises to meet the total energy of the vibrator. Now obviously these conditions will lead to the vibrator spending a proportionally greater part of its time near the well walls than anywhere else. This is of course what we have in mind when we say that the Franck-Condon principle favors transitions at the classical turning points.

An ab initio calculation of the potential curve in its repulsive portions would call for a precise determination of the molecular electronic wave function at the internuclear separation of interest. Although the simplest (generally) portion of the potential involved in such a calculation is the Coulomb interaction between the two nuclei, this nuclear portion contributes substantially and increases with increasing atomic number as does the exchange reaction contributing in this region. However, the nuclear repulsion will be essentially the same in both upper and lower state, while

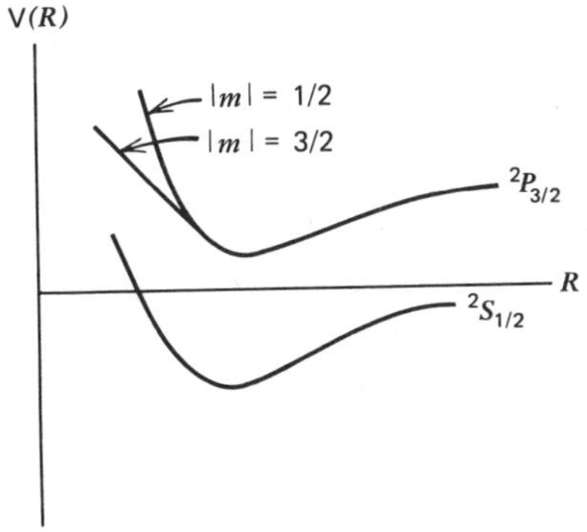

Figure 46.1 After Klein and Margenau (163).

the exchange contribution will increase more rapidly with atomic number in the ground than in the excited state. At least KM view it thus. This means that the classical turning point will move toward larger internuclear separation with increasing atomic number. As we can see from Fig. 46.1, such an increase will reduce parent-satellite separation. KM further remark that the difference in exchange energy between the two components of the split upper state and hence the splitting increases with atomic size. This effect, however, is presumed to be overshadowed by the movement of the turning point.

4 KM obtain good agreement with the Ch'en observation that the intensity of the violet satellites is 10^{-3} to 10^{-4} that of the parent. Insofar as relative intensity is concerned, the intensity of the satellite will be proportional to the probability for the presence of a satellite-producing perturber in the interaction sphere multiplied by the probability that the alkali is in a ${}^2P_{1/2}$ state for which $|m|=\frac{1}{2}$. The relative intensity of the resonance line, on the other hand, will be made up of two contributions:

a when no perturber is present it will include a contribution from the ${}^2P_{3/2}\ |m|=\frac{1}{2}$

b when a perturber is present it will be limited to the ${}^2P_{3/2}\ |m|=\frac{3}{2}$.

Therefore

$$\frac{I_s}{I_p}=\frac{\left(\frac{1}{3}\tau d e^{-\tau d}\right)}{\left[\frac{1}{3}+\frac{1}{3}\tau d e^{-\tau d}\right]}=10^{-3}$$

for a d of 1 amagat.

Blue Satellites Based on the fact that no red satellites had been observed at high pressure, KM took them to be low-pressure phenomena, which certainly seems straightforward enough. Low pressure in turn would appear to favor the outer or asymptotic parts of the potential curve and militate against the split-curve type of explanation. The authors therefore look to the vibrational structure of the collision molecule for their explanation.

We will probably agree that, given upper and lower state potential wells which yield the appropriate vibrational structure, a large variety of satellite spectra could be produced. The location of the satellites will of course depend on the relative separation of the appropriate vibrational levels in the two electronic states. The relative intensity of the satellites will depend on the number of collision pairs trapped in the wells and the vibrational transition probability. Although experimental data was not available on these wells, KM were able to ultilize some experimental results—van der Waals constants, for example—and resonable estimation in order to arrive at approximations for the intensities of the satellites. The agreement is persuasive if not ultimately conclusive.

As a footnote to the considerations of this section we remark that Royer (237), in a statistical study, pointed out that the singularity at $\omega=0$, which was dismissed by KM as an extraneous feature, is actually a satellite which should not be identified with the resonance line itself. That split levels and the vibrational levels of collision molecules, although of importance, are insufficient to account for all satellite phenomena we shall see in what follows.

47 Broadening of an Alkali Doublet by a Noble Gas

KM developed a detailed treatment of the broadening of the two members of an alkali doublet by a noble gas atom, a problem which Takeo and Ch'en (264) had previously considered less comprehensively. It is true that the treatment produces no satellites, but, with the simple sort of asymptotic force presumed, this should hardly come as a surprise. The value of the calculation lies in the groundwork it lays for the qualitative applications of the statistical theory that we have already encountered in the previous section.

We begin with the dipole-dipole interaction potential of an alkali in a p state surrounded by n foreign-gas atoms:

$$\mathsf{V}=e^2\sum_{i=1}^{n}\mathbf{r}\cdot\bar{\mathbf{r}}_iR_i^{-3}-3e^2\sum_{i=1}^{n}\mathbf{r}\cdot\mathbf{R}_i(\bar{\mathbf{r}}_i\cdot\mathbf{R}_i)R_i^{-5} \tag{47.1}$$

where $\mathbf{r}$ refers to the alkali electron, $\bar{\mathbf{r}}_i$ the dipole coordinate of the ith perturber, and $\mathbf{R}_i$ to the separation of the radiator from the ith perturber.

The alkali functions are familiar and here designated as $\psi(^2P_{jm})$ for $j=\frac{1}{2}$, $\frac{3}{2}$, $m=\pm\frac{3}{2}$, $\pm\frac{1}{2}$. The perturber functions are taken as $\phi(r)=\Pi_i u_{10}(i)$ (spin), the system function written as $\psi(^2P_{1/2,\pm1/2})\phi(r)=\psi_{\pm1/2}(r)\phi(r)$ for the unperturbed case. Since the problem is to be carried through second order in ordinary perturbation theory, the objective is to find the combination of these functions that diagonalizes $\Sigma_i[\mathsf{V}_{\nu i}\mathsf{V}_{i\lambda}/(E_\lambda^0-E_\nu^0)]$, V given by Eq. (47.1), the expression of course being that for the second-order energy. Now this second-order energy may be dealt with rather easily if the denominator can be removed from the summation sign. This is effected by taking the energy difference as essentially ΔF, the ionization potential of the perturbing atom alone. Thus we obtain $(1/\Delta F)(\mathsf{V}^2)_{\nu\lambda}$. That this is diagonal in the chosen wave functions is argued as follows.

V^2 as given by Eq. (47.1) will not change sign when its Cartesian coordinates are reflected through the origin; thus its parity is even. In addition to V^2, the factor $\psi_{1/2}\psi_{-1/2}$ will appear in the off-diagonal matrix elements over the radiator ground state. The reader may wish to use Eqs. (2.18) in order to convince himself that the parity of this wave function product is odd. However, a matrix element, since it is a scalar, must have even parity. Ergo, V^2 is diagonal in the $^2P_{1/2}$ states. Such is not the case for the $^2P_{3/2}$ states, however.

Our argument of the last paragraph tells us that, insofar as V^2 is concerned, matrix elements for which $\Delta m=\pm1$ are forbidden. Similar arguments based on V and elementary matrix multiplication tells us that $\Delta m=0, \pm 2$ elements exist. The off-diagonal elements, $(V^2)_{3/2,-1/2}$, $(V^2)_{1/2,-3/2}$, and so on are all equal, and $(V^2)_{3/2,3/2}=(V_2)_{-3/2,-3/2}$, $(V^2)_{1/2,1/2}=(V^2)_{-1/2,-1/2}$. The resulting determinantal equation has a solution involving two energies,

$$E_{+-}=\Big\{(V^2)_{3/2,3/2}+(V^2)_{1/2,1/2}$$

$$\pm\Big[(V^2)_{3/2,3/2}-(V^2)_{1/2,1/2}+4\{(V^2)_{3/2,-1/2}\}^2\Big]^{1/2}\Big\}/2\Delta F \quad (47.2)$$

which energies depend on (1) the spatial distribution and (2) the number of perturbers included in V.

Simplifying matters somewhat by (1) fixing all active perturbers at a distance, $\bar{R}$, from the radiator, (2) distributing them at the six corners of an octahedron, and (3) restricting the perturbers considered to a volume τ surrounding the radiator, KM obtain, for the $^2P_{1/2}$ state,

$$E=(2e^4n/\Delta F\bar{R}^6 3)(\bar{r}^2)_{00}\{\tfrac{2}{3}(r_{12})^2+\tfrac{1}{3}(r_{10})^2\}, \quad (47.3a)$$

and for the $^2P_{3/2}$,

$$E_{+-}=-\left(\frac{e^4}{\Delta F\bar{R}^6}\right)(\bar{r}^2)_{00}\Big\{2n\big[\tfrac{2}{3}(r_{12})^2+\tfrac{1}{3}(r_{10})^2\big]/3$$

$$\pm\Big[\sum_{j=1}^{n}\{(\tfrac{3}{90})^2(r_{12})^4+(\tfrac{3}{18})^2(r_{10})^4\}\{(\delta_{\lambda_j,x(y)}$$

$$-2\delta_{\lambda_j,z})^2+3(\delta_{\lambda_j,x}-\delta_{\lambda_j,y})^2\}\Big]^{1/2}/3\Big\} \quad (47.3b)$$

In Eq. (47.3b) $(\bar{r}^2)_{00}$ is the matrix element of the squared dipole moment coordinate of the perturber, r_{12} the radiator radial matrix element for $l=1\rightarrow l=2$ while λ_j is the jth perturber's Cartesian coordinate.

KM consider the $P_{1/2}$ state to be spherically symmetric, and hence the spatial distribution of the perturbers is irrelevant, their separation of course not being so. The perturbation of the $P_{3/2}$ state, on the other hand, will

Table 47.1

n	Average van der Waals energy $^2P_{3/2}$
1	$(G/R^6)(K\pm 2J)$
2	$(G/R^6)(2K\pm 2.9J)$
3	$(G/R^6)(3K\pm 3.5J)$
4	$(G/R^6)(4K\pm 4.4J)$
5	$(G/R^6)(5K\pm 5.0J)$
6	$(G/R^6)(6K\pm 5.8J)$

Source: After Klein and Margenau (163).

depend on just where the n perturbers are located. KM take the probability of occurrence of each perturber distribution to be proportional to the number of permutations of n objects taken p at a time, p the number of perturbers on each axis. This probability is used to weight the energies obtained from Eq. (47.3b) for the various perturber distributions within the interaction sphere, the result for up to six perturbers being reproduced as Table 47.1.

The symbols appearing in the table are defined as follows:

$$G=\frac{e^4(\bar{r}^2)_{00}}{\Delta F} \tag{47.4a}$$

$$K=\frac{2\left[2(r_{12})^2/3+(r_{10})^2/3\right]}{3} \tag{47.4b}$$

$$J=\left[(1/90)^2(r_{12})^4+(1/18)^2(r_{10})^4\right]^{1/2} \tag{47.4c}$$

In terms of these quantities the $^2P_{1/2}$ state energy is

$$E=GR^{-6}Kn \tag{47.4d}$$

For purposes of calculation it is convenient to rewrite these quantities as

$$K=\mathrm{m}^{-1}\left[\frac{f_{12}}{\Delta E_{12}}+\frac{f_{10}}{\Delta E_{10}}\right] \tag{47.4b$'$}$$

$$J=(4\mathrm{m})^{-1}\left[(10)^{-2}\left(\frac{f_{12}}{\Delta E_{12}}\right)^2+\left(\frac{f_{10}}{\Delta E_{10}}\right)^2\right]^{1/2} \tag{47.4c$'$}$$

where

$$\alpha=\frac{2e^2(\bar{r})^2_{00}}{3\Delta F} \tag{47.4e}$$

$$f_{12}=\tfrac{4}{9}\mathrm{m}\Delta E_{12}(r_{12})^2 \tag{47.4f}$$

$$f_{10}=\tfrac{2}{9}\mathrm{m}\Delta E_{10}(r_{10})^2 \tag{47.4g}$$

Here α is the perturber polarizability.

KM now consider the specific case of an alkali interacting with Ar, and we shall follow them through on this. They take $\alpha=1.63$ Å^3, $f_{12}/\Delta E_{12}\doteq 1.5/2$ eV, $f_{10}/\Delta E_{10}\doteq 0.52$, so that

$$^2P_{1/2}\text{:}\ E=-267.4n/\bar{R}^6 \tag{47.5a}$$

$$\begin{aligned} ^2P_{3/2}\ E_{+-} &= -(267.4/R^6)\{n\pm(1/16)[2\delta_{n1}+2.9\delta_{n2} \\ &\quad +3.5\delta_{n3}+4.4\delta_{n4}+5.0\delta_{n5}+5.8\delta_{n6}]\} \end{aligned} \tag{47.5b}$$

The intensity, $I(\Delta\omega)$, at a frequency $\Delta\omega=\omega-\omega_0$ in the spectral line will be equal to the probability for a configuration of perturbers corresponding to that frequency, that is

$$I(\Delta\omega)=\sum_n P(n)\delta(\omega_n-\Delta\omega) \tag{47.6a}$$

where

$$P(n)=(\gamma d)^n e^{-rd}/n! \tag{47.6b}$$

Here $P(n)$ is the Poisson distribution and tells us the probability of finding n particles of a gas of density d within a sphere of volume r. Note that $\bar{n}=rd$ is the most probable number within r. The ω_n are the energies of Table 47.1 minus the corresponding ground-state perturbation energies.

We may perhaps have forgotten that all our machinations of the last few paragraphs have related to the excited p state of the radiating alkali. The same sort of thing is done for the spherically symmetric ground s state with the result

$$E=G_0\bar{R}^{-6}n \tag{47.7a}$$

$$G_0=\alpha e^2 3f_{10}/2\mathrm{m}\Delta E_{10} \tag{47.7b}$$

and $f_{10}/\Delta E_{10} \doteq \frac{1}{2}$. Therefore

$$^2P_{1/2}\colon \omega_n=-134\bar{R}^{-6}n \Rightarrow$$

$$^2P_{3/2}\colon \omega_n=-134\bar{R}^{-6}\{n\pm 8^{-1}[2\delta_{n1}+\cdots+5.8\delta_{n6}]\} \tag{47.8a}$$

$$^2P_{1/2}\colon I(\omega)=\sum_{n=1}^{6}(rd)^n e^{-rd}\frac{\delta[134R^{-6}n-\omega]}{n!}$$

$$^2P_{3/2}\colon I(\omega)_{\pm}=\sum_{n=1}^{6}(rd)^n e^{-rd}\delta\{134R^{-6}\left[n\pm\tfrac{1}{8}(2\delta_{n1}+\cdots+5.8\delta_{n6})\right]-\omega\} \tag{47.8b}$$

$$I(\omega)=I(\omega)_{+}+I(\omega)_{-}$$

Here the two $\frac{3}{2}$ states have equal weight so that simple addition of intensity distributions follows as shown. What have we arrived at?

We have arrived at a statistical line shape for the two members of an alkali doublet broadened by Ar. KM carried out a large number of numerical calculations of linewidth, which they compared with observations of Rb–Ar, and line shift, which they compared with observations of both Rb–Ar and Na–Ar. The agreement the reader may judge for himself from Fig. 47.1.

As is apparent, both doublet components are shifted to the red, the $^2P_{3/2}$ being broadened more than the $^2P_{1/2}$. The authors remark that by adjusting the radius of the interaction volume the shift may be made to fit the experimental curve exactly, but the difference in the broadening of the two doublet components becomes too small by a factor of about one fifth. It is to be recalled that this entire treatment has been based on the simple, asymptotic form of the interaction, that is, on cr^{-6}. Under these conditions, the results do seem rather reasonable.

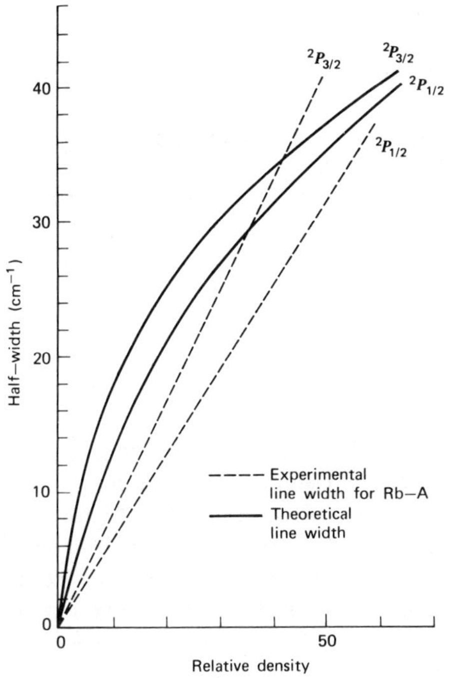

(a)

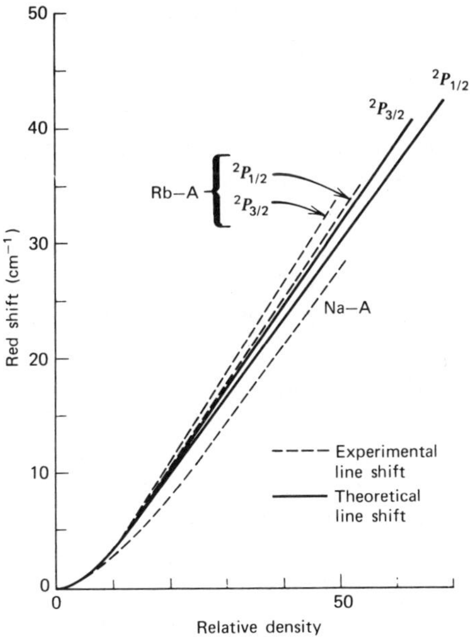

(b)

Figure 47.1 After Klein and Margenau (163).

48 Quantum Oscillations in the Line Wing

Mies (201) and Royer (237) have utilized what we may classify as the Jablonski theory of broadening in order to bring out a number of interesting aspects of the satellite problem. In a series of papers of which we remark the first (152) and the last (154), Jablonski developed a theory of spectral line broadening [cf. Chapter 3 of (9)] which we mentioned in Chapter 1 but which we now remark on in somewhat greater detail.

Jablonski began by considering the radiator and n perturbers as an $(n+1)$-atomic molecule. Practically, and rather quickly, the n reduces to one so that we might simply begin with a diatomic molecule. Then the problem of determining the intensity distribution in the broadened spectral line constitutes an analogy to that of determining the intensity in an electronic-vibrational band system. In dealing with the intensity of an electronic-vibrational band it has long proven fruitful to consider this intensity as the product of an intensity for the electronic transition and a squared vibrational overlap integral or a so-called Franck-Condon factor. A great many such calculations have been carried out over the years by Nicholls (cf. 210), and, although certain modifications of the approach (for example, the so-called r-centroid) have occasionally proved fruitful, the hypothesis to the effect that the intensity varies from band to band with the square of the overlap integral has proved generally good. (In a precise calculation one would take the dipole operator and integrate it over electronic and vibrational coordinates. If the operator is integrated first over the electronic coordinates, we obtain what has been called an electronic "transition moment." It has been found that this moment, which is now the operator for the vibrational integration, is often essentially independent of which pair of vibrational wave functions are under consideration. Therefore, the relative intensity from band to band is equivalent to the vibrational overlap integral from band to band.)

Therefore, we might suppose that some sort of analogous calculation in the line broadening situation might prove of value. It isn't, but it is. That is to say, the broadening analogy is not very precise, but the ideas are of value. Let us take the time to discuss the situation elementarily with reference to Fig. 48.1.

We are going to suppose that a spectral line absorbed by atom X in its transition from state i to state f is broadened by a collision with atom Y. When atom X is in electronic state i it could combine with atom Y in its ground state to form the set of vibrational levels I. (All of this of course is being simplified to the level of schematism.) An analogous statement may be made with respect to states f and F. When the situation is such that the states I obtain, the transition may take place, its relative intensity determinable by integrating the product of the appropriate vibrational wave functions in I and F over all internuclear separations, R, and squaring the result. This is the *modus operandi* for dealing with molecular electronic-vibrational intensities.

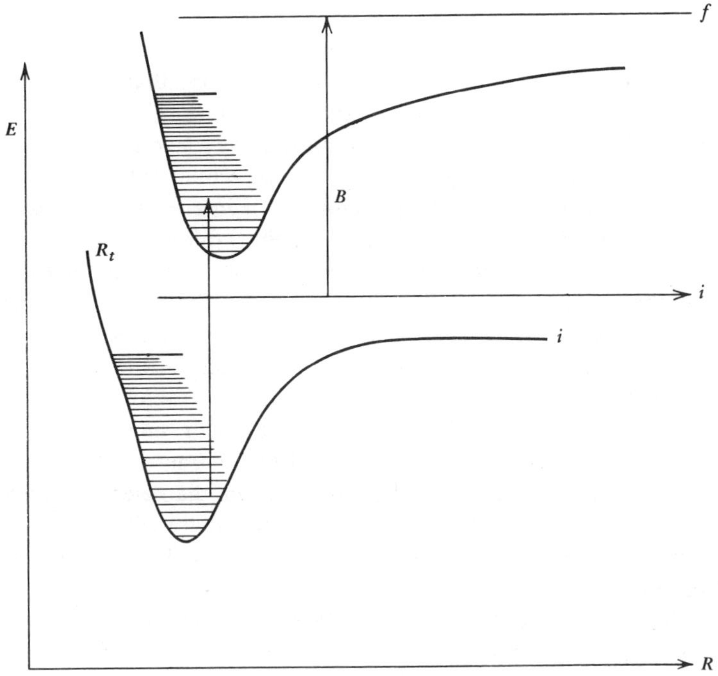

Figure 48.1

Now we say we are going to deal with broadened atomic transitions "analogously."

This means, first, that we take the collision molecule wave function as a product of the electronic wave function for the radiating atom and the vibrational wave function for the dissociated collision pair, this pseudo-vibrational function describing the translational motion of the system. It means, second, that the intensity at a particular frequency in the spectral line corresponds to the integral of a pair of such functions over the vibrational coordinates of those upper and lower state vibrational functions appropriate to the frequency in the line and the initial translational energy, the electronic integration being assumed to yield a constant. It means, third, that the integral over the vibrational coordinates contains no operators; it is simply an overlap integral that we justify by the arguments of band theory.

Take the transition B in Fig. 48.1. The intensity corresponding to the frequency associated with such a separation will be proportional to the integral over the product of the vibrational wave functions in the upper and lower states. These wave functions of: (1) the angular momentum of the colliding pair, which is of course dependent on the distance of closest approach; (2) the thermal energy of the colliding pair. Having obtained the overlap integral, one finally obtains the intensity at the corresponding frequency in the spectral line by averaging over initial angular momentum and thermal energy. It is perhaps worthwhile to emphasize the dissociated nature of the diatomic under consideration. Suppose that our collision diatomic were captured into the well in the lower state. This would of course immediately eliminate the momentum and energy averaging, and, as we moved thus from the analogy to the original, we would enter a realm having but a tenuous connection with the phenomena of interest to us. This bring us to a consideration of the statistical nature of the Jablonski theory.

This theory is a statistical theory that is not applicable in the central portion of the spectral line, a fact that is obviously of no consequence to satellite studies since satellites are a wing phenomena. Such is the case because of the importance to the central portion of the line of distant collisions, distant collisions which would render the Jablonski theory somewhat bizarre. Through, say, Eq. (17.5b) the angular momentum quantum number l, will be related to the optical collision diameter. Therefore, corresponding to whatever maximum optical collision diameter is chosen as a consequence of our exclusion of distant collisions, there will be a maximum l in the angular momentum average. At this point, recall the form the vibrational functions may take.

As we have discussed in some detail (§ 37 of 9) and as is obvious in any event we may choose a vibrational wave function

$$\psi_k = \sum_{l=0}^{\infty} a_l P_l(\cos\vartheta)\frac{G_{kl}(r)}{r} \tag{48.1a}$$

which, when substituted into the Schrödinger equation yields

$$\frac{d^2G_{kl}}{dR^2} + \left[2\mu E_k - 2\mu \mathrm{U}(r) - 2\mu l(l+1)R^{-2}\right]G_{kl} = 0 \tag{48.1b}$$

where μ is the reduced mass of the collision pair. The Wentzel-Kramers-Brillouin (WKB) solution to this equation is

$$G_{kl}(R) = \begin{cases} c(p_{kl}(r))^{-1/2}\exp\left(-\int_R^{R_t}|p_{kl}(R')|\,dR'\right) & \text{for } R<R_t \\ c(p_{kl}(r))^{-1/2}2\cos\left[\int_R^{R_t}p_{kl}(R')\,dR' - \frac{1}{4}\pi\right] & \text{for } R>R_t \end{cases} \tag{48.1c}$$

We remark that R_t, indicated in Fig. 48.1, is the classical turning point where $p_{kl} = (2\mu E_k - 2\mu\mathrm{U}(r) - 2\mu l(l+1)R^{-2})^{1/2} = 0$. Quantum mechanically the solution for $R<R_t$ is the familiar one corresponding to percolation through a potential barrier. However, the interesting solution from our point of view is that for $R>R_t$, this is an appropriate point at which to establish contact with Mies.

Tanaka and Yoshino (267) had observed bandlike structure in the continuous radiation emitted by He ($1s2s\,{}^1S_0$) in collision with ground state He which Mies and Smith (202) quantitatively explained. Their method of explanation is essentially what Mies (201) later improved on somewhat and applied to spectral line broadening in general. What Mies said was this:

The Jablonski theory and Eq. (48.1c) tell us that the intensity in a spectral line is proportional to the R-integral over $\cos\beta'\cos\beta''$, β' and β'' being obvious from Eq. (48.1c). Here the single prime refers to the upper state and hence p_{kl}, E_k and $\mathsf{U}(r)$ there; the double prime refers to the lower state. The Jablonski argument given by Breene (9) in his Eqs. (3.77) to (3.79) is a quite simple one and leads to the following result for the overlap integral:

$$\cos\left\{\beta'(R_c)-\beta''(R_c)\pm\tfrac{1}{4}\pi\right\}=\cos\alpha \tag{48.2}$$

Therefore, the intensity in the spectral line would be proportional to the square of an angular function were it not for the angular momentum and kinetic energy averaging. It is worthwhile to follow through Mies determination of when such averaging will not obliterate oscillations by its smoothing action.

Mies considers only the case where the initial state is attractive as this is the sole continuum-continuum transition wherein we may anticipate observing these fluctuations. His expression for the spectral line intensity is

$$I(kT)\sim\int_0^\infty d\epsilon' \exp\left(\frac{-\epsilon}{RT}\right)M(\epsilon') \tag{48.3a}$$

where

$$M(\epsilon')\sim\sum_{l'l''} S_{l''\Lambda''}^{l'\Lambda'} A^{\Lambda'\Lambda''}\cos^2\alpha \tag{48.3b}$$

Here we have simply followed Herzberg (22), as did Mies, in using Λ for the angular momentum number associated with the electronic structure of the collision molecule. ($\Sigma\sim\Lambda=0, \Pi\sim\Lambda=1, \Delta\sim\Lambda=2$, etc.) Only the Hönl-London factors, S, are l dependent. These Hönl-London factors [cf. p. 208 of (22)] of course allow $l''=l'$, $l'\pm1$ and we sum over l'' to $(2J'+1)$ so that Eq. (48.3b) becomes

$$\sum_{l'=\Lambda'}^{l_{\max}} (2l'+1)A^{\Lambda'\Lambda''}\cos^2\alpha \tag{48.3b$'$}$$

Perhaps this is an appropriate place at which to emphasize a point with reference to the matter of angular momentum. We have just pointed out that there is angular momentum associated with the electronic structure of the collision molecule. We know there is what amounts to a molecular rotational momentum by virtue of the motion of, say, the perturber past the emitter. These two momenta, as in a stable diatomic, will combine to yield a total angular momentum which we have designated l. We have already remarked on the necessity for an upper limit on l imposed by the necessity for excluding distant collisions. The psuedo-rotational momentum obviously dominates these consideration since the Λ contribution to l is a mere 0, 1, or 2. However, the Λ contribution establishes the lower limit on possible values of l as is indicated in the summation of Eq. (48.3b′).

Mies establishes the upper limit on this summation with a classical criterion. For an attractive potential there will be an internuclear separation at which a rotational barrier will exist. Classically at least, this barrier will prevent a perturber having greater than barrier separation from approaching the critical separation, that is, the Franck-Condon separation, R_c. This would seem a quite sensible way of establishing a maximum R and hence a maximum l. This barrier will then correspond to a particular l value, its height specificable as ϵ_l. Obviously $\epsilon_{l_{max}} = \epsilon'$, the energy of the initial state. Thus the criterion. The summation over l is, predictably, replaced by an integration. Since the energy is proportional to $l(l+1)$, it is qualitatively obvious that Eq. (48.3b′) may be approximately rewritten as

$$\int_0^{E_{max}} dE_J \cos^2\alpha \tag{48.3b″}$$

Mies shows by numerical calculation that, for ϵ' small compared to the dissociation energy of the collision molecule, α is, to a very good approximation, linearly dependent on E_J; thus $\alpha = \alpha_0 - \alpha_1 E_J$. With this Eq. (48.3b″) becomes

$$\tfrac{1}{2}E_{max}\left[1 + \left\{\frac{\sin(\alpha_1 E_{max})}{\alpha_1 E_{max}}\right\}\cos(2\alpha_0 - \alpha_1 E_{max})\right] \tag{48.3b‴}$$

What Mies calls Jablonski's random-phase approximation [cf. Eq. (3.93) and the preceding equations of (9)] amounts to the replacement of $\cos^2\alpha$ by $\frac{1}{2}$, that is, the neglect of the second term within the square bracket of Eq. (48.3b‴). Thus the magnitude of what Mies calls the quantum oscillations depends on

$$A(\epsilon') = \sin(\alpha_1 E_{max})/\alpha_1 E_{max} \tag{48.4}$$

Obviously $A(\epsilon')$ varies from 1 at very small E_{max} or, indirectly, incident kinetic energies to 0 for very large E_{max} or incident kinetic energies. Since the incident kinetic energies depend on kT, the oscillations will surely disappear for sufficiently high temperature. This result must in turn be averaged over a Boltzmann distribution which, as we are well aware, broadens with temperature increases from zero. Thus at or near zero the quantum oscillations will be pronounced, smoothing out with increases in temperature. Because of insufficient knowledge of the interaction potentials, Mies was of course unable to carry out a precise calculation for a particular transition. However, he was able to carry out what appears to be a quite reasonable estimate for the He broadened by He case, from which he concluded that the maximum temperature at which these oscillations could be observed was of the order of 300° K. As he points out, Tanaka et al. (265, 266) observed a considerable resolution on cooling the discharge from 300 to 77° K, which is strong support for the Mies conclusion. Further, Tanaka et al. observed the same sort of oscillations in Ne and Ar.

Mies is enthusiastic about the possibility of determining interaction potentials from these oscillations—for example, the continuum vibrational wave function immediately above the well and the "last" bound vibrational wave function are essentially identical—but we would be straying too far afield in any serious consideration of these potentials. Our final remark, though obvious, is worth setting down: the quantum oscillations considered in this section almost certainly exist and can be important to the broad category, "satellites." However, they cannot be looked to for an explanation of the wide range of satellite phenomenon nor, certainly, did Mies intend them for such an explanation.

49 Path Integrals, the Rainbow and Interference Effects

It has not been uncommon to consider the Jablonski theory as treating the paths of the collision pair quantally as opposed to the classical paths of other treatments. "Not so," says Royer (238), for the WKB functions which, in the final analysis, really describe the translational motion imply classical trajectories. Royer describes the result of all this as the semiclassical spectrum. It is of considerable basic interest for us to follow through his demonstration of this assertion.

We may recall that, in the Feynmanian formulation of quantum theory (14), a probability amplitude is associated with each possible path a system may take in proceeding from an initial to a final state. The total transition amplitude is then the sum of all these amplitudes for a given initial and final state. For the initial state the translational state of an incident particle having a particular velocity vector, and for the final state, that of an outgoing particle having a velocity vector differing by whatever the scattering angle of interest may be, we are dealing with a quantal situation and hence amplitude. On the other hand, the classical situation and hence amplitude for a particular initial velocity is completely determined by the angular momentum or optical collision diameter. We join the multitude in defining therefore the semiclassical amplitude for scattering through a particular angle as the sum over those amplitudes corresponding to the optical collision diameters capable of producing scattering through the angle of interest. With this we follow through the Royer obtention of an amplitude using the semiclassical definition and find that, indeed, this is what our considerations of the preceding sections have been.

First we outline a typical path for a broadening collision as follows. This path is to connect the states (i, ϵ_i, l) and (f, ϵ_f, l) where ϵ_i and ϵ_f are the energies of the initial and final states, respectively, and l is the angular momentum. The perturber proceeds from infinity to a collision pair separation r and, in doing so, follows the trajectory prescribed by ϵ_i, l, and $\mathsf{U}_i(r)$. While the perturber is at r, the transition i to f occurs instantaneously. The perturber then proceeds to infinity following the trajectory prescribed by ϵ_f, l, and $\mathsf{U}_f(r)$. The amplitude for this sequence of events will be the product of three amplitudes for the three subsequences. We now work this out in detail.

The amplitude associated with perturber motion from r_1 to r_2 as prescribed by ϵ_e, l, and $\mathsf{U}_e(r)$ is designated as $\langle r_1 \to r_2 \rangle_e$. That the amplitude and phase are related as

$$\langle r_1 \to r_2 \rangle \sim \exp\left[\mathrm{i} \left| \int_{r_1}^{r_2} dr\, k_e(r) \right| \right] \tag{49.1}$$

may be seen from the fact that, for $\mathsf{U}_e = 0$, Eq. (49.1) simply reduces to the product of the translational wave functions at points r_1 and r_2. (At this point Royer drops specific reference to potential—$k_e \to k$, etc.—and so, as Mr. Dunkinfield was wont to remark, shall we.) From Eq. (49.1) the reader will agree that, if r_3 lies between r_1 and r_2,

$$\langle r_1 \to r_2 \rangle = \langle r_1 \to r_3 \rangle \langle r_3 \to r_2 \rangle \tag{49.2}$$

Finally, if the perturber passes through the turning point, r_t, a phase factor, $\exp[\mathrm{i}\pi/2]$, a direct consequence of quantum barrier penetration, is picked up; thus

$$\begin{aligned} \langle r_1 \to r_t \to r_2 \rangle &= \langle r_1 \to r_t \rangle e^{\mathrm{i}\pi/2} \langle r_t \to r_2 \rangle \\ &= \exp\left[\int_{r_t}^{r_1} k(r)\, dr + \mathrm{i}\pi/2 + \mathrm{i} \int_{r_t}^{r_2} k(r)\, dr \right]. \end{aligned} \tag{49.3}$$

The amplitude associated with perturber presence at r will be the sum of the amplitudes associated with the two paths to r: (1) the perturber proceeds from infinity to r ($\langle \infty \to r \rangle$); (2) the perturber proceeds from infinity to the turning point, is there reflected, and continues to r ($\langle \infty \to r_t \to r \rangle$). Therefore, the amplitude for the perturber to be at r is

$$A(r) = (\langle \infty \to r \rangle + \langle \infty \to r_t \to r \rangle)[k/k(r)]^{1/2} \tag{49.4}$$

We have inserted a factor $[k/k(r)]^{1/2}$ with the following thoughts in mind: the separation r will correspond to a particular frequency in the spectral line; the square of $A(r)$, to the intensity at that frequency. Just as the Condon point corresponds to the most probable transition separation of the molecular constituents because the molecule spends more time there, so the intensity at a given frequency will to some extent depend on the time spent by the collision pair at the corresponding separation. That $k/k(r)$ corresponds to this time explains its insertion.

Since Eq. (49.2) allows us to write $\langle \infty \to r \rangle$ as $\langle \infty \to r_t \rangle \langle r \to r_t \rangle^{-1}$, Eq. (49.4) becomes

$$\begin{aligned} A(r) &= \langle \infty \to r_t \rangle [\langle r \to r_t \rangle^{-1} + e^{\mathrm{i}\pi/2} \langle r_t \to r \rangle][k/k(r)]^{1/2} \\ &= \langle \infty \to r_t \rangle 2 e^{\mathrm{i}\pi/4} \psi_{\mathrm{WKB}}(r) \end{aligned} \tag{49.5}$$

by Eq. (49.1). Here ψ_{WKB} is Eq. (49.1c) for $r > r_t$, admittedly the only portion of the WKB function used in the Jablonski theory. With Eq. (49.5) we have completed the Royer proof that this portion of the WKB theory is a semiclassical one. In order to show that the Jablonski theory itself is semiclassical we develop the transition amplitude.

Now we have said that the transition amplitude is the sum of the amplitudes over the various paths connecting the initial state (i, ϵ_i, l) with the final state (f, ϵ_f, l). There are four such paths:

1 The perturber proceeds from infinity to some point $r > r_t$ at which the electronic transition occurs. Its radial velocity immediately before collision is $\dot{r} = -k_i(r)/\mathrm{m}$. The radial velocity immediately after collision is:

a $-k_f(r)/\mathrm{m}$ so that the perturber proceeds to the turning point, then back to infinity,

b $k_f(r)/\mathrm{m}$ so that the perturber proceeds directly back to infinity.

2 The perturber proceeds from infinity to the turning point, then to some point $r > r_t$ at which point the electronic transition occurs. Its radial velocity immediately before collision is $r = k_i(r)/\mathrm{m}$. The radial velocity immediately after collision is:

a $k_f(r)/\mathrm{m}$ so that the perturber proceeds directly to infinity.

b $-k_f(r)/\mathrm{m}$ so that the perturber proceeds to the turning point, then to infinity.

The amplitudes associated with these paths are

(1a) $\langle \infty \to r \rangle_i T_{if}(r) \langle r \to r_{tf} \to \infty \rangle_f$

(1b) $\langle \infty \to r \rangle_i T_{if}(r) \langle r \to \infty \rangle_f$

(2a) $\langle \infty \to r_{ti} \to r \rangle_i T_{if}(r) \langle r \to \infty \rangle_f$

(2b) $\langle \infty \to r_{ti} \to r \rangle_i T_{if}(r) \langle r \to r_{ft} \to \infty \rangle_f$

Royer makes use of

$$\langle \infty \to r \rangle_i = \langle \infty \to r_{ti} \rangle_i \langle r \to r_{ti} \rangle_i^{-1}$$

$$\langle \infty \to r_{ti} \rangle_i = \langle \infty \to r_{ti} \rangle_i e^{\mathrm{i}\pi/2} \langle r_{tf} \to \infty \rangle_f$$

and two similar expressions, drops the constant phase factor, $\langle \infty \to r_{ti} \rangle_i \langle r_{tf} \to \infty \rangle_f$, and utilizes Eq. (49.3) in order to obtain the amplitudes

$$(1\mathrm{a}) = A(-,-,1) \tag{49.5a}$$

$$(1\mathrm{b}) = A(-,+,0) \tag{49.5b}$$

$$(2\mathrm{a}) = A(+,+,1) \tag{49.5c}$$

$$(2\mathrm{b}) = A(+,-,1) \tag{49.5d}$$

where

$$A(\pm,\pm,n) = T_{if}(r)\exp\left[\mathrm{i}\int_{r_{ti}}^{r} (\pm) k_i(r')\,dr' - \mathrm{i}\int_{r_{tf}}^{r} (\pm) k_f(r')\,dr' + \tfrac{1}{2} i n \pi\right]$$

$$= T_{if}(r)\exp\left[(\pm)\mathrm{i}\alpha - (\pm)\mathrm{i}\beta + \tfrac{1}{2} i n \pi\right] \tag{49.5e}$$

If we now sum over these amplitudes and integrate over r, we should obtain the intensity in the spectral line. There is a restriction on the integration over r in that its lower limit is taken as r_t^* where r_t^* is the greater of r_{ti} and r_{tf}. The proper overlap integral is thus obtained if $T_{if}(r)$ is identified with $D(r)(k_i k_f/k_i(r)k_f(r))^{1/2}$, where $D(r)$ is the electronic transition moment. (Here the most general case, that of D a function of r, is taken. The often useful approximation that D is a constant could of course replace this.) This means that $|D(r)|^2$ is to be interpreted as a transition probability per unit time, $k_i/k_i(r)$ as the relative time spent at r, and $k_f/k_f(r)$ as the density of final states.

We drop the integration over r and write out the summation over paths, grouping paths (1a) and (2a) and paths (1b) and (2b) together with the result:

$$T_{if}\left\{\left[e^{-i\alpha+i\beta}+e^{i\alpha-i\beta}\right]e^{i\pi/2}+\left[e^{-i\alpha-i\beta}+e^{i\alpha+i\beta+i\pi}\right]\right\}$$

$$=T_{if}\left\{e^{i\pi/2}\cos(\alpha-\beta)+e^{i\pi/2}\cos\left(\alpha+\beta+\tfrac{1}{2}\pi\right)\right\} \tag{49.3b}$$

The main contributions to the r-integral will obviously arise from the first of the two cosines which infers further affirmation of the Franck-Condon principle since this is equivalent to $k_i(r)=k_f(r)$. It means too that those paths for which the direction of perturber motion reverses itself at the electronic transition are comparatively unimportant. When the second term in Eq. (49.6) is dropped we obtain the Jablonski result with which we became familiar in the preceding section. The important point, of course, is that, in obtaining this result, we have seen the specific correspondences between the various classical perturber paths and the various factors in the Jablonski theory.

Thus what Royer calls the semiclassical theory produces the Jablonski result. The term "semiclassical" would appear to infer the existence of a "classical" treatment, as indeed it does, and to this we now turn our attention.

We begin by simplifying the semiclassical spectrum by taking the kinetic energy difference in initial and final states

$$\left[k_f^2(r)-k_i^2(r)\right]/2\mathrm{m}=\epsilon_f-\epsilon_i-\mathrm{U}(r), \qquad \mathrm{U}(r)=\mathrm{U}_f-\mathrm{U}_i \tag{49.6}$$

as small compared to $k_i^2(r)/2\mathrm{m}$, the original kinetic energy. We may then expand the final kinetic energy in a series, retaining only the first two terms, thus

$$k_f(r)=\left\{k_i^2(r)+2\mathrm{m}\left[\epsilon_f-\epsilon_i-\mathrm{U}(r)\right]\right\}^{1/2} \tag{49.7a}$$

$$=k_i\left\{1+\left[\epsilon_f-\epsilon_i-\mathrm{U}(r)\right]2\mathrm{m}/k_i^2(r)\right\}^{1/2}$$

$$=k_i+\left[\epsilon_i-\epsilon_f-\mathrm{U}(r)\right]\mathrm{m}/k_i \tag{49.7b}$$

and $r_{tf}\doteq r_{ti}$.

Equations (49.7) are used in the phase of the WKB approximation $k_f = k_i$ in its amplitude. Thus the overlap integral inferred by Eq. (49.6) becomes

$$(k_i k_f)^{1/2} \int_{r_{ti}}^{\infty} D(r) k_i^{-1}(r)\, dr \cos\left\{ \int_{r_{ti}}^{r} dr' \frac{\left[\epsilon_f - \epsilon_i - \mathsf{U}(r')\right]\mathsf{m}}{k_i(r')} \right\}$$

$$= (k_i k_f)^{1/2} \mathsf{m}^{-1} \int_0^{\infty} dt\, D(r(t)) \cos\left\{ \int_0^t dt' \left[\epsilon_f - \epsilon_i - \mathsf{U}(r(t'))\right] \right\}, \quad (49.8)$$

where $\dot{r} = k_i(r)/\mathsf{m}$, $r_t(t) = r(0)$, and $r(t)$ is the perturber trajectory.

Equation (49.8) is squared and substituted into the intensity expression that Royer is using. This expression involves a $\delta(\omega - \epsilon_f + \epsilon_i)$, ω being frequency separation from line center, and integration over k_i and k_f. The result includes the absolute square of

$$\exp\left\{ \mathrm{i} \int_0^t dt' \left[\omega - \mathsf{U}(r(t', v, \rho))\right] \right\}$$

$$= \lim_{t_1 t_2 \to \infty} \int_{-t_1}^{t_2} dt \exp\left[-\mathrm{i} \int_{-t_1}^{t} dt' E_f(t') \right] D(t) \exp\left[-\mathrm{i} \int_t^{t_2} dt' E_i(t') \right] \quad (49.9)$$

In Eq. (49.9) $E_i(t)$ is the total energy of the radiator and the radiation field before the radiative transition; $E_f(t)$ is the total energy after the transition. E_f differs from E_i by the energy of the absorbed photon plus the atomic energy difference $\mathsf{U}(t)$. Once again we have encountered amplitudes for the system.

The first exponential in Eq. (49.9) is the amplitude for the system to evolve from $-t_1$ to t. $D(t)$ is the amplitude for the radiative transition to occur at t. The last exponential is the amplitude for the evolution of the system from t to t_2 in the final state.

Based on the fact that the equation may alternately be (classically) interpreted as the spectrum radiated by a classical oscillator of frequency $\mathsf{U}(t)$ and oscillation amplitude $D(t)$, which amplitude varies with time due to the influence of the perturber, Royer terms this result the (fully) classical spectrum. This may be an appropriate point at which to insert Table 49.1 (Royer's Table I), a rather interesting display of the parallelism he has demonstrated between the classical spectrum and the scattering cross section.

Table 49.1

	Classical spectrum	Cross secion
Observed quantity φ	Frequency ω	Scattering angle ϑ
Variable parametrizing different paths	Transition time t	Angular momentum l
"Path differential" dP	dt	$(l+1)^{1/2}\, dl$
Action $A(P, \varphi)$	$\omega t - \int_0^t \mathsf{U}(t')\, dt'$	$\vartheta l - \int_0^l \theta(l')\, dl'$
	U = potential difference	θ = classical deflection angle

Source: After Royer (237).

We have seen, in Eq. (49.8) for example, that the path integral for the classical spectrum is of the form $\int dP \exp[\mathrm{i}A(P,\varphi)]$, the action A given in Table 49.1. The main contribution to the integral arises from the Condon point, the separation at which the perturber velocity is unchanged by the electronic transition since there $\omega - \mathsf{U}(r) = 0$. Royer carries out an expansion of the action about this separation, keeping only the first two terms:

$$A(r \text{ or } t,\omega) \doteq A_c(\omega) - \left[\tfrac{1}{2}(t-t_c)^2\dot{\mathsf{U}}(t_c) \text{ or } \tfrac{1}{2}(r-r_c)^2\mathsf{U}'(r_c)/\dot{r}(r_c)\right] \qquad (49.10a)$$

where

$$\mathsf{U}' = \frac{d\mathsf{U}}{dr'}, \qquad \dot{\mathsf{U}} = \mathsf{U}'\dot{r}, \qquad \dot{r} = \frac{k_i(r)}{\mathsf{m}} \qquad (49.10b)$$

Now $\dot{\mathsf{U}}(t_c)$ is of course constant for a given Condon point insofar as t is concerned. Therefore, the path integral is readily evaluable with change in variable from t to $t' = t - t_c$:

$$\int dP\, e^{\mathrm{i}A} = e^{\mathrm{i}\eta_c}\int dt \exp\left[-\tfrac{1}{2}(t-t_c)^2\dot{\mathsf{U}}(t_c)\right] = e^{\mathrm{i}\eta_c}\int_0^\infty dt'\, e^{-t'^2\dot{\mathsf{U}}/2}$$

$$= \frac{\text{const } e^{\mathrm{i}\eta_c}}{(\dot{r}\mathsf{U}')^{1/2}}$$

Putting in the electronic transition moment and summing over Condon points yields the path integral given by Royer:

$$\sum_c e^{\mathrm{i}\eta_c} D_c |\dot{r}\mathsf{U}'|_c^{-1/2}, \qquad \eta_c = A_c \pm \tfrac{1}{4}\pi \qquad (49.11)$$

where the sign on $\pi/4$ is the sign of $-\dot{\mathsf{U}}$ at the Condon point. The absolute square of Eq. (49.11),

$$\sum_c |D_c|^2 |\dot{r}\mathsf{U}'(r)|_c^{-1} + \sum_{c\neq c'} D_c D_{c'} \cos[\eta_c(\omega) - \eta_{c'}(\omega)]\, |\dot{r}\mathsf{U}'|_{c'}^{-1/2} |r\mathsf{U}'|_c^{-1/2}$$

$$(49.12)$$

when averaged over angular momenta and initial energy, will yield the sought after spectrum.

If we neglect the second term on the right of Eq. (49.12), this term telling us the effects of interference between different paths, the momentum and energy averages may be carried out in order to obtain

$$I(\omega) = \sum_c 4\pi r_c^2 e^{-\mathsf{U}_i(r_c)/kT} |D(r_c)|^2 |\mathsf{U}'(r_c)|^{-1}, \qquad \mathsf{U}(r_c) = \omega \qquad (49.13)$$

Eq. (49.13) is essentially what Ben-Reuven obtains as his Eq. (163), the result of taking the slow collision limit of the Liouvillian formulation considered in Section 13. It certainly is "immediately apparent," as Royer remarks, that a local extremum, for which of course $\mathsf{U}' = 0$, at r_m will yield a local maximum or satellite at the frequency $\omega = \mathsf{U}(r_m)$. This is the analog of rainbow scattering. A remark is assuredly in order on this qualitative observation.

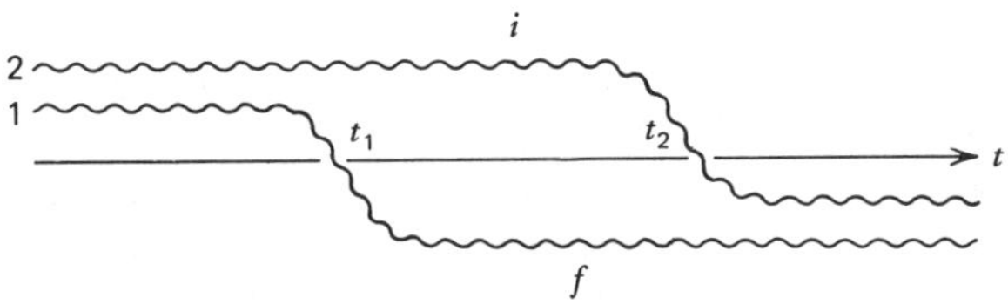

Figure 49.1 After Royer (238).

An extremum would, according to Eq. (49.13), yield a divergent spike at the minimum of, say, a Lennard-Jones potential. What this really means is that the expression Eq. (49.10a) has not been carried far enough. Royer found that carrying the expansion a term further yielded a result in terms of Airy functions [cf. Chapter 10 of (3)] and a finite rainbow spike.

The intensity maximum associated with the minimum of a Lennard-Jones type potential Royer concluded to be responsible for the red satellite observed by many authors. The violet satellites he attributed to the interference effects of which we have already considered the Mies study in the preceding section.

Although we have ruminated à la Mies, Royer tells us how this interference arises due to an interesting interference between the—in this case—two paths involved in the broadening phenomenon. This is worth considering more specifically.

In Fig. 49.1 the temporal axis increases from left to right, the upper half-plane corresponding to the radiator in its initial electronic state, the lower half-plane to the final. The times t_1 and t_2 correspond to the times at which perturber initial and final velocities are the same and transitions are to be expected. Then path 1 corresponds to perturber motion from infinity to r_c, transition, and perturber motion on to the turning point, then back to infinity. Path 2 corresponds to perturber motion from infinity to the turning point, back to r_c, transition, and then perturber motion on out to infinity. The interference between these two paths is precisely illustrated by the development leading to Eq. (49.6).

We conclude by remarking that there is no apparent conflict between the Mies and Royer treatments of the interference effect. We probably will not be able to see these maxima at temperatures much above 300° K.

50 Certain Unified Theories of Line Broadening

We may, albeit with a modicum of disagreement (177), describe a unified theory as one that reduces to the interruption theory near line center, the statistical theory in the line wing. Of course it is supposed to yield the proper results between these two limits. As we have already remarked, Smith, Cooper, and Vidal (249) were one of the first groups to apply the nomenclature to the attempt. We have also already enunciated the principles involved in this theory. Nevertheless, it will be of some value for us to remark on the details of two of these theories vis à vis satellites, that of Jacobson and that of Szudy and Baylis. Both are basically of an interruption-Baranger variety, the latter having Jablonskiesque overtones.

Jacobson and his collaborators (124, 59, 60, 155) have carried out a number of calculations that have perhaps been more concerned with the potentials between the collision partners than with the general broadening theory. Therefore, although Jacobson's results have been of considerable interest, we shall consider his work rather briefly.

Fox and Jacobson (FJ) (124) developed what they called a general pressures method of computing spectral line shapes, and they referenced the considerably earlier work of Bloom and Margenau (80). We have discussed [§ 3.15 of (9)] the work of these latter authors and shall not essay another discussion. Indeed, the reader will find it a familiar procedure to proceed from the Fourier transform of the average over $\mathbf{d}(0)\cdot\mathbf{d}(t)$ to

$$I(\omega)=\int\Phi(t)\exp[\mathrm{i}\omega t]\,dt \tag{50.1a}$$

$$\Phi(t)=\int\exp[\mathrm{i}\omega t]\left[\exp\left\{\mathrm{i}\int_0^t[\mathsf{V}_{Tf}(t')-\mathsf{V}_{Ti}(t')]\,dt'-\frac{\epsilon_T}{2kT}\right\}\right]_{\mathrm{av}} \tag{50.1b}$$

for an isolated line. Here the subscript "T" refers to the total of all perturbing molecules; ω to the frequency separation from line center. The average is over collisions. When the interactions are of the form with which we have already dealt, $\mathsf{V}=\Sigma\mathsf{V}_{ij}$, the correlation function becomes

$$\Phi(t)=A^{-n}\left[\exp\left\{\mathrm{i}\int_0^t[\mathsf{V}_f(t')-\mathsf{V}_i(t')]\,dt'-\frac{\epsilon_i}{2kT}\right\}\right]_{\mathrm{av}}^n \tag{50.1b$'$}$$

where n is the total number of perturbers, A a normalization constant. The average is to be carried out over the initial positions and velocities of the perturbers.

At this point the initial positions and velocities are reexpressed in plane polar coordinates. We take m as perturber mass, v_r as perturber radial velocity, and v_φ as perurber azimuthal velocity. Then $\epsilon_i=\mathsf{m}v_r^2+\mathsf{m}v_\varphi^2+\mathsf{V}_i(r)$, and we obtain

$$\Phi(t)=\frac{2\mathsf{m}}{kT}\int_0^\infty r^2\,dr\int_{-\infty}^\infty dv_\varphi$$
$$\times\exp\left[-\left(\frac{\mathsf{m}}{2kT}\right)\left(v_r^2+v_\varphi^2\right.\right.$$
$$\left.\left.+\frac{2\mathsf{V}_i(r)}{\mathsf{m}}\right)\right]\left[1-\exp\left(i\int_0^t[\mathsf{V}_f(t')-\mathsf{V}_i(t')]\,dt'\right)\right] \tag{50.1b$''$}$$

That we are dealing with the classical path goes without saying. The potentials that Jacobson and his associates used were of the Lennard-Jones type, $\mathsf{V}=C_n/r^n(t)-C_p/r^p(t)$, the constants being different for the upper and lower states. Next these authors develop these potentials as power series in the time t. This is accomplished by equating what may be considered a differential equation for the orbit to a particular form of expansion,

$$V(r(t)) = (C_n a_{in} - C_p \alpha_{ip}) t^i$$

for the potential. If we take the original speed of the perturber—or relative speed of the collision pair—as v_0, the original translational energy being $\frac{1}{2} m v_0^2$, then by subtracting the instantaneous and azimuthal energies from this we will obtain the instantaneous potential. By abstracting the mass from this expression for the potential then we may, with FJ, say $V(r)$ is proportional to $v_0^2 - \dot{r}^2 - v_0^2 \rho^2 / r^2 = \Sigma D_i t^i$. Taking $D_0 = v_0^2 - a_1^2 - v_0^2 \rho^2 / R^2$ we may obviously develop a recursion relation for the a_i and proceed to an evaluation of the line shape expansion. Jacobson and his associates used terms in the expansion through those quadratic in the time. A large number of the calculations reported by these authors related to Cs radiation broadened by Ar. For the ground state a Lennard-Jones (6-12) potential was utilized, its constants evaluated from molecular beam data. The upper state potential presented the problems one would anticipate. The authors in one instance (FJ) took the radial dependence of the potential in the upper state to be the same as that of the ground state with coefficients such as to reproduce the experimental widths and shifts for the $6^2P_{1/2}$–$6^2S_{1/2}$ line.

Jacobson does not claim that this treatment will provide a detailed description of all satellite phenomena, but a number of valuable descriptions have certainly been provided by this work. Further, that author apparently had no intention of accounting for Doppler effects nor, indeed, is there any cogent reason for always including them. However, there are certain situations wherein their inclusion is indicated, and, for such, one might turn to the work of Nienhuis (211, 212), who not only treated Doppler effects but also so cast the theory that detailed balance and Kirchhoff's law were maintained. His latter work (212) comprehends the additional detail to be anticipated from the specifics of multiplet spectra. We now turn to the unified theory of Szudy and Baylis (SB) (261, 262).

We saw in Eq. (14.5) how the one-particle correlation function may be written in terms of $g(t)$. Let us conform to the SB notation by writing Eq. (14.4b) as

$$\varphi(t) = \sum_{if} P_i^{(1)} |\langle f^{(1)} | i^{(1)} \rangle|^2 e^{\mp i \omega_{fi}^{(1)} t} \tag{14.4b$'$}$$

where $P_i^{(1)}$ is the probability of finding the single perturber in $|i\rangle$ and $\exp[\mp i \omega_{fi}^{(1)} t]$, the perturber energy difference exponential, in emission (+) or absorption (−). From Eqs. (14.4b′) and (14.5) we will agree that

$$g(t) = V[1 - \varphi(t)] = V \sum_{if} P_i^{(1)} |\langle f^{(1)} | i^{(1)} \rangle|^2 (1 - e^{i \omega_{fi} t}) \tag{50.2}$$

where we have dropped the ± sign. Now what SB are going to do is use an adiabatic Jablonski treatment to obtain a useful and specific expression for Eq. (50.2). This amounts to making the Born-Oppenheimer approximation in order to substitute a specific expression for the overlap integral and to insert a particular probability function. As we might anticipate, the summation will go over to an integration.

If we suppose V_i and V_f to be central, we may write the wave functions as $\psi_{i(f)l}(r)\mathsf{Y}^{lm}(r)/r$ where the radial portions of the wave functions are solution of

$$\left[\frac{d^2}{dr^2}+k^2_{i(f)}(r)\right]\psi_{i(f)l}(r)=0 \tag{50.3a}$$

where

$$k^2_{i(f)}(r)=2\mathsf{m}\left[E_{i(f)}-\mathsf{V}_{i(f)}(r)\right]-l(l+1)/r^2 \tag{50.3b}$$

$$\int_0^R dr\,|\psi_{i(f)l}(r)|^2=1, \qquad \psi_{i(f)l}(0)=\psi_{i(f)l}(R)=0 \tag{50.3c}$$

and

$$\begin{aligned}\langle f^{(1)}|i^{(1)}\rangle &=\delta_{l_il_f}\delta_{m_im_f}\int_0^\infty dr\,\psi^*_{fl}(r)\psi_{il}(r)\\ &=\delta_{l_il_f}\delta_{m_im_f}A_l(\omega_{fi})\end{aligned} \tag{50.4}$$

The probability of a particular linear momentum k and a particular angular momentum l is taken in product form

$$P^{(1)}_i=P(k_i)Q_l \tag{50.5a}$$

Q_l being given by Eq. (3.88) of Breene (9) as

$$Q_l=3(2l+1)/2k_i^2R^2 \tag{50.5b}$$

We are therefore in a position to write out Eq. (50.2) as

$$g(t)=\left\langle(2\pi R/k_i^2)\sum_{k_fl}(2l+1)|A_l(\omega_{fi})|^2(1-e^{i\omega_{fi}t})\right\rangle \tag{50.6}$$

where the brac operator is the average, $\langle\cdots\rangle=\Sigma_{\mathrm{k_i}}\mathrm{P(k_i)}\cdots$, over the initial linear momenta. The sum in Eq. (50.6) is replaced by the integral $\int dE_f v_{lm}(E_f)$ as is appropriate to the limit of large V. Here $v_{lm}(E_f)\,dE_f$ is the number of discrete k_f values for which $k_f^2/2\mathsf{m}$ lies in dE_f, that is, for which $k_f^2/2\mathsf{m}=dE_f$. Considering the asymptotic form of the radial wave function,

$$\psi_{fl}(r)=2\cos(\delta_l+k_fr)/R$$

and ignoring δ_l, the boundary condition, Eq. (50.3c), tells us that $\psi_{fl}(R)=0$ for $(2n-1)\pi/2=k_fR$. Therefore,

$$v_{lm}(E_f)\,dE_f=k_fR/\pi+\tfrac{1}{2}=k_fR\Rightarrow v_{lm}(E_f)=2\mathsf{m}R/\pi k_f \tag{50.7}$$

so that Eq. (59.6) becomes

$$g(t)=\left\langle\left(\frac{4R^2}{E_i}\right)\sum_{l=0}^{\infty}(2l+1)\int_{-\infty}^{\infty}d\omega_{if}k_f^{-1}(1-e^{i\omega_{fi}t})|A_l(\omega_{fi})|^2\right\rangle \quad (50.8)$$

where

$$E_i=k_i^2/2\mathrm{m}, \qquad \omega_{if}=E_i-E_f\Rightarrow dE_f=d\omega_{if}$$

Eq. (50.8) is what SB call the starting point for their unified Franck-Condon line shape. From this point reasonably straightforward manipulations lead to

$$I(\omega)=\frac{N}{\pi}\left[(\omega-\omega_0-\Delta)-\mathrm{i}\left(\tfrac{1}{2}\delta\right)\right]^{-2}\xi^2 J(\xi) \quad (50.9\text{a})$$

where

$$\xi=\omega-\omega_0-N\kappa \quad (50.9\text{b})$$

$$\kappa=\left\langle(4R^2/E_i)\sum_{l=0}^{\infty}(2l+1)\int_{-\infty}^{\infty}d\omega_{fi}k_{if}^{-1}|A_l(\omega_{if})|^2\omega_{fi}\right\rangle \quad (50.9\text{c})$$

$$J(\xi)=\left\langle(\pi R^2/E_ik_f)\sum_{l=0}^{\infty}(2l+1)|A_l(\xi)|^2\right\rangle \quad (50.9\text{d})$$

A_l is to be evaluated for the energy difference $E_f-E_i=\omega_{fi}=\xi$.

Eqs. (50.9) describe what SB dubbed the unified Franck-Condon line shape. These authors proceeded to utilize the WKB approximation to evaluate the overlap integral in Eq. (50.9d), complete evaluation of the line profile then being possible depending on the form presumed for the interaction potentials. [A somewhat similar procedure has been utilized by Sando and Wormhoudt (247) and Sando (246).] Their results reduced to the interruption and statistical ones in the appropriate limits. Now from, say, Eqs. (48.1) we may determine the phase in a WKB treatment as

$$\eta(r)=\int_{r_t}^{r}dr'\left[k_i(r')-k_f(r')\right]=\int_{r_t}^{r}dr'\frac{\left[k_i^2(r')-k_f^2(r')\right]}{\left[k_i(r')+k_f(r')\right]}$$

$$=\int_{r_t}^{r}dr'\frac{[\xi-\Delta\mathrm{V}]}{\left[k_i(r')+k_f(r')\right]} \quad (50.10)$$

by Eqs. (50.3b) and (50.9b) and where $\mathrm{V}=\mathrm{V}_f-\mathrm{V}_i$. At or near the Condon points will be regions of stationary phase such that $\partial\eta/\partial r=0$ which condition will rather obviously be guaranteed by

$$\xi=\Delta\mathrm{V} \quad (50.11)$$

There may be one or several real Condon points, determinable as solutions to Eq. (50.11). For these an asymptotic solution for Eq. (50.9d) leads to the statistical limit

$$I(\omega)=\left\{4\pi\xi^2/\left[\xi^2+\left(\tfrac{1}{2}\delta\right)^2\right]\right\}\sum r_c^2\frac{\exp[-\mathrm{V}_i(r_c)/kT]}{|\Delta\mathrm{V}'(r_c)|} \tag{50.12}$$

where the sum is over the Condon points satisfying Eq. (50.11). By V′ we refer to the derivative of $\mathrm{V}(r_c)$ so that, for extrema of $\Delta\mathrm{V}(r_c)$ we will encounter satellites. The location of these satellites is of course

$$\xi_s=\Delta\mathrm{V}(r_0)=\omega_s-\omega_0-\Delta \tag{50.13}$$

the point here being the shift of the satellite location with parent line shift.

SB also consider the complex solutions to Eq. (50.11), that is, the complex Condon points. In this case they obtain what may be called the antistat as compared to the statistical distribution. This distribution is essentially that of Eq. (50.12) except that it is diluted by a factor proportional to $\exp[-|z_c|]$, where

$$z_c=-|(2v_0)^{-1}r_c^2\sin\vartheta\Delta\mathrm{V}'(r_c)|^{2/3}$$

Here the Condon radius $r_c\equiv|r_c|\exp[i\vartheta]$ and $v_0=(2kT/\mathrm{m})^{1/2}$. As does the statistical profile, the antistat profile diverges as a satellite is approached.

Let us suppose that $\Delta\mathrm{V}(r)$ does not change sign for a pair of potentials, V_i and V_f. Then we see that solutions for Eq. (50.11) are real for one sign of $(\omega-\omega_0)$ and hence for one side of the spectral line; they are complex for the other sign and hence the other side of the line. Thus one of the two wings is statistical, the other of antistat form. The possibility of antistat intensity distributions was apparently first predicted by Holstein (24). Lizengevich and Formin (185), for example, had also encountered the possibility of either a strong asymmetry or a symmetrical statistical distribution in the wings of the spectral line. For further discussion of this symmetry and comparison with experiment we may consult Royer and Allard (38), Kielkopf (160), and Szudy and Baylis (262).

51 Born-Oppenheimer Breakdown and Satellites

Several of the theories considered in this chapter have treated the broadened spectral line as a reasonably precise analogy to an electronic-vibrational band system. However, they have gone a step further than this, for they have supposed the wave functions to be products of the electronic and vibrational wave functions for the collision pair, that is to say, they have presumed the applicability of the so-called Born-Oppenheimer approximation. Now we know that, when there is no term in the Hamiltonian describing the interaction of two types of motion, the total wave function is indeed a product of the wave functions for the two types of motion. Equivalently, when there are no pairs of states between which such a "perturbing" Hamiltonian can act, the Born-Oppenheimer approximation is unassailable. What all this means to the preceding ruminations is the following.

When a spectral line is absolutely isolated, that is, when the collision pair possesses one and only one upper electronic state, the theories presented above may account for all satellite phenomena. We now consider the work of Breene (86) who showed how the presence of a second electronic upper state —not that, under the appropriate circumstances, a lower state could not have the same effect—may introduce additional effects.

The spectral lines emitted by the alkali Cs provide familiar examples of the satellite phenomenon when perturbed by the presence of noble-gas atoms such as Ar. The second doublet of the principal series of Cs arises from transitions having the pair of upper levels $7p^2P_{1/2,3/2}$ (30). Since the collision partner is a noble-gas atom, we shall assume it to be in its ^1S state. Therefore, the Wigner-Wittmer rules (285) tells us that the diatomic molecule formed by the collision pair will have the states $^2\Sigma^+$ and $^2\Pi$ when Hund's coupling case (a) is assumed. Cs has a $6d^2D$ level some 600 wave numbers above its $7p^2P$ which would, in the isolated case, have nothing to do with the resonance doublet. When a collision occurs with another atom in a ^1S state, however, the $^2\Sigma^+$, $^2\Pi$, and $^2\Delta$ diatomic states are formed, and the situation is radically altered. Let us suppose that the ^2D: $^2\Sigma^+$ state is attractive. Then we shall anticipate its proximity to the ^2P: $^2\Sigma^+$ as indicated schematically in Fig. 51.1. (The two $^2\Sigma^+$ curves are not required to intersect in order for the phenomenon that we shall discuss to occur, although some intersection could of course take place.) Insofar as the wells are concerned, Bernstein and Muckerman (76) tell us that the experimentally determined depth of the Cs–Ar well

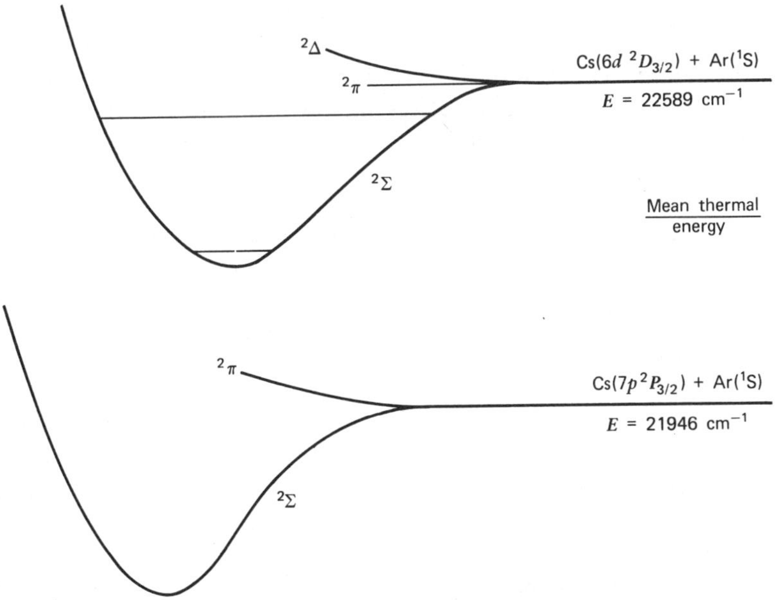

Figure 51.1 After Breene (86).

associated with the ground electronic state is 50 cm^{-1}. Therefore, the roughly 300 cm^{-1} depth of the upper well indicated here would not appear unreasonable. Obviously, the lower of the two pictured wells could have any depth, even zero depth, without affecting the treatment. For reasons that will later be obvious, Breene supposed the upper potential well to contain two vibrational levels as is indicated in the figure. Finally, in order to complete our explanation of the figure, we remark that the observations of Gilbert and Ch'en (133) for the Cs–Ar collision pair, for example, were carried out at a temperature of about 440° K. The mean continuous vibrational energy of the Cs–Ar molecule for such a temperature is indicated.

Let us consider the complete wave function for the CsAr molecule and suppose that the solution may be assumed to be of product form,

$$\Psi_{mk} = \Theta_m R_{mk}/r \tag{51.1}$$

where Θ_m is a function of the electronic and nuclear rotational coordinates, and R_{mk} is the discrete or continuous function of the nuclear vibrational coordinates. If we suppose the electronic wave function to be only parametrically dependent on the internuclear separation, the Born-Oppenheimer approximation, as Rice (230) has shown, allows us to neglect the following terms in the Hamiltonian:

$$\mathsf{V}\Psi_{mk} = -\frac{R_{mk}(r)}{2\mathsf{m}r}\frac{\partial^2\Theta_m}{\partial r^2} - (\mathsf{m}r)^{-1}\frac{dR_{mk}(r)}{dr}\frac{\partial\Theta_m}{\partial r} \tag{51.2}$$

where m is the reduced mass of the collision pair.

Now suppose that, if we carry out the calculation requisite to obtaining an extended configuration-interaction linear-combination-of-atomic-orbitals-molecular-orbital (LCAO-MO) wave function, Fig. 51.1 will result from taking matrix elements of these electronic wave functions over the Hamiltonian, which includes the elecrically dependent portions plus the nuclear repulsion. (We assume ground rotational states.) The determination of bound and continuous vibrational wave functions for the two potentials completes the calculation under the Born-Oppenheimer approximation. In energy regions such as those corresponding to the "mean thermal energy," this approximation breaks down, however, since electronic-vibrational wave functions must now be combinations of continuous functions associated with the lower of the two potential curves and bound functions associated with the upper potential curves. The Born-Oppenheimer neglected terms of Eq. (51.2) provide the perturbing portion of our Hamiltonian while the Fano configuration-interaction approach provides the method for the treatment of the perturbation, this treatment being originally concerned with autoionization (114).

In Section II of his paper Fano deals with the situation postulated here—one discrete state and one continuum. He begins by assuming that the wave function corresponding to an energy E may be expressed as a linear combination of the wave functions φ for the bound state and $\psi_{E'}$ for the continuous state,

$$\Psi_E = a\varphi + \int dE' \, b_{E'} \psi_{E'} \tag{51.3}$$

and proceeds to determine expansion coefficients, which we need not write down. The actual evaluation of these expansion coefficients would require accurate knowledge of both potential curves and wave functions. Suffice it to say that these expansion coefficients give rise to a resonance in the intensity that corresponds to the bound vibrational level of the upper state embedded in the continuum of the lower.

Gilbert and Ch'en, for example, have observed a satellite at about 240 cm^{-1} (which the close lying lower level of the last paragraph would be unable to account for) in Cs broadened by an Ar line. Such a satellite may be accounted for by including a second bound vibrational state in the configuration-mixed wave function. In such a situation the intensity behavior near unperturbed line center is again controlled by, say, the configuration mixing of the ground vibrational state with the continuum. As we move into the line wing, we are proceeding to energies such that the mixing effect is of little consequence. As our energy continues to increase and as we approach the energy of the first ($v=1$) bound level, however, the mixing effects again begin to make themselves felt, the violet satellite which subsequently appears being roughly a measure of the bound vibrational level separation. (It is important to emphasize here that the this mixing phenomenon is completely different from capturing the collision molecule into a potential well.) Thus the violet satellite observed by Gilbert and Ch'en would correspond to a level separation of about 250 cm^{-1}. In order to illustrate the increase in intensity anticipated in the neighborhood of the higher vibrational level, we partially write down Fano's expression for the overlap integral; one of Fano's states is a configuration mixture of one continuum and several discrete states [Eq. (65)]:

$$\langle \psi_E | i \rangle = \cos\Delta \sum_\nu \tan\Delta_\nu \frac{\langle \varphi_\nu | i \rangle}{\pi V_{\nu E}} + \cdots \tag{51.4a}$$

$$-\sum_\nu \tan\Delta_\nu = \sum_\nu \frac{\pi |V_{E\nu}|^2}{(E-E_\nu)} \tag{51.4b}$$

It is obvious that the factor $(E-E_\nu)^{-1}$ will lead to an intensity maximum in the neighborhood of the higher bound vibrational state.

In sum then, Breene's work demonstrated that in addition to the satellite origins discussed earlier in this chapter, satellites may occur due to the configuration embedding of bound vibrational levels in the continuum as a consequence of the breakdown of the Born-Oppenheimer approximation.

Bibliography

1a Fuhr, J. R., W. L. Weise, and L. Z. Roszman. 1972. *Bibliography on Atomic Line Shapes and Shifts*. NBS Special Publication 366. Washington.

1b Fuhr, J. R., L. Z. Roszman, and W. L. Weise. 1974. *Bibliography on Atomic Line Shapes and Shifts*. NBS Special Publication 366. Supplement 1. Washington.

1c Fuhr, J. R., G. A. Martin, and B. J. Specht. 1975. *Bibliography on Atomic Line Shapes and Shifts*. NBS Special Publication 366. Supplement 2.

1d Fuhr, J. R., B. J. Miller, and G. A. Martin. 1978. *Bibliography on Atomic Line Shapes and Shifts*. NBS Special Publication 366. Supplement 3.

Books Reports and Theses

2 Abragam, A. 1961. *Low Temperatures Physics*. New York:

3 Abramowitz, M. and I. A. Stegun (Editors). 1964. Handbook of Mathematical Functions. Washington: NBS Applied Mathematics Series No. 55.

4 Abrikosov, A. A., L. P. Gorkov, and I. E. Dzyaloshinski. 1963. *Methods of Quantum Field Theory in Statistical Physics*. Englewood Cliffs, N.J.: Prentice-Hall.

5 Allen, C. W. 1963. *Astrophysical Quantities*. 2nd ed. London: University of London: Athlone.

6 Beals, R. 1973. *Advanced Mathematical Analysis*. New York: Springer-Verlag.

7 Bethe, H. A. and E. E. Salpeter. 1957. *Quantum Mechanics of One- and Two-Electron Atoms*. Berlin: Springer-Verlag.

8 Bezzerides, B. 1966. *Radiation Absorption Phenomena in Gases*. (Thesis) University of California.

9 Breene, R. G. Jr. 1961. *The Shift and Shape of Spectral Lines*. Oxford, England: Pergamon.

10 Condon, E. U. and G. H. Shortley. 1935. *The Theory of Atomic Spectra*. Cambridge, England: Cambridge University.

11 Cramer, H. 1946. *Mathematical Methods in Statistics*. Princeton, N.J.: Princeton University.

12 Dirac, P. A. M. 1947. *The Principles of Quantum Mechanics*. 3rd ed. Oxford, England: Oxford University.

13 Edmonds, A. R. 1957. *Angular Momentum in Quantum Mechanics*. Princeton, N.J.: Princeton University.

14 Feynman, R. P. and A. R. Hibbs. 1965. *Quantum Mechanics and Path Integrals*. New York: McGraw-Hill.

15 Friedrichs, K. O. 1973. *Spectral Theory of Operators in Hilbert Space*. New York: Springer-Verlag.

16 Goldberger, M. L. and K. M. Watson. 1964 *Collision Theory*. New York: Wiley.

17 Goldstein, H. 1950. *Classical Mechanics*. Cambridge, Mass.: Addison-Wesley.

18 Griem, H. A. 1964. *Plasma Spectroscopy*. New York: McGraw-Hill.

19 ________. 1974. *Spectral Line Broadening by Plasmas*. New York: Academic.

20 Hausdorf, F. 1962. *Set Theory*. New York: Chelsea (Translated from 3rd German ed.).

21 Heitler, W. 1954. *The Quantum Theory of Radiation*. 3rd ed. Oxford, England: Oxford University.

22 Herzberg, G. 1950. *Molecular Spectra and Molecular Structure. I. Spectra of Diatomic Molecules*. New York: Van Nostrand.

23 Herzberg, G. 1945. *Molecular Spectra and Molecular Structure. II. Infrared and Raman Spectra of Polyatomic Molecules*. New York: Van Nostrand.

24 Holstein, T. 1953. *Pressure Broadening of Spectral Lines*. (Report) Pittsburgh: University of Pittsburgh.

25 Huang, K. 1963. *Statistical Mechanics*. New York: Wiley.

26 Jauch, J. M. and F. Rohrlich. 1976. *The Theory of Photons and Electrons*. 2nd ed. New York: Springer-Verlag.

27 Källén, G. 1972. *Quantum Electrodynamics*. (translated by G. K. Iddings and M. Mizushima) New York: Springer-Verlag.

28 Kirzhnits, D. A. 1967. *Field Theoretical Methods in Many-Body Systems*. (translated by A. J. Meadows) Oxford, England: Pergamon.

29 Mattuck, R. D. 1976. *A guide to Feynman diagrams in the many-body problem*. 2nd ed. New York: McGraw-Hill.

30 Moore, C. E. 1958. *Atomic Energy Levels*. Vol. III, NBS Circular No. 467. Washington, D.C.: U.S. GPO.

31 Morse, P. M. and H. Feshbach. 1953. *Methods of Theoretical Physics*. 2 vols. New York: McGraw-Hill.

32 Mott, N. F. and H. S. W. Massey, 1949. *The Theory of Atomic Collisions*. 2nd ed. Oxford, England: Oxford University.

33 Mott, N. F. and I. N. Sneddon. 1948. *Wave Mechanics and Its Applications*. Oxford, England: Oxford University.

34 Murnaghan, F. D. 1962. *The Unitary and Rotation Groups*. Washington, D. C.: Spartan.

35 von Neumann, J. 1955. *Mathematical Foundations of Quantum Mechanics*. (translated by R. T. Beyer) Princeton, N.J.: Princeton University.

36 Roney, L. R. 1973. *On the Theory of Combined Doppler and Binary Collision Foreign Gas Broadening*. (Thesis) Université Libre de Bruxelles.

37 Ross, D. W. 1964. *Quantum Theory of the Electromagnetic Propertier of Matter: Broadening of Spectral Lines*. (Thesis) Harvard University.

38 Royer, A. and N. F. Allard. 1975. *Classical Pressure Broadening Theory*. (Report) Centre de Recherches Mathematique.

39 Schiff, L. I. 1955. *Quantum Mechanics*. 2nd ed. New York: McGraw-Hill.

40 Schweber, S. S. 1961. *An Introduction to Relativistic Quantum Field Theory*. Evanston, Ill.: Row Peterson.

41 Slater, J. C. 1960. *Quantum Theory of Atomic Structure*. 2 vols. New York: McGraw-Hill.

42 Traving, G. 1960. *Über die Theorie der Druckverbreiterung von Spektrallinien*. Karlsruhe: Verlag G. Braun.

43 Unsold, A. 1955. *Physik der Sternatmosphären*. Berlin: Springer-Verlag.

44 Van Vleck, J. H. 1932. *The Theory of Electric and Magnetic Susceptibilities*. Oxford, England: Oxford University.

45 Wentzel, G. 1949. *Quantum Theory of Fields*. (translated by C. Houtermans and J. M. Jauch) New York: Interscience.

46 Weyl, H. No date. *The Theory of Groups and Quantum Mechanics*. (translated by H. P. Robertson) New York: Dover.

47 Whittaker, E. T. and G. N. Watson. 1927. *A Course of Modern Analysis*. 4th ed. Cambridge, England: Cambridge University.

48 Wigner, E. P. 1959. *Group Theory*. (translated by J. J. Griffin) New York: Academic.

49 Wilson, E. B. Jr., J. C. Decius, and P. C. Cross. 1955. *Molecular Vibrations*. New York: McGraw-Hill.

50 Yutsis, A. P., I. B. Levinson, and V. V. Vanagas. 1962. *Theory of Angular Momentum*. (translated by A. Sen and R. N. Sen) NASA TT F-98. Washington, D. C.: U.S. GPO.

51 Zaidi, H. R. 1967. *Application of Many-Body Theory to the Problem of Spectral Line Shapes*. (Thesis) University of Tennessee.

52 Ziman, J. M. 1969. *Elements of Advanced Quantum Theory*. Cambridge, England: Cambridge University.

Articles

53 Albers, J. and I. Oppenheim. *Physica* **59**, 161(1972).

54 ________. *Physica* **59**, 187(1972).

55 Anderson, P. W. *Phy. Rev.* **76**, 647(1949).

56 Armstrong, B. H. *JQSRT* **7**, 61(1967).

57 Arnous, E. *Helv. Phys. Acta* **25**, 631(1952).

58 Arnous, E. and W. Heitler. *Proc. R. Soc. London A* **220**, 290(1953).

59 Atakan, A. K. and H. C. Jacobson. *JQSRT* **12**, 289(1972).

60 Atakan, A. K. *Phys. Rev. A* **7**, 1452(1973).

61 Bali, L. M. and R. B. Higgins. *Phys. Lett. A* **35**, 95(1971).

62 Balling, L. C., R. J. Hanson, and F. Pipkin. *Phys. Rev.* **133**, A607(1964).

63 Baranger, M. *Phys. Rev.* **111**, 481(1958).

64 ________. *Phys. Rev.* **111**, 494(1958).

65 ________. *Phys. Rev.* **111**, 855(1958).

66 Baym, G. and L. P. Kadanoff. *Phys. Rev.* **124**, 287(1961).

67 Ben-Reuven, A. *Phys. Rev.* **141**, 34(1966).

68 ________. *Adv. At. Mol. Phys.* **5**, 201(1969).

69 ________. *Phys. Rev. A* **4**, 753(1971).

70 ________. *Phys. Rev. A* **4**, 2115(1971).

71 ________. *Adv. Chem. Phys.* **33**, 235(1975).

72 Berman, P. R. *JQSRT* **12**, 1331(1972).

73 Berman, P. R. and W. E. Lamb, Jr. *Phys. Rev.* **187**, 221(1969).

74 Berman, P. R. *Phys. Rev. A* **2**, 2435(1970).

75 Berne, B. T., J. Jortner, and R. G. Gordon. *J. Chem. Phys.* **47**, 1600(1967).

76 Bernstein, R. B. and J. T. Muckerman. 1967. in *Intermolecular Forces*. (edited by J. O. Hirschfelder) New York: Interscience.

77 Besombes, F., J. Granier, and R. Granier. *Opt. Commun.* **1**, 161(1969).

78 Bethe, H. A. and J. Goldstone. *Proc. R. Soc. London A* **238**, 551(1957).

79 Bezzerides, B. *Phys. Rev.* **159**, 3(1967).

80 Bloom, S. and H. Margenau. *Phys. Rev.* **90**, 791(1953).
81 Bottcher, C. *J. Phys B* **4**, L99(1971).
82 Branson, D. *Phys. Rev.* **135**, B1255(1964).
83 Brechot, S. and H. van Regemorter. *Ann. Astrophys.* **27**, 432(1964).
84 Brechot, S. *Ann. Astrophys.* **27**, 739(1964).
85 Breene, R. G. Jr. *J. Mol. Spectrosc.* **26**, 465(1968).
86 ________. *Phys. Rev.* **2**, A1164(1970).
87 ________. *Nuovo Cimento B* **4**, 1(1971).
88 Breit, G. *Phys. Rev.* **34**, 553(1929).
89 Buckingham, A. D. *Adv. Chem. Phys.* **12**, 107(1967).
90 Byron, F. W. and H. M. Foley. *Phys. Rev. A* **134**, 625(1964).
91 Byron, F. W. Jr., M. N. McDermott, and R. Novick. *Phys. Rev. A* **134**, 615(1964).
92 Cattani, M. *Rev. Bras. Fis.* **1**, 351(1971).
93 ________. *Lett. Nuovo Cimento* **4**, 346(1970).
94 Callen, H. B. and T. A. Welton. *Phys. Rev.* **83**, 34(1951).
95 Cantor, G. F. L. P. Math. Ann. 46, 481(1895).
96 ________. *Math. Ann.* **49**, 207(1897).
97 Castex, M. *CR Acad. Sci.*, **268**, 552(1969).
98 Ch'en, S. Y. and C. W. Fountain. *JQSRT* **4**, 323(1964).
99 Ch'en, S. Y. and A. T. Longseth. *Phys. Rev. A* **3**, 946(1971).
100 Ch'en, S. Y. and R. V. Phelps. *Phys. Rev. A* **7**, 470(1973).
101 Ch'en, S. Y. and M. Takeo. *Rev. Mod. Phys.* **29**, 20(1957).
102 Ch'en, S. Y., M. R. Atwood, and T. H. Warnock. *Physica* **27**, 1170(1961).
103 Cherkasov, M. R. *Opt. Spectrosc.* **40**, 3(1976).
104 Cooper, J. *Rev. Mod. Phys.* **39**, 167(1967).
105 Czuchaj, E. *Acta Phys. Pol. A* **45**, 731(1974).
106 ________. *Acta Phys. Pol. A* **45**, 97(1974).
107 ________. Z. Phys. **276**, 85(1976).
108 Davies, J. T. and J. M. Vaughan. *Astrophys. J.* **137**, 1302(1963).
109 Dicke, R. H. *Phys. Rev.* **89**, 472(1953).
110 Dillon, T. A., E. W. Smith, J. Cooper, and M. Mizushima. *Phys. Rev. A* **2**, 1839(1970).
111 Exton, R. J. *JQSRT* **15**, 1141(1975).
112 Fadeev, L. D. *Sov. Phys. JETP* **12**, 1014(1961).
113 Fano, U. *Rev. Mod. Phys.* **29**, 74(1957).
114 ________. *Phys. Rev.* **124**, 1866(1961).
115 ________. *Phys. Rev.* **131**, 259(1963).
116 Faxen, H. and J. Holtsmark. Z. Phys. **45**, 307(1927).
117 Fermi, E. *Rev. Mod. Phys.* **4**, 105(1932).
118 Feynman, R. P. *Phys. Rev.* **76**, 749(1949).
119 Fiutak, J. *Acta Phys. Pol.* **26**, 919(1964).
120 ________. *Acta Phys. Pol.* **27**, 753(1965).
121 ________. *J. Math. Phys. Chem.* **8**, 39(1968).
122 Foley, H. M. *Phys. Rev.* **69**, 616(1946).
123 Fonda, L., G. C. Ghirardi, T. Weber, and A. Rimini. *J. Math. Phys.* **7**, 1643(1966).
124 Fox, R. L. and H. C. Jacobson. *Phys. Rev.* **188**, 232(1969).

125 Frohlich, H. *Nature* **157**, 478(1946).

126 Furry, W. H. *Phys. Rev.* **81**, 115(1951).

127 Futrelle, R. P. *Phys. Rev. A* **5**, 2162(1972).

128 Galatry, L. *Phys. Rev.* **122**, 1218(1961).

129 Garstens, M. A. *Phys. Rev.* **93**, 1228(1954).

130 Gerbal, D. and M. Prud'Homme. *JQSRT* **14**, 351(1974).

131 Gersten, J. I. and H. M. Foley. *J. Opt. Soc. Am.* **58**, 933(1968).

132 di Giacomo, A. and F. Feo. *Nuovo Cimento B* **25**, 730(1975).

133 Gilbert, D. E. and S. Y. Ch'en. *Phys. Rev.* **188**, 40(1969).

134 Girardeau, M. *J. Math. Phys.* **4**, 1096(1963).

135 Godfrey, C. *Philos. Trans. R. Soc. London A* **195**, 329(1901).

136 Godfrey, J. T., C. R. Vidal, E. W. Smith and J. Cooper. *Phys. Rev. A* **3**, 1543(1971).

137 Grainer, R. *Ann. Phys. (Paris)* **4**, 383(1969).

138 Granier, R., M. C. Castex, J. Granier, and J. Romand. *C. R. Acad. Sci. Ser. B* **264**, 778(1967).

139 Gwinn, J. A., P. M. Thomas, and J. F. Kielkopf. *J. Chem. Phys.* **48**, 568(1968).

140 Herbert, F. *JQSRT* **14**, 943(1974).

141 Herman, R. M. and R. G. Breene, Jr. *J. Mol. Spectrosc.* **23**, 343(1967).

142 Hess, S. *Physica* **61**, 80(1972).

143 Hindmarsh, W. R. and J. M. Farr. *J. Phys. B* **2**, 1388(1969).

144 Holtsmark, J. *Ann Phys.* **58**, 577(1919).

145 ________. *Z. Phys.* **34**, 722(1925).

146 Hood, R. J. and G. P. Reck. *J. Chem. Phys.* **56**, 4053(1972).

147 Huber, D. L. and J. H. Van Vleck. *Rev. Mod. Phys.* **38**, 187(1966).

148 van de Hulst, H. C. and J. J. M. Reesink. *Astrophys. J.* **106**, 121(1947).

149 Hummer, D. G. *Mem. R. Astron. Soc.* **70**, 1(1965).

150 Hynne, F. *JQSRT* **14**, 437(1974).

151 Irving, J. H. and R. W. Zwanzig. *J. Chem. Phys.* **19**, 1173(1951).

152 Jablonski, A. *Acta Phys. Pol.* **6**, 371(1937).

153 ________. *Physica* **7**, 541(1940).

154 ________. *Phys. Rev.* **68**, 78(1945).

155 Jacobson, H. C. *Phys. Rev. A* **4**, 1368(1971).

156 Jefimenko, O. *J. Chem. Phys.* **39**, 1556(1959).

157 ________. *J. Chem. Phys.* **42**, 205(1965).

158 Jefimenko, O. and S. Y. Ch'en. *J. Chem. Phys.* **26**, 913(1957).

159 Karplus, R. and J. Schwinger. *Phys. Rev.* **73**, 1020(1948).

160 Kielkopf, J. F. *J. Phys. B* **9**, 1601(1976).

161 Kielkopf, J. F. and J. A. Gwinn. *J. Chem. Phys.* **48**, 5570(1968).

162 Kielkopf, J. F., J. F. Davis, and J. A. Gwinn. *J. Chem. Phys.* **53**, 2605(1970).

163 Klein, L. and H. Margenau. *J. Chem. Phys.* **30**, 1556(1959).

164 Kolb, A. C. and H. Griem. *Phys. Rev.* **111**, 514(1958).

165 Kronig, R. deL. *Physica* **5**, 65(1938).

166 Kubo, R. 1959. in *Lectures in Theoretical Physics.* (edited by W. E. Britten and L. G. Dunham) New York: Interscience.

167 ________. *J. Phys. Soc. Jn.* **17**, 1100(1962).

168 Kuhn, H. *Philos. Mag.* **18**, 987(1934).
169 ________. *Proc. R. Soc. London A* **158**, 212(1937).
170 Kuhn, H. and E. L. Lewis, *Proc. Phys. Soc. London A* **299**, 423(1967).
171 Kuhn, H. and O. Oldenberg. *Phys. Rev.* **41**, 72(1932).
172 Kuhn, H. and J. M. Vaughan. *Proc. R. Soc. London A* **277**, 297(1963).
173 Lehmberg, R. H. *Phys. Rev.* **181**, 32(1969).
174 Lenitzky, I. R. *Phys. Rev.* **119**, 670(1960).
175 ________. *Phys. Rev.* **124**, 642(1961).
176 ________. *Phys. Rev.* **131**, 2827(1963).
177 Lee, R. *J. Phys. B* **4**, 1640(1971).
178 Lenz, W. *Z. Phys.* **80**, 423(1933).
179 Leslie, D. C. M. *Philos. Mag.* **42**, 37(1951).
180 Lewis, M. *Phys. Rev.* **130**, 666(1963).
181 Lindholm, E. *Ark. Mat. Astron. Fys.* **28B**, No. 3(1942).
182 ________. *Ark. Mat. Astron. Fys.* **32A**, No. 17(1946).
183 Lippmann, B. and J. Schwinger. *Phys. Rev.* **79**, 469(1950).
184 Lisitsa, V. S. and S. I. Yakovlenko. *Sov. Phys. JETP* **41**, 233(1975).
185 Lizengevich, A. I. and V. V. Formin. *Opt. Spectrosc.* **34**, 277(1973).
186 Lorentz, H. A. *Proc. R. Acad. (Amsterdam)* **8**, 591(1906).
187 Low, F. *Phys. Rev.* **88**, 53(1952).
188 Margenau, H. *Phys. Rev.* **40**, 387(1932).
189 ________. *Phys. Rev.* **44**, 931(1933).
190 ________. *Phys. Rev.* **48**, 755(1935).
191 ________. *Phys. Rev.* **76**, 1423(1949).
192 ________. *JQSRT* **3**, 445(1963).
193 Margenau, H. and M. Lewis. *Rev. Mod. Phys.* **31**, 569(1959).
194 Margenau, H. and W. W. Watson. *Rev. Mod. Phys.* **8**, 22(1936).
195 Majorana, E. *Nuovo Cimento* **9**, 43(1930).
196 Mead, C. A. *Phys. Rev.* **120**, 854(1960).
197 ________. *Phys. Rev.* **120**, 860(1960).
198 ________. *Phys. Rev.* **128**, 1753(1962).
199 ________. *Int. J. Theor. Phys.* **1**, 317(1968).
200 Michelson, A. A. *Astrophys. J.* **11**, 251(1895).
201 Mies, F. H. *J. Chem. Phys.* **48**, 482(1968).
202 Mies, F. H. and A. L. Smith. *J. Chem. Phys.* **45**, 994(1966).
203 Minkowski, R. *Z. Phys.* **36**, 839(1926).
204 Mizushima, M. *JQSRT* **7**, 505(1967).
205 ________. *JQSRT* **11**, 471(1971).
206 ________. D. Robert and L. Galatry. *J. Phys.* **26**, 194(1965).
207 Mollow, B. R. and M. M. Miller. *Ann. Phys.* (N.Y.) **52**, 464(1969).
208 Moore, H. R. *Science* **66**, 543(1927).
209 Morozov, V. A. and P. P. Shorygin. *Opt. Spectroc.* **19**, 289(1965).
210 Nicholls, R. W. and A. L. Stewart. 1962. in *Atomic and Molecular Processes*. (edited by D. R. Bates) New York: Academic.
211 Nienhuis, G. *Physica* **66**, 245(1973).

212 ________. *Physica* **74**, 157(1974).

213 Nowotny, H. *Phys. Lett. A* **36**, 481(1971).

214 Ny, T. Z. and S. Y. Ch'en. *Nature* **138**, 1055(1936).

215 Ny, T. Z. *C. R. Acad. Sci.* **203**, 242(1936).

216 Oldenberg, O. *Z. Phys.* **47**, 184(1928).

217 ________. *Z. Phys.* **55**, 1(1929).

218 Pasmanter, R. A. and A. Ben-Reuven. *JQSRT* **13**, 57(1973).

219 Posener, D. W. *Aust. J. Phys.* **12**, 184(1959).

220 Preston, W. M. *Phys. Rev.* **51**, 298(1937).

221 Primas, H. *Helv. Phys. Acta.* **34**, 331(1961).

222 Rautian, S. G. and I. I. Sobelman. *Sov. Phys. Usp.* **9**, 701(1967).

223 Rebane, V. N. *Opt. Spectrosc.* **42**, 123(1977).

224 Reck, G. P. *JQSRT* **9**, 1419(1969).

225 Reck, G. P. and R. J. Hood. *J. Chem. Phys.* **56**, 1230(1972).

226 Reck, G. P., H. Takebe, and C. A. Mead. *Phys. Rev.* **137**, A683(1965).

227 van Regemorter, H. *C. R. Acad. Sci.* **257**, 63(1963).

228 ________.1971. in *Atoms and Molecules in Astrophysics*. Amsterdam: North-Holland.

229 Rice, O. K. *Phys. Rev.* **33**, 748(1929).

230 ________. *Phys. Rev.* **35**, 1551(1930).

231 Robin, S. and S. Robin. *Rev. Opt. Theor. Instrum.* **37**, 161(1958).

232 Roney, P. L. *JQSRT* **15**, 361(1975).

233 ________. *JQSRT* **15**, 181(1975).

234 ________. *JQSRT* **15**, 301(1975).

235 ________. *JQSRT* **15**, 706(1975).

236 Ross, D. W. *Ann. Phys.* (N.Y.) **36**, 458(1966).

237 Royer, A. *J. Chem. Phys.* **50**, 1906(1969).

238 ________. *Phys. Rev. A.* **4**, 499(1971).

239 ________. *Phys. Rev. A* **6**, 1741(1972).

240 ________. *Phys. Rev. A* **7**, 1078(1973).

241 ________. *Can. J. Phys.* **52**, 1816(1974).

242 ________. *Can J. Phys.* **53**, 2470(1975).

243 ________. *Can J. Phys.* **53**, 2477(1975).

244 Sakata, S. *Prog. Theor. Phys.* **16**, 686(1956).

245 Sahal-Brechot, S. *Astron. Astrophys.* **1**, 91(1969).

246 Sando, K. M. *Phys. Rev. A.* **9**, 1103(1974).

247 Sando, K. M. and J. C. Wormhoudt. *Phys. Rev. A* **7**, 1889(1973).

248 Smith, E. W. and C. F. Hooper, Jr. *Phys. Rev.* **157**, 126(1967).

249 Smith, E. W., J. Cooper, and C. R. Vidal. *Phys. Rev.* **185**, 140(1969).

250 Smith, E. W., C. R. Vidal, and J. Cooper. *J. Res. Natl. Bur. Stand. Sec. A* **73**, 389(1969).

251 Smith, E. W., J. Cooper, W. R. Chappell, and T. Dillon. *JQSRT* **11**, 1547(1971).

252 Sobelman, I. I. *Fortschritte Physik* **5**, 175(1957).

253 Spitzer, L. Jr. *Phys. Rev.* **55**, 699(1939).

254 ________. *Phys. Rev.* **56**, 39(1939).

255 ________. *Phys. Rev.* **58**, 348(1940).

256 Srivastava, R. P. and H. R. Zaidi. *Can. J. Phys.* **53**, 84(1975).

257 Stapp, H. P. *Phys. Rev.* **125**, 2139(1962).

258 Stettin, G. 1964. in *The Equilibrium Theory of Classical Fluids.* (edited by H. L. Frisch and J. L. Lebowitz) New York: Benjamin.

259 Strutt, J. W. (Lord Rayleigh) *Philos. Mag.* **27**, 298(1889).

260 Snider, R. F. *J. Chem. Phys.* **32**, 1051(1960).

261 Szudy, J. and W. E. Baylis. *JQSRT* **15**, 641(1975).

262 Szudy, J. *JQSRT* **17**, 681(1977).

263 Takeo, M. *Phys. Rev. A* **1**, 1143(1970).

264 Takeo, M. and S. Y. Ch'en. *Phys. Rev.* **93**, 420(1954).

265 Tanaka, Y. *J. Opt. Soc. Am.* **45**, 710(1955).

266 ________. *J. Opt. Soc. Am.* **48**, 304(1958).

267 Tanaka, Y. and K. Yoshino. *J. Chem. Phys.* **39**, 3081(1963).

268 Tip, A. *Physica* **52**, 493(1971).

269 Tsao, C. J. and B. Curnutte. *JQSRT* **2**, 41(1962).

270 Vainshtein, L. A. *Sov. Phys. JETP* **18**, 1383(1964).

271 ________. *Proc. Phys. Soc. London* **89**, 511(1966).

272 Van Vleck, J. H. *Phys. Rev.* **71**, 413(1947).

273 Van Vleck, J. H. and V. F. Weisskopf. *Rev. Mod. Phys.* **17**, 227(1945).

274 Vaughan, J. M. *Phys. Rev.* **166**, 13(1968).

275 Voigt, W. K. *Bayer. Akad. München, Ber.* 603(1912).

276 Voslamber, D. *Phys. Lett. A* **40**, 266(1972).

277 ________. *Z. Naturforsch. A* **27**, 1783(1972).

278 Weingeroff, M. *Z. Phys.* **67**, 679(1931).

279 Weisskopf, V. F. *Z. Phys.* **75**, 287(1932).

280 ________. *Z. Phys.* **77**, 398(1932).

281 Weisskopf, V. F. and E. P. Wigner. *Z. Phys.* **63**, 54(1930).

282 Weisskopf, V. F. *Z. Phys.* **65**, 18(1930).

283 Wigner, E. P. *Phys. Rev.* **40**, 749(1932).

284 ________. *Phys. Rev.* **98**, 145(1955).

285 Wigner, E. P. and E. E. Witmer. *Z. Phys.* **51**, 859(1928).

286 Wittke, J. P. and R. H. Dicke. *Phys. Rev.* **103**, 620(1956).

287 Yakimets, V. V. *Sov. Phys. JETP* **24**, 990(1967).

288 Zaidi, H. R. *Phys. Rev.* **173**, 123(1968).

289 ________. *Can. J. Phys.* **50**, 1175(1972).

290 ________. *Can. J. Phys.* **50**, 2791(1972).

291 ________. *Can. J. Phys.* **50**, 2801(1972).

292 ________. *Can. J. Phys.* **53**, 76(1975).

293 Zwanzig, R. 1961. in *Lectures in Theoretical Physics.* (edited by W. E. Brittin) New York: Interscience.

Author Index

Subject Index